D1123360

# FRONTIER ORBITALS AND PROPERTIES OF ORGANIC MOLECULES

**ELLIS HORWOOD BOOKS IN ORGANIC CHEMISTRY**
*Series Editor:* Dr J. MELLOR, University of Southampton

|  |  |
|---|---|
| Belen'kii, L.I. | **CHEMISTRY OF ORGANOSULFUR COMPOUNDS: General Problems** |
| Beyer, H. & Walter, W. | **ADVANCED ORGANIC CHEMISTRY** |
| Blazej, A. & Kosik, M. | **PHYTOMASS: A Raw Material for Chemistry and Biotechnology** |
| Brown, S.S. & Kodama, Y. | **TOXICOLOGY OF METALS: Clinical and Experimental Research** |
| Cervinka, O. | **ENANTIOSELECTIVE REACTIONS IN ORGANIC CHEMISTRY** |
| Corey, E.R., Corey, J.Y. & Gasper, P.P. | **SILICON CHEMISTRY** |
| Dunn, A.D. & Rudorf, W.D. | **CARBON DISULPHIDE IN ORGANIC CHEMISTRY** |
| Foá, V., Emmet, E.A., Moroni, M. & Colonksi, A. | **OCCUPATIONAL AND ENVIRONMENTAL CHEMICAL HAZARDS: Cellular and Biochemical Indices of Monitoring Toxicity** |
| Fresenius, P. | **ORGANIC CHEMICAL NOMENCLATURE** |
| Hudlicky, M. | **CHEMISTRY OF ORGANIC FLUORINE COMPOUNDS: A Laboratory Manual, 2nd (Revised) Edition** |
| Hudlicky, M. | **REDUCTIONS IN ORGANIC CHEMISTRY** |
| Kova, J., Krutosikova, A. & Kada, R. | **CHEMISTRY OF HETEROCYCLIC COMPOUNDS** |
| Lukevics, E., Pudova, O. & Sturkovich, R. | **MOLECULAR STRUCTURE OF ORGANOSILICON COMPOUNDS** |
| Maksic, Z.B. | **MODELLING OF STRUCTURE AND PROPERTIES OF MOLECULES** |
| Manitto, P. | **BIOSYNTHESIS OF NATURAL PRODUCTS** |
| Mason, J.M. | **PRACTICAL SONOCHEMISTRY: User's Guide to Applications in Chemistry and Chemical Engineering** |
| Neidleman, S.L. & Geigert, J. | **BIOHALOGENATION: Principles, Basic Roles and Applications** |
| O'Hagan, D. | **THE POLYKETIDE METABOLITES** |
| Pizey, J.S. | **SYNTHETIC REAGENTS: Volume 3: Diborane; 2,3-Dichloro-5,6-Dicyanobenzoquinone DDQ; Iodine; Lead Tetra-acetate Lithium Aluminium Hydride** (reprinted from Synthetic Reagents, Volume 1) |
| Sakurai, H. | **ORGANOSILICON AND BIOORGANOSILICON CHEMISTRY: Structure, Bonding, Reactivity and Synthetic Application** |
| Smith, E.H. | **COMPARATIVE OXIDATION OF ORGANIC COMPOUNDS** |
| Smith, K. | **SOLID SUPPORTS AND CATALYSTS IN ORGANIC SYNTHESIS** |
| Traven, V.F. | **FRONTIER ORBITALS AND PROPERTIES OF ORGANIC MOLECULES** |
| Trinajstic, N., Knop, J.V., Müller, W.R. & Szymanski, K. | **COMPUTATION CHEMICAL GRAPH THEORY: Generation of Structures by Computer Methods** |
| Tundo, P. | **CONTINUOUS FLOW METHODS IN ORGANIC SYNTHESIS** |
| Vernin, G. | **THE CHEMISTRY OF HETEROCYCLIC FLAVOURING AND AROMA COMPOUNDS** |
| Lukevicś, E. & Zablocka, A. | **NUCLEOSIDE SYNTHESIS: Organosilicon Methods** |

# FRONTIER ORBITALS AND PROPERTIES OF ORGANIC MOLECULES

**VALERY F. TRAVEN**
Professor of Organic Chemistry, D.I. Mendeleev Institute of Chemical Technology, Moscow, Russia

***Translation Editor & Series Editor:***
Dr JOHN MELLOR
Department of Chemistry, University of Southampton

**ELLIS HORWOOD**
NEW YORK LONDON TORONTO SYDNEY TOKYO SINGAPORE

First published in 1992 by
**ELLIS HORWOOD LIMITED**
Market Cross House, Cooper Street,
Chichester, West Sussex, PO19 1EB, England

A division of
Simon & Schuster International Group
A Paramount Communications Company

***Translated by:***
ALEXANDER A. ROSINKIN
D.I. Mendeleeu Institute Moscow



Printed and bound in Great Britain
by Bookcraft, Midsomer Norton

British Library Cataloguing in Publication Data

A Catalogue Record for this book is available from the British Library

ISBN 0–13–327487–X

Library of Congress Cataloging-in-Publication Data

Available from the publishers

# Table of contents

**Foreword: Frontier orbitals versus resonance structures** ix
**Preface** xi
**Abbreviations used in this work** xiii

PART I METHODS OF STUDYING ELECTRON STRUCTURE OF MOLECULES

**1 Methods based on classical physics** 3

**2 Methods of quantum chemistry** 13
2.1 Some principles of quantum mechanics 13
2.2 Concepts of molecular orbital theory 20
2.2.1 General characteristics of calculation methods 20
2.2.2 Symmetry of molecules, orbitals and states 32
2.3 Experimental methods 36
2.3.1 Photoelectron spectroscopy 37
2.3.2 Electron transmission spectroscopy 41
2.3.3 Polarographic methods 44
2.3.4 Electron spin resonance spectroscopy 46
2.3.5 Electron absorption spectroscopy 48
2.4 Estimation of orbital interactions 53
2.4.1 Fragment (group) orbitals 53
2.4.2 Orbital interaction rules 55
2.4.3 Parametrization of orbital interactions by PE spectroscopy 57

PART II ORBITAL STRUCTURE AND PROPERTIES OF ORGANIC MOLECULES

**3 Basic concepts** 67

3.1 Orbital structure and reactivity 67
3.2 Orbital structure and biological activity 75
3.3 Orbital structure and conformational effects 80
3.4 Orbital structure, colour and phototropy 88
3.5 Effect of self-association and medium on orbital structure of molecules 95

**4 Alkanes** 102
4.1 Acyclic alkanes 102
4.1.1 Methane 102
4.1.2 Ethane 106
4.1.3 Medium acyclic alkanes 109
4.2 Cycloalkanes 114
4.3 Some features of the electron structure of alkanes and their organoelement analogues 117
Problems 123

**5 Alkenes and Alkynes** 124
5.1 Alkenes 124
5.1.1 Ethylene 124
5.1.2 Linear and cyclic alkenes 127
5.2 Alkynes 130
5.3 Electron structure of polyenes and Woodward–Hoffmann rules 132
5.3.1 $4n\pi$ electron systems: butadiene-1,3 134
5.3.2 $(4n + 2)\pi$ electron system: hexatriene-1,3,5 139
5.3.3 Cyclohexadiene-1,3 and cyclohexadiene-1,4 140
5.3.4 Norbornadiene and related compounds 141
Problems 146

**6 Aromatic hydrocarbons** 147
6.1 Annulenes 147
6.1.1 [4]Annulene (cyclobutadiene) 148
6.1.2 [6]Annulene (benzene) 148
6.1.3 [10]Annulenes 155
6.2 Alkylbenzenes 158
6.2.1 Toluene 158
6.2.2 tert-Butylbenzene, neopentylbenzene and their organoelement analogues 162
6.2.3 α-(Alkyl)phenylcyclopropanes 167
6.2.4 Styrene and stilbene 167
6.2.5 Ethynylbenzenes 172
6.3 Polycyclic aromatic hydrocarbons 174
6.3.1 Biphenyl, terphenyl, paracyclophanes 174
6.3.2 Naphthalene 179
6.3.3 Anthracene and phenanthrene 184
6.3.4 Higher polycyclic hydrocarbons 187

6.3.5 Frontier orbitals and reactivity of polycyclic aromatic hydrocarbons 199
Problems 204

**7 Halogenohydrocarbons** 205
7.1 Haloalkanes 205
7.2 Haloethylenes 211
7.3 Halobenzenes 212
Problems 218

**8 Oxygen and sulphur compounds** 219
8.1 Aliphatic alcohols, ethers, and sulphides 219
8.1.1 Methanol and dimethyl ether 220
8.1.2 Organoelement analogues of dimethyl ether 225
8.1.3 Linear and cyclic ethers and sulphides 226
8.1.4 Crown ethers and their analogues 228
8.2 Disulphides 230
8.3 Vinyl ethers and sulphides 232
8.4 Aromatic hydroxy compounds and ethers 235
8.4.1 Phenol 235
8.4.2 Anisole and its homologues 238
8.4.3 Organoelement analogues of anisole 239
8.5 Aldehydes and ketones 242
8.5.1 Formaldehyde 243
8.5.2 Acetaldehyde 244
8.5.3 Acetone and related compounds 245
8.5.4 Acrolein 251
8.5.5 Benzaldehyde 254
8.5.6 Acetophenone 254
8.5.7 Phenalenone 255
8.5.8 Benzanthrone 257
8.5.9 Diketones 260
8.5.10 Benzoquinone, 1,4,-naphthoquinone, 9,10-anthraquinone 261
8.6 Carboxylic acids 263
8.6.1 Acetic acid 263
8.6.2 Benzoic acid 265
Problems 267

**9 Nitrogen compounds** 269
9.1 Amines 269
9.1.1 Ammonia and its analogues 269
9.1.2 Trimethylamine and its organoelement analogues 271
9.1.3 Trialkylamines and azocycloalkanes 272
9.1.4 Transannular effects in bicyclic amines and diamines 274
9.1.5 Enamines 280
9.1.6 Aniline 283

9.1.7 N,N-dimethylaniline and its organoelement analogues 284
9.1.8 Other arylamines 287
9.2 Hydrazo compounds 289
9.2.1 Aliphatic hydrazo compounds 289
9.2.2 Hydrazobenzenes 292
9.3 Azo compounds 294
9.3.1 Azomethane and other azoalkanes 295
9.3.2 Diazirines 298
9.3.3 Cyclic azoalkanes 299
9.3.4 Azobenzenes 301
9.3.5 Electron absorption spectra of azo compounds 305
9.4 Nitroso and nitro compounds 310
9.4.1 Nitrosomethane 310
9.4.2 Nitromethane 311
9.4.3 Nitrosobenzene 313
9.4.4 Nitrobenzene 314
9.5 Nitriles 316
9.5.1 Acetonitrile 316
9.5.2 Cyanoalkenes 317
9.5.3 Benzonitrile 317
Problems 319

**10 Orbital-controlled properties of substituted aromatic hydrocarbons** 321
Problems 329

**11 Heterocyclic compounds** 331
11.1 Five-membered heterocycles 331
11.1.1 Pyrrole 331
11.1.2 Furan 332
11.1.3 Thiophene, selenophene, tellurophene 334
11.2 Six-membered heterocycles 336
11.2.1 Pyridine 336
11.2.2 Other heterocyclic analogues of benzene 339
11.2.3 Diazines and sym-triazines 342
11.3 Polycyclic heteroaromatic compounds 342
11.4 Orbital-controlled properties of heteroaromatic molecules 348
Problems 354

**Conclusion** 355

**Answers to problems** 357

**Appendix** 363

**References** 371

**Index** 387

# Foreword: Frontier orbitals versus resonance structures

Since the discovery of electrons, chemists have been looking for the most effective way to visualize how electrons are spread in organic molecules and how they participate in chemical reactions.

G. Lewis proposed that each pair of electrons forming a chemical bond might be shared by a pair of atoms. Electrons may be shifted along the bond depending on the difference in the electronegativity of those atoms. Although the shift might be rather definite, nevertheless, in terms of the inductive effect, electrons should still be located between two atoms.

R. Robinson developed the concept of mesomerism. According to this concept, electrons can be shared by more than two atoms in molecules containing double bonds. In conjunction with the theory proposed by L. Pauling, resonance structures and resonance hybrids were suggested as the best way to show the distribution of electrons in organic molecules and the corresponding intermediates. However, the resonance structures seem to be rather artificial or even useless in many cases. One should note the need for at least seven resonance structures of the intermediate to explain the preferability of the α-attack in electrophilic substitution of naphthalene. An explanation of the results of substitution in polycyclic aromatic hydrocarbons such as anthracene or benz[a]anthracene in terms of the resonance theory is far more complicated.

Quantum mechanics suggests a different way of understanding how electrons are spread in organic molecules. Even simple calculations by the Hückel method afford energies and coefficients of molecular orbitals, which, in turn, allow the evaluation of the electron densities in unsaturated organic molecules. K. Fukui, R. Woodward and R. Hoffmann have developed the concepts explaining many reactions in terms of molecular orbitals.

Applications are not restricted to well-known pericyclic reactions. Electrophilic aromatic substitution can be considered in terms of frontier orbitals. Following

K. Fukui, one can see that the highest occupied molecular orbital of naphthalene, with its higher coefficients at the $\alpha$-positions, determines the character of its interaction with electrophiles. The same explanation seems to be quite reasonable for the substitution in polycyclic aromatic hydrocarbons as well.

Recently, physicists and physical chemists have been very successful in showing the reliability of orbital parameters obtainable by quantum chemical calculations of organic molecules. Experimental data about occupied electron levels of molecules are, in particular, available from photoelectron spectra. These methods have been called 'experimental quantum chemistry'. Therefore, both energies and shapes of molecular orbitals of key organic compounds are quite trustworthy at the present time and can be used as definite parameters, just as is the case with the geometric parameters of molecules.

Orbital parameters available from quantum chemical calculations and electron spectroscopy are applicable not only to the gas phase. In many cases, these parameters are confirmed, for example, by ESR spectra of the corresponding ion radicals recorded in solution. In this sense, 'experimental quantum chemistry' seems to be one of the methods of structural chemistry. We can follow Dunitz and be reasonably confident that any particular arrangement of atoms observed in a molecular crystal cannot be far from an equilibrium structure of the isolated molecule' [1].

In this approach, the orbital concepts may be more extensively used even in undergraduate courses of organic chemistry since the undergraduate has a reasonable basis for understanding of chemical bonding in terms of orbitals from the introductory courses in chemistry and courses in structural chemistry.

# Preface

Organic chemists who are not interested in evaluation of molecular orbitals do not have to read Part I of this book. They can see how energies and coefficients of orbitals contribute to the understanding of organic chemistry in Part II, where the frontier orbitals of molecules are mostly discussed since they are predominantly involved in chemical interactions.

Part II begins with the survey of the properties of organic molecules which can be understood in terms of the orbital concepts. Appropriate definitions and generalizations are given in Chapter 3 of the book.

To make the spectral and calculation data more attractive (and useful!) for organic chemists, problems are given at the end of each chapter of Part II. The effectiveness of the frontier orbital concept compared with the octet rule, the model of hybridized orbitals and resonance structures can be seen when solving these problems. Answers to the problems are given at the end of the book. Some data of experimental quantum chemistry and HMO calculations, which might be useful in solving the problems, are given in the Appendix.


**ACKNOWLEDGEMENTS**

I am indebted to Professor Boris Stepanov of Moscow D. Mendeleev Institute of Chemical Technology, Professor Nikolai Zefirov and Professor Yurii Ustynyuk of the Moscow University, and to Professor Dennis J. Sardella of Boston College, USA, who reviewed some parts of the manuscript, for their useful criticism and suggestions. It is, however, the author alone who is responsible for any errors or inaccuracies in the final draft. I am also thankful to Tatyana Chibisova, Vladimir Redchenko, Andrei Safronov, and Sergei Khaljapin for their help in preparing this manuscript for print.

I also thank the various authors and editors who have so kindly given permission to reproduce figures, tables and diagrams from their papers. The corresponding references are made in the appropriate parts of the text.


Valery F. Traven

# Abbreviations used in this work

| | |
|---|---|
| AO | Atomic orbital |
| CI | Configurational interaction |
| CNDO | Complete neglect of differential overlap |
| CT | Charge transfer |
| DZ | Double-zeta basis |
| EAS | Electron absorption spectroscopy |
| EHMO | Extended Hückel molecular orbital method |
| ESR | Electron spin resonance |
| ETS | Electron transmission spectroscopy |
| GF | Gaussian function |
| GTO | Gaussian type orbital |
| HAM/3 | Hydrogen atom in molecule |
| HFI | Hyperfine interaction |
| HMO | Hückel molecular orbital method |
| HOMO | Highest occupied molecular orbital |
| HSAB | Hard and soft acids and bases |
| INDO | Incomplete neglect of differential overlap |
| INDO/S | Incomplete neglect of differential overlap with spectroscopic parametrization |
| LC | Liquid crystal |
| LCAO | Linear combination of atomic orbitals |
| LCFO | Linear combination of fragment orbitals |
| LUMO | Lowest unoccupied molecular orbital |
| MINDO | Modified incomplete neglect of differential overlap |
| MNDDO | Modified neglect of diatomic differential overlap |
| MO | Molecular orbital |
| NDDO | Neglect of diatomic differential overlap |
| PAH | Polycyclic aromatic hydrocarbon |
| PES | Photoelectron spectroscopy |

| | |
|---|---|
| PPP | Parr–Pariser–Pople method |
| PRDDO | Partial retention of diatomic differential overlap |
| SCF | Self-consistent field |
| SPINDO | Spectroscopic potential adjusted for INDO |
| STO | Slater type orbital |
| STO4-31G | DZ basis in which the inner (non-valence) orbitals are approximated near the nucleus by four GTO and far from the nucleus by one GTO, while valence orbitals are approximated by three and one GTO |
| STO6-31G | DZ basis extended at the expense of addition of polarization d-AO for all atoms of the 2nd period elements |
| TB effect | Through bonds effect |
| TCNE | Tetracyanoethylene |
| TS effect | Through space effect |
| UEP | Unshared electron pair |
| UV spectroscopy | Ultra-violet spectroscopy |

# Part I

# Methods of studying electron structure of molecules

# 1

# Methods based on classical physics

The electron structure of organic molecules has become the subject of studies since early this century. In 1897, it was established that 'indivisible' atoms are actually composed of positively charged nuclei and negatively charged electrons. In his experiments, J. Thomson showed that electron shells of atoms of various elements differ in their stability. The electron shells of the noble gases are especially stable. The atoms of other elements can, under certain conditions, lose or accept electrons, and the number of electrons in their outer shells can thus become the same as in the atoms of the noble gases: 2, 8, 18, etc. Based on these facts, A. Kossel (1916) explained the formation of the **ionic bond** as follows: as the atoms of different elements combine, the atom of one element gives off its electrons while the atom of the other element accepts them, and both atoms thus acquire stable electron shells.

In the same year of 1916, G. Lewis proposed a concept of formation of covalent bonds that are characteristic of organic molecules, and formulated the **octet rule**. According to this concept, the combining atoms share an electron pair rather than transfer of the electron from one to the other. The electron shell of one atom is completed to the stable shell of a noble gas atom by the electrons of the other atom: the atom tends to form bonds until its outermost shell becomes a complete octet of electrons (for the elements of the second period).

The octet rule has played an important role in the development of the structural concepts in organic chemistry. Dashes in structural formulas of chemical compounds were interpreted by chemists as pairs of electrons. The concept of electron pairs that bond covalently the atoms in organic molecules (concept of **covalent bond**) and move from one atom to the other under various effects has proved quite fruitful, and it still remains valid within the framework of the classical theory of structure [2–6].

The electron effects in organic molecules depend on the differences in the electronegativity of the elements that form these molecules. Several scales of **electronegativity** of chemical elements have been proposed. Mulliken gave an adequate definition: the degree of electronegativity of an atom $\chi_m$ is an arithmetic mean of its first ionization potential $IP_1$ and its electron affinity $EA_1$ [7]:

$$\chi_m = (IP_1 + EA_1)/2$$

According to Pauling, electronegativity is estimated by the energy of bonds. Pauling's electronegativity values are more commonly used now. (Electronegativities of some elements according to Pauling are given in Table 1.1).

**Table 1.1.** Pauling electronegativity values for some elements (parenthesized are electronegativities according to Mulliken)

| IA | IIA | IIIA | IVA | VA | VIA | VIIA |
|---|---|---|---|---|---|---|
| H<br>2.1(−) | | | | | | |
| Li<br>1.0(0.8) | Be<br>1.5(2.2) | B<br>2.0(1.9) | C<br>2.5(2.5) | N<br>3.0(3.7) | O<br>3.5(3.5) | F<br>4.0(3.9) |
| Na<br>0.9(0.7) | Mg<br>1.2(1.2) | Al<br>1.5(1.6) | Si<br>1.8(2.2) | P<br>2.1(2.8) | S<br>2.5(2.6) | Cl<br>3.0(3.0) |
| K<br>0.8(0.8) | Ca<br>1.0(−) | Ga<br>1.6(1.9) | Ge<br>1.8(−) | As<br>2.0(1.8) | Se<br>2.4(2.2) | Br<br>2.8(2.8) |

The different electronegativity of various atoms explains the shift of electron pairs in the bonds and polarity of the molecule on the whole. Several mechanisms of electron shifts in molecules have been proposed, such as the inductive effect, the field effect, conjugation, etc.

The **inductive effect**, which is a shift of an electron pair due to the different electronegativities of atoms in a molecule, can be estimated, e.g. by the change in the acidity of carboxylic acids. Thus, for example, monochloroacetic acid is about 100 times stronger than acetic acid, while trichloroacetic acid is 10 000 times as strong (Table 1.2) due to the higher electronegativity of chlorine compared to the hydrogen atom.

**Table 1.2.** Values of $pK_a$ for some carboxylic acids R-COOH

| R | $pK_a$ | R | $pK_a$ | R | $pK_a$ |
|---|---|---|---|---|---|
| $CF_3$ | 0.23 | $CH_2SO_2CH_3$ | 2.36 | $CH_2Cl$ | 2.85 |
| $CCl_3$ | 0.66 | $CH_2CN$ | 2.47 | $CH_2Br$ | 2.89 |
| $CHCl_2$ | 1.25 | $CH_2SCN$ | 2.58 | $CH_2I$ | 3.16 |
| $CH_2NO_2$ | 1.68 | $CH_2F$ | 2.57 | | |

(From Albert, and Serjeant, (1973), *Determination of ionization constants*, Chapman and Hall, London [8])

The inductive effects of atoms and substituents are universal: while effecting acidity, they can also change the basicity of organic compounds. Thus, for example, the increasing electronegativity of substituent R decreases the basicity of the amine $R—NH_2$ (see Table 1.3).

The **field effect** is also determined by the differences in the electronegativity of atoms, but as distinct from the inductive effect, it acts not through the bond but through space.

The **conjugation** effect in organic molecules is illustrated by the following examples.

**Table 1.3.** Values of $pK_a$ for conjugated acids of amines $R\text{-}NH_2$

| R | $pK_a$ | R | $pK_a$ | R | $pK_a$ |
|---|---|---|---|---|---|
| $CH_3$ | 10.62 | $C(CH_3)_3$ | 10.45 | $CH_2Ph$ | 9.34 |
| $CH_2CH_3$ | 10.63 | $CH_2COOCH_3$ | 7.66 | $CH_2CH{=}CH_2$ | 9.69 |
| $CH(CH_3)_2$ | 10.63 | $CH_2CONH_2$ | 7.93 | $CH_2CN$ | 5.34 |

(From Albert, and Serjeant, (1973), *Determination of ionization constants*, Chapman and Hall, London [8])

Conjugation of double bonds in a molecule of buta-1,3-diene accounts for a considerable reduction in the heat of hydrogenation of this diene as compared with alkenes and alkadienes, in which double bonds are not conjugated:

| Hydrocarbon | $H^0$, kJ/mol [5] |
|---|---|
| $CH_3CH_2CH{=}CH_2$ | −126.4 |
| $CH_2{=}CH{-}CH_2{-}CH{=}CH_2$ | −252.8 |
| $CH_2{=}CH{-}CH_2{-}CH_2{-}CH{=}CH_2$ | −251.2 |
| $CH_2{=}CH{-}CH{=}CH_2$ | −236.6 |

Delocalization of the electron density with involvement of the conjugation effects is especially important for the interaction of substituents with the adjacent charged centre. Thus, the higher acidity of carboxylic acids compared with alcohols is explained by delocalization of the negative charge over three atoms in the carboxylate ion (due to the conjugation effect) rather than fixation on one atom in the alcoholate ion:

$$\left[R{-}C({=}O){-}O^- \longleftrightarrow R{-}C(O^-){=}O\right] \equiv R{-}C(\cdots O)(\cdots O)^-$$

This delocalization was confirmed by direct independent experiment. Electron diffraction findings and x-ray structural analysis of the molecule of formic acid and its sodium salt [5] demonstrate complete equivalence of the CO bonds in the anion (nm):

$$R{-}C({=}O)_{0.123}({-}OH)_{0.136} \qquad \left[R{-}C(\cdots O)_{0.127}(\cdots O)_{0.127}\right]^- \; Na^+$$

This effective delocalization is not feasible with the $R{-}O^-$ anions that are formed by acid dissociation of alcohols. These anions are far less stable, while alcohols are, as a result, much weaker acids.

Heteroatoms with unshared electron pairs and methyl groups are capable of effective conjugation with an electron-deficient centre to form corresponding carbocations. Thus, the methoxymethyl cation has been prepared in the form of a stable solid complex $[CH_3OCH_2^+][SbF_6^-]$ [5,9].

The energy of resonance stabilization of the carbo-cation increases with the number of methyl groups at a positively charged carbon atom due to the hyperconjugation effect:

| $CH_3—CH_2^+$ | $(CH_3)_2CH^+$ | $(CH_3)_3C^+$ |
|---|---|---|
| 150.7 kJ/mol | 276.3 kJ/mol | 351.7 kJ/mol |

It is difficult to differentiate qualitatively between the electron effects of substituents in organic molecules. Nevertheless, attempts were made within the framework of the $\sigma\rho$-method, or correlation analysis.

While studying the dependence of dissociation of meta- and para-substituted benzoic acids $X—C_6H_4—COOH$ on substituent X, G. Hammett (1935) formulated the concept of the **σρ-method** and proposed the equations that were given his name [10]:

$$\begin{aligned} \log k_x &= \log k_0 + \rho\sigma \\ \log K_x &= \log K_0 + \rho\sigma \end{aligned} \tag{1.1}$$

where $k_x$ and $K_x$ are rate and equilibrium constants of benzene derivatives with X substituents respectively; $k_0$ and $K_0$ are the same constants for the derivatives of benzene without the substituent X. Constant $\sigma$ of the substituent X in these equations measures the polar (electron) effect of the hydrogen atom substitution for X. Its value depends on the substituent position ($\sigma_{meta}$ and $\sigma_{para}$ constants respectively) but does not depend on the nature of the reaction or its conditions. The **electron-withdrawing** substituents have positive values of $\sigma$ and the **electron-donating** substituents negative ones.

The **reaction constant** $\rho$ depends on the nature of the reaction (and also on the conditions of the reaction, such as the solvent, temperature, etc.) and is the measure of sensitivity of a given reaction to electron effects of substituents. For example, the reactions with positive $\rho$ values are accelerated on the introduction of an electron-withdrawing substituent and decelerated by substitution of the hydrogen atom by the electron-donating groups.

Using only two tables (which give the $\sigma$ and $\rho$ constants) one can calculate the relative rates of hundreds of reactions. Using, for example, the value of $\sigma$ for the m-$NO_2$ group (+0.710) and the value of $\rho$ for ionization of benzoic acid at 25°C (1.0), it is easy to find that the dissociation constant of m-nitrobenzoic acid is 6.13 times higher than that of benzoic acid. Using the same values of $\sigma$ and $\rho$ for the acid-catalysed hydrolysis of benzamide in 60% alcohol at 80°C (−0.298) it is easy to determine that the rate of hydrolysis of m-nitrobenzamide is only 0.615 of the rate of hydrolysis of unsubstituted benzamide.

Equations similar to Hammett's are commonly used because of their simplicity. Jaffe used the simple Hammett equation for more than 400 reaction series and found out that it held to an accuracy of 15% for most of them. This can be explained by the fact that Hammett's equation expresses objectively the principle of linearity of free energies of reactions for the related compounds.

At the same time, Jaffe's statistics showed that many reactions do not obey the simple correlation equations of the Hammett's equation type. Besides, it was shown that the $\sigma_{para}$ constants account for the summary effects of substituents that are transmitted to the reaction centre by two mechanisms: the resonance mechanism (the conjugation effect) which is characterized by the constant $\sigma_R$ and the field mechanism, characterized by the constant $\sigma_I$:

$$\sigma_{para} = \sigma_R + \sigma_I.$$

Further development of the quantitative theory of organic reactions gave birth to many new constants [11–13]. It has thus been established that the reactions that proceed with substitution of hydrogen at the carbon atom of the benzene ring through an electron-poor or electron-rich transition state, should be better described by the substituent constants designated $\sigma^+$ and $\sigma^-$ respectively. It was found that the resonance constants $\sigma_R$ also depend on the type of the reaction. In order to separate the overall electron effect into its component parts, the $\sigma^0$ constants were determined by the ionization constant for substituted phenylacetic acids. These constants $\sigma^0$ do not include the effect of direct interaction of the $\pi$ system of the phenyl group with the reaction centre. A new set of $\sigma_R^0$ constants characterizing the resonance effects of substituents ($\sigma_R^0 = \sigma_{para}^0 - \sigma_I$) was thus derived. More than ten different constants can now be used to describe various substituent effects at the para position of the benzene ring (Table 1.4).

Swain and Lupton analysed more that 40 types of various constants and showed that they were not independent. They showed that linear combinations of $F$ values (characterizing the contribution of the field effect) and $R$ values (the share of the resonance effect) characterize satisfactorily at least 43 types of constants $\sigma$. Each type is given in the form:

$$\sigma = fF + rR$$

where $f$ and $r$ are the respective weight coefficients.

Table 1.5 gives the parameters of $F$ and $R$ for some substituents. This approach can be used to estimate quantitatively the contribution, e.g., of the resonance effect to various constants: 22% to the $\sigma_{meta}$ constant, 53% to the $\sigma_{para}$ constants, 66% to the $\sigma_{para}^+$, and 92% to the $\sigma_{para} - \sigma_{meta}$ difference.

In the terms of the $\sigma\rho$-method, numerous attempts were made to separate the electron and steric constituent effects. Determining substituent constants at the ortho position of the benzene ring relative to the reaction centre turned out to be more difficult than determining the substituent constants at the meta- and para positions. The difficulties were explained by the so-called **ortho-effect**, i.e. the effect that includes partly a steric influence of the substituent. The analysis of some reaction series made it possible to estimate the steric effects of substituents. Acid-catalyzed hydrolysis of carboxylic esters in aqueous acetone turned out to be a suitable reaction: it is insensitive to polar effects of substituents (see Table 1.6).

At a certain stage of evolution of the quantitative theory of organic reactions, an

**Table 1.4.** Constant $\sigma$ values for some substituents

| Substituent | $\sigma_{para}$ | $\sigma_{meta}$ | $\sigma^+_{para}$ | $\sigma^+_{para}$ | $\sigma^-_{para}$ | $\delta_I$ | $\delta_R$ | $\sigma^0_{meta}$ | $\sigma^0_{para}$ | $\sigma^0_R$ | $\sigma^+_R$ | $\sigma^-_R$ |
|---|---|---|---|---|---|---|---|---|---|---|---|---|
| $NH_2$ | −0.66 | −0.16 | −1.3 | −0.16 | — | 0.19 | −0.79 | −0.09 | −0.30 | −0.48 | −1.61 | — |
| $N(CH_3)_2$ | −0.60 | −0.21 | −1.7 | — | — | — | — | — | — | — | — | — |
| $O^-$ | −0.52 | −0.71 | — | — | — | — | — | — | — | — | — | — |
| OH | −0.36 | −0.002 | −0.92 | — | — | 0.25 | −0.63 | 0.02 | −0.20 | −0.43 | −1.20 | — |
| $OCH_3$ | −0.27 | −0.12 | −0.78 | 0.05 | — | 0.27 | −0.61 | 0.10 | −0.14 | −0.45 | −1.02 | — |
| $C(CH_3)_3$ | −0.20 | −0.12 | −0.26 | −0.06 | — | — | — | — | — | — | — | — |
| $CH_3$ | −0.17 | −0.07 | −0.31 | −0.10 | — | −0.05 | −0.11 | −0.06 | −0.14 | −0.10 | −0.25 | — |
| H | 0 | 0 | 0 | 0 | 0 | 0 | 0 | 0 | 0 | 0 | — | — |
| $C_6H_5$ | 0.01 | 0.22 | −0.18 | 0.11 | — | — | — | — | — | — | — | — |
| F | 0.06 | 0.34 | −0.07 | 0.35 | — | 0.50 | −0.45 | −0.34 | 0.15 | −0.34 | −0.57 | — |
| Cl | 0.23 | 0.37 | 0.11 | 0.40 | — | 0.46 | −0.23 | 0.37 | 0.24 | −0.23 | −0.36 | — |
| Br | 0.23 | 0.39 | 0.15 | 0.41 | — | 0.44 | −0.20 | 0.37 | 0.26 | −0.19 | −0.30 | — |
| COOH | 0.27 | 0.36 | 0.42 | 0.32 | 0.73 | — | — | — | — | — | — | — |
| I | 0.28 | 0.35 | 0.14 | 0.36 | — | 0.39 | −0.17 | 0.34 | 0.28 | −0.14 | −0.25 | — |
| CHO | 0.45 | 0.36 | — | — | 1.13 | — | — | — | — | — | — | — |
| $COCH_3$ | 0.52 | 0.31 | — | — | 0.87 | 0.28 | 0.20 | 0.36 | 0.47 | 0.16 | — | 0.56 |
| CN | 0.63 | 0.68 | 0.66 | 0.56 | 1.00 | — | — | — | — | — | — | — |
| $NO_2$ | 0.78 | 0.71 | 0.79 | 0.67 | 1.27 | 0.65 | 0.15 | 0.71 | 0.81 | 0.15 | — | 0.60 |
| $N(CH_3)_3{}^+$ | 0.86 | 0.90 | 0.41 | 0.36 | — | — | — | — | — | — | — | — |

(From March, J. (1977), *Advanced Organic Chemistry*, McGraw-Hill, N.Y. [5] and Shorter, J. (1982), *Correlation analysis of organic reactivity with particular reference to multiple regression*, Res. Stud. Press Chichester [13]).

**Table 1.5.** *F* and *R* values for some substituents

| Substituent | *F* | *R* | Substituent | *F* | *R* |
|---|---|---|---|---|---|
| $C(CH_3)_3$ | −0.104 | −0.138 | $CF_3$ | 0.631 | 0.186 |
| $CH_2CH_3$ | −0.065 | −0.114 | I | 0.672 | −0.197 |
| $CH_3$ | −0.052 | −0.141 | Cl | 0.690 | −0.161 |
| H | 0 | 0 | F | 0.708 | −0.336 |
| $NH_2$ | 0.037 | −0.681 | Br | 0.727 | −0.176 |
| Ph | 0.139 | −0.088 | CN | 0.847 | 0.184 |
| $OCH_3$ | 0.413 | −0.500 | $NO_2$ | 1.109 | 0.155 |
| OH | 0.487 | −0.643 | $\overset{+}{N}(CH_3)_3$ | 1.460 | 0 |
| COOH | 0.552 | 0.140 | $\overset{+}{N}_2$ | 2.760 | 0.360 |

(From March, J. (1977), *Advanced Organic Chemistry*, McGraw-Hill, NY [5])

**Table 1.6.** Steric constants $E_s$ for some substituents

| Substituent | $E_s$ | Substituent | $E_s$ | Substituent | $E_s$ |
|---|---|---|---|---|---|
| H | 1.24 | $CH_2I$ | −0.37 | $C(CH_3)_3$ | −1.54 |
| $CH_3$ | 0 | $CH_2OPh$ | −0.38 | Neopentyl | −1.85 |
| $CH_2CH_3$ | −0.07 | $CH(CH_3)_2$ | −0.47 | $CCl_3$ | −2.06 |
| $CH_2Cl$ | −0.24 | Cyclohexyl | −0.79 | $CBr_3$ | −2.43 |
| $CH_2Br$ | −0.27 | $CF_3$ | −1.16 | $C(CH_2CH_3)_3$ | −3.8 |

(From March, J. (1977), *Advanced Organic Chemistry*, McGraw-Hill, NY [5]).

illusion appeared that the $\sigma$ constant could be used to estimate all types of electron effects of substituents in organic molecules. The replacement of the simple Hammett and similar equations by multi-parameter equations [12,13]:

$$\begin{aligned}\log(k/k_0) &= \rho_i\sigma_i + \rho_\pi^+\sigma_\pi^+ + \rho_\pi^-\sigma_\pi^-,\\ \log(k/k_0) &= \alpha\sigma_{I,X} + \beta\sigma_{R,X} + \psi v_X + h \text{ etc.}\end{aligned} \tag{1.2}$$

seems to be a natural development of the correlation analysis. It was believed that such multi-parameter correlation equations could evaluate various electron effects. It appeared, however, that their practical use is much more complicated than the use of equations of the (1.1) type.

Improved correlation equations (in addition to Hammett's equation, also proposed were the Taff, Edwards, Swain–Scott, and other equations [11]) made them applicable to many new reactions that had not earlier been described quantitatively, but demonstrated also disadvantages of the $\sigma\rho$-method. The difficulties multiplied greatly when an attempt was made to describe quantitatively the reactivity not only of benzene derivatives but also of polycyclic hydrocarbons and their heteroatomic analogues. While, for example, the constant $\sigma$ should be determined only for the three positions (ortho, meta, and para) of the substituent relative to the reaction centre of the benzene derivatives, 14 values of this constant should be determined for each possible substituent position with respect to the reaction centre in the series of substituted naphthalenes: $\sigma_{12}$, $\sigma_{13}$, $\sigma_{14}$, $\sigma_{15}$, $\sigma_{16}$, $\sigma_{17}$, $\sigma_{18}$, $\sigma_{21}$, $\sigma_{23}$, $\sigma_{24}$, $\sigma_{25}$,

$\sigma_{26}, \sigma_{27}, \sigma_{28}$. (The first figure in the subscript designates the position of the reaction centre, and the other figure the position of the substituent).

Corresponding measurements were partly made for the naphthalene series [14, 15]. From the calculated basicity of nitronaphthylamines (assume that the reaction constant $\rho$, that had been earlier calculated for the meta-X-anilines was suitable for the protonation of naphthalene derivatives as well) the following ten values of the $\sigma$ constants were found for the nitro group [15]:

$$\sigma_{13} = 0.625, \quad \sigma_{14} = 1.21, \quad \sigma_{15} = 0.39, \quad \sigma_{16} = 0.34, \quad \sigma_{17} = 0.44$$
$$\sigma_{24} = 0.58, \quad \sigma_{25} = 0.37, \quad \sigma_{26} = 0.52, \quad \sigma_{27} = 0.34, \quad \sigma_{28} = 0.48$$

It is easy to imagine the tremendous range of measurements and calculations that would be necessary for quantitative description of the interactions between the reaction centre and variable substituents in aromatic hydrocarbons with more than two benzene rings in the molecule. No wonder therefore that there are only few reports on futile attempts of such descriptions in the fluorene, anthracene and phenanthrene series [12].

Estimation of the electron structure of molecules within the framework of the $\sigma\rho$ method (with separation of the total substituent effect into components) by the determination of rate and equilibrium constants of the corresponding organic reactions is also limited by some factors. Such estimations describe only 'collective' contribution of all electrons of the molecule to any component of the total electron effect. This is due to the fact that the correlation equations are based on the principle of linearity of free energies, while the value of free energy is the 'collective' property of the corresponding molecular system.

Estimations of the electron effects in organic molecules on the basis of the electron concepts and the $\sigma\rho$ method by measuring dipole moments, the Kerr effect, chemical shifts in the NMR spectra, and force constants of bonds in the IR spectra have the same limitations, since they also evaluate collective effects of all electrons of the molecule [16].

Thus the electric dipole moment of the molecule, which is the product of the radius-vector $\boldsymbol{l}$, directed away from the centre of gravity of the positive charge toward the gravity centre of the negative charge, and the absolute value of the charge $q$:

$$\boldsymbol{\mu} = q\boldsymbol{l} \tag{1.3}$$

depends on the electron shift and is an integral property of the molecule, because it is the vector sum of moments of separate bonds.

The separation of the total electron effects into components by dipole moments within the framework of the electron concepts is based on the selection of suitable models. Thus, from the comparison of the dipole moments of nitromethane and nitrobenzene:

| $CH_3—X$ | $C_6H_5—X$ |
|---|---|
| $X = NO_2, \quad \mu = 3.10$ D | $X = NO_2, \quad \mu = 4.01$ D |
| $X = NH_2, \quad \mu = 1.46$ D | $X = NH_2, \quad \mu = 1.53$ D |

the negative mesomeric effect of the nitro group is evaluated as $-0.76$ D, while the positive mesomeric effect of the amino group (1.02 D) is evaluated from the comparison of the $\mu$ of methylamine and aniline [17].

The use of IR spectroscopy in the estimation of the electron structure depends also on the charge characteristics of the atoms in the molecule, since its ability to absorb IR radiation is based on the interaction of its oscillating electric dipole moment with the oscillating electric component of radiation [16]. The energy difference of two adjacent oscillating levels $E_n$ and $E_{n+1}$ for the harmonic oscillator is determined by the equation:

$$\Delta E = E_{n+1} - E_n = h\nu = \left(\frac{h}{2\pi}\right)\left(\frac{k}{m}\right)^{1/2}, \quad \nu = \left(\frac{1}{2\pi}\right)\left(\frac{k}{m}\right)^{1/2}, \tag{1.4}$$

where $k$ is the force constant of the bond between fragments $A$ and $B$; and $m$ is the mean mass of the fragments $A$ and $B$.

The equations (1.4) relate unambiguously the recorded frequencies of the corresponding oscillations with the force constants and are therefore used for estimation of the electron effects in molecules by IR spectra. For example, the dependence of frequency of valence oscillations of the carbonyl group on the electron properties of the substituent is illustrated by the following $\nu_{C=C}$ values for some p-substituted acetanilides $X—C_6H_4NHCOCH_3$: greater $\nu_{C=O}$ values correspond to a stronger electron acceptor X:

| X | $OCH_3$ | $C_6H_5$ | $CH_3$ | Br | CN | $NO_2$ |
|---|---|---|---|---|---|---|
| $\nu_{C=O}$ cm$^{-1}$ | 1648 | 1689 | 1691 | 1698 | 1706 | 1712 |

The use of NMR spectroscopy is based on the dependence of the nuclear magnetic moment on the electron distribution in the molecule. In the absence of the external magnetic field, all orientations of the nucleus moment are degenerate. The degeneration is removed in the applied external magnetic field $H_0$. The signal can be seen in the NMR spectrum due to a nuclear transition between the corresponding energy states.

Since the electrons screen the nucleus, the magnetic field on the nucleus $H_n$ differs from the $H_0$ of the applied magnetic field:

$$H_n = H_0(1 - \sigma) \tag{1.5}$$

where $\sigma$ is the screening constant.

With this understanding of the NMR experiment, NMR chemical shift $\sigma$ determined from the spectrum is also a 'collective' property of the electrons even with differentiation between the diamagnetic and paramagnetic contributions to this property. This property depends on the electrons of a given atom and its immediate surroundings (for more detail see [16]).

NMR spectroscopy is also used to estimate substituent constants within the framework of the correlation analysis. Thus, constants $\sigma_I$ were estimated for many X substituents by measuring chemical shifts $\sigma_F$ of the fluorine atom in NMR spectra of meta-substituted fluorobenzenes [11]:

$$\sigma_F^{\text{m-X}} = -7.10\sigma_I + 0.60 \tag{1.6}$$

The mentioned methods of evaluation of electron effects in organic molecules based on classical physics do not extend beyond the framework of the classical theory of structure, electron concepts, and the $\sigma\rho$ method.*

The traditional approaches cannot however explain, on the whole, many aspects of reactivity and properties of organic molecules. These difficulties are objective because the processes occurring in molecules with involvement of electrons cannot be understood on the basis of the laws of classical physics (see, for example, [15]).

* J. Kochi *et al.* and V. Koptyug *et al.* have undertaken a new step in the use of the $\sigma\rho$ method for the study of the electron structure of molecules. The results of their studies, based on the orbital concepts, are discussed in detail in Chapter 10.

# 2

# Methods of quantum chemistry

The methods of quantum chemistry have offered new possibilities in the study of the electron structure of organic molecules. The processes occurring in a molecule with involvement of electrons are interpreted by these methods from the standpoints of quantum mechanics, which is the only method applicable to the study of processes occurring in the world of microparticles. The quantum chemistry methods can be used to calculate not only the 'collective' properties of electrons in the molecule (charges on the atoms, bond orders, etc.) but also the energy and wave function of each electron in a molecule.

The fundamental concepts of quantum chemistry are discussed in much detail in many special publications and we shall therefore limit ourselves to the discussion of those aspects that are useful for understanding of the subject matter of this monograph [18–21].

We discuss here only some concepts of quantum mechanics, quantum chemistry, group theory, electron spectroscopy and polarography, which are mainly used now in the study of the electron structure of organic molecules in terms of molecular orbitals and molecular states.

## 2.1 SOME PRINCIPLES OF QUANTUM MECHANICS

The introduction of quantum mechanics became inevitable owing to advances made in physics in the past century. First Balmer (1885) and then Rydberg (1889) discovered the regularities in the position of lines in the emission spectrum of the hydrogen atom. They proposed to determine the frequency $\nu$ of emission from the following equation:

$$\nu = (R/n_1^2) - (R/n_2^2) \qquad (2.1)$$

where R is the Rydberg constant, 109 677.58 cm$^{-1}$; and $n_1$ and $n_2$ are integers.

The simplicity of this empirical formula for the description of atomic spectra provided a convincing proof of the existence of a universal mechanism of the atom interaction with radiation. Since only certain frequencies were observed in the spectrum, it was suggested that the atoms could exist only in certain states. After discovery of electrons, it appeared quite reasonable to associate the various states of the atom with various electron orbits on the assumption that only certain orbits are allowed.

These were Bohr's postulates that laid the foundation for the quantum mechanical theory of atomic structure. In 1913, he formulated the **postulates** that were not derived from classical physics and even contradicted some of its fundamental principles:

(1) each electron of an atom in the ground state moves by its own stable orbit; each orbit is assigned a particular energy state of the atom;
(2) the electron found in one of these orbits does not emit energy;
(3) energy is emitted only when the electron is excited from one orbit to another; the energy is emitted and absorbed in quanta; e.g. as an electron moves from a stable orbit (with the energy $E_1$) to another orbit (with a smaller energy $E_2$), a quantum is emitted; its energy is:

$$h\nu = E_1 - E_2 \qquad (2.2)$$

According to Bohr, the permitted orbitals are circular and their size satisfies the quantum requirement for multiplicity of the momentum of quantity of electron movement to the $h/2\pi$ value (where $h$ is Planck's constant, equal to $6.626 \times 10^{-34}$ J s). Bohr calculated the energy levels of the hydrogen atom for each $n$:

$$E = hR/n^2 \qquad (2.3)$$

and differences of the corresponding energy levels for each $n_1$ and $n_2$ pair:

$$h\nu = \frac{hR}{n_1^2} - \frac{hR}{n_2^2} \qquad (2.4)$$

The most stable orbit is the one with the minimum radius $n = 1$; the electron of the hydrogen atom in the ground state is found in this orbit. The empirical Balmer's formula was thus explained.

Bohr's theory overcame the contradictions between the probability of finding an electron in stable orbits and the emission of certain energy by the excited atom that were unsolved by classical physics. At the same time, despite the apparent advantages, Bohr's theory failed to explain some delicate effects in the spectrum of the hydrogen atom. Precise measurements showed that the spectral lines have fine structures. For example, all lines in the Balmer series ($n_1 = 2$; $n_2 = 3, 4, 5, \ldots$) are doublets. It turned out that orbits with the same principal quantum number $n$ can differ by shape. They can, for example, be elliptic. These differences are supplemented by the differences in their energies, the limited number of lines in the spectrum corresponding to the limited number of elliptic orbits.

It was later established that the lines in the emission spectrum are split to a greater extent when the excited atoms are placed in a magnetic field. The cause of this splitting was explained by a different spatial orientation of the orbits: this splitting was also characterized by the number of lines.

It was thus necessary to admit that the state of the electron in the atom is characterized, in addition to the principal quantum number $n$, by two other quantum numbers, namely, the azimuthal quantum number $l$ and the magnetic quantum number $m$. The quantum mechanical principles used in the description of electrons in atoms were emphasized by using the term 'orbital'.

The **principal quantum number** $n$ characterizes the size of the orbital: the greater the number $n$, the greater the probability of finding the electron farther from the nucleus. The level of the electron energy of one-electron atom depends on the value of $n$.

The angular momentum, or the **azimuthal quantum number** $l$ is associated with the shape of the orbital. The value of $l = 0$ corresponds to a spherically symmetric shape. Other $l$ values are associated with various non-spherical shapes. The $l$ value does not substantially affect the energy of a one-electron atom. But the effect of this number on the energy of multi-electron atoms is considerable.

The **magnetic quantum number** $m$ characterizes the spatial orientation of atomic orbitals. It is evident that a spherically symmetric orbital ($l = 0$) can have only one orientation ($m = 0$). Less symmetric orbitals have various orientations, e.g. an orbital characterized by $l = 2$ may have five different orientations associated with $m$ values of $0, \pm 1, \pm 2$.

The **spin quantum number** was later introduced into quantum mechanics. This fourth quantum number was labelled by s. Doublet splitting of the lines in atomic spectra were interpreted as a result of rotation of the electron about its own axis (clockwise and counter-clockwise). Thus, the spin quantum number $s$ can only have two values: $+\frac{1}{2}$ or $-\frac{1}{2}$. It is noteworthy that the quantum numbers can vary only within certain limits and thus form all permitted combinations. Each of these **permitted combinations of the quantum numbers** corresponds to the permitted energy state of the electron in the atom and characterizes the orbital in which the electron is found (Table 2.1).

**Table 2.1.** Some permitted combinations of the quantum numbers

| Orbital | $n$ | $l$ | $m$ | $s$ | Number of combinations for a given $n$ value |
|---|---|---|---|---|---|
| $1s$ | 1 | 0 | 0 | $\pm\frac{1}{2}$ | 2 |
| $2s$ | 2 | 0 | 0 | $\pm\frac{1}{2}$ | |
| $2p_x$ | 2 | 1 | $-1$ | $\pm\frac{1}{2}$ | |
| $2p_y$ | 2 | 1 | 0 | $\pm\frac{1}{2}$ | 8 |
| $2p_z$ | 2 | 1 | $+1$ | $\pm\frac{1}{2}$ | |

The important rules for the atomic orbitals were formulated:

(1) atomic orbitals may be occupied by one or two electrons, or remain unoccupied: **occupancy** $g_\mu$ **of an atomic orbital** $\mu$ can be 0, 1 or 2;

(2) an atom cannot have two electrons with the same combination of quantum

numbers; if two electrons are found in the same orbital, they have three equal quantum numbers ($n$, $l$, $m$) and a different spin quantum number $s \pm \frac{1}{2}$;

(3) each electron occupies the lowest possible orbital, i.e. the lowest energy orbitals are first occupied by electrons.

Occupied orbitals with the same principal quantum number $n$ form an **electron shell**. The outer electron shell of a neutral atom is called a **valence shell**, and the corresponding orbitals and electrons are respectively called valence orbitals and electrons.

One of the most important principles of quantum mechanics determines the wave properties of an electron. In the general form, this principle is formulated by the equation proposed by Louis de Broglie (1924):

$$\lambda = h/mv \tag{2.5}$$

According to this principle, any particle with mass $m$, moving at velocity $v$ is associated with the wavelength $\lambda$. This property, which can practically be disregarded for macroparticles, has been proved for microparticles.

Another fundamental principle of quantum mechanics is the **uncertainty principle** (Heisenberg, 1926). According to classical mechanics, one can predict accurately the position, velocity and energy of all particles in a system on the condition that their accurate position, velocity and energy at the previous moment of time (and also the nature of the interacting forces) are known. Accurate prediction of planet position in the orbit at any moment of time is an illustration of the validity of classical mechanics. Prediction for microparticles is confined within the framework of the uncertainty principle, according to which it is impossible to predict an accurate position and velocity of a subatomic particle simultaneously. Instead, only a probability of finding a microparticle, e.g. an electron within a given volume of space, can be determined.

The methods for solution of problems of microparticles were also determined by quantum mechanics postulates.

Any state of a wave-like particle (of an electron, in the first instance) is described by function $\Psi(q_1, q_2, \ldots, q_n, \tau)$ which depends on its coordinates and time. The function $\Psi$ is called the wave function and is interpreted not as an amplitude function that is commonly used to describe waves in classical mechanics, but as a measure of probability of an event, in particular a measure of probability of finding an electron in a definite point of space. Wave functions $\Psi$ can assume only the permitted eigenvalues.

The problems for particles with wave properties can be solved by the **Schrödinger wave equation** (1926):

$$H\Psi = E\Psi \tag{2.6}$$

One of the quantum mechanical postulates suggests that operator $H$ can be found for any observed property of a system described by function $\Psi$; this operator acts on the permitted value of $\Psi$ to give $\Psi$ multiplied by constant $E$. This constant corresponds to the eigenfunction $\Psi$ of the system. It follows from the Schrödinger equation that all relations between the values of classical mechanics are replaced in

**Table 2.2.** Some quantum mechanical operators

| Property | Classical expression | Quantum mechanical operator |
|---|---|---|
| Coordinate | $x$ | $x$ |
| Momentum of particle with mass $m$ | $p = mv$ | $\left(\frac{h}{2\pi i}\right)\left(\frac{\mathrm{d}}{\mathrm{d}x}\right)$ |
| Kinetic energy of particle with mass $m$ moving along axis $x$ | $T = \frac{1}{2}mv^2 = \left(\frac{1}{2m}\right)p_2$ | $-\left(\frac{h^2}{8\pi^2 m}\right)\left(\frac{\mathrm{d}^2}{\mathrm{d}x^2}\right)$ |
| Potential energy in the vicinity of nucleus with charge $z$ | $U = -ze^2/r$ | $-ze^2r$ |
| Potential energy of two-electron interaction | $U = e^2/r_{12}$ | $e^2/r_{12}$ |
| Total energy of electron moving along axis $x$ | $E = T + U$ | $H = -\left(\frac{h^2}{8\pi^2 m}\right)\cdot\left(\frac{\mathrm{d}^2}{\mathrm{d}x^2}\right) + U$ |

quantum mechanics by the relations between operators. For a better understanding of the operator 'concept', Table 2.2 gives some properties, their expression in classical physics and the corresponding quantum mechanical operators.

Energy operators are most important for the solution of problems in chemistry. Table 2.2 gives full energy operator $H_x$ for an electron moving across the axis $x$. As all the three coordinates change, the electron operator $H$ assumes the following form:

$$\mathscr{H} = -\left(\frac{h^2}{8\pi^2 m}\right)\left(\frac{\partial^2}{\partial x^2} + \frac{\partial^2}{\partial y^2} + \frac{\partial^2}{\partial z^2}\right) + U = -\left(\frac{h^2}{8\pi^2 m}\right)\Delta + U \tag{2.7}$$

where $\Delta$ is Laplace operator:

$$\left(\frac{\partial^2}{\partial x^2} + \frac{\partial^2}{\partial y^2} + \frac{\partial^2}{\partial z^2}\right)$$

The Schrödinger equation (2.6) with operator (2.7), which is called the Hamiltonian (energy) operator, is thus a differential equation describing the dependence between the electron energy and spatial coordinates of the electron. It can be solved accurately only for one-electron systems, since the energy operator depends in this case on the coordinates of only one electron.

The solutions of the Schrödinger equation are atomic functions expressed in the polar system of coordinates $(r, q, \varphi)$:

$$\chi_{nlm} = R_{nl}(r)Y_{lm}(\theta, \varphi) \tag{2.8}$$

The radial part $R_{nl}$ in these functions is determined by the quantum numbers $n$ and $l$, while the angular part $Y_{lm}$, by the numbers $l$ and $m$.

Given below are some functions $\chi_{nlm}$ obtained by accurate solution of the Schrödinger equation for the hydrogen atom (as an example of a one-electron system):

$$\chi_{100} = \frac{1}{\sqrt{\pi}}\left(\frac{z}{a_0}\right)^{3/2} e^{-\rho} \qquad 1s\text{-orbital}$$

$$\chi_{200} = \frac{1}{4\sqrt{\pi}}\left(\frac{z}{a_0}\right)^{3/2} (2-\rho)e^{-\rho/2} \qquad 2s\text{-orbital}$$

$$\chi_{210} = \frac{1}{4\sqrt{\pi}}\left(\frac{z}{a_0}\right)^{3/2} \rho e^{-\rho/2}\cos(\theta) \qquad 2p\text{-orbital}$$

$$\chi_{211} = \frac{1}{4\sqrt{\pi}}\left(\frac{z}{a_0}\right)^{3/2} e^{-\rho/2}\sin(\theta)\cos(\rho) \qquad 2p_x\text{-orbital}$$

$$\rho = zr/a_0, \qquad a_0 = \frac{h^2}{4\pi^2 me^2}$$

According to the principles of quantum mechanics, the given analytical expressions for atomic orbitals do not determine the accurate position of an electron in the atom. They only suggest that the electron can most probably be found in the vicinity of the nucleus, while this probability decreases with increasing distance from the nucleus.

In a three-dimensional space the **electron density** is the product of $\chi^2$ and the area of the spherical shell $4\pi r^2$. In the coordinates $4\pi\tau^2\chi^2 - \tau$ (Fig. 2.1), the maximum on the curve for the 1$s$ orbital corresponds accurately to the classical Bohr's radius: the highest probability of finding the electron is at a distance of 0.053 nm from the nucleus. Another conclusion is, however, even more noteworthy: the atom has no definite boundaries. The density of the 1$s$ electron is appreciable even at relatively great distances. Figure 2.1 shows that the probability of finding the 2$s$ electron is the highest at $r = 0.25$ nm, of the 3$s$ electron at $r = 0.6$ nm, etc. At the same time, in the states corresponding to 2$s$ and 3$s$ orbitals, the electron can also be found in the vicinity of the nucleus at a certain moment of time.

If all $s$ orbitals are spherically symmetric, the $p$ orbitals are only symmetric relative to appropriate axes.

Objective estimation of orbital parameters for various electron states of the hydrogen atom within the framework of quantum mechanics, makes them applicable to other one-electron systems [21]. Thus, the expression for the determination of electron energy of the hydrogen atom:

$$E_n = 2\pi^2 mz^2e^4/n^2h^2 \qquad \text{or} \qquad E_n = -13.6z^2 \tag{2.9}$$

can be used to estimate the energy of the electron in the corresponding ions of multi-electron atoms. According to (2.9), the energy of the electron in a singly charged helium ion ($z = 2$) is $E_2(\mathrm{He}) = -13.6 \times 4 = 54.5$ eV, in a doubly charged lithium ion

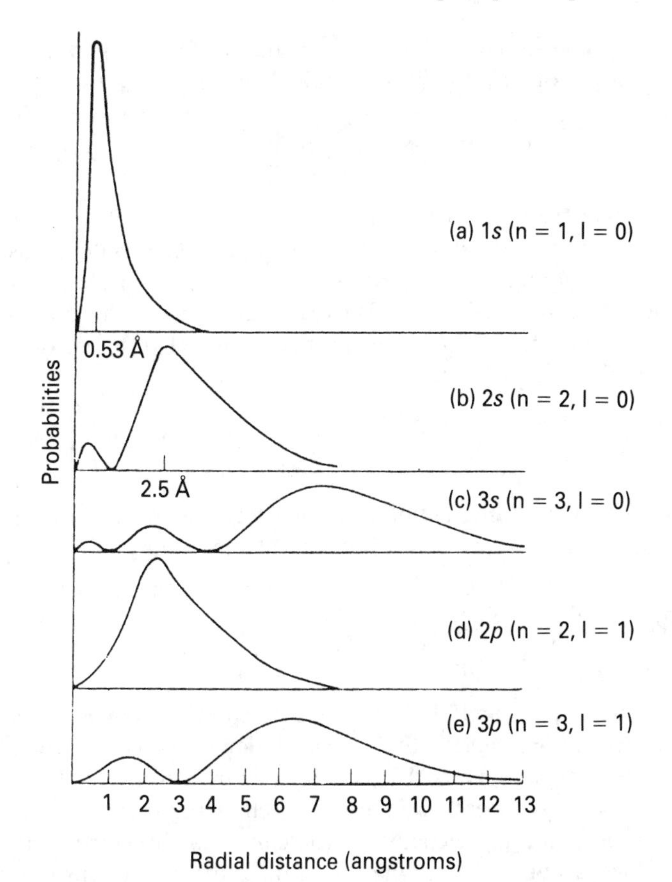


Fig. 2.1. Dependence of $4\pi r^2\chi^2$ electron densities probabilities as determined by the functions of the hydrogen-like atom on distance $r$. (From Lehmann [19])

($z = 3$) it is $E_3(\text{Li}) = -13.6 \times 9 = -122$ eV, in a triply charged beryllium ion ($z = 4$), $E_3(\text{Be}) = -13.6 \times 16 = -218$ eV, and in a quadruply charged boron ion, $E_4(\text{B}) = -13.6 \times 25 = -340$ eV.

For the purpose of comparison , given below are experimentally found values of the corresponding ionization potentials:

$$IP_2(\text{He}) = 54.416 \text{ eV} \qquad IP_3(\text{Li}) = 122.451 \text{ eV}$$
$$IP_4(\text{Be}) = 217.713 \text{ eV} \qquad IP_5(\text{B}) = 340.217 \text{ eV}$$

Wave functions of the hydrogen atom cannot however be used to describe the electron states in multi-electron atoms, because the energy operator $H$ (2.7) (which was used for solution of the hydrogen atom) does not account for the inter-electron interactions.

In order to determine the wave functions for multi-electron systems, the total energy operators should be used in the general case. Thus, the energy operator accounting for the **inter-electron repulsion** $e^2/r_{ij}$ has the following form (determination of other terms of the operator is illustrated in Table 2.2):

$$\mathscr{H} = \left[ -\left(\frac{h^2}{8m\pi^2}\right) \sum_{i=1}^{n} \Delta_i - \sum_{i=1}^{n} \frac{ze^2}{r_i} + \sum_{i \neq j}^{n} \sum^{n} \frac{e^2}{r_{ij}} \right] \tag{2.10}$$

When used for estimation of the inter-electron effects, such operators, however, depend on the coordinates of more than one electron simultaneously. Accurate solution of the Schrödinger equation thus becomes impossible.

Hartree and Fock proposed more effective energy operators for quantum mechanical calculations. The Hartree–Fock operator recommended for the determination of the energy of multi-electron systems:

$$F = \left[ -\left(\frac{h^2}{8m\pi^2}\right) \sum_{i=1}^{n} \Delta_i - \sum_{i=1}^{n} \frac{ze^2}{r_i} + \sum_{i \neq j}^{n} \int \frac{e^2 \varphi_j^2(j)}{r_{ij}} \mathrm{d}\tau_j - K_{ij} \right] \tag{2.11}$$

accounts for the Coulomb interaction $I_{ij}$ of the electrons by introducing the effective Hartree potential. This potential is formed by all electrons (except the $i$th electron) that act on the $i$th electron:

$$\sum_{i \neq j}^{n} \sum^{n} \frac{e^2}{r_{ij}} \longrightarrow \sum_{i \neq j}^{n} \int \frac{e^2 \varphi_j^2(j)}{r_{ij}} \mathrm{d}\tau_j$$

The Hartree–Fock operator (2.1) accounts also for the **exchange energy** $K_{ij}$ that reduces the **electrostatic repulsion** due to **correlation of electron motion**. Accounting for the exchange energy suggests that electrons cannot, in principle, be assigned to particular orbitals: they appear as if constantly changing their orbitals.

The solution of quantum mechanical problems with the Hartree–Fock operator (2.11) ensures high-accuracy determination of the atomic functions of multi-electron atoms (Fig. 2.2). The Hartree–Fock atomic functions have, however, a serious disadvantage: it is impossible to present them in the analytic form. Hence the difficulties arising in their use as the basis functions in quantum chemical calculations.

## 2.2 CONCEPTS OF MOLECULAR ORBITAL THEORY

### 2.2.1 General characteristics of calculation methods

***Simple methods***

The theory of molecular orbitals (MO) has laid the foundation for the wide use of quantum mechanical methods for the study of the electron structure of organic molecules. The concepts of this theory were largely formulated by Hund, Mulliken and Hückel (1927–1937). The main concept of the MO theory suggests that the electrons in a molecule are not located between the atoms but are rather found on **delocalized molecular orbitals**. A **molecular orbital** $\varphi$ **is a linear combination** of **atomic orbitals** (MO LCAO):

$$\varphi = c_1\chi_1 + c_2\chi_2 + \cdots + C_\mu\chi_\mu + \cdots + c_n\chi_n \tag{2.12}$$

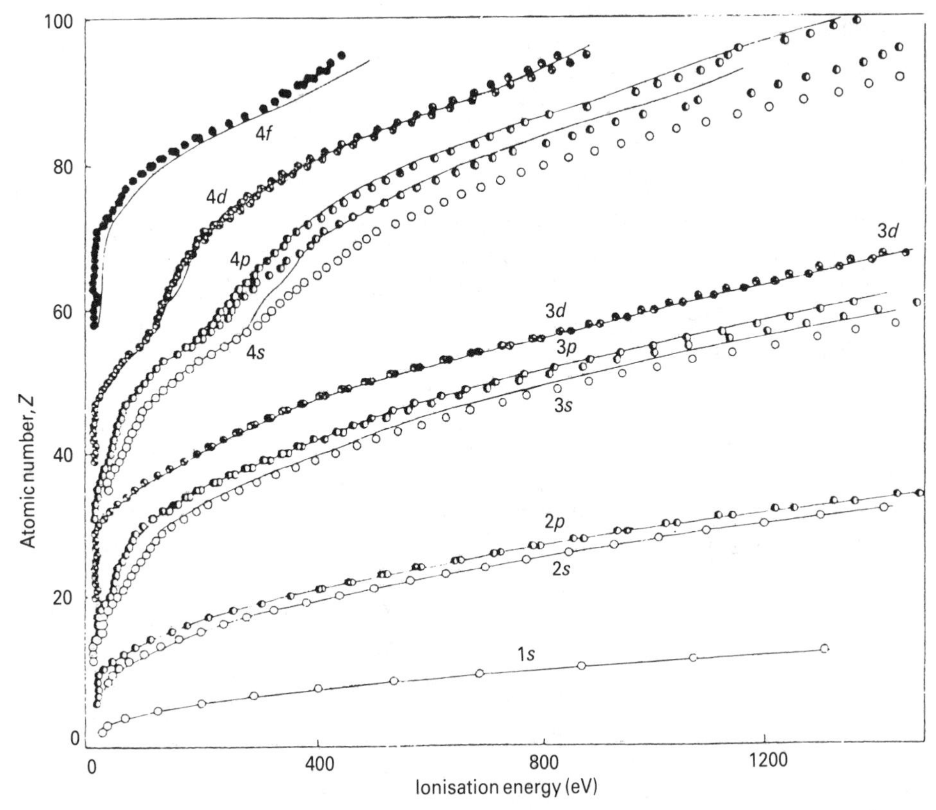


Fig. 2.2. Energies of atomic orbitals calculated by the Hartree equation (solid lines) and ionization energies measured by x-ray photoelectron spectroscopy. (From Ballard [22])

The atomic orbitals $\chi$, which are included into MO, are called **basis atomic orbitals**. The equation (2.12) suggests that $\chi_\mu$ orbitals can belong to different atoms of the molecule, each atom contributing only definite AO to the MO.

Molecular orbitals are typical wave functions and have all their properties:

(1) the square of $c_\mu$ coefficient at the $\chi_\mu$ atomic orbital describe the probability of finding the electron occupying the $\varphi$ molecular orbital in a given atomic orbital; the coefficients $c_1, c_2, \ldots, c_\mu, \ldots, c_n$ are called **eigencoefficients** of MO.
(2) each MO is characterized by a certain energy $\varepsilon_1$, which is known as **eigenvalue** or **eigenenergy** of MO; this parameter determines the energy of electrons occupying appropriate orbital of the molecule;
(3) $n$ atomic orbitals form $n$ molecular orbitals:

$$\begin{aligned}\varphi_1 &= c_{11}\chi_1 + c_{12}\chi_2 + \cdots + c_{1n}\chi_n \\ \varphi_2 &= c_{21}\chi_1 + c_{22}\chi_2 + \cdots + c_{2n}\chi_n \\ &\vdots \\ \varphi_n &= c_{n1}\chi_1 + c_{2n}\chi_2 + \cdots + c_{nn}\chi_n\end{aligned} \tag{2.13}$$

or, in short,

$$\varphi_i = \sum_{\mu=1}^{n} c_{i\mu}\chi_\mu \tag{2.14}$$

where $i = 1, 2, \ldots, n$; the subscript $i$ (or $j$) usually designates molecular orbitals, while $\mu$ (or $\nu$, $\lambda$, $\sigma$) are atomic orbitals; eigencoefficients have twin subscripts $c_{i\mu}$ (the first coefficient relates to MO and the second one to AO).

**Variational method*** modifies the quantum mechanical problem of determining eigenenergies $\varepsilon_i$ and eigencoefficients $c_{i\mu}$ of the wave function (2.12) in the framework of the Schrödinger equation:

$$H\varphi = \varphi E$$

to the mathematical problem. A system of linear equations (2.14) relative to $c_\mu$ values should be solved. In the general case for the wave function (2.12), these equations have the following form:

$$\begin{aligned}
(H_{11} - \varepsilon S_{11})c_1 + (H_{12} - \varepsilon S_{12})c_2 + \cdots + (H_{1n} - \varepsilon S_{1n})c_n &= 0\\
(H_{21} - \varepsilon S_{21})c_1 + (H_{22} - \varepsilon S_{22})c_2 + \cdots + (H_{2n} - \varepsilon S_{2n})c_n &= 0\\
&\vdots\\
(H_{n1} - \varepsilon S_{n1})c_1 + (H_{n2} - \varepsilon S_{n2})c_2 + \cdots + (H_{nn} - \varepsilon S_{nn})c_n &= 0
\end{aligned} \tag{2.15}$$

where $H_{\mu\nu} = \int \chi_\mu H \chi_\nu \, d\tau$ are the **integrals of energy operators** and $S_{\mu\nu} = \int \chi_\mu \chi_\nu \, d\tau$ are **overlap integrals**.

The solution of systems of equations similar to (2.14) requires preliminary estimation of the energy operator integrals $H_{\mu\nu}$ and the overlap integrals $S_{\mu\nu}$. These integrals are either calculated from the corresponding atomic functions or replaced by the appropriate parameters.

Equations (2.15) are called **secular**. As distinct from the case where all $c_{i\mu}$ coefficients are zero, they can be only solved if the appropriate **secular determinant** is zero:

$$\begin{vmatrix}
(H_{11} - \varepsilon S_{11}) & (H_{12} - \varepsilon S_{12}) \ldots & (H_{1n} - \varepsilon S_{1n})\\
(H_{21} - \varepsilon S_{21}) & (H_{22} - \varepsilon S_{22}) \ldots & (H_{2n} - \varepsilon S_{2n})\\
\vdots & \vdots & \vdots\\
(H_{n1} - \varepsilon S_{n1}) & (H_{n2} - \varepsilon S_{n2}) \ldots & (H_{nn} - \varepsilon S_{nn})
\end{vmatrix} = 0 \tag{2.16}$$

It can be seen that the integrals $S_{\mu\nu}$ and $H_{\mu\nu}$ form the corresponding square matrices $[S_{\mu\nu}]$ and $[H_{\mu\nu}]$. If all elements of these matrices are known (e.g. on the condition that $S_{\mu\nu} = \sigma_{\mu\nu}$, where $\sigma_{\mu\nu} = 1$ at $\mu = \nu$, and $\sigma_{\mu\nu} = 0$ at $\mu \neq \nu$), the solution of the characteristic equation of $H$ matrix of the general form

* For a detailed discussion of the variational method see, e.g. [18] or [21].

$$\begin{vmatrix} (H_{11}-\varepsilon) & H_{12} & \dots & H_{1n} \\ H_{21} & (H_{22}-\varepsilon) & \dots & H_{2n} \\ \vdots & \vdots & & \vdots \\ H_{n1} & H_{n2} & \dots & (H_{nn}-\varepsilon) \end{vmatrix} = 0 \quad (2.17)$$

consists in finding eigenvalues $\varepsilon$ of the matrix $H$. The methods of matrix algebra suggest the use of matrix diagonalization for the purpose. By one of these methods, the initial matrix $H$ is transformed by similar transformations (multiplying by matrices $B$ and $B^{-1}$) into the diagonal matrix $D = B^{-1}HB$ [18]. The specific feature of the **diagonal matrix** $D$ is that its non-diagonal elements are equal to zero, while the diagonal elements are its eigenvalues, and hence the eigenvalues of the initial matrix $H$ are at the same time eigenvalues of the unknown wave function $\varphi$, since the latter, in turn, are eigenfunctions of the energy operator. Introduction of thus found $n$ values of $\varepsilon_i$ into the system of equations (2.15) and solution of this system give $n$ vectors $\mathbf{c}_{i\mu}$ of the sought wave function.

The solution of a quantum mechanical problem within the framework of the variational method in $\pi$ **approximation** includes the introduction into calculations of only atomic orbitals with $\pi$ symmetry (see Section 2.2.2). The diagonal elements $H_{11}, H_{22}, \dots, H_{66}$ of the characteristic equation (2.17) are assumed to be equal to $\boldsymbol{\alpha}$, and are called **Coulomb integrals**. Non-diagonal elements $H_{12}, H_{21}, \dots, H_{\mu\nu}$ are assumed to be equal to $\beta$ (if $\mu$ and $\nu$ belong to adjacent atoms) and equal to 0 (if $\mu$ and $\nu$ belong to the non-adjacent atoms); the $\beta$ integrals are called the **resonance integrals**. This method of calculation of molecules is known as the **Hückel method** of molecular orbitals (HMO method). Below follow exemplary calculations by the HMO method for ethylene, butadiene-1,3 and benzene.

*Ethylene*. The solution of the quantum chemical problem for ethylene by the HMO method includes the determination of eigenvalues of the characteristic equation (2.17) of the second order.

By dividing all terms of this equation by $\beta$, and replacing $(\alpha - \varepsilon)/\beta$ by $x$, we have:

$$\begin{vmatrix} x & 1 \\ 1 & x \end{vmatrix} = 0$$

whence $x_{1,2} = \pm 1$, $\varepsilon_1 = \alpha + \beta$, $\varepsilon_2 = \alpha - \beta$*.

The introduction of eigenvalues of $\varepsilon_1$ and $\varepsilon_2$ into the system of equations (2.14), which has the following form for ethylene:

$$(\alpha - \varepsilon)c_1 + \beta c_2 = 0$$

$$\beta c_1 + (\alpha - \varepsilon)c_2 = 0$$

and accounting for the normalization of eigencoefficients of the resultant MO:

* Eigenenergy, expressed as a result of calculation of the molecule by the HMO method in units of $\beta$, is designated by the Greek letter $\lambda$ (e.g. for ethylene, $\lambda_1 = 1$; $\lambda_2 = -1$). In this case

$$\varepsilon_i = \alpha + \lambda_i \beta.$$

$$c_1^2 + c_2^2 = 1$$

gives

$$\lambda_1 = 0.707\chi_1 + 0.707\chi_2; \qquad \varepsilon_1 = \alpha + \beta$$
$$\lambda_2 = 0.707\chi_1 - 0.707\chi_2; \qquad \varepsilon_2 = \alpha - \beta$$

Formation of molecular $\pi$ orbitals of ethylene from basis AOs is shown graphically in the correlation diagram (Fig. 2.3). Molecular orbitals are **occupied** by electrons in the same way as the atomic orbitals. The permitted numbers are 0, 1 or 2. Since the ethylene molecule has two $\pi$ electrons, only the $\varphi_1$ orbital is occupied. This is the **bonding orbital** since the energy of its electrons is lower than in the initial $\chi_1$ and $\chi_2$ AOs. The molecular $\pi$ orbital $\varphi_2$ is **unoccupied**; this is the **anti-bonding orbital**. Insertion of an electron into this MO from the bonding MO as a result of photo-excitation or formation of an anion particle, upsets stability of the molecule. Other ways of representation of $\pi$ MOs of ethylene are given in Fig. 2.3: these are dumb-bells (side view of the molecule) and circles (top view). The shaded lobes of the dumb-bells are associated with positive regions, and clear lobes with negative regions. The dimensions of dumb-bells and circles are proportional to absolute values of the eigencoefficients:

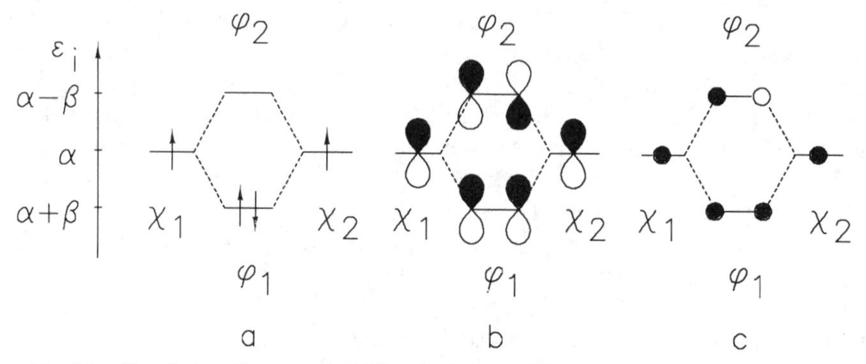


Fig. 2.3. Correlation diagram of $\pi$MOs of ethylene; different graphic representation of $\pi$MOs (a, b and c) are shown.

*Buta-1,3-diene.* Similar HMO calculations with characteristic determinant and the $H$ matrix for buta-1,3-diene are given below.
The solution of the buta-1,3-diene molecule by the HMO method gives four molecular $\pi$ orbitals*:

$$\varepsilon_1 = \alpha + 1.618\beta; \quad \varphi_1 = 0.372\chi_1 + 0.602\chi_2 + 0.602\chi_3 + 0.372\chi_4$$
$$\varepsilon_2 = \alpha + 0.618\beta; \quad \varphi_2 = 0.602\chi_1 + 0.372\chi_2 - 0.372\chi_3 - 0.602\chi_4$$
$$\varepsilon_3 = \alpha + 0.618\beta; \quad \varphi_3 = 0.602\chi_1 - 0.372\chi_2 - 0.372\chi_3 + 0.602\chi_4$$
$$\varepsilon_4 = \alpha - 1.618\beta; \quad \varphi_4 = 0.372\chi_1 - 0.602\chi_2 + 0.602\chi_3 - 0.372\chi_4$$

* Eigenenergy expressed as a result of molecule calculation by the HMO method in $\beta$ units designated by the Greek letter $\lambda_1$ (e.g. for benzene $\lambda_1 = 2$; $\lambda_2 = \lambda_3 = 1$, etc.); hence:

$$\varepsilon_i = \alpha + \lambda_i\beta.$$

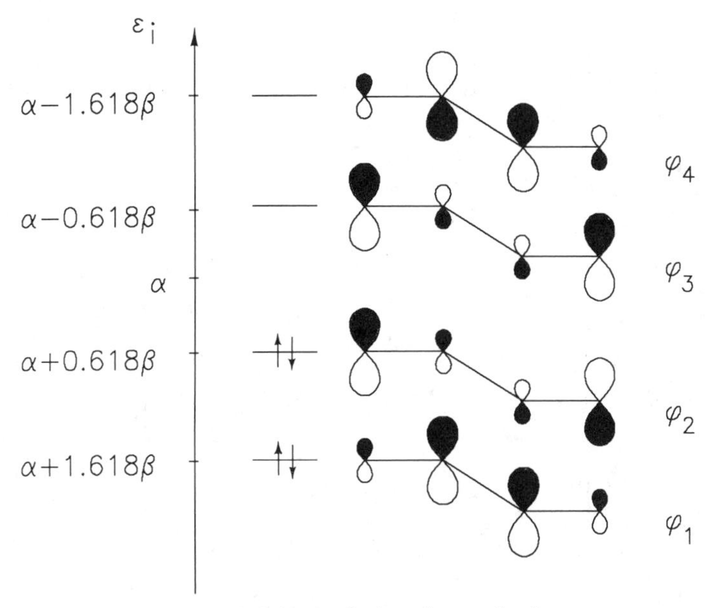


Fig. 2.4. $\pi$ Orbitals of a butadiene molecule

The shapes of these $\pi$ orbitals are shown in Fig. 2.4).

The $\varphi_1$ and $\varphi_2$ are occupied MOs; the $\varphi_2$ is the Highest Occupied MO (HOMO). The $\varphi_3$ and $\varphi_4$ are unoccupied (vacant) MOs; the $\varphi_3$ is the Lowest Unoccupied MO (LUMO).

HOMO and LUMO are Frontier Orbitals of the Molecule (FMO). These MOs are especially important in the molecule: they are the first to be involved in the interaction with electrophilic and nucleophilic agents or activated on exposure to radiation with formation of appropriate intermediates. Thus, the electron jumps into the LUMO of a neutral molecule to form an anion radical $M^{\dot{-}}$ (2.18):

$$\begin{array}{lccc} & \text{———} & & \text{———} \\ \text{LUMO} & \text{———} & & \text{—}\uparrow\text{—} \\ \text{HOMO} & \text{—}\uparrow\downarrow\text{—} & + \ e^- \longrightarrow & \text{—}\uparrow\downarrow\text{—} \\ & \text{—}\uparrow\downarrow\text{—} & & \text{—}\uparrow\downarrow\text{—} \\ & M & & M^- \end{array} \qquad (2.18)$$

Similarly, the electron is detached from HOMO of a neutral molecule to form a cation-radical $M^{\dot{+}}$ (2.19).

*Benzene.* The characteristic determinant and $H$ matrix for calculation of the benzene molecule by the HMO method are given below:

$$\begin{array}{lcccc} & \text{———} & & & \text{———} \\ \text{LUMO} & \text{———} & -\ e^{-} & \longrightarrow & \text{———} \\ \text{HOMO} & \uparrow\downarrow & & & \uparrow \\ & \uparrow\downarrow & & & \uparrow\downarrow \\ & \text{M} & & & \text{M}^{\dot{+}} \end{array} \qquad (2.19)$$

$$\begin{vmatrix} (\alpha-\varepsilon) & \beta & 0 & 0 & 0 & \beta \\ \beta & (\alpha-\varepsilon) & \beta & 0 & 0 & 0 \\ 0 & \beta & (\alpha-\varepsilon) & \beta & 0 & 0 \\ 0 & 0 & \beta & (\alpha-\varepsilon) & \beta & 0 \\ 0 & 0 & 0 & \beta & (\alpha-\varepsilon) & \beta \\ \beta & 0 & 0 & 0 & \beta & (\alpha-\varepsilon) \end{vmatrix} = 0 \qquad (2.20)$$

$$\begin{bmatrix} \alpha & \beta & 0 & 0 & 0 & \beta \\ \beta & \alpha & \beta & 0 & 0 & 0 \\ 0 & \beta & \alpha & \beta & 0 & 0 \\ 0 & 0 & \beta & \alpha & \beta & 0 \\ 0 & 0 & 0 & \beta & \alpha & \beta \\ \beta & 0 & 0 & 0 & \beta & \alpha \end{bmatrix}$$

Diagonalization of the $H$ matrix gives six eigenenergies $\varepsilon_1$–$\varepsilon_6$ for the benzene molecule while the solution of the system of linear equations (2.14) gives six sets of eigencoefficients according to the number of molecular $\pi$ orbitals:

$$\begin{aligned} &\varepsilon_1 = \alpha + 2\beta; && \varphi_1 = 0.408(\chi_1 + \chi_2 + \chi_3 + \chi_4 + \chi_5 + \chi_6) \\ &\varepsilon_2 = \alpha + \beta; && \varphi_2 = 0.577(\chi_1 - \chi_4) + 0.289(\chi_2 - \chi_3 - \chi_5 + \chi_6) \\ &\varepsilon_3 = \alpha + \beta; && \varphi_3 = 0.500(\chi_2 + \chi_3 - \chi_5 - \chi_6) \\ &\varepsilon_4 = \alpha - \beta; && \varphi_4 = 0.500(-\chi_2 + \chi_3 - \chi_5 + \chi_6) \\ &\varepsilon_5 = \alpha - \beta; && \varphi_5 = 0.577(\chi_1 + \chi_4) - 0.289(\chi_2 + \chi_3 + \chi_5 + \chi_6) \\ &\varepsilon_6 = \alpha - 2\beta; && \varphi_6 = 0.408(\chi_1 - \chi_2 + \chi_3 - \chi_4 + \chi_5 - \chi_6) \end{aligned}$$

Three $\pi$ orbitals $\varphi_1$, $\varphi_2$, and $\varphi_3$ of benzene are occupied. Two molecular orbitals $\varphi_2$ and $\varphi_3$ are **degenerate** because they have equal energies; they are frontier occupied molecular orbitals of benzene.

Three $\pi$ orbitals $\varphi_4$, $\varphi_5$, and $\varphi_6$ of benzene are unoccupied. Two of them ($\varphi_4$ and $\varphi_5$) are frontier unoccupied MOs; they are degenerate orbitals as well.

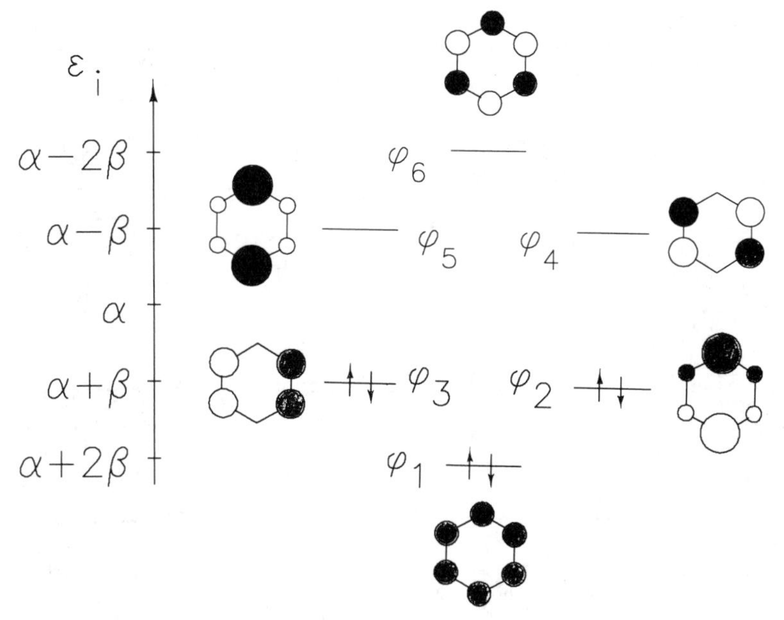


Fig. 2.5. π Orbitals of a benzene molecule

All π MOs of benzene are shown in Fig. 2.5.

It can be seen from Fig. 2.5 that the doubly degenerate HOMOs and LUMOs of benzene differ substantially by their coefficients: $\varphi_2$ and $\varphi_5$ orbitals are called **symmetric** π MOs, while $\varphi_3$ and $\varphi_4$ are **antisymmetric** MOs. These differences are important for removal of degeneracy of the frontier MOs on the insertion of substituent X into the benzene molecule (Fig. 2.6).

Within the framework of the MO theory, the occupied MOs with equal energies form an electron shell. The benzene molecule thus has two π electron shells. Two electrons in the $\varphi_1$ orbital with the energy $\varepsilon_1 = \alpha + 2\beta$ form the first shell, and four electrons in the $\varphi_2$ and $\varphi_3$ orbitals with the same energy $\varepsilon_2 = \varepsilon_3 = \alpha + \beta$ form the second shell.

In the discussion of the results of benzene molecule calculations by the HMO method, it is necessary to mention another property of its electron structure which is also confirmed by the most accurate quantum chemical calculations. Being a typical **alternant hydrocarbon** (all atoms in the molecules of these conjugated hydrocarbons are divided into two alternating types, viz., labelled and unlabelled), benzene obeys the **Pairing theorem**. From the above calculations it follows that all π orbitals of benzene are symmetric relative to the zero level ($\varepsilon = \alpha$), i.e. each bonding MO with the energy $\varepsilon_i = \alpha + \lambda_i\beta$ corresponds to the anti-bonding MO with the energy $\varepsilon_j = \alpha - \lambda_j\beta$. The coefficients in the anti-bonding MOs of labelled atoms coincide in values and signs with the coefficients in the bonding MOs of the same atoms, while the coefficients in the bonding and anti-bonding MOs of the unlabelled atoms have opposite signs.

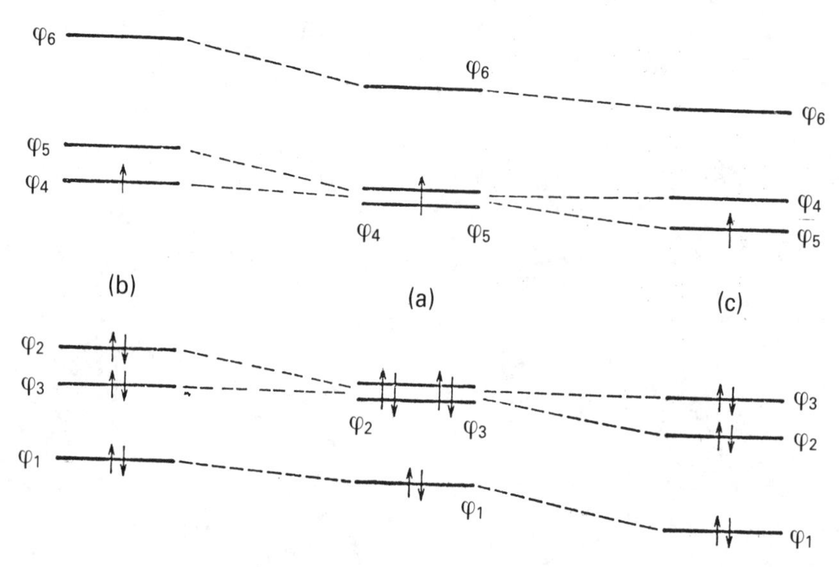


Fig. 2.6. Frontier orbitals of substituted benzenes $C_6H_5X$: (a) X = H; (b) X is electron donor; (c) X is electron acceptor.

Eigencoefficients $c_{i\mu}$ and eigenenergies $\varepsilon_i$ derived by quantum chemical calculations are the basis for the determination of the parameters of the electron structure of a molecule within the framework of the MO theory. Each $c_{i\mu}$ value estimates objectively the localization of the $i$th MO in the $\mu$th atomic orbital, in other words, the contribution of the $\mu$th AO to the $i$th molecular orbital. The value of the coefficient $c_{i\mu}$ equal to 1 indicates that a given MO is completely localized in the $\mu$th atomic orbital. Such an MO is known as a non-bonding molecular orbital. Non-bonding MOs are characteristic of molecules that contain heteroatoms with an unshared electron pair (nitrogen, oxygen, the halogens, etc.); the resultant MOs are designated n-orbitals.

The square of the coefficient $c_{i\mu}^2$ of an occupied molecular orbital estimates the **partial electron density** in the $\mu$th basis orbital in the $i$th MO:

$$(\rho_\mu)_i = g_i c_{i\mu}^2$$

The electron density in the $\mu$th basis orbital in the $i$th MO occupied by one electron ($g_i = 1$) is called **spin density**.

The product $c_{i\mu}c_{i\nu}$ of coefficients in the $i$th occupied MO characterizes the $\mu$–$\nu$ **bond order**, i.e. the electron density in the region of overlapping of the $\mu$th and $\nu$th orbitals in the $i$th MO:

$$(\rho_{\mu\nu})i = g_i c_{i\mu} c_{i\nu}$$

Summation of electron densities $(\rho_\mu)_i$ and $(\rho_{\mu\nu})_i$ in all occupied MOs gives **total electron densities** in the appropriate centres:

$$\rho_\mu = \sum_i g_i c_{i\mu}^2, \qquad \rho_{\mu\nu} = \sum_i g_i c_{i\mu} c_{i\nu}$$

Summation of electron densities in all AOs of the atom in the corresponding MOs estimates the charge $Z$ on the atom as calculated by the quantum chemical method. If, in the simplest case, atom $\mu$ contributes its sole AO (with the number of electrons in it equal to $\eta\mu$) to the MO, the charge $Z_\mu$ on the atom is determined by the formula $Z_\mu = \eta_\mu - \rho_\mu$. Since electron densities are summed up with accounting for MO occupancy, only occupied MOs contribute to the interatomic bonding in the molecule.

In calculation of organic molecules in the valence basis set, the number of the resultant MOs increases in proportion to the size of the basis set. For example, twelve AOs should be accounted in the calculation of ethylene in this approximation: four AOs of each of the two carbon atoms ($2s$, $2p_x$, $2p_y$ and $2p_z$) and one AO of each of the four hydrogens. Twelve resultant MOs are available by these calculations.

By placing 12 electrons of the ethylene molecule into its MOs one can obtain six occupied and six unoccupied molecular orbitals: $\varphi_6$-HOMO, and $\varphi_7$-LUMO.

In calculations accounting for orbital overlapping (e.g. by the Extended Hückel method), estimation of electron densities and bond order includes the overlap integral $S_{\mu\nu}$. The calculated values are called **orbital populations**. For example, the population of the orbital $\mu$ in the atom A (orbital population) is calculated as $P_{\mu\mu}$:

$$P_{\mu\mu} = \sum_i g_i \left( c_{i\mu}^2 + \sum_\nu c_{i\mu} c_{i\nu} S_{\mu\nu} \right)$$

while the population $P_{\mu\nu}$, which characterizes bonding between two AOs ($\mu$ and $\nu$), and which is shared by atoms A and B, is calculated as follows:

$$P_{\mu\nu} = \sum_i g_i c_{i\mu} c_{i\nu} S_{\mu\nu}$$

The symbol $\mu$ in the expressions for both $P_{\mu\mu}$ and $P_{\mu\nu}$ is used to designate the orbitals of the atom A, while the symbol n designates the orbitals of the atom B.

The main feature of the **simple methods** of the MO theory is that the interelectron effects in a definite form are disregarded. This simplifies significantly calculations of matrix elements of the energy operator $H$, and the system of equations (2.15) need thus be solved only once. The simplicity of calculations within the framework of simple methods practically removes limitations associated with the size of the basis set and allows calculations of organic molecules of any complexity.

These calculations are widely used in $\pi$ approximation in particular. It is most important that the parameters of molecular $\pi$ orbitals calculated, e.g., by the HMO method practically coincide with the parameters that are available on the basis of the more accurate *ab initio* and semi-empirical calculations. Besides, errors due to the approximations of simple methods can partly be compensated by empirical parametrization, although such parametrization in relative units $\beta$ has not been substantiated (see Section 2.4). In selecting parameters, it is assumed that the diagonal elements $\alpha\mu$ of the $H$ matrix estimate, in the first approximation, the energy of the electron in an isolated 2p$\pi$ orbital of the $\mu$ atom, while non-diagonal elements $\beta_{\mu\nu}$

estimate the energy of the electron in the region of overlapping of $2p\pi$ orbitals of the adjacent $\mu$ and $v$ atoms [18,23].

**Ab initio** ***methods***

In *ab initio* calculations, the orbital structure of organic molecules is determined by solving the system of **Roothaan equations** (1951):

$$\sum_{\mu=1}^{n} (F_{\mu v} - \varepsilon_i S_{\mu v})c_{i\mu} = 0, \qquad i = 1, 2, \ldots, n \tag{2.21}$$

In these calculations, all elements of the $F$ and $S$ matrices are calculated. Matrix elements of the Fock energy operator, Fockian $F_{\mu v}$, for example, are calculated as the sum of the integrals $H_{\mu v}$ and the integrals estimating the interelectron effects:

$$F_{\mu v} = H_{\mu v} + \sum_i \sum_\lambda \sum_\sigma P_{\lambda\sigma}[(\mu v \,|\, \lambda\sigma) - \tfrac{1}{2}(\mu\lambda \,|\, v\sigma)] \tag{2.22}$$

The interelectron interactions are estimated by the density matrix $P_{\mu\sigma}$, whose elements are the unknown solutions of the Roothaan equations. For this reason, calculations with application of the Fock energy operator are possible only within the framework of the self-consistent field technique, i.e. by successive approximations, the number of which is determined by the wanted accuracy of calculations (**SCF method**).

Since it is impossible to use the Hartree–Fock functions as basis orbitals in calculations of molecules by the MO LCAO method, the appropriate analytical expressions, which accurately approximate the mentioned functions, were derived. Thus, for approximation of radial parts of **atomic functions**, the **Slater** formulae were proposed:

$$R_{nl}(r) = r^{n-1} \exp(-\xi r) \tag{2.23}$$

in which the orbital exponent $\xi$ is determined as a quotient resulting from division of the effective nuclear charge $Z_{\text{eff}}$ by the principal quantum number $n$. In accordance with Eq.(2.23), and with accounting for the normalization factors, the radial parts of carbon AOs can be expressed as follows [18]:

$$\chi(1s) = 7.66\text{e}^{10.8r},\ \chi(2s) = 2.06\text{e}^{-3.07r},\ \chi(2p_x) = 3.58r\cos\theta^{-3.07r}$$

where $r$ is the distance from the nucleus (Å), and $\theta$ is the angle at the coordinate $x$.

The corresponding integrals can be calculated more easily if the radial parts of AO are represented by **Gaussian functions**:

$$R_{nl}(r) = r^{n-1} \exp(-\xi r^2) \tag{2.24}$$

The necessary accuracy of approximation is attained by using a linear combination of Gaussian functions rather than a single Gaussian function [24].

Minimal basis sets approximate reliably only atomic orbitals of the second period elements from boron to fluorine, several Gaussian functions determining one atomic orbital. For example, in one of the 'split valence' basis sets 4-31GF, each $1s$ orbital of the second period elements is approximated by four Gaussian functions. The valence shell orbitals are expressed by the inner and outer components, each of which is represented by three and one Gaussian functions respectively. The internal electron shells are approximated to a greater accuracy in another basis set, 6-31GF. The 'polarization' sets account for the effect produced on each atomic orbital by its surroundings in the molecule. For example, in the polarization basis sets, said atomic functions are supplemented by vacant orbitals. Thus, $3d$ orbitals can be added to the basis set of the oxygen atom.

*Ab initio* methods are widely used to calculate small organic molecules. The electron shells of methane, formaldehyde, ethylene and other simple organic molecules, as well as their cations, anions and radicals, were calculated in various basis sets by *ab initio* methods [24–26]. There were determined algorithms for calculation of charge distribution, dipole moments, molecule symmetry, and internal rotation energy. Calculated were the surfaces of potential energy of model reactions with involvement of simple molecules [26] (including catalytic reactions [27]).

The importance of *ab initio* calculations should be definitely emphasized. These methods can be used to evaluate the electron state of particles inaccessible to direct experimental studies, such as highly reactive intermediate complexes (including those superfast reactions), interstellar space particles, etc.

***Semi-empirical methods***

The calculations by *ab initio* quantum chemical methods become rather bulky with increasing number of basis functions. It especially concerns the determination of two-electron four-centre integrals $(\lambda v | \lambda\sigma)$ and $(\mu\mu | v\sigma)$. In **semi-empirical quantum chemical methods**, each atomic orbital is approximated by only one Slater function. The scope of calculations decreases because of the decreased number of integrals to be calculated.

In one approximation, for example, all integrals $(\mu v | \lambda\sigma)$ are neglected except those in which $\mu$ and $v$ belong to the atom A, while $\lambda$ and $\sigma$ belong to the atom B. The overlap integral $S$ is equated to the Kronecker symbol $\sigma_{\mu v}$, which is equal to 1 when $\mu = v$, and zero when $\mu \neq v$. Appropriate approximation is called the neglect of diatomic differential overlap approximation (**NDDO approximation**). If the approximation is based on complete neglect of differential overlap (**CNDO approximation**), the number of the calculated integrals becomes even smaller: the integrals $(\mu v | \lambda\sigma)$ are assumed to be zero in all cases except when $\mu = v$ and $\lambda = \sigma$. In this approximation, two-electron integrals are replaced by parameters of $\gamma_{AB}$ estimating the mean repulsion between the electron in the valence shell of the atom A and the electron in the valence shell of the atom B.

The semi-empirical methods of calculations are described in detail in special literature [28,29]; these calculations are less cumbersome and more suitable even in the study of electron structure of complicated molecules and their associates. To summarize, it is necessary to note only that each semi-empirical quantum chemical method is usually parametrized thoroughly for estimation of a property. For example, the **Parr–Pariser–Pople method (PPP)** can be used for reliable quantum chemical estimation of electron absorption spectra of aromatic hydrocarbons and dyes [30–32]. Assigning the bands of photoelectron spectra is more objective within the framework of calculations by **HAM/3**, **CNDO/S3** and **Xα** methods [20, 23, 33, 34]. Inner rotational barriers are calculated reliably by the MNDO method.

### 2.2.2 Symmetry of molecules, orbitals and states

The properties of symmetry are important for the solution of problems associated with the study of electron structure of molecules, such as selection of basis set and determination of eigenvalues of the $H$ matrix in quantum chemical calculations, classification of molecular states and determination of the probability of transitions between these states. Thus, the properties of symmetry underlie the methods of calculations in $\pi$ approximation: only atomic orbitals with $\pi$ symmetry are regarded as the basis set in the HMO and PPP methods.

Symmetry is analysed by the **Group Theory** [35]. The symmetry of a molecule, orbital or state is determined by the set of symmetry elements inherent in a given system, i.e. the combinations of rotations and reflections that match it with itself. Appropriate axes and planes of symmetry are called the **elements** of symmetry. Each symmetry element is responsible for the corresponding **transformation** of symmetry or **operation** of symmetry. An operation of symmetry of a system causes transformation of a system into a position coinciding with the initial one. The corresponding operations are called point operations: all point operations of symmetry must leave at least one unchanged point of the system.

All point operations can be derived from the following two main types of transformation:

(a) **Rotation** around the axis through angle $2\pi/n$, by means of which the system repeats an identical pattern; this operation is designated $C_n$ and is called an axis of symmetry of order $n$;
(b) **Reflection** in plane $\sigma$; if the axis of symmetry is perpendicular to the plane of reflection, the symbol $\sigma$ has a subscript $h$ ($\sigma_h$); if the axis of symmetry lies in the reflection plane, the symbol $\sigma$ has the subscript $v$ ($\sigma_v$).

The operation of **mirror rotation** $S_n$ is, for example, a combination of rotation about the axis $C_n$ through an angle of $2\pi/n$ and reflection in the plane. Furthermore, the point of intersection of the axis $C_h$ and the plane $\sigma_h$ forms the **centre of symmetry**. The corresponding operation is designated $i$ and is called the **operation of inversion**.

A molecule of water, for example, has the following elements of symmetry: identity transformation $E$ (the operation that leaves the molecule unchanged); the axis of rotation of the second order $C_2^z$ (the main axis of symmetry always coincides with the axis $z$); planes $\sigma_v^{xz}$ and $\sigma_v^{yz}$. All these elements obey the rules of group multiplication:

$$(C_2^z) \times (C_2^z) \equiv E,$$

$$(\sigma_v^{xz}) \times (\sigma_v^{xz}) \equiv E$$

$$(C_2^z) \times (\sigma_v^{xz}) \equiv \sigma_v^{yz}, \text{ etc.}$$

and are connected by the table of group multiplication and form the **point group** designated in the group theory as $C_{2v}$.

**Table 2.3.** Group multiplication of symmetry elements of a water molecule

| | $E$ | $C_2^z$ | $\sigma^{xz}$ | $\sigma^{yz}$ |
|---|---|---|---|---|
| $E$ | $E$ | $C_2^z$ | $\sigma^{xz}$ | $\sigma^{yz}$ |
| $C_2^z$ | $C_2^z$ | $E$ | $\sigma^{yz}$ | $\sigma^{xz}$ |
| $\sigma^{xz}$ | $\sigma^{xz}$ | $\sigma^{yz}$ | $E$ | $C_2^z$ |
| $\sigma^{yz}$ | $\sigma^{yz}$ | $\sigma^{xy}$ | $C_2^z$ | $E$ |

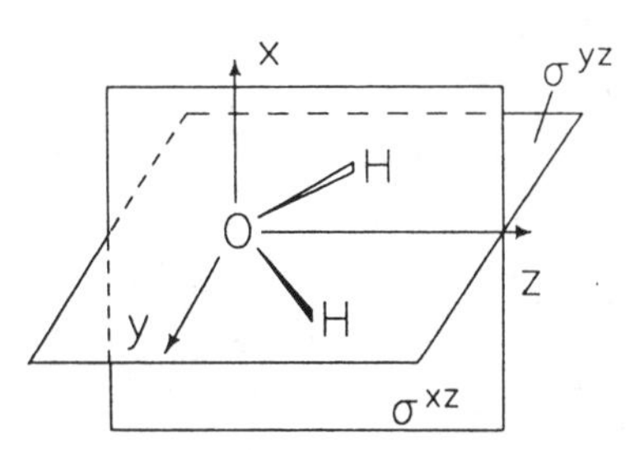


The transformations of a system (molecule, orbital or state) effected by the symmetry elements are determined by the table of characters of the corresponding point group. The point group $C_{2v}$, for example, has the table of characters as shown in Table 2.4.

**Table 2.4.** Table of characters of point group $C_{2v}$

| $C_{2v}$ | $E$ | $C_2^z$ | $\sigma_v^{xy}$ | $\sigma_v^{yz}$ | $f(j)$ |
|---|---|---|---|---|---|
| $A_1$ | +1 | +1 | +1 | +1 | $z, x^2, y^2$ |
| $A_2$ | +1 | +1 | −1 | −1 | $xy$ |
| $B_1$ | +1 | −1 | +1 | −1 | $x, xz$ |
| $B_2$ | +1 | −1 | −1 | +1 | $y, yz$ |

Each figure in the table is the character (trace) of the matrix for operation which it represents. The last column gives some functions $f(j)$ that are transformed according to the corresponding type of symmetry. The type of symmetry symbols are usually given in the first column, in the appropriate line of the table of characters. The letters $A$ and $B$, for example, designate non-degenerate, the letter $E$, doubly degenerate, and the letter $T$, triply degenerate types of symmetry. The point of group symmetry to which a given molecule belongs determines the possibility of the presence of degenerate MOs in this molecule: doubly and triply degenerate MOs may be found only in molecules of the point group of symmetry having the $E$ and $T$ types of symmetry in the table of characters respectively. In all point groups, the symbol $A$ corresponds to the types of symmetry having symmetric (relative to the main axis) transformation (for all types of $A$, in the tables of characters for the operation of the main axis, the character is equal to +1). The symbol $B$ corresponds to the types of symmetry having antisymmetrical (relative to the main axis) transformation (for all types of $B$, in the tables of characters for the operation of the main axis, the character is equal to $-1$). The subscripts 1 and 2 in designations of types of symmetry indicate, respectively, symmetry and antisymmetry relative to the axis of symmetry that is different from the main axis (if another axis in a given point group is absent, the plain of reflection is used as the reference). The subscripts $g$ and $u$ indicate symmetric and antisymmetric transformations to the operation of inversion $i$.

About 50 point groups of symmetry are known. A special nomenclature is used to designate them. The main component of each designation is the capital letter: $C$ (cyclic), $D$ (dihedral), $T$ (tetrahedral), $O$ (octahedral). The subscripts used to designate point groups depend on the symmetry operation: the subscript 1 is assigned to the point group including only one symmetry operation $E$; the subscript $S$ is given to the point group including the rotoflection axis of the first order $S_1$, which is equivalent to the reflection plane $\sigma_h$, etc.

**Table 2.5.** Characters of some point groups

| $C_1$ | $E$ | $f(j)$ |
|---|---|---|
| $A$ | +1 | any |

| $C_2$ | $E$ | $C_2^z$ | $f(j)$ |
|---|---|---|---|
| $A$ | +1 | +1 | $z, x^2$ |
| $B$ | +1 | −1 | $x, y$ |

| $C_s$ | $E$ | $\sigma_h$ | $f(j)$ |
|---|---|---|---|
| $A'$ | +1 | +1 | $x, y, xy$ |
| $A''$ | +1 | −1 | $z, xz, y$ |

| $C_{2h}$ | $E$ | $C_2^z$ | $i$ | $\sigma_h$ | $f(j)$ |
|---|---|---|---|---|---|
| $A_g$ | +1 | +1 | +1 | +1 | $xy$ |
| $B_g$ | +1 | −1 | +1 | −1 | $zy, zx$ |
| $A_u$ | +1 | +1 | −1 | −1 | $z$ |
| $B_u$ | +1 | −1 | −1 | +1 | $x, y$ |

| $D_{2h}$ | $E$ | $C_2^z$ | $C_2^y$ | $C_2^x$ | $i$ | $\sigma^{xy}$ | $\sigma^{xz}$ | $\sigma^{yz}$ | $f(j)$ |
|---|---|---|---|---|---|---|---|---|---|
| $A_g$ | +1 | +1 | +1 | +1 | +1 | +1 | +1 | +1 | $x^2, y^2, z^2$ |
| $B_{1g}$ | +1 | +1 | −1 | −1 | +1 | +1 | −1 | −1 | $xy$ |
| $B_{2g}$ | +1 | −1 | +1 | −1 | +1 | −1 | +1 | −1 | $xz$ |
| $B_{3g}$ | +1 | −1 | −1 | +1 | +1 | −1 | −1 | +1 | $yz$ |
| $A_u$ | +1 | +1 | +1 | +1 | −1 | −1 | −1 | −1 | — |
| $B_{1u}$ | +1 | +1 | −1 | −1 | −1 | −1 | +1 | +1 | $z$ |
| $B_{2u}$ | +1 | −1 | +1 | −1 | −1 | +1 | −1 | +1 | $y$ |
| $B_{3u}$ | +1 | −1 | −1 | +1 | −1 | +1 | +1 | −1 | $x$ |

(From Flurry [35])

The characters of some point groups of symmetry are given in Table 2.5. The selection of the point group of symmetry for the analysis of the electron structure of a molecule depends on the approximation of quantum chemical calculations. Anisole, for example, can be calculated in several approximations: (a) in the $\pi$ approximation, assuming that the methoxy group is a pseudoheteroatom; its molecule can then be attributed to the group of symmetry $C_{2v}$; (b) and in the valence basis approximation, taking into account all orbitals of the valence shell that form the anisole molecule;

its symmetry is then attributed to the group of symmetry $C_1$:

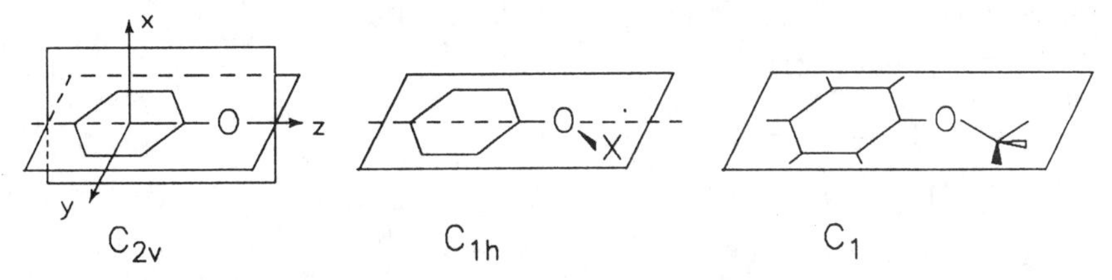


**Table 2.6.** Conversions of buta-1,3-diene MOs ($C_{2h}$ point group)

| $C_{2h}$ | $E$ | $C_2^z$ | $i$ | $\sigma_h^{xy}$ | Symmetry type |
|---|---|---|---|---|---|
| $\varphi_1$ | +1 | +1 | −1 | −1 | $a_u$ |
| $\varphi_2$ | +1 | −1 | +1 | −1 | $b_g$ |
| $\varphi_3$ | +1 | +1 | −1 | −1 | $a_u$ |
| $\varphi_4$ | +1 | −1 | +1 | −1 | $b_g$ |

Symmetry of the wave functions, e.g. of the molecular orbitals, is determined in the same way. Shown below are $\pi$ MOs of buta-1,3-diene and the types of their symmetries, while Table 2.6 gives the types of their transformations within the framework of the point group $C_{2h}$, to which the butadiene molecule belongs:

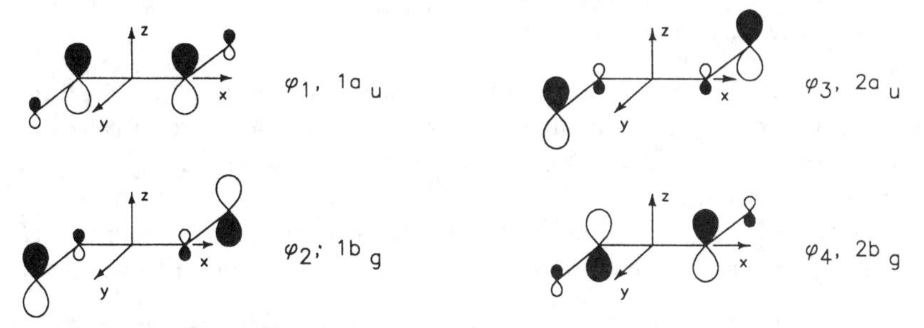


According to these data, the HOMO $\varphi_2$ of butadiene has $b_g$ symmetry, while LUMO $\varphi_3$ of butadiene belongs to the $a_u$ symmetry type. Molecular orbitals having the same type of symmetry are enumerated in the order of their increasing energies. Thus, $\pi$ MOs of butadiene are designated as follows:

$$1a_u, 1b_g, 2a_u, 2b_g.$$

The $S_k$ electron states of the molecules are classified by their symmetry as well.

The symmetry of the electron state $S_k$ is determined as the direct product of the characters of transformation matrices of all occupied MOs $\varphi_i$, taking account of $g_i$, the number of electrons in the MO:

$$\Psi(S_k) = \varphi_1^{g_1}\varphi_1^{g_2}\varphi_3^{g_3}\cdot\varphi_n^{g_n}$$

The wave function, e.g. of the ground state of butadiene, has the form:

$$\Psi(S_0) = (1a_u)^2(1b_g)^2$$

while its symmetry is $A_g$ (MO symmetry types are usually designated by lower-case characters and the symmetry types of states by capitals):

$$\left.\begin{aligned} E &= (+1)^2\ \ (+1)^2 = +1 \\ C_2^z &= (+1)^2(-1)^2 = +1 \\ i &= (-1)^2\ \ (+1)^2 = +1 \\ \sigma_h^{xy} &= (-1)^2(-1)^2 = +1 \end{aligned}\right\} Ag$$

The wave function of the ground state of the cation radical $SS_0^{+}$ of butadiene has the symmetry $B_g$:

$$\Psi(S_0^+) = (1a_u)^2(1b_g)^1 \qquad (B_g \text{ state})$$

The total wave function of the ground state of the anion radicals $S_0^{+}$ of butadiene has the symmetry $A_u$:

$$\Psi(S_0^-) = (1a_u)^2(1b_g)^2(2a_u) \qquad (A_u \text{ state}).$$

The wave function of the first excited state $S_1$ of butadiene has the $B_u$ symmetry ($^1B_u$—singlet or $^3B_u$—triplet):

$$\Psi(S_1) = (1a_u)^2(1b_g)^1(2_u)^1 \qquad (B_u \text{ state}).$$

## 2.3 EXPERIMENTAL METHODS

At the present time, the results of quantum chemical calculations within the framework of the MO theory can be estimated experimentally. Coincidence of calculated and measured parameters makes it possible to describe quantitatively the electron structure of a molecule.

According to Koopmans' theorem*, the values of ionization potentials $IP_i$, e.g., are equal to the absolute values of the energies of the corresponding occupied MOs:

$$IP_i = -\varepsilon_i \quad \text{(the subscript } i \text{ is used for occupied MOs)}$$

while the values of electron affinity $EA_j$ are equal to the absolute values of the energies of the corresponding unoccupied MOs:

$$EA_j = -\varepsilon_j \quad \text{(the subscript } j \text{ is used for the unoccupied MOs).}$$

Some methods of electron spectroscopy are in this situation justly regarded as 'experimental' methods of quantum chemistry. First of all, they include methods of photoelectron (PE) spectroscopy, which is used to estimate the energies of occupied electron levels [36–43]. Besides, information on the occupied electron levels of molecules can be obtained by electron absorption spectroscopy of the corresponding charge-transfer (CT) complexes [44]. The method of electron transmission (ET) spectroscopy estimates the energies of unoccupied electron levels of the molecule [45, 46]. Information on the frontier occupied and vacant electron levels can be obtained

* The Koopmans' theorem is described in more detail in Sections 2.3.1 and 2.3.2.

by polarography [47–49] and by electron spin resonance spectroscopy of the corresponding ion-radical [16–50]. The differences in the energies of the ground electron state and various excited states of organic molecules can be evaluated from the data of one-configurational electron transitions by the method of electron absorption spectroscopy.

The basic principles of each method and their applicability to the study of electron structure of organic molecules are discussed in detail in the relevant monographs. In this chapter we shall give only a short review of them.

### 2.3.1 Photoelectron spectroscopy

Photoelectron spectroscopy (PES) is widely used to study the electron structure of organic molecules [22, 33, 36–39]. The method is based on the photoelectric effect:

$$\mathrm{M} + h\nu \rightarrow M^{\dot{+}} + e^{-}$$

As a substance is irradiated, the energy of a photon absorbed in the molecule is spent in removing the electron from the molecule (this part of the energy determines the ionization potential $IP_i$) with a transfer of the corresponding kinetic energy to the ionized electron:

$$h\nu = IP_i + E_i^{\mathrm{kin}} \tag{2.25}$$

Equation (2.25) explains why the intensity of the electron flow (not electron energy) can only be increased by increasing the irradiation intensity. The energy of photoelectrons can be changed by alternating the frequency of the irradiating photons. Low-energy sources are used in the study of valence shells of organic molecules: He(I) (21.2 eV), He(II) (40.8 eV), Ne(I) (16.8 eV), etc.*

Vapour of a substance is ionized in a deeply evacuated chamber of a photoelectron spectrometer. The ionization chamber is connected with an analyser where photoelectrons are differentiated by their kinetic energies. Since the relationship between the ionization potential and kinetic energy of the electron is described unambiguously by Eq. (2.25), the abscissa of the photoelectron spectrum is expressed directly in the $IP_i$ values (Fig. 2.7a). The quantity of electrons per unit of energy per second is the measure of the intensity of the photoelectron current which is plotted against the axis of ordinates and designed as cps (counts per second). Photoelectron spectroscopy thus estimates energy spectra of electrons emitted by the molecules irradiated by monochromatic UV- or X-rays.

In fact, electromagnetic radiation interacts with a molecule by a mechanism that is more complicated than described by Eq. (2.25): during photo-ionization, excess energy of electromagnetic radiation is not only converted into kinetic energy of a photoelectron but is also spent for excitation of the molecular cation-radical $\mathrm{M}^{\dot{+}}$.

In the general case, the resulting ion $\mathrm{M}^{\dot{+}}$ can be found in various electron, vibrational, and rotational states. Changes in the energies of rotational states induced by photoionization are usually disregarded. According to the law of conservation of

* A new method of ionization of rarefied vapour of substances by excited atoms of helium and neon (He* and Ne*) has been widely used in recent years. The method is known as Penning ionization.

momentum, the changes in the kinetic energy of the ion during photo-ionization are also neglected since the mass of the ion $M^+$ is much greater than that of the electron. The mass of a hydrogen molecule, for example, is 3672 times higher than the electron mass; accordingly, the kinetic energy of the molecular ion $H_2^{\dot{+}}$ that is formed by ionization of hydrogen is only 0.00027 of the energy of the photoelectron removed from the molecule.

Various electron states of the **cation radical** $M^{\dot{+}}$ can be designated $S_0^{\dot{+}}$, $S_1^{\dot{+}}$, $S_2^{\dot{+}}$, etc. The symbol $S_0^{\dot{+}}$ indicates the ion in the ground state that corresponds to the removal of the electron from the highest occupied molecular orbital; the symbol $S_1^{\dot{+}}$ designates the first excited state of the cation radical corresponding to the removal of the electron from the highest occupied MO, etc. Energy of the $i$th electron state of the cation radical is designated $E(S_i^{\dot{+}})$. This corresponds to the removal of the electron from the $i$th occupied molecular orbital. In order to account for possible changes in the vibrational energy of the ion, the second subscript is used. The symbol $S_{0,0}^{\dot{+}}$ is used to designate the cation radical in the ground electron state and the ground vibrational state, while the symbol $S_{1,n}^{\dot{+}}$ designates the ion in the first excited electron and $n$th vibrational state.

The energy conservation during photoionization can be described as:

$$h\nu + E(S_{0,0}) = E(S_{i,n}^{\dot{+}}) + E_i^{\text{kin}} \tag{2.26}$$

The equation suggests that the initial state of a molecule can be described as the ground electron and ground vibrational state:

$$E(S_{i,n}^{\dot{+}}) - E(S_{0,0}) = h\nu - E_i^{\text{kin}}$$

Comparison of this equation with Eq. (2.25) shows that the ionization potential $IP_i$, or the ionization energy of the molecule $M$, is the difference of the energies of the cation radical formed by ionization and the initial state of the molecule:

$$IP_i = E(S_{i,n}^{\dot{+}}) - E(S_{0,0}) \tag{2.27}$$

The processes occurring on photoexcitation are shown in Fig. 2.7. It can be seen that various ionization potentials ($IP_1, IP_2, \ldots, IP_i, \ldots, IP_n$) correspond to singly charged molecular cation radicals formed by withdrawal of an electron from HOMO, (HOMO-1), (HOMO-2), etc., respectively. The second, third, etc. ionization potentials differ by their meaning from the corresponding ionization potentials of atoms for which $IP_2$ means the energy which is spent to remove the second electron from a singly charged ion, and $IP_3$ means the energy spent to remove the third electron from a doubly charged ion, etc. The diagram explains also the formation of **vibrational structure** in the photoelectron spectrum bands. The ionization band with a permitted vibrational structure appears in the form of a series of peaks, the distance between peaks being equal to the difference of the energies between the corresponding vibrational levels of the ion, and can be used to calculate the frequency of vibrational movement in it.

Since the transformation of excitation in an electron transition occurs in a shorter time than required to change the energy of the vibrational state, the most probable vibrational state of the ion is that which corresponds to the minimal changes in the

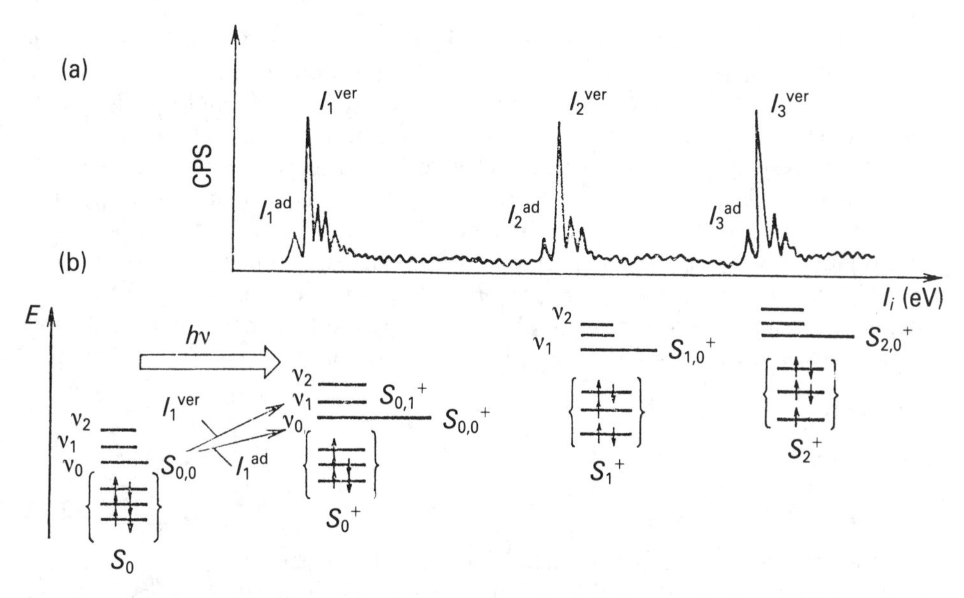


Fig. 2.7. (a) Photoelectron spectrum and (b) formation of vibrational structure of photoionization bands. (From Ghosh [33])

geometry of the initial molecule and whose energy is therefore minimal. The first peak of the photoelectron spectrum band characterizes the value of the **adiabatic potential** $IP_i^{ad}$ and corresponds to the zero vibrational state of the $S_{i,0}^+$ ion. The second peak is usually the highest. It characterizes the value of the **vertical ionization potential** $IP_i^{vert}$ and relates to the so-called **Franck–Condon transition**. It corresponds to the first vibrational state of the cation radical $S_{i,1}^+$. As a rule, only the vertical ionization potentials are measured and analysed because the differences between vertical and adiabatic ionization potentials are quite small ($\simeq 0.1$ eV).

The photoelectron spectrum band is designated by the state symmetry of the cation radical formed by photoionization. The state symmetry is determined unambiguously by the symmetry of the molecular orbital in which an unpaired electron remains as a result of photoionization. Since, for example, three occupied MOs of formaldehyde have symmetries $b_2$, $b_1$, and $a_1$ respectively, the first three ionization potentials determined from the photoelectron spectrum should be assigned by the symmetry of these orbitals [38]:

| | | | |
|---|---|---|---|
| Ionization potential (eV) | 10.88 | 14.5 | 16.0 |
| MO number and symmetry | $2b_2$ | $1b_1$ | $5a_1$ |
| Symmetry of cation-radical state (parenthesized is an electron configuration) | $^2B_2$ $(b_2^1 b_1^2 a_1^2)$ | $^2B_1$ $(b_2^2 b_1^1 a_1^2)$ | $^2A_1$ $(b_2^2 b_1^2 a_1^1)$ |

Quantitative characteristics of the photoelectron spectrum, such as the band shape, the presence of vibrational structure, the values of the Jahn–Teller and spin-orbital effects, intensity and position of the bands are usually sufficient for reliable assignment of the spectrum bands of comparatively simple molecules such as HBr, $H_2S$, ethylene, etc. The photoelectron spectrum of a polyatomic molecule is usually more complicated. Strict assignment of the bands is, as a rule, based on Koopmans' theorem, suggesting the use of quantum chemical calculations for the purpose.

According to Fig. 2.7(b), photoelectron spectroscopic bands and ionization potentials determined from their position, correspond to the differences in the energies of the ground state of a neutral molecule $S_{0,0}$ and various doublet states of the ions $S^{+}_{i,n}$ produced by photoionization [see Equation (2.27)]. Assuming that photoionization of an electron from a molecular orbital is not attended by considerable transformation of the other occupied MOs, one can obtain:

$$E(S_{0,0}) - E(S^{+}_{i,1}) = \varepsilon_i^{\mathrm{SCF}} \tag{2.28}$$

It then turns out that:

$$IP_i = -\varepsilon_i^{\mathrm{SCF}} \tag{2.29}$$

This relationship is called **Koopmans' theorem** [40]: measured vertical ionization energies $IP_i$ are equal to the negative values of calculated orbital energies $-\varepsilon_i$. The corresponding configurations of ionic states $S^{+}_{i,1}$ formed by photoionization and described by wave functions calculated for the ground state of a neutral molecule are called the Koopmans configurations.

It is Koopmans' theorem (2.29) that determines applicability of photoelectron spectroscopy to the study of electron structure within the framework of the orbital approach. This theorem is used universally for interpretation of photoelectron spectra. Within the framework of Koopmans' approximation one can calculate the orbital structure of the atom or molecule and use thus obtained wave functions for both neutral structure in the ground state and various states of cation radicals formed by photoionization.

Figure 2.2 shows the comparison of the measured ionization potentials of atoms of various elements with *ab initio* quantum chemical calculations. It can be seen that Koopmans' relationship holds true for all electron states of the atoms:

$$IP_i(\Psi_i) = -\varepsilon^{\mathrm{SCF}}(\Psi_i)$$

Koopmans' theorem is also widely used for the interpretation of photoelectron spectra of molecules. Photoelectron spectroscopic data for some hydrocarbons are given below by way of illustration:

| | Benzene | Naphthalene | Phenanthrene | Anthracene | Tetracene |
|---|---|---|---|---|---|
| $\varepsilon_{\mathrm{HOMO}}^{\mathrm{HMO}}$ | $\alpha + \beta$ | $\alpha + 0.618\beta$ | $\alpha + 0.4141\beta$ | $\alpha + 0.414\beta$ | $\alpha + 0.295\beta$ |
| $IP_i$ (eV) | 9.24 | 8.15 | 7.86 | 7.41 | 6.97 |

In the general case, however, molecular orbitals of the cation radical $M^{\dot{+}}$ can differ substantially from the orbitals of a neutral molecule. This can be explained by possible transformation of the remaining occupied electron levels after photoionization. This effect is called **electron reorganization** and is estimated by the

**reorganization energy** $E_{reorg}$. Equation (2.29) can also be complicated by the fact that while photoionization changes the number of electrons in the molecule, the magnitude of interelectron effects and, primarily, of the **correlation energy** ($\Delta E_{corr}$) is inevitably changed.

Detailed analysis shows that the effects following photoionization can be disregarded: in most cases the different effects compensate one another [41–42]. The reorganization energy resulting from photoionization of some compounds can, however, be 20% of the measured ionization potential. Besides, the photoionization of localized orbitals is attended by high relaxation effects ($\Delta E_{rel}$): the higher the orbital localization, the higher the relaxation energy. Transition metal complexes (e.g. ferrocene) are a convincing example.

The difference between orbital energy calculated by the *ab initio* method (Hartree–Fock) and the experimentally found vertical ionization potential is called the **Koopmans defect** $\Delta K_i$:

$$\Delta K_i = -(\varepsilon_i^{SCF} + IP_i^{vert}) \tag{2.30}$$

**Table 2.7.** Measured ionization potentials $IP_i$, calculated MO energies $\varepsilon_i$ and effects following photoionization of water in the gas phase (eV)

| MO number and symmetry | $IP_i$ | $-\varepsilon_i$ (*ab initio*) | $E_{reorg}$ | $\Delta E_{corr}$ | $E_{rel}$ |
|---|---|---|---|---|---|
| $1a_1$ | $540.2 \pm 0.2$ | 559.5 | −20.4 | 0.5 | 0.5 |
| $2a_1$ | $32.2 \pm 0.2$ | 36.8 | −2.5 | −1.9 | 0.0 |
| $1b_2$ | $18.6 \pm 0.2$ | 19.5 | −1.9 | 1.3 | 0.0 |
| $3a_1$ | $14.7 \pm 0.1$ | 15.9 | −2.5 | 1.4 | 0.0 |
| $1b_1$ | 12.6 | 13.9 | −2.8 | 1.4 | 0.0 |

(From Schäfer [26]; Hohlneicher [43])

Assignment of PE spectrum bands of water (Table 2.7) shows that in ionization of the inner (localized) shells, Koopmans' defect $\Delta K_1$ is sufficiently high: reorganization energy $E_{reorg}$ can exceed the change in the correlation energy $E_{corr}$ more than 40 times. Conversely, in ionization of the higher (delocalized) electron shells, Koopmans' defect is small: the contribution of the reorganization energy and changes in the correlation energy are closer in their absolute values and partly compensate one another since they have opposite signs.

### 2.3.2 Electron transmission spectroscopy

Photoelectron spectroscopy can be used to obtain objective information on theenergies of occupied molecular orbitals. The energies of vacant electron levels are no less important for the discussion of behaviour of organic compounds in chemical and photochemical reactions.

The method of electron transmission spectroscopy (ETS) [45, 46] can be used to estimate the energies of vacant MOs. An electron transmission spectrum is recorded by measurement in the gas phase: a monochromatic beam of electrons is passed through vapour of the analysed substance. The molecules absorb some electrons in their vacant orbitals, and this causes changes in the intensity (number of electrons per second) of the electron beam as it passes through the measuring chamber. Recording these changes is the operating principle of electron transmission spectroscopy.

On recording the electron transmission spectrum, a signal is formed which is proportional to the section of electron scattering, i.e. to the sum of elastic or inelastic scattering. This gives very high sensitivity to the recording of anion radicals by this method.

The electron transmission experiment is very simple. Electrons are emitted by a thorium-coated iridium electrode; they pass the monochromator and enter the working chamber. Vapour of the examined substance is rarefied in the chamber to a density sufficient to scatter about 60% of the electrons that enter the chamber. As the beam of electrons leaves the chamber, it enters a collector where its intensity is measured. Since the life span of intermediate anion radicals is short, the electrons released on their decay are returned into the beam. But these 'secondary' electrons have a non-oriented velocity vector, and they do not enter the collector. Unscattered 'primary' electrons alone are therefore measured in the collector.

A typical curve describing electron scattering by atoms or molecules in the coordinates – intensity $I$ of scattered electrons vs electron energy – is shown in Fig. 2.8.

A special method has been developed by which rapid changes in the intensity of the electron beam can be measured to high accuracy. This method is based on differential recording of the signal.

Resonance changes in the intensity (induced by inelastic electron scattering) are especially marked in differential recording against the insignificantly changed non-resonance (elastic) scattering. The resonance in the electron transmission spectrum is detected as a point of transition of minimum transmitted electron current into its maximum as the energy of the electrons directed into the chamber increases. For asymmetric signals, the point on the vertical line, midway between the maximum and minimum, is assumed conventionally as the energy of electron insertion into the vacant MO, and is called electron affinity $EA$ (in eV) of the molecule:

$$M + e^- \rightarrow M^{\dot{-}}$$

ETS findings are also commensurable with the results of quantum chemical calculations within the framework of Koopmans' theorem. Enlargement of, for example, the conjugated $\pi$ system not only decreases the first ionization potential but also (according to the pairing theorem) increases the first value of electron affinity, i.e. reduces the energy of LUMO:

| | Ethylene | buta-1,3-diene | hexa-1,3,5-triene |
|---|---|---|---|
| $\varepsilon_{LUMO}^{HOMO}$ | $\alpha + \beta$ | $\alpha - 0.618\beta$ | $\alpha - 0.445\beta$ |
| $EA_1$ (eV) | $-1.78$ | $-0.62$ | $>0$ |

The symmetry of unoccupied MO of a neutral molecule to which the electron is attached determines the symmetry of the electron state of the anion radical formed.

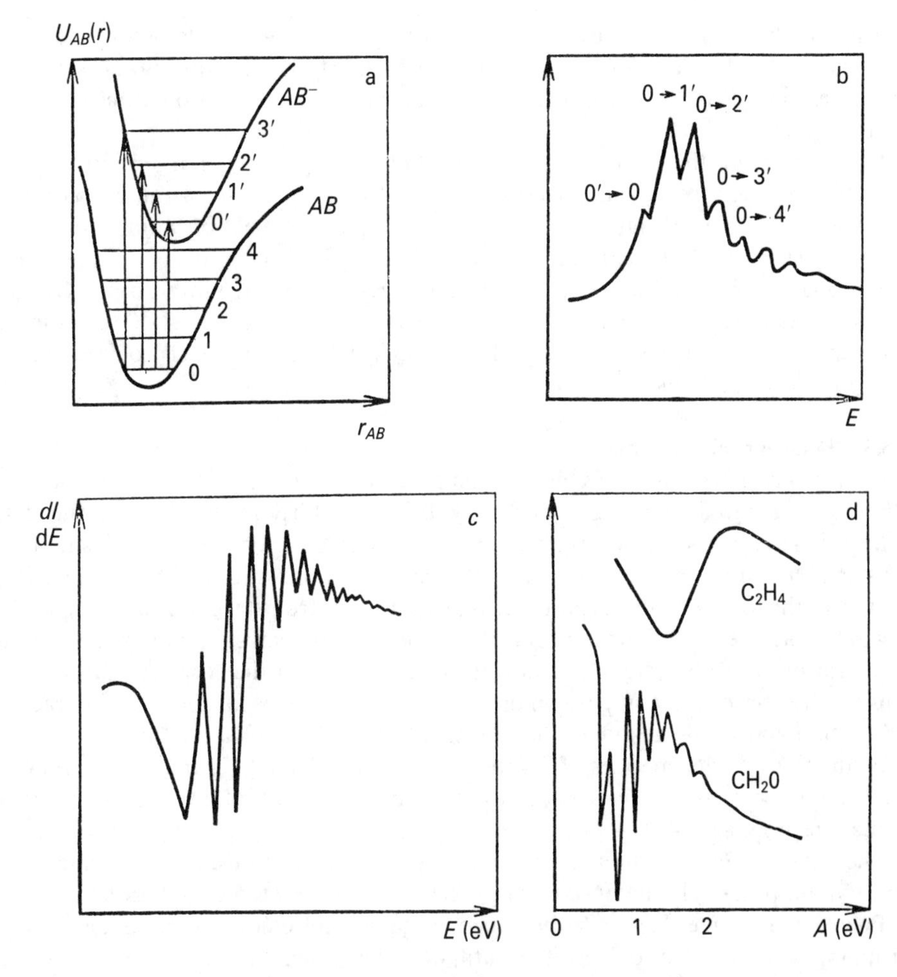


Fig. 2.8. Recording electron transmission spectrum: (a) potential energy of a hypothetical diatomic molecule and its intermediate anion radical AB-; (b) section of electron scattering by compound AB; (c) derivative of passed flow as electron energy function; (d) ETS of ethylene and ETS of $CH_2O$ (From Jordan and Burrow [46])

Two resonance signals in the ET spectrum of butadiene ($-0.62$ eV and $-2.80$ eV) are thus assigned to $\pi$ MOs $2a_u$ and $2b_g$ or to the symmetry states $A_u$ and $B_g$ respectively.

Possible observation of **vibrational structure of signals** in the spectrum depends on the life span of the anion radical formed. It can be assumed tentatively that if the average life span of the anion radical is longer than the period of nucleus vibration, a well-resolved vibrational structure of the resonance signal appears. During observation of the vibrational structure, the **adiabatic energy of the electron affinity** of a molecule $EA_j^{\text{adiab}}$ is determined by the first peak of the signal. The ground state of, for example, the anion radical $CH_2O^{\dot{-}}$, is lower than of the anion radical $C^2H_4^{\dot{-}}$.

A distinct vibrational structure is therefore seen in the ET spectrum of formaldehyde (Fig. 2.8). If the life span of the anion radical is short and the vibrational structure of the resonance signal is not discovered, **vertical energy of electron affinity** ($EA_j^{vert}$) is determined.

If the energy of the main vibrational level of the negatively charged ion is lower than the energy of the ground state of a neutral molecule, a stable anion radical is formed. It is assumed that such a molecule has positive electron affinity. Electron affinity of such molecules cannot be measured by the EST method: energies of higher vibrational states of anion radicals can be measured only in favourable cases. Oxygen is a good example; its molecule in the gas state has $EA = +0.44$ eV [45]. Accordingly, resonance peaks, corresponding only to high vibrational levels of the anion $\nu(O_2^{\dot{-}}) \geq 4$, can be found in the ET spectrum.

### 2.3.3 Polarographic methods

Polarography makes it possible to compare ionization potentials and electron affinities (measured in the gas phase by PES and ETS methods) with the data concerning the behaviour of organic molecules in solvents. The relationships that are thus established are of great interest for at least the following reasons. Firstly, they lead us to the following important conclusions: the electron structures of complicated organic molecules and their ion radicals do not undergo substantial changes during their transition from individual states in the gas phase to the solvated states in the liquid phase. Secondly, comparison of the formation energies for the same ion radical in solution and in the gas phase helps estimation of their solvation energies.

Estimations of the energies of frontier orbitals of solvated organic molecules can, for example, be derived from the polarographic measurements of reversible oxidation–reduction processes [47].

The polarographic method consists in measuring the relationships between current density and potential difference during electrolysis of the studied solution.

**Polarographic one-electron reduction** of organic molecules is carried out with a dropping-mercury electrode as the cathode. The potential across the cathode is changing gradually. The other electrode is the so-called non-polarizable electrode whose potential remains constant during the reduction process. If the potential of the dropping-mercury electrode is low and its value is insufficient to induce an electrochemical reaction, current is almost absent in a polarographic cell. As the potential across the cathode grows to the level that induces an electrochemical reduction, the molecule of an organic substance in a polarographic cell accepts an electron:

$$M + e^- \rightarrow M^{\dot{+}}$$

The generated current is recorded. As the potential increases at the cathode, the current intensity increases but only to a certain limit, until all molecules at the cathode are reduced. The limiting current is called the **diffusion current** because it only depends on the rate of diffusion of the molecules of the reduced substance toward the electrode surface. The potential corresponding to the midpoint on the diffusion current curve (current–potential curve) is called the **half-wave potential** of

**polarographic reduction**, $E_{1/2}^{\mathrm{red}}$, and is measured in volts. Since the surface of the dropping-mercury cathode is renewed continuously, the method of polarographic reduction can be used for accurate measurements of $E_{1/2}^{\mathrm{red}}$.

Estimations made in reversible reduction processes (cyclic voltammetry ) are especially reliable. Their quantum chemical interpretation is based on the assumption that the electron captured by an organic molecule is attached to the frontier unoccupied MO. Half-wave potential of a reversible polarographic reduction $E_{1/2}^{\mathrm{red}}$ thus characterizes the energy of LUMO of the organic molecule. Consider, e.g., $E_{1/2}^{\mathrm{red}}$ potentials for some 1,4-benzoquinones: electron donating methyl groups increase the energy of LUMO (and hence the absolute value of the potential) while the chlorine atoms, having electron acceptor properties, decrease the energy substantially:

| | $H_3C$, $CH_3$ | (unsubstituted) | Cl, Cl |
|---|---|---|---|
| $E_{1/2}^{\mathrm{red}}$, V | −0.67 | −0.51 | −0.18 |

Using Eq. (2.20) for the MO energy, one can write the following relationships:

$$\varepsilon_j = \beta\lambda_j + \alpha \qquad E_{1/2}^{\mathrm{red}} = b\lambda_j + C \tag{2.31}$$

The parameter $C$ depends on the solvation energy of a hydrocarbon molecule and its anion-radical. It is assumed to be constant for some compounds with a similar structure. Since the dependence is linear, it can be used to determine the effective value for the parameter $\beta$ by the angle of inclination of the line in the coordinates $E_{1/2}^{-\lambda}$. Thus, the following relationships of the type (2.31) are known for various groups of organic compounds [18]:

for linear conjugated polyene hydrocarbons

$$E_{1/2}^{\mathrm{red}} = 2.23\lambda_j - 0.94$$

for aromatic polycyclic hydrocarbons

$$E_{1/2}^{\mathrm{red}} = 2.41\lambda_j - 0.40$$

for aromatic ketones

$$E_{1/2}^{\mathrm{red}} = 2.15\lambda_j - 0.68$$

for nitrobenzenes

$$E_{1/2}^{\mathrm{red}} = 2.28\lambda_j + 0.24$$

**Polarographic one-electron oxidation** of organic molecules is associated with removal of one electron from the molecule with formation of a cation-radical:

$$\mathrm{M} - \mathrm{e}^- \rightarrow \mathrm{M}^{\dot{+}}$$

Since on oxidation of the molecule, the electron is first of all removed from the frontier occupied molecular orbital (e.g. from $\varphi_2$ and $\varphi_3$ orbitals of a benzene molecule), half-wave potentials of polarographic oxidation of related compounds depend on the energies of their HOMOs:

$$E_{1/2}^{ox} = \beta\lambda_i + C \tag{2.32}$$

or, according to Koopmans' theorem, on ionization potentials measured by the PES method. The parameter $C$ is determined by the difference in solvation energies of a neutral molecule and the corresponding cation radical. The following relationship has been derived for many (about 60) organic compounds of different classes [48]:

$$E_{1/2}^{ox} = 0.92IP - 6.20.$$

Almost the same parameters were obtained for a series of aromatic hydrocarbons:

$$E_{1/2}^{ox} = 0.98IP - 6.04$$

and only slightly different parameters were derived for strained cyclic saturated hydrocarbons [49]:

$$E_{1/2}^{ox} = 0.78IP - 4.88$$

### 2.3.4 Electron spin resonance spectroscopy

Polarographic methods are suitable for estimation of the energies of frontier orbitals, but they give no information on the distribution of electron density in these orbitals. The method of electron spin resonance (ESR) spectroscopy is very informative for the purpose. It determines the splitting constants of an unpaired electron on the nuclei of the corresponding atoms in ion radicals formed chemically or electrochemically from neutral molecules [16, 50]. In the absence of the external magnetic field, the magnetic moment of the electron is not manifest because of random spatial orientations of spins. But only two orientations of magnetic moments can be realized in a stationary external magnetic field $H$: field- and counter-field orientations. The energy of particles whose spins are field-oriented is lower than the energy of particles in the absence of the external magnetic field. Conversely, the particles whose spins are counter-field oriented have higher energies. The energy difference of the formed levels determines the condition for resonance in ESR spectroscopy:

$$\Delta E = h\nu = g\beta H \tag{2.33}$$

where $g$ is a dimensionless factor 2.002313 for the 'free' electron, which only slightly changes in atoms and molecules; $\beta$ is Bohr's magneton, equal to $0.92732 \times 10^{-27}$ J/gauss; and $H$ is the magnetic field strength.

If an unpaired electron in an organic radical (or ion-radical) interacts only with the external magnetic field, a single line can be seen in the ESR spectrum. But the nuclei of some atoms of organic molecules have non-zero spins (e.g. H, P, $C^{13}$). Spin states of an unpaired electron in a molecule can be split by the interaction of the electron magnetic moment with nuclear magnetic moments. This effect is called **hyperfine interaction** (HFI). The magnitude of HFI is characterized by the **splitting constant**. Splitting, e.g. on a proton, is designated $a_H$. The $a_H$ value is proportional to spin density on the hydrogen atom which, in turn, is determined by electron density $\rho_\mu$ on the carbon atom $\mu$, bound with this hydrogen. The dependence is known as the McConnely equation:

$$a_H = Q_{C-H} \cdot \rho_\mu \cdot \tag{2.34}$$

The parameter $Q_{C-H}$ measures the efficiency with which the unpaired electron in the p$\pi$ orbital of carbon can induce an unpaired spin on the C—H proton. This parameter will also be designated $Q$. The absolute value of $Q$ is between 20 and 30 Oe.

As an unpaired electron interacts with two equivalent protons in the radical particle, the splitting pattern changes. The ESR spectrum of, for example, radical $CH_2OH$ containing the only pair of equivalent protons (hyperfine splitting on protons of the OH group can be disregarded under certain conditions), is a triplet with line intensity distribution of 1:2:1. If an unpaired electron interacts in the radical with $n$ equivalent protons (or other nuclei with non-zero spin $Z$), the number of lines in the ESR spectrum is $n + 1$ (in the general case, $2nZ + 1$).

If a molecule (from which a radical is formed) contains non-equivalent protons or nuclei with different spins, the ESR spectrum becomes more complicated. Accurate interpretation of the spectrum is based on its comparison with the results of quantum chemical calculations. By determining the $a_H$ constants from the spectrum, and calculating $\rho_\mu^{exp}$ from Eq. (2.34), one can determine the parameters of the electron structure that can be compared with the results of quantum chemical calculation of spin density $\rho_\mu^{theor}$.

The ESR spectrum of, for example, buta-1,3-diene anion radical is a quintuplet with a relative intensity of 1:4:6:4:1, which corresponds to unpaired electron splitting on four equivalent protons. Each component of the quintuplet is split into a triplet (1:2:1) corresponding to the interaction of an unpaired electron with two equivalent protons. The measured values of the HFI constants (7.62 Oe and 2.79 Oe respectively) agree well with the results of calculations by the HMO method: the ratio of the measured unpaired electron densities on the atoms 1(4) and 2(3) in the butadiene anion radical is $7.62/2.79 = 2.73$, while the ratio of the calculated densities is $0.6015^2/0.3717^2 = 2.61$, in accordance with its LUMO coefficients:

$$\varphi_3 = 0.6015_{\chi 1} - 0.3717_{\chi 2} - 0.3717_{\chi 3} + 0.6015_{\chi 4}$$

The analysis of the spectra of ion radicals of substituted benzenes is an interesting example of application of ESR spectroscopy to the study of the electron structure of organic molecules. Insertion of a substituent into the benzene molecule removes degeneracy of both the frontier occupied ($\varphi_2$ and $\varphi_3$) and of the frontier unoccupied ($\varphi_4$ and $\varphi_5$) orbitals, as shown in Fig. 2.6. The character and effectiveness of removal of degeneracy depends on the substituent nature. Both electron donating and electron withdrawing substituents should probably, first of all, change the energies of symmetric orbitals $\varphi_2$ and $\varphi_5$, while the energies of antisymmetric orbitals $\varphi_3$ and $\varphi_4$ should, in the first approximation, remain unchanged: these orbitals have zero coefficients at the site of substitution [41]. It explains how the distribution of spin density, determined from the ESR spectrum of the anion radical of a substituted benzene $C_6H_5X$, can characterize the substituent X interaction with the $\pi$ system of the

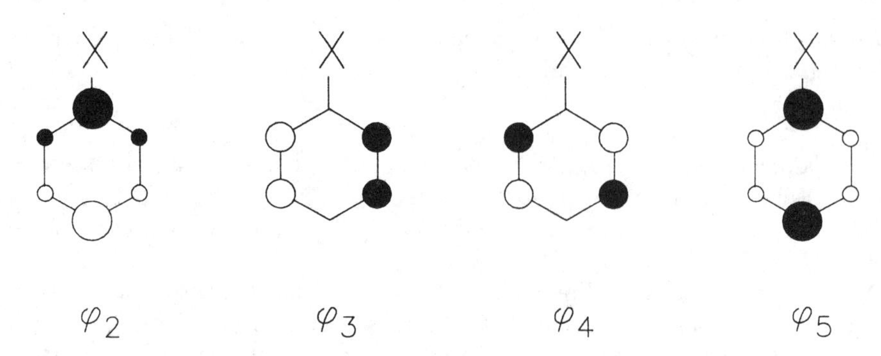


benzene ring. The HFI constants, determined from the ESR spectrum, e.g. for the toluene anion radical, are $a_{H(CH_3)} = 0.79$ Oe, $a_{H(ortho)} = 5.12$ Oe, $a_{H(meta)} = 4.50$ Oe and $a_{H(para)} = 0.59$ Oe. These values can be compared with the values of spin densities ($Q = 22.5$ Oe):

$$\rho_{\text{ortho}}^{\text{exp}} = 0.23, \qquad \rho_{\text{meta}}^{\text{exp}} = 0.20, \qquad \rho_{\text{para}}^{\text{exp}} = 0.03$$

calculated for the location of an unpaired electron in an antisymmetrical orbital $\varphi_4$ of the benzene ring. According to the diagram in Fig. 2.6 this fact suggests the electron donating effect of the $CH_3$ group that increases the energy of the $\varphi_5$ orbital, so that the benzene orbital $\varphi_4$, which becomes the frontier LUMO of the toluene molecule, is practically unchanged. From the benzene molecule calculation it follows that the densities of unpaired electron in the $\varphi_4$ orbital are

$$\rho_{\text{ortho}}^{\text{theor}} = 0.25, \qquad \rho_{\text{meta}}^{\text{theor}} = 0.25, \qquad \rho_{\text{para}}^{\text{theor}} = 0.00$$

These results well agree with the experimentally found values (see above).

The values of splitting constants of an unpaired electron on ring protons in the anion radical of trimethyl(phenyl)silane $C_6H_5SiMe_3$:

$$a_{\text{H(ortho)}} = 2.66 \text{ Oe}, \qquad a_{\text{H(meta)}} = 1.06 \text{ Oe}, \qquad a_{\text{H(para)}} = 8.42 \text{ Oe}$$

illustrate another case of removal of degeneracy of the $\varphi_4$ and $\varphi_5$ orbitals of the benzene molecule. HFI constants correspond to the situation where the $\varphi_5$ orbital of the benzene ring becomes the frontier LUMO in this silane. According to Fig. 2.6, the low energy of this orbital (compared with the $\varphi_4$ orbital) is a direct indication of the electron withdrawing effect of the $SiMe_3$ group.

ESR spectra of cation radicals give similar information on the electron structure of organic molecules. (For more detail see subsequent chapters.)

### 2.3.5 Electron absorption spectroscopy

Electron absorption spectroscopy is also important for understanding the electron structure of molecules within the framework of the molecular orbitals and electron state concepts. As the absorption spectrum is recorded, electrons absorb energy and leave occupied molecular orbitals, but they still remain in the molecule since they transfer into some unoccupied orbitals. This method can sometimes be used to estimate properties of vacant and occupied MOs.

The importance of electron absorption spectroscopy for the study of the electron structure of molecules is explained by the very short excitation time (as in other electron spectroscopic methods) which does not exceed $10^{-13}$ s. This time is much shorter than required to change the geometry of the molecule: the transitions in electron absorption spectroscopy obey the Franck–Condon principle.

If electron levels alone were involved in electron excitation, the electron absorption spectrum would consist mostly of separate lines. But an experimental electron absorption spectrum includes wide bands, which often have fine structure: changes in the electron energy of the molecule are accompanied by changes in its vibrational energy.

The determination of electron transition probability on the basis of multiplicity and symmetry selection rules is especially informative as regards study of the electron structure of molecules. Transitions that fulfil the selection rules, are called *permitted*. Transitions for which the rules do not hold are called *forbidden*. This is practically important for the band intensity: highly intensive bands correspond to the most probable transitions.

The **multiplicity selection rule** reads: permitted electron transitions occur only between states with the same multiplicity. The multiplicity of the state of a molecule is determined by the expression $2s + 1$, where $s$ is the total spin quantum number, calculated as the sum of electron spins in all MOs. The total spin quantum number of a neutral molecule in the ground state $S_0$ with doubly occupied MOs is always equal to zero, while in the excited states $S_1$, $S_2$, $S_3$, etc., it can be either 0 or 1. The values of $s$ correspond to the states with multiplicity of 1 (these states are called singlets and designated by $^1S$) or 3 (these states are called triplets and designated $^3S$). In accordance with the multiplicity selection rule, the transitions $^1S_0 \rightarrow {}^1S_1$ and $^1S_0 \rightarrow {}^1S_2$ in buta-1,3-diene are permitted, while the transitions $^1S_0 \rightarrow {}^3S_1$ and $^1S_0 \rightarrow {}^3S_2$ are forbidden. Electron transitions in the buta-1,3-diene molecule are shown in Fig. 2.9: each transition is considered as an electron transition configuration.*

According to the **symmetry selection rule**, the probability of an electron transition is estimated objectively by the value of the transition moment.

The **transition moment** describes the charge migration on the transition of the molecule from one electron state to another:

$$M_{ik} = \int \Psi_i^* \hat{M} \Psi_k \, d\tau \tag{2.35}$$

where $\Psi_k$ and $\Psi_i$ are wave functions of the initial and final electron states of the molecule; $\hat{M}$ is the transition moment operator.

The value of $M_{ik}$ determines the intensity of absorption according to the following expression for the **oscillator strength**:

$$f_{ik}^{\text{theor}} = 1.085 \times 10^{-3} \nu_{ik} (|M_{ik}|^2/e^2) \tag{2.36}$$

* If the absorption band corresponds to one configuration, this band is supposed to be due to one-configurational transition. If the band is assigned to several configurations, the band is believed to be due to multi-configurational transition.

For electron transitions permitted by symmetry, the value of $M_{ik}$ should be other than zero, while for the forbidden transitions it is zero.

The transition moment is zero if the integrand is transformed as a total symmetry type, i.e. it relates to the symmetry type in the table of characters of the point group in which all characters are equal to +1 (in the table of characters of the point group, $C_{2h}$, to which the butadiene molecule belongs, $A_g$ is a total symmetry type). The expression for the transition moment can then be modified:

$$M_{ik} = \int \Psi_i^* M \Psi_k \, d\tau = \int \Psi_i^* \times \Psi_k \, d\tau + \int \Psi_i^* y \Psi_k \, d\tau + \int \Psi_i^* z \Psi_k \, d\tau \quad (2.37)$$

Consider estimation of electron transition probabilities in a buta-1,3-diene molecule. It was shown in Section 2.2.2 that the ground state of butadiene is the $A_g$ symmetry type, while the first excited state belongs to the $B_u$ symmetry type. Then, the expression for the dipole moment of the $S_0 \rightarrow S_1$ transition in the butadiene molecule is:

$$\begin{array}{cccc} M_{S_1,S_0} = & \int \varphi_{S_1}^* x \varphi_{S_0} \, d\tau + & \int \varphi_{S_1}^* y \varphi_{S_0} \, d\tau + & \int \varphi_{S_1}^* z \varphi_{S_0} \, d\tau = 0 \\ & B_u B_u A_g & B_u B_u A_g & B_u A_u A_g \\ & A_g & A_g & B_g \end{array}$$

According to this expression, the $S_0 \rightarrow S_1$ transition is permitted by the symmetry selection rule, since the subintegral components $M_x$ and $M_y$ are transformed by the total symmetry type $A_g$. This transition is then polarized in the direction of the $x$ and $y$ axes and is forbidden in the direction of the $z$ axis. The same holds for the $S_0 \rightarrow S_4$ transition.

Conversely, the $S_0 \rightarrow S_2(a_u^1 b_g^2 a_u^1)$ and $S_0 \rightarrow S_3(a_u^2 b_g^1 b_g^1)$ transitions are forbidden by the symmetry selection rule: not a single component of the transition moment has its integrand transformed by the total-symmetry type $A_g$. The corresponding integrals are therefore equal to zero:

$$\begin{array}{cccc} M_{S_2,S_0} = & \int \varphi_{S_2}^* x \varphi_{S_0} \, d\tau + & \int \varphi_{S_2}^* y \varphi_{S_0} \, d\tau + & \int \varphi_{S_2}^* z \varphi_{S_0} \, d\tau = 0 \\ & A_g B_u A_g & A_g B_u A_g & A_g A_u A_g \\ & B_u & B_u & A_u \end{array}$$

The value of $M_{ik}$ determined by quantum chemical calculations can, according to Eq. (2.36), be used for theoretical estimation of the oscillator strength $f_{ik}^{\text{theor}}$.

The oscillator strength can also be determined experimentally by measuring the area circumscribed by the absorption curve and the wave number axis:

$$f_{ik}^{\text{exp}} = 4.32 \times 10^{-9} \int_{\nu_1}^{\nu_2} \varepsilon \, d\nu, \quad (2.38)$$

where $\varepsilon$ is the molar extinction coefficient and $\nu$ is the wave number.

The symmetric band integral in the first approximation (2.38) can be replaced by the product $\varepsilon \Delta\nu_{1/2}$, where $\Delta\nu_{1/2}$ is the half-width of the absorption band, i.e. the width at the height $\varepsilon_{\max}/2$.

For a totally permitted transition $f = 1$; the values of $f$ from 0.1 to 1.0 correspond to the values of $\varepsilon$ equal to 10 000–100 000. Coincidence of the theoretical and experimental estimations of oscillator strength $f$, as well as the energies of electron transition, can be regarded as confirmation of correctness of the absorption band assignment.

The scheme in Fig. 2.9 shows applicability of the **one-configurational model** for interpretation of UV spectra of compounds in which the frontier molecular orbitals are at a sufficiently great distance from the adjacent vacant and occupied MOs. If

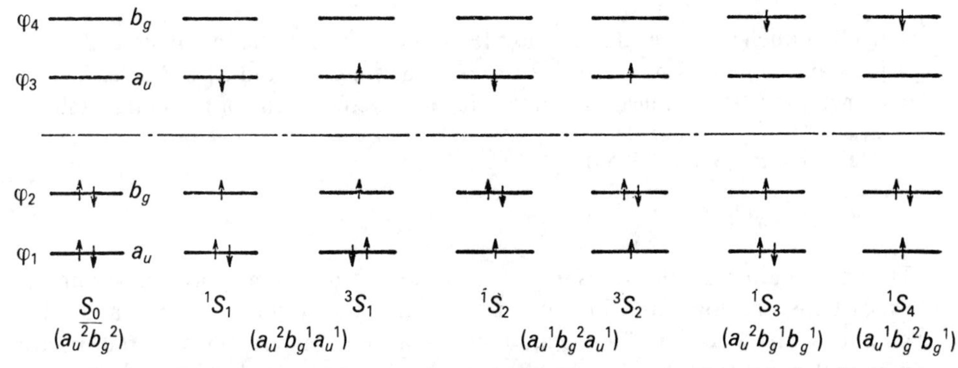


Fig. 2.9. Configurations of the ground and excited electron states of buta-1,3-diene ($S_0$ is the ground state; $^1S_1$, $^1S_2$, $^1S_3$, and $^1S_4$ are excited singlet states; $^3S_1$ and $^3S_2$ are excited triplets).

HOMOs and LUMOs of organic molecules are degenerate or close to degeneracy, their electron absorption spectra should be analysed taking into account the interaction of configurations of the corresponding electron transitions, the so-called **multi-configurational interaction.**

The estimation of the energy of long-wave transition in a benzene molecule follows below. In one-configurational approximation, the said energy should be determined as the difference of energies of the LUMO and HOMO:

$$\Delta E = E_{\text{LUMO}} - E_{\text{HOMO}} = 2\beta$$

In fact, the long-wave transition in a benzene molecule can have at least four different configurations of the same energy:

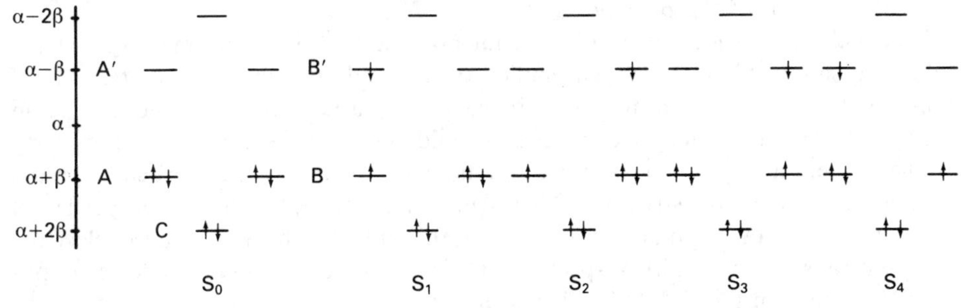

α−2β
α−β A′ B′
α
α+β A B
α+2β C

$S_0$ $S_1$ $S_2$ $S_3$ $S_4$

Best agreement with experimental findings is attained for such molecules with the quantum chemical calculations of electron spectra that account for the multi-configurational interaction. The interaction of configurations obeys the same rules as orbital interactions:

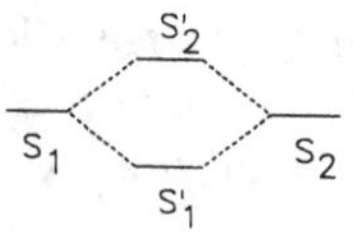

$$\begin{vmatrix} (ES_1 - E) & H_{12} \\ H_{21} & (ES_2 - E) \end{vmatrix} = 0$$



It actively influences the energies of excited electron states: the initial states $S_1$ and $S_2$ with similar energies mix better. The resultant states $S_2'$ and $S_1'$ are described by wave functions that are linear combinations of wave functions of the initial states:

$$\Psi(S_1') = a\Psi(S_1) + b\Psi(S_2)$$

$$\Psi(S_2') = a\Psi(S^1) - b\Psi(S_2)$$

The absorption band positions in the electron spectra of some polycyclic aromatic hydrocarbons are compared in Section 6.3.4 with the results of HMO and PPP calculations which account for the multi-configurational interactions. In the spectra of naphthalene, anthracene and naphthacene, the longest-wave bands are, first of all, assigned to the electron transitions between the frontier orbitals: in the molecules of these hydrocarbons, these orbitals are sufficiently well separated from other MOs and are not degenerate. On the contrary, it has already been said that benzene has degenerate frontier molecular orbitals; the low-energy electron transition in it cannot therefore be a one-configurational transition.

In our acquaintance with electron absorption spectroscopy, attention should be paid to designation of electron transitions and the corresponding bands in the absorption spectra. The symmetry of electron states, between which the transition occurs, is used to designate electron transitions. Since the symmetry of a molecule in the ground state with doubly occupied MOs is always a total-symmetry type, the bands in the electron absorption spectra are often designated only by the symmetry of the wave function of the excited state into which the molecule is converted. Thus, the band in the UV spectrum of a butadiene molecule that corresponds to the transition from the ground state $S_0$ to the first excited state can be designated as the $A_g \rightarrow B_u$ band, or as the $B_u$ band.

To conclude the short review of 'experimental' methods in quantum chemistry, the following should be noted. When applied to the study of orbital structure, each of these methods is based on quantum-chemical calculations. It was stated in Section 2.2.1 that a definite method of calculation should preferably be used for each property studied. There are, however, universal methods that can be used to calculate many parameters of electron structure. They can objectively estimate the properties of molecules in both ground and excited states. In Table 2.8, the data of photoelectron, electron transmission, and UV-spectroscopy for benzene are compared, for example, with the results of the INDO/S calculation.

**Table 2.8.** The measured values of ionization potentials $IP_i$, electron affinities $EA_1$, UV spectrum bands of benzene and assignment of spectral data by the INDO/S calculations

| | Spectral data | Assignment | Calculated values |
|---|---|---|---|
| $IP_i$ (eV) | 9.31 | $1e_{1g}(\pi)$ | 9.33 |
| | 11.42 | $2e_{2g}$ | 12.00 |
| | 12.11 | $1a_{2u}(\pi)$ | 13.78 |
| | 13.82 | $2e_{1u}$ | 15.37 |
| | 14.70 | $1b_{2u}$ | 15.90 |
| | 15.40 | $1b_{1u}$ | 17.88 |
| | 16.90 | $2a_{1g}$ | 22.82 |
| $EA_1$ (eV) | −1.15 | $1e_{2u}(\pi)$ | −0.06 |
| Band energies in UV spectrum | 4.77 | $B_{2u}$ | 4.82 |
| ($f$ values) | (0.00) | | (0.00) |
| | 6.20 | $B_{1u}$ | 6.24 |
| | (0.04) | | (0.00) |
| | 6.88 | $E_{1u}$ | 6.77 |
| | (0.88) | | (1.00) |

(From Millefiori and Millefiori [51])

## 2.4 ESTIMATION OF ORBITAL INTERACTIONS

### 2.4.1 Fragment (group) orbitals

Although Roothaan's equations are more accurately solved within the framework of *ab initio* methods, their use is limited by at least the following factors.

Firstly, the size of the basis set and the potential abilities of computers impose limitations to the wide use of the most perfect schemes for *ab initio* calculations of complicated organic molecules.

Second, the results of the corresponding calculations (especially for complicated molecules) are difficult to interpret in the traditional (for an organic chemist) terms, which suggest estimations of mutual effects of groups of atoms or molecular fragments (substituents, functional groups, etc.), rather than of separate atoms or separate atomic orbitals. For this reason, approaches to the estimation of electron structures of organic molecules within the framework of the simplified quantum chemical calculations are now widely used.

One such approach consists in the use of the so-called **fragment** or **group** orbitals in the basis set of quantum chemical calculations. These orbitals are supposed to be localized not over atoms but over a group of atoms or molecular fragments. They are formed in the same way as molecular orbitals, i.e. as linear combinations of appropriate atomic orbitals of the fragment. The calculations using fragment orbitals in the basis set are also known in *ab initio*, semi-empirical, and simple quantum chemical calculations. This approach simplifies both calculations and analysis of their results [52–54]. According to Heilbronner and Maier [54], the possibility of using fragment orbitals in calculations depends largely on the fact that the canonical molecular orbitals ($\varphi_i$) calculated by the MO LCAO method are the same conventional pictures of the electron structure of the molecule as is the case with the other molecular orbitals $\{\varphi'\}$, obtained by appropriate unitary transformations [54]:

$$\{\varphi_i'\} = U\{\varphi_i]$$

While selecting the matrix for a unitary transformation $U \equiv L$, one can transform the canonical $\varphi_i$ orbitals into fragment orbitals $\lambda_i$:

$$\{\lambda_i\} = L\{\varphi_i\}$$

Using the appropriate matrix of transformation it is possible to obtain the correspondingly transformed Hartree–Fock matrix:

$$F_\lambda = (F_{\lambda,ij}) = LF_\Psi L^{\mathrm{T}}$$

It is quite natural that matrix elements differ in their absolute values, but they become quite versatile for compounds within the framework of an appropriate model.

Thus obtained fragment orbitals $\lambda_i$ can be used in the basis set for calculation of another molecule, e.g. in the Hückel approximation:

$$\varphi_i = \sum c_{i\mu}\lambda_i; \qquad \langle\lambda_i|\mathcal{H}|\lambda_i\rangle = A_i; \qquad \langle\lambda_i|\mathcal{H}|\lambda_j\rangle = B_{ij};$$

The first equation is a linear combination of fragment orbitals, the second one describes eigenenergies of fragment orbitals, and the third equation determines the integrals of the interaction of these orbitals. The corresponding scheme for calculations is called MO LCFO (molecular orbital is a linear combination of fragment orbitals).

Fragment orbitals are common wave functions; they have all properties of these functions. Like atomic and molecular orbitals, fragment orbitals have the same $g_i$ values, which are equal to 0, 1, or 2. The overlapping of fragment orbitals is determined by calculation of the corresponding overlap integral:

$$S_{\mu\nu} = \int \chi_\mu \chi_\nu \,\mathrm{d}\tau$$

Wave functions made on the basis of fragment orbitals are also determined by solving Roothaan equations. The elements of the energy operator matrix cannot be determined either by appropriate calculations using atomic orbitals in the basis set (e.g. by the *ab initio* method), or by appropriate physical experiment, i.e. by measuring ionization potentials of model compounds using photoelectron spectroscopy.

Some fragment orbitals can be considered as examples. In the study of the electron structure of benzene derivatives with various functional groups, $\pi$ orbitals of the unsubstituted benzene are usually regarded as the fragment orbitals $\pi_1$, $\pi_2$, $\pi_3$. They are shown graphically in Section 2.2.1.

While estimating orbital interactions with involvement of heteroatoms or substituents, the orbitals located mainly on these heteroatoms or substituents are often regarded as fragment orbitals. Such orbitals, e.g. in compounds of sulphur, oxygen or nitrogen, can only tentatively be regarded as unshared electron pair (lone pair) orbitals and are designated by *n*. In fact, lone pair orbitals are to a certain degree mixed with the orbitals of groups attached to the heteroatom, and are thus fragment orbitals. (An example of a fragment orbital presented in an analytical form is given in Section 4.1.1.)

### 2.4.2 Orbital interaction rules

Efficiency of interaction or mixing of orbitals (atomic, fragment, or molecular) depends on their relative energy, spatial overlapping and symmetry [52–5].

*Energies of mixing orbitals.* This factor depends directly on the basis set. As a rule, atomic orbitals of the valence shell are dealt with only in calculations of organic compounds. For the hydrogen atom, this is the 1*s* orbital, while for the carbon atom these are 2*s*-, $2p_x$, $2p_y$, and $2p_z$ orbitals.

Confinement of the basis for calculations to the valence shell orbitals turns out reasonably due to substantial differences in the energies of the electron in the inner and outer electron shells. For the purpose of comparison, given below are the values of the ionization potentials (in eV) of the electrons of some atomic orbitals [33]:

| | B | C | N | O | F |
|---|---|---|---|---|---|
| 1*s* | 192 | 288 | 403 | 538 | 694 |
| 2*s* | 12.93 | 16.59 | 20.33 | 28.48 | 37.85 |
| 2*p* | 8.30 | 11.26 | 14.50 | 13.62 | 17.42 |

The high values of the ionization potentials of the inner electrons interfere with active involvement of their orbitals in the formation of MOs: the inner electrons remain mostly localized over the atoms. It can be seen that the electrons of the valence shell are also ionized to a different degree; e.g. ionization potentials of 2*p* electrons are much lower than those of 2*s* electrons. This difference is sufficiently obvious: the orbitals of organic molecules with the highest energy are usually formed only by atomic *p* orbitals; in planar systems these are $\pi$ MOs.

Spatial conditions for orbital overlap is another factor that is important for effective mixing of orbitals. It relates to the geometry of a molecule. Those orbitals can only mix, for which the overlap integral $S_{\mu\nu}$ is greater than zero:

$$S_{\mu\nu} > 0$$

This condition is first of all fulfilled for the orbitals of the adjacent atoms and for the orbitals of the atoms that are found close to one another in space. The integrals of the orbital overlap for the adjacent atoms in a buta-1,3-diene molecule $\chi_1$ and $\chi_2$, as well as $\chi_2$ and $\chi_3$, $\chi_3$ and $\chi_4$ are definitely other than zero. Overlapping of the $\chi_1$ and $\chi_4$ orbitals in the *trans* conformation can be neglected, while it is 0.2 in the *cis* conformation [18].

Examples of orbital mixing will be discussed in much detail in other chapters of this book. Much attention is given to the $\pi$(C = C)-orbital interaction in organic molecules. In this chapter we shall only note effective interaction of these orbitals in a molecule of cyclohexa-1,3-diene: while the ionization potential of electrons in the $\pi$(C = C) orbital of the cyclohexene molecule is 9.12 eV, the ionization potentials of two $\pi$(C = C) orbitals of cyclohexa-1,3-diene differ substantially from this value and are 8.80 and 9.80 eV respectively [54].

*Symmetry of mixing orbitals.* The third factor that determines the orbital mixing is their symmetry: orbitals of the same symmetry can only be mixed.

Consider the following example illustrating the role of the symmetry factor in mixing of atomic orbitals:

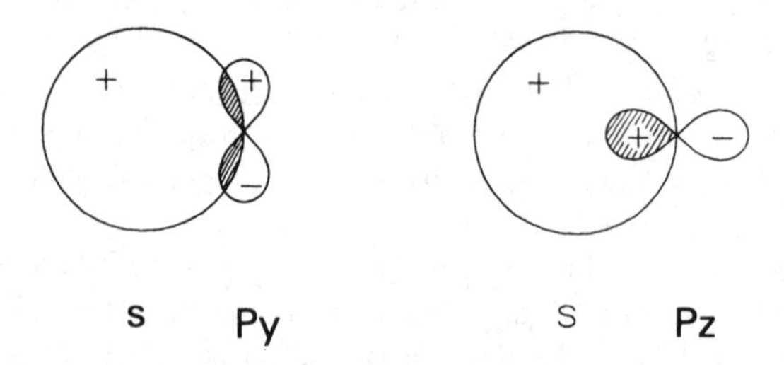


The overlapping elements of the $s$ and $p_y$ orbitals have the same signs on one side of the $p_y$ orbital and the opposite signs on its other side. These orbitals have different symmetry relative to the bond axis ($\sigma$ and $\pi$ respectively) and are called **orthogonal.** They have zero overlap. Their interaction by the $\pi$ type is excluded. Conversely, overlapping of the $s$ and $p$ orbitals by the $\sigma$ type is not ruled out. Thus, the overlapping of the $s$ and $p_z$ orbitals (right) is not zero. Such orbitals have the same symmetry relative to the bond line; they can mix and are called **non-orthogonal.**

The orbital with lower energy interacts with a higher-energy orbital and is thus stabilized, while the higher-energy orbital mixes with a lower-energy orbital and is

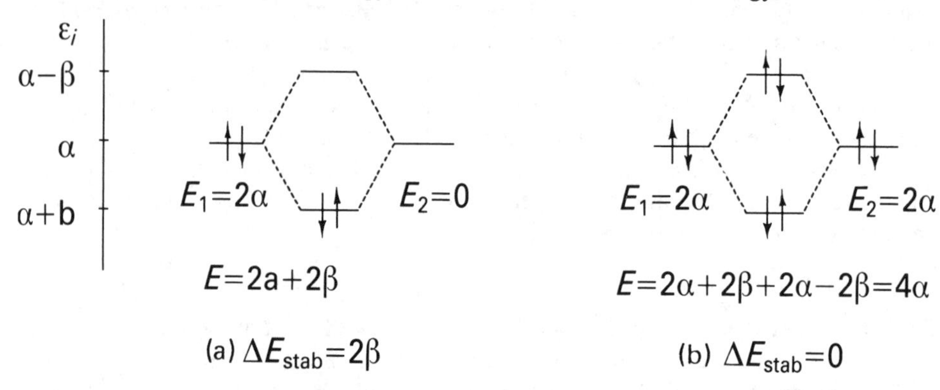


Fig. 2.10. Energy stabilization $\Delta E_{stab}$ due to mixing of (a) occupied and unoccupied orbitals; (b) two occupied orbitals.

thus destabilized (Fig. 2.10). It is important to note that the higher orbital is always destabilized slightly more than the lower orbital. These rules explain why mixing of doubly occupied orbitals cannot stabilize the system, and actually destabilizes it as a rule. On the contrary, it can be seen from the diagram that the interaction between the occupied and vacant orbitals produces a stabilizing effect.

It should finally be noted that the interactions between orbitals are additive because the final result does not depend on the sequence in which these interactions are considered.

The mentioned orbital mixing rules are illustrated in Fig. 2.11 by the formation of ethylene MOs. In the left part of the correlation diagram, one $\pi$ and five $\sigma$ occupied MOs with various contributions of the CH and CC orbitals are shown as a result of mixing of 12 basis AOs (for two carbon and four hydrogen atoms). The sequence of MOs is determined by the relative energies of the basis AOs ($C_{2p} > H_{1s} \gg C_{2s}$) and the number of sign alternations in the resultant MOs [53].

In the right part of the diagram of Fig. 2.11, MOs of ethylene are formed from the orbitals of two $CH_2$ fragments. The fragment orbitals $\sigma(CH_2)$ and $\pi(CH_2)$ form bonding (+) and antibonding (−) combinations respectively:

$$\sigma^+(CH_2), \sigma^-(CH_2), \pi^+(CH_2) \quad \text{and} \quad \pi^-(CH_2).$$

Besides, the group of bonding orbitals includes also $\sigma(CC)$ and $\pi(CC)$ orbitals. The $\sigma^+(CH_2)$ and $\sigma(CC)$ orbitals are fully symmetric (have the $a_g$ symmetry within the point group $D_{2h}$) and can mix, while all other orbitals have different types of symmetry. When estimating the MO sequence, it is understood that the energy of the $\sigma(CH)$ bond is about 420 kJ/mol, the energy of the $\sigma(CC)$ bond is about 350 kJ/mol, while the $\pi$ orbital should be located above the corresponding $\sigma$ orbitals [52].

The diagrams of ethylene MOs obtained are compared with the photoelectron spectrum of ethylene. The first band (having a fine vibrational structure) is assigned to the ionization of the $\pi(CC)$ orbital, the second band to the $\sigma(CH)$ orbitals and the third band (14.5 eV) to the $\sigma(CC)$ bonding orbital.

It should be noted that the sequence of ethylene MOs, determined both in the basis of AOs and in the fragment orbital basis, has been confirmed by *ab initio* quantum chemical calculations (see Section 5.1.1).

Group orbitals can be used not only to estimate the electron structure of organic molecules. The correlation diagram in Fig. 2.12 shows, for example, a preferable conformation of ethyl cation. The interaction between the unoccupied p orbital of the planar $CH_2^+$ fragment and $\pi(CH_3)$ orbital is shown for the eclipsed and staggered conformations of the ethyl cation. Since the $\pi^x(CH_3)$ and $\pi^y(CH_3)$ orbitals are degenerate and their overlap with $p_x$ and $p_y$ AOs of carbon is similar, both conformations are equivalent as regards their energy, while the calculated rotational barrier for ethyl cation has the minimum value.

If the hydrogen of the methyl group is substituted by the atom X, $\pi(CH_2X)$ orbitals become non-degenerate: depending on a particular X, the $\pi^x(CH_3)$ orbital either decreases or increases its energy. The interaction with the vacant $p$ orbital of the $CH_2^+$ fragment in the staggered conformation can thus be strengthened or weakened. If X is $CH_3$, the staggered conformation turns out to be most advantageous, while at X = F, the eclipsed conformation is preferable [52].

For a detailed description of the orbital interaction effects on conformation of neutral molecules and ions see Section 3.3.

### 2.4.3 Parametrization of orbital interactions by PE spectroscopy

According to Koopmans' theorem, photoelectron spectroscopy is most suitable for parametrization of quantum chemical calculations.

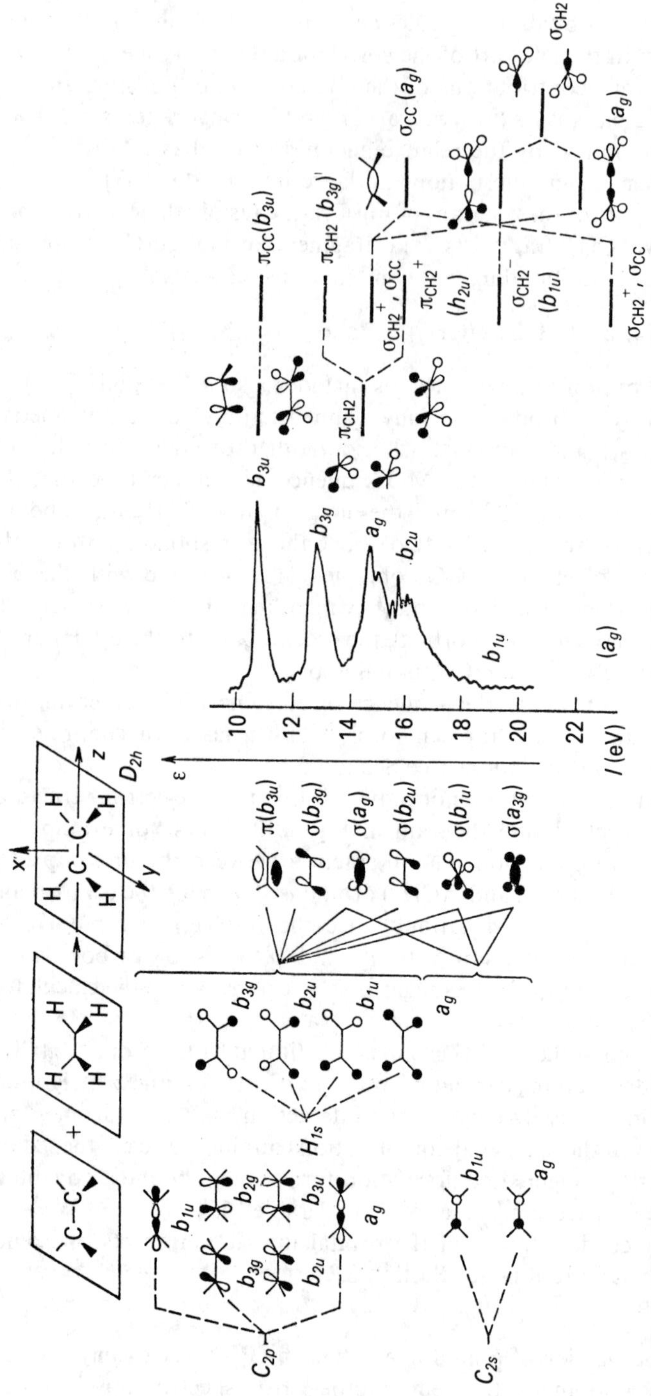


Fig. 2.11. Formation of MOs of ethylene and its photoelectron spectrum left – formation of MOs from AO; right – formation of MOs from fragment orbitals. (From Bock and Ramsey [53])

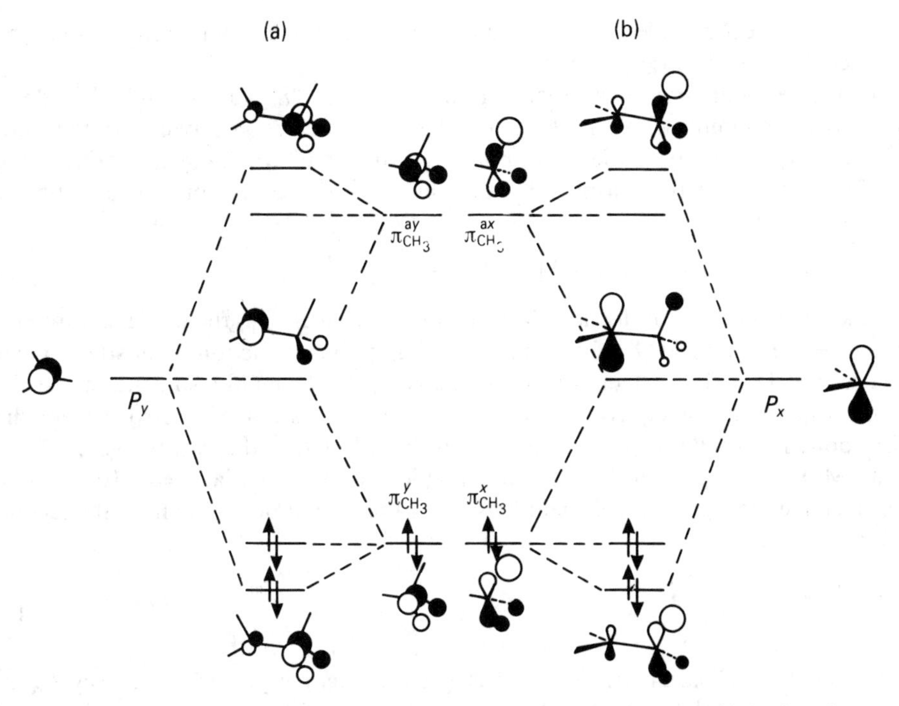


Fig. 2.12. Formation of ethyl cation MOs: (a) eclipsed conformation; (b) bisected conformation. (From Klessinger [52])

Numerous attempts have been made to use PES data for estimation of orbital parameters, i.e. for parametrization. Mixing of the $3p_\pi$ orbital of the sulphur atom with the $\pi(C = C)$ orbital, for example, was estimated from the PE spectrum of *trans*-1,2-di(methylthio)ethylene [56]. The energy of electrons in the $3p_\pi$ orbital of sulphur was determined from the spectrum of 1,1-di(methylthio)ethylene by the 8.80 eV band corresponding to the (−)-combination of $3p_\pi$ orbitals of two sulphur atoms: this combination is not mixed with the $\pi(C{=}C)$ orbital (for symmetry considerations) but is influenced by its inductive effect. The energy of the electrons in the $\pi(C{=}C)$ orbital was determined by the first ionization potential of unsubstituted ethylene.

In the estimation of the effect of interaction between the corresponding fragment orbitals in the molecule of para-bis(methylthio)benzene with involvement of the benzene $\pi$(sym)orbital, the inductive substituent effect on this orbital was also determined from the position of the band assigned to the ionization of $\pi$(asym) orbital that is incapable of mixing with substituent orbitals [57].

Consider in more detail the method of parametrization in the basis of fragment orbitals (MO LCFO). It is based on the determination of the $F$ elements of the Fock energy operator matrix which are necessary for the solution of the system of Roothaan equations

$$\sum_{\mu=1}^{n} (F_{\mu\nu} - \varepsilon_i S_{\mu\nu}) c_{i\mu} = 0$$

The photoelectron spectra of the corresponding model compounds should be analysed for this purpose [52, 58].

For the determination of the diagonal matrix element $F_{\mu\mu}$, use should be made of the model compound whose molecular states are described as having orbitals with prevalent contribution of the corresponding $\mu$th (atomic or fragment) orbital; the ionization of the electrons from this $\mu$th orbital should be presented as a distinctly identifiable band

$$IP_i(\mu) = E[M_i^+ + (\mu)] - E[M_0] = -F_{\mu\mu} \tag{2.39}$$

For the determination of the non-diagonal matrix element $F_{\mu\nu}$ the model compound should be used, some MOs of which are formed by the interaction of mostly $\mu$th and $\nu$th orbitals. The PE spectrum of this model compound should also contain distinct bands which could unequivocally be assigned to ionization of electrons from the corresponding resultant ($\mu\nu$) orbitals. Then, by estimating the eigenvalue $\varepsilon_i$ of the ($\mu\nu$)th MO by the value of the corresponding ionization potential $\varepsilon_i = -\mathrm{IP}_i(\mu\nu)$, one can calculate the $F_{\mu\nu}$ matrix element by solving the secular determinant of the second order

$$\begin{vmatrix} (F_{\mu\mu} - \varepsilon_i) & F_{\mu\nu} \\ F_{\nu\mu} & (F_{\nu\nu} - \varepsilon_i) \end{vmatrix} \tag{2.40}$$

In this method of parametrization, all effects, such as reorganization energy $E_{\text{reorg}}$, changes in the correlation energy $\Delta E_{\text{corr}}$, and the relaxation energy $\Delta E_{\text{relax}}$ (the defect of Koopmans' theorem), are taken into account in the values of the $F_{\mu\mu}$ and $F_{\mu\nu}$ matrix elements [see (2.30)]:

$$I_i(\mu) = -\varepsilon_i^{\text{SCF}}(\mu) + E_{\text{reorg}} + \Delta E_{\text{corr}} + \Delta E_{\text{relax}} = -F_{\mu\mu} \tag{2.41}$$

The integrals determined by the proposed scheme account for the total energy of the electron in the $\mu$th and ($\mu\nu$)th orbitals and can in general be designated by the Fock matrix elements $F_{\mu\mu}$ and $F_{\mu\nu}$.

The specific feature of the parametrization scheme discussed consists in that the interelectron effects $(\mu\nu|\lambda\sigma)$ and $(\mu\lambda|\nu\sigma)$ in the matrix $F_{\mu\nu}$ elements of the Roothaan equation:

$$F_{\mu\nu} = H_{\mu\nu} + \sum_i \sum_\mu \sum_\nu c_{i\mu}c_{i\nu}[2(\mu\nu|\lambda\sigma) - (\mu\lambda|\nu\sigma)]$$

are not calculated (as is the case with *ab initio* methods), or found from empirical formulae (as with semi-empirical methods), but are measured directly from the PES data.

It is quite natural that quantum chemical calculations with such integrals need no self-consistent field technique and can be performed within the framework of the simple method. This, in turn, justifies designation of the corresponding integrals by the symbols of the $H$ matrix elements: $H_{\mu\mu}$ and $H_{\mu\nu}$. In parametrization of calculations in the $\pi$ approximation, the $H_{\mu\mu}$ integral estimates, e.g. the energy of the electron in the $\mu$th $p_\pi$ orbital, and can be designated by the symbol of the element alone, e.g. $H_O$, $H_S$, $H_N$, etc.

The proposed scheme of parametrization is illustrated below by estimation of the occupied $\pi$MOs of organic compounds of chalcogens [58]. Thus the $H_S$ matrix element evaluating the energy of the electron in $3p_\pi$ orbitals of the sulphur atom in alkyl and aryl sulphides should not probably be determined by the ionization potential of the $3p$ shell of the isolated suphur atom (its value can be assumed to be 10.36 eV [7]), or by the ionization potential of the orbital $b_1$ of the $H_2S$ molecule, although the $3p_\pi$ orbital of the sulphur atom in the $H_2S$ molecule is not mixed with other AOs and is the only $b_1$ type orbital (Table 2.9).

**Table 2.9.** Symmetry transformations of basis AOs in the $H_2S$ molecule ($C_{2v}$)

| $C_{2v}$ | $E$ | $C_2^z$ | $\sigma^{xz}$ | $\sigma^{yz}$ | Symmetry type |
|---|---|---|---|---|---|
| $3p_z$ | +1 | +1 | +1 | +1 | $a_1$ |
| $3p_x$ | +1 | −1 | +1 | −1 | $b_1$ |
| $3p_y$ | +1 | −1 | −1 | +1 | $b_2$ |
| $3_s$ | +1 | +1 | +1 | +1 | $a_1$ |
| (+)1$s$ | +1 | +1 | +1 | +1 | $a_1$ |
| (−)1$s$ | +1 | −1 | −1 | +1 | $b_2$ |

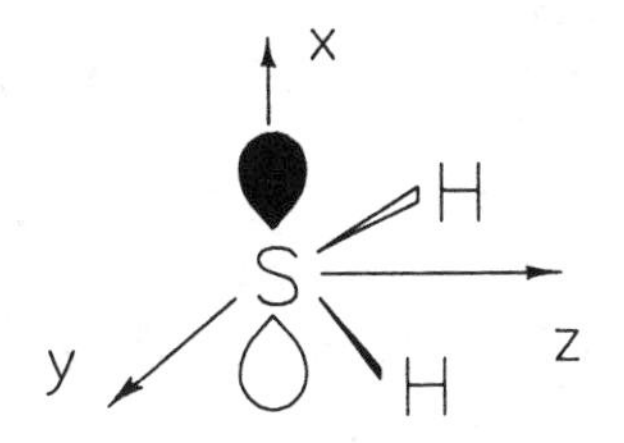


The nearest surroundings of the sulphur atom in organic sulphides are probably best of all modelled by mixing its orbitals in the dimethyl sulphide molecule. Despite the fact that in addition to the $3p_\pi$ orbital of sulphur, the type $b_1$ orbital set of dimethyl sulphide includes also (+) combination of $2p_\pi$ orbitals of the carbon atoms, the contribution of the 3p$\pi$ orbital of sulphur to the population of the HOMO of this compound is, according to the extended Hückel calculations, actually not less than 75%. This corresponds to the narrow band assigned to the first ionization at 8.67 eV in the PE spectrum [59]. The value of the $H_S$ integral can thus be determined from the PE spectrum of $(CH_3)_2S$: $H_S = -8.67$ eV.

It is hardly reasonable to use a more complicated model for the evaluation of the $H_S$ parameter [60–62]. The $H_S$ matrix found, for example, for 1,4-bis(methylthio) benzene from the PE spectrum is about the same: $H_S = -8.80$ eV [52].

It can be seen from Table 2.10 that the $3p_\pi$ orbital of the sulphur atom can be mixed by symmetry ($b_1$) with the $\pi$ (sym) orbital of benzene within the framework of the symmetry group $C_{2v}$ in which thioanisole can conventionally be considered (see Fig. 2.5 and Section 2.2.2). The results of this mixing is splitting of the highest occupied $\pi$ level of the benzene molecule, and the presence of well resolved bands in

**Table 2.10.** Characters of transformation of thioanisole fragment orbitals

| $C_{2v}$ | $E$ | $C_2^z$ | $\sigma_v^{xz}$ | $\sigma_v^{yz}$ | Symmetry type |
|---|---|---|---|---|---|
| $\pi_1$ | +1 | −1 | +1 | −1 | $b_1$ |
| $\pi_2$ (symm) | +1 | −1 | +1 | −1 | $b_1$ |
| $\pi_3$ (asymm) | +1 | +1 | −1 | −1 | $a_2$ |
| $3p_z$ | +1 | +1 | +1 | +1 | $a_1$ |
| $3p_y$ | +1 | −1 | −1 | +1 | $b_2$ |
| $3p_x$ | +1 | −1 | +1 | −1 | $b_1$ |

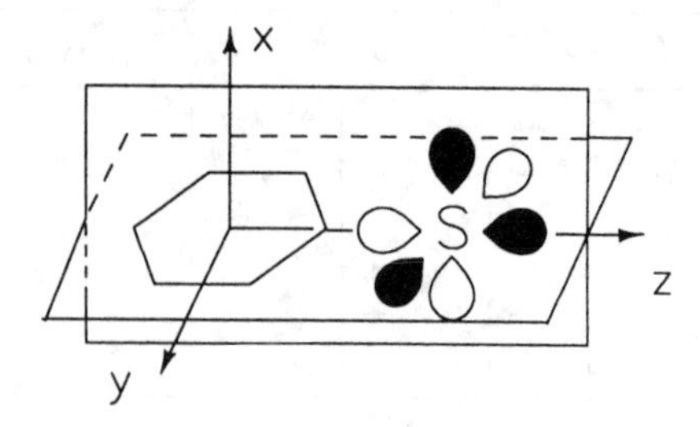


the region of low ionization potentials of the PE spectrum of thioanisole [60]. The value of the $H_{S\pi(sym)}$ integral is determined by solving the determinant:

$$\begin{vmatrix} (H_S - \varepsilon_1) & H_{S\pi(sym)} \\ H_{S\pi(sym)} & (H_S - \varepsilon_1) \end{vmatrix} = 0 \tag{2.42}$$

$$H_{\pi(sym)} = -IP_1(C_6H_6) = -9.24\text{ eV}; \quad \varepsilon_1 = -IP_1(C_6H_5SMe) = -8.07\text{ eV}$$

$$H_{S\pi(sym)} = \sqrt{[(-8.67 + 8.07)(-9.04 + 8.07)]} = -0.84\text{ eV}$$

The $H_{S\pi(sym)}$ parameter changes only insignificantly if the $\pi_1(b_1)$ orbital is also taken into account in the calculation [61]:

$$\begin{matrix} (H_{\pi(sym)} - \varepsilon_1) & H_{S\pi(sym)} & 0 \\ H_{S\pi(sym)} & (H_S - \varepsilon_1) & H_{S\pi_1} \\ 0 & H_{S\pi_1} & (H_{\pi_1} - \varepsilon_1 \end{matrix} = 0 \tag{2.43}$$

The values of the $H_{S\pi(sym)}$ integrals found by solution of these determinants (2.42) and (2.43) taking into account the calculation error are not statistically meaningful. The corresponding data are given in Table 2.11 (assuming chalcogens are included).

Furthermore, the involvement of the $\pi_1$ orbital of benzene in the formation of HOMOs of hetero-analogues of anisole $C_6H_5EMe$ is insignificant; in a molecule of thioanisole, for example, it is only 1.2%.

The value of the $H_{CS}$ integral, determining effectiveness of the $\pi$ type interaction between $p$ orbitals of the adjacent carbon and sulphur atoms, can be found from the relation $H_{CS} = \sqrt{3}H_{S\pi(symm)}$ taking into account for the $1/\sqrt{3}$ of the coefficient of benzene $\pi$ (sym) orbital at the site of substituent attachment.

**Table 2.11.** Parametrization of quantum chemical calculations from PES data. Values of $H_{E\pi\,(\mathrm{symm})}$ integrals

| Element $E$ | $H_{E\pi\,(\mathrm{symm})}$ | | $\pi_1$ share in HOMO (%) |
|---|---|---|---|
| | not accounting for $\pi_1$ | accounting for $\pi_1$ | |
| O | $1.37 \pm 0.04$ | $1.29 \pm 0.08$ | 2.0 |
| S | $0.84 \pm 0.04$ | $0.78 \pm 0.06$ | 1.2 |
| Se | $0.72 \pm 0.05$ | $0.67 \pm 0.06$ | 1.0 |
| Te | $0.69 \pm 0.06$ | $0.64 \pm 0.06$ | 0.8 |

(From Rodin *et al.* [61])

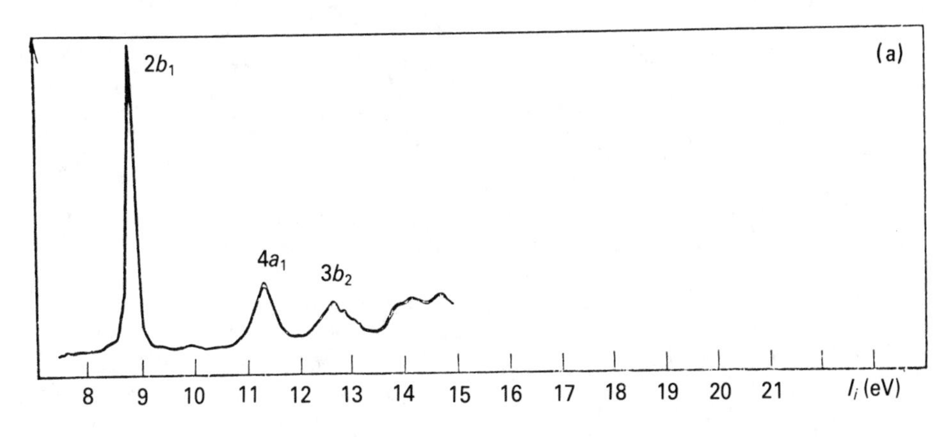


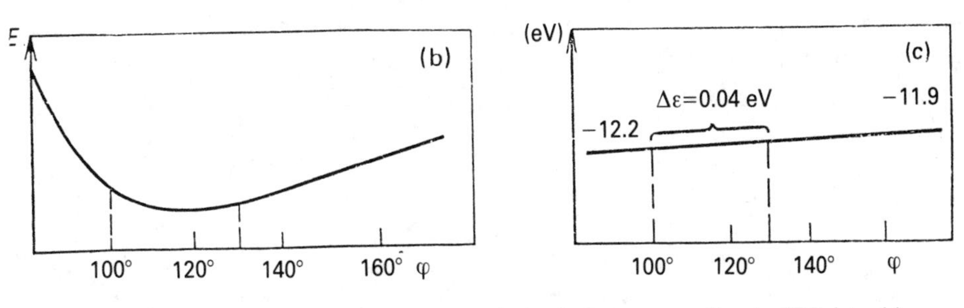


Fig. 2.13. Parametrization of quantum chemical calculations according to PES data: (a) PE spectrum of dimethyl sulphide; (b) dependence of total energy (EHM data) of dimethyl sulphide on the CSC valence angle; (c) dependence of energy of dimethyl sulphide HOMO on the CSC valence angle. (From Traven *et al.* [58])

Figure 2.13 shows the dependence of total energy $E$ and the HOMO energy of dimethyl sulphide on the value of the CSC angle as determined by the EHM method. In the series of $Me_2E$ compounds with E varying from O to S, and further to Te, the dependence of the energy of HOMO on variations in the CSC valence angle definitely decreases. As the angle increases from 70 to 140°, the energy of HOMO of dimethyl ether increases by 1 eV, of dimethyl sulphide by 0.23 eV and of dimethyl telluride only by 0.06 eV. And conversely, the energies of $\sigma$ orbitals localized in the CEC plane are rather sensitive to the same changes in the CEC angle in all $Me_2E$ molecules. Thus, the energy of the $a_1$ orbital of dimethyl telluride increases almost by 2 eV, while the energy of the $b_2$ orbital decreases by 1.3 eV. These data convincingly show why the PES parametrization is best of all suitable for simple quantum chemical calculations in the $\pi$ approximation.

# Part II

# Orbital structure and properties of organic molecules

# 3

# Basic concepts

## 3.1 ORBITAL STRUCTURE AND REACTIVITY

Methods of physics can be used to study the ionization of organic molecules in the gas, liquid and solid phases. The study of these processes is, however, important not only because it gives information of the electron structure of neutral molecules, but also because it can help estimation of the properties of organic molecules in chemical interactions. The first stages of these interactions usually depend on the frontier orbitals, i.e. the highest occupied and lowest unoccupied orbitals of the reactant molecules.

The concept of frontier orbitals proposed by Fukui [63–7] was among the first concepts that established the relationships between the molecular orbitals and the behaviour of molecules in chemical reactions.

Reactions of unsaturated and aromatic hydrocarbons are a good example of the application of this concept. Many aspects of their reactivity are unexplainable by the traditional concepts of electron effects. It is not clear, for example, why naphthalene, the closest analogue of benzene with sufficiently high energy of $\pi$ electron delocalization, is attacked by electrophilic agents almost exclusively at the $\alpha$ position, and only to an insignificant degree at the $\beta$ position. Why are many aromatic hydrocarbons attacked by both nucleophilic and electrophilic agents at the same atoms?

These and many other questions cannot be answered by considering only total electron densities (or total charges) on atoms.

While postulating the importance of valence electrons in the formation of molecules from atoms, Fukui tried to estimate the role of not only the total electron densities but the role of electron levels in the molecules. This approach includes the discussion of the reactivity in terms of molecular orbitals. According to Fukui [64,66], chemical interaction is attended by electron transitions between the interacting molecules. Mixing of the reactant orbitals obeys the same rules as mixing of atomic and fragment orbitals:

- in a reaction with an electrophilic agent, the most susceptible to attack is that

position of the substrate at which the electron density in the HOMO is the highest;
- in a reaction with a nucleophilic agent, the most susceptible to attack is that position of the substrate at which the coefficient of LUMO is the maximum.

According to Fukui, the frontier orbitals, viz. the highest occupied MO of one reagent and the lowest unoccupied MO of the other reactant, play an especially important role in chemical reactions. The reaction is accompanied by the transfer of electron density from the HOMO to the LUMO and depends largely on the conditions of their overlap. Thus the rate of an electrophilic substitution reaction in aromatic hydrocarbons is proportional to the electron density in various HOMO positions. A naphthalene molecule is preferably attacked by an electrophile at positions 1, 4, 5, and 8, because the highest coefficients (electron densities) are at the 1st, 4th, 5th and 8th carbon atoms in its HOMO:

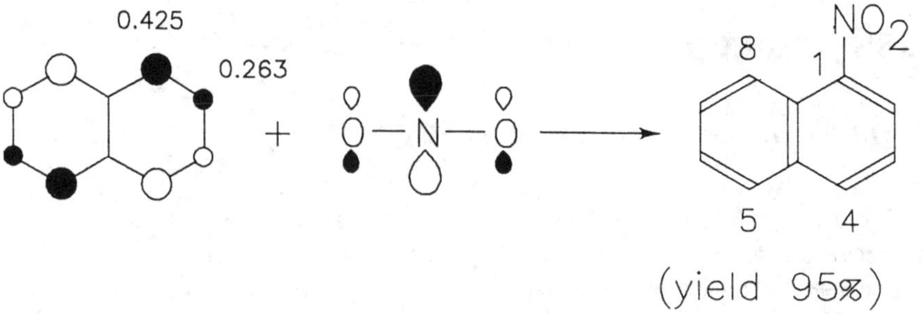

The frontier orbitals concept was at first criticized (see the review in Ref. [68]), but new facts were discovered that showed the importance of the frontier orbitals of molecules. Among the decisive evidence were the data concerning the charge-transfer interactions that occur on formation of the so-called 'weak' complexes of organic molecules by donor D and acceptor A [36]:

$$D + A \leftrightharpoons [D, A] \leftrightharpoons D^{\dot{+}}A^{\dot{-}}$$

Complexes formed with involvement of aromatic hydrocarbons (donors) have been studied in more detail. Their donor properties in such interactions were confirmed by direct experiment: excitation by a laser showed that the complexation reactions between aromatic hydrocarbons and $\pi$ acceptors resulted in formation of the cation radical $ArH^{\dot{+}}$ and the anion radical $A^{\dot{-}}$ [69]:

$$[ArH, A] \rightarrow [ArH^{\dot{+}}A^{\dot{-}}].$$

The role of the frontier orbitals of the donor and acceptor in charge-transfer interactions is determined unambiguously by direct dependence of the energy of the longest wavelength charge-transfer band in the electron absorption spectrum of the complex on the first ionization potential of the donor $IP_1^D$ and the first value of electron affinity of the acceptor $EA_1^A$

$$h\nu_{CT} = IP_1^D - EA_1^A - W \tag{3.1}$$

where $W$ is the dissociation energy of the excited state of the complex.

The dependence (3.1) is simplified when a series of related donors is studied with the same acceptor:

$$h\nu_{CT} = aIP_1 - b$$

where $a$ and $b$ are constants.

The following dependences have been established for the tetracyanoethylene (TCNE) complexes with substituted benzenes [70–72]:

$$h\nu_{CT} = 0.83IP_1 - 4.42 \quad (3.2)$$

$$h\nu_{CT} = 0.80IP_1 - 4.16 \quad (3.3)$$

Charge-transfer complexes were studied not only for aromatic hydrocarbons and their derivatives. Similar dependences were established for TCNE complexes with n-donors as well, e.g. with dialkyl sulphides [73, 74]:

$$h\nu_{CT} = 0.45IP_1 - 1.38 \quad (3.4)$$

and even with $\sigma$ donors, saturated compounds of the group IVA elements [75]:

$$h\nu_{CT} = 0.84IP_1 - 4.25 \quad (3.5)$$

Later studies showed that the low differences between the energies (energy gap) of the frontier MOs of the reacting molecules is a decisive condition for orbital control of appropriate reactions.

The rules for concerted cycloaddition reactions proposed by Woodward and Hoffmann [76–8] were a further development of the frontier orbitals concept. These highly stereospecific reactions obey the rule of conservation of orbital symmetry: the symmetry of the corresponding molecular orbitals remains unaltered during the reaction. Using the symmetry of orbitals of the reacting molecules one can explain and predict the direction of concerted reactions of many unsaturated compounds.

The frontier orbitals concept can, for example, explain why thermal cyclodimerization of ethylene does not occur while a photochemical reaction is feasible. In the thermal process, the HOMO of one molecule ($\pi'_1$) and the LUMO of the other ($\pi''_2$) should be regarded as frontier orbitals. It can be seen that mixing of these orbitals is forbidden by symmetry: the interaction between a pair of carbon atoms in structure I is antibonding:

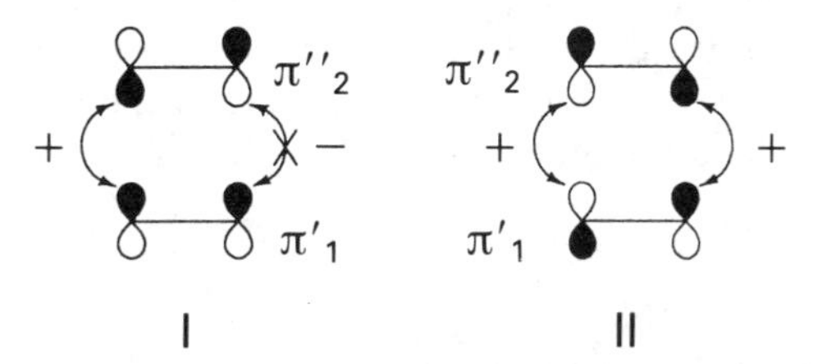


In a photochemical reaction, the $\pi_2$ orbital of ethylene is involved in cyclization with acceptance of the excited electrons. It can be seen that the interaction of two $\pi_2$ orbitals is permitted by symmetry (structure II).

The Woodward–Hoffmann rules for pericyclic reactions of unsaturated hydrocarbons are discussed in detail in Section 5.3. In this chapter we shall only note that when determining the direction of the reaction, not only the $\pi$ orbitals can be the decisive frontier orbitals. The Fukui concept was applied successively to the reaction of hydrogen detachment from aliphatic hydrocarbons and also to the reactions of $S_N2$ and E2 in the series of haloalkanes [66]. Many examples showing the importance of frontier orbitals in reactions of saturated compounds of the group IVA elements have been studied by Kochi [79]:

$$R_4Pb + [Ir^{4+}Cl_6]^{2-} \rightarrow R_4Pb^{\dot{+}} + [Ir^{3+}Cl_6]^{3-}$$

$$R_4Pb^{\dot{+}} - \text{rapid} \rightarrow R^{\cdot} + R_3Pb^{+}$$

$$R^{\cdot} + [Ir^{4+}Cl_6]^{2-} - \text{rapid} \rightarrow RCl + [Ir^{3+}Cl_5]^{2-}$$

The linear dependence of the reaction rates ($\log k$) of the compounds $PbMe_nEt_{4-n}$ on ionization potentials (measured in the gas phase) and on one-electron oxidation potentials (measured in solution) suggests the mechanism of electron transfer between the molecules of tetraalkyl plumbane and $[IrCl_6]^{2-}$. The electron transfer with involvement of the $\sigma$ orbitals occurs not only with organometal compounds. Charge-transfer processes were also studied between TCNE and saturated hydrocarbons [80, 81] (see Section 4.3).

It is important to emphasize that application of the frontier orbitals concept to qualitative estimation of reactions is, by no means, the result of the simplified approach to estimation of reactivity. The direction and rate of reactions of organic molecules depend on various kinetic and thermodynamic factors and can only be estimated objectively within the framework of perfect quantum chemical calculations of potential energy surfaces of reaction complexes, i.e. calculations that are now only applicable to comparatively simple reactions occurring in the gas phase [82].

Klopman [68,83] and Salem [84] used the MO perturbation method to illustrate the role of the frontier orbitals in chemical reactions. The change in energy attainable in pericyclic reaction has been determined as the sum of three terms:

$$\Delta E = -E^{\text{repuls}}_{\text{occ/occ}} + E_{\text{q}} + E^{\text{attr}}_{\text{occ/unocc}} \tag{3.6}$$

The first term is negative because on the interaction of two occupied orbitals the system is partly destabilized (for more detail see 2.4.2, Fig. 2.10)

The second term of Eq. (3.6) is important only for charged molecules since it depends on their electrostatic interactions. The third term has the positive sign: on the interaction between the occupied and vacant orbitals, the system is stabilized (see Section 2.4.2, Fig. 2.10). In the general form, the value of the third term is determined by the equation:

$$E^{\text{attr}}_{\text{occ/unocc}} = \sum_{i}^{\text{occ}} \sum_{j}^{\text{unocc}} \frac{2\left(\sum_{\mu\nu} c_{i\mu} c_{j\nu} \beta_{\mu\nu}\right)^2}{E_i - E_j}$$

where $c_{i\mu}$ is the coefficient of atomic orbital $\mu$ in the $i$th MO of the donor; $c_{j\nu}$ is the coefficient of the atomic orbital $\nu$ in the $j$th MO of the acceptor; $\beta_{\mu\nu}$ is the resonance integral between $\mu$ and $\nu$ AOs; and $E_i$ and $E_j$ are energies of the $i$th MO and $j$th MO respectively.

It follows from the equation that the prevalent contribution to the stabilization energy should belong to the interaction between the HOMO of the donor and the LUMO of the acceptor. The lower the energy gap between the frontier orbitals, the higher the energy stabilization. The third term of Eq.(3.6) can be transformed into Eq. (3.7), while estimates the role of interaction of the frontier orbitals:

$$E = 2\Sigma(c_{\mathrm{HOMO}\mu} c_{\mathrm{LUMO}\nu} \beta_{\mu\nu})^2/(E_{\mathrm{HOMO}} - E_{\mathrm{LUMO}}) \tag{3.7}$$

The most probable is thus the reaction occurring with formation of a bond between the atoms with the maximum coefficients in the frontier orbitals.

The importance of the frontier orbitals becomes especially overt if we follow the transformations of separate MOs that occur when the reactants approach each other: various MOs undergo deep but different changes [68]. The electron density of most MOs decreases gradually in the region of formation of a new bond, while the electron density of one of them (the frontier MO), on the contrary, increases in the same

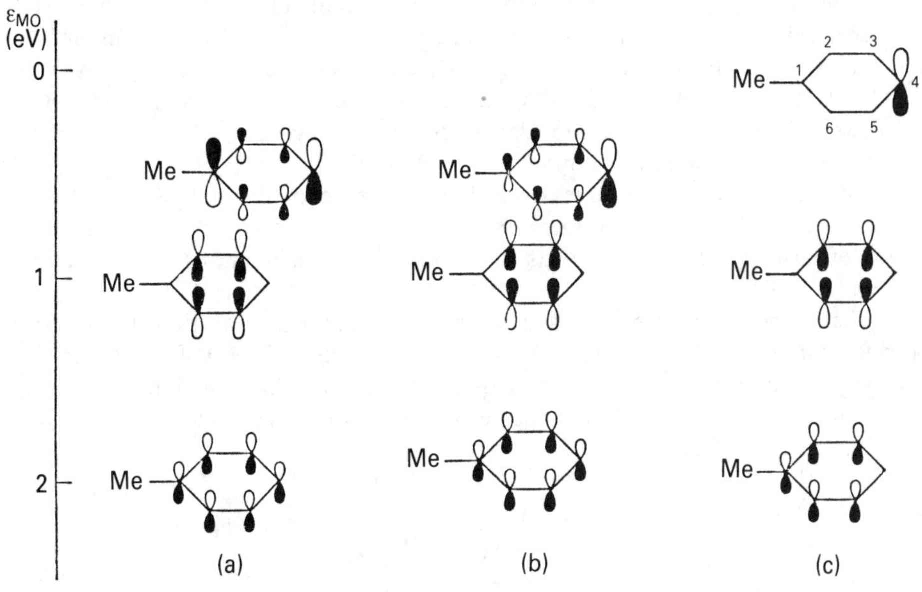


Fig. 3.1. Transformation of toluene MOs on the interaction with an electrophilic agent: (a) $\beta_{3\text{-}4} = \beta_{4\text{-}5} = 1.0$; (b) $\beta_{3\text{-}4} = \beta_{4\text{-}5} = 0.5$; (c) $\beta_{3\text{-}4} = \beta_{4\text{-}5} = 0$. (From Klopman [68])

region. Figure 3.1 illustrates these processes on a diagram showing transformation of toluene MOs caused by an electrophilic attack: as the electrophilic agent approaches, the electron density in the HOMO of toluene becomes concentrated at the para position to the methyl group, while nodes are formed at the site of attachment of the electrophilic agent in other occupied MOs. The dependence of the isomer composition on hardness of the electrophilic agent also agrees with the frontier orbital concept: toluene is brominated mostly at the para position and chlorinated at the ortho position. Quantum chemical calculations show that the bromonium ion, being a soft electrophile, is oriented in accordance with the distribution of electron density in toluene's HOMO, mostly at the para position, while a harder chloronium ion in chlorination is oriented at the ortho position of toluene (in accordance with the total charge determined by summation of charges in all occupied MOs).

Another example emphasizing the importance of frontier orbitals in the reaction of electrophilic reagents with aromatic hydrocarbons is provided by the modified INDO study of the mechanism of benzene nitrosation by the nitrosonium cation in the gas phase [82,85]

$$C_6H_6 + NO^+ \rightarrow C_6H_5NO + H^+$$

It follows from calculations of the potential energy surface that at the initial stage of the reaction, when the distance between the electrophile and the substrate exceeds the sum of van der Waals' radii, i.e. 0.3–0.4 nm, the most probable orientation of the nitrosonium cation is along the $C_6$ axis. The maximum overlap of $e_{1g}$ symmetry HOMO of benzene and $e$ symmetry LUMO of the cation is achieved in this geometry of the intermediate. According to the correlation diagram in Fig. 3.2, collision between the benzene molecule and the nitrosonium cation results in the electron transfer from the hydrocarbon HOMO to the LUMO of the attacking reagent: $C_6H_6 + NO^+ \rightarrow C_6H_5^{+}NO$. In agreement with this diagram, the energies of the corresponding electron levels were determined: the first ionization potential of benzene is 9.24 eV, while the electron affinity of the nitrosonium cation is 9.27 eV.*

The one-electron transfer stage was also discovered experimentally for substituted benzenes. It is induced by the nitronium cation in the gas phase [86], and by the action of various electrophilic agents (chlorine, bromine, mercury trifluoroacetate) in solvents [87,88].†

In the proposed mechanism of electrophilic aromatic substitution, the electron transfer from the HOMO of aromatic hydrocarbon to the LUMO of the electrophile unambiguously determined the isomer distribution during the reaction.

It can be seen from Fig. 3.2 that deeper orbitals of the reactants also interact when the intermediate III is formed:

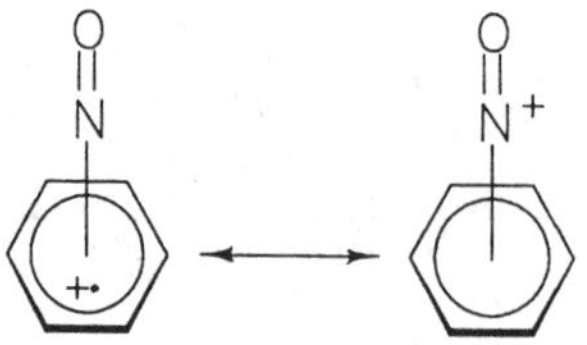

III

* According to the modified INDO/3 calculations, these values in isolated molecules are 9.22 and 8.87 eV respectively. But as the reactants approach each other to a distance of O.3 nm, the orbitals are brought close to one another [85].

† One-electron transfer was also discovered in reactions of electrophilic agents with polycyclic aromatic hydrocarbons [89].

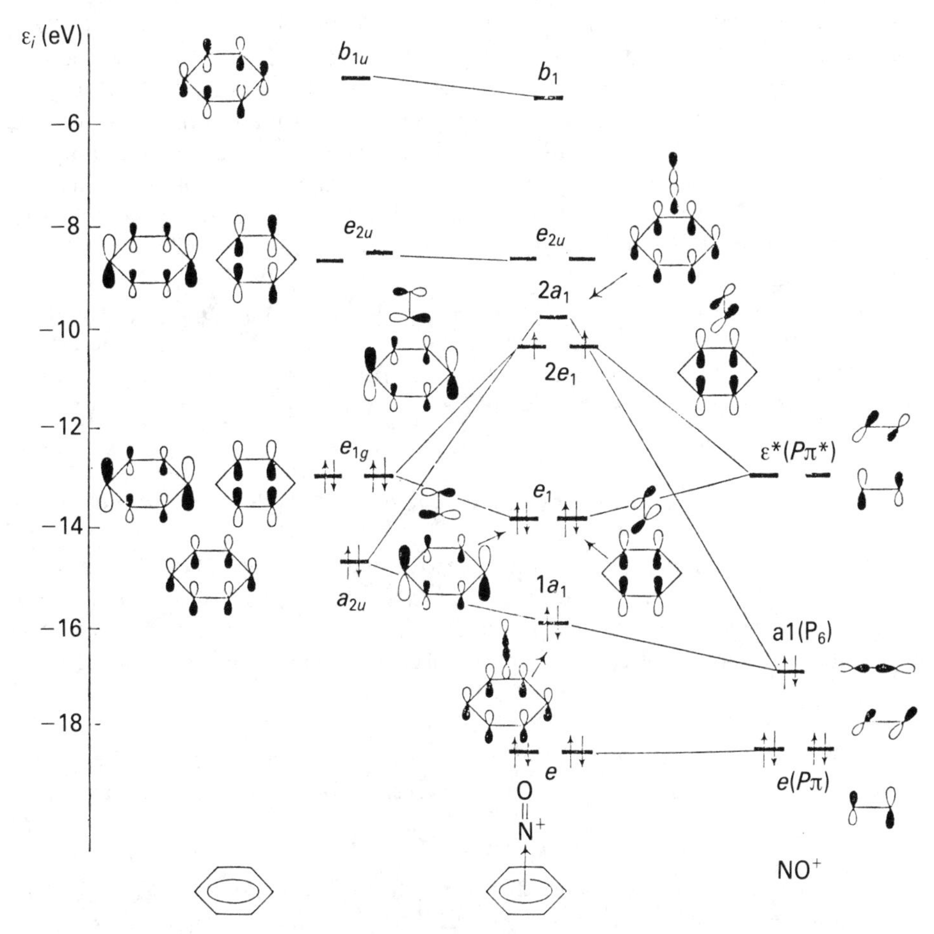


Fig. 3.2. Orbital interactions between nitrosonium cation and benzene molecule: unperturbed benzene MOs (left) and unperturbed electrophile MOs (right). (From Minkin *et al.* [85])

but the contribution of the frontier orbitals is decisive.

The reactions of nucleophilic aromatic substitution are also controlled by the parameters of the frontier molecular orbitals: regioselectivity is controlled by the overlapping conditions of the HOMO of the nucleophilic reagent with the LUMO of the aromatic substrate, while their total rate is controlled by electron affinity, i.e. by the energy of the aromatic substrate LUMO (see, e.g., Section 3.5).

The new data support the Fukui concept and promote better understanding of the other orbital concepts that were proposed to explain much of organic chemistry.

The analysis of changes in the first ionization potentials of monosubstituted benzenes, on the basis of Koopmans' theorem, for example, brings a fresh understanding of $\sigma^+$ constants in terms of molecular orbitals (for more detail see Ch. 10).

The parameters of the molecular orbitals have been used to develop the principle of hard and soft acids and bases (HSAB). Hard acids ($H^+$, $Na^+$, $AlCl_3$, $RCO^+$) presumably react with hard bases ($H_2O$, $OH^-$, $F^-$, $R_2O$, $NH_3$, $RNH_2$) while soft acids ($Ag^+$, $BH_3$, carbenes) preferably interact with soft bases ($R_2S$, $R_3P$, $C_2H_4$, $C_6H_6$). Since it was impossible to determine accurately hardness of the molecule and its quantitative parameters, the HSAB principle was, till recent times, used only for qualitative estimations. In full agreement with this principle, complexes of alkenes and aromatic hydrocarbons (soft bases) with $Ag^+$, $Pt^{2+}$, $Hg^{2+}$ (soft acids) were identified in many cases; complexes of soft bases with $Na^+$ and $A^{3+}$ (hard acids) were identified only in rare cases.

Pearson analyzed the concepts of electronegativity and hardness of chemical reagents (atoms, ions, radicals, molecules). As a result, the limitations of the HSAB principle have been markedly lessened within the framework of the orbital approach [90]. The concept of **electron chemical potential** $\mu$ was proposed:

$$\mu = (\partial E/\partial N) \tag{3.8}$$

where $E$ is the total electron energy of the system and $N$ is the number of electrons in the system.

The value of $\mu$, estimating the absolute electronegativity $\chi^0_M$, can be determined for any molecule M from experimentally found ionization potentials $IP$ (during ionization, the number of electrons in a molecule decreases from $N - 1$ to $N$) and from electron affinity $EA$ (the number of electrons in a molecule increases from $N$ to $N + 1$):

$$-\mu = \chi^0_M = -(\partial E/\partial N) = (IP + EA) \tag{3.9}$$

If two particles, A and B, react with formation of AB in the gas phase, the change in the electron energy of the system can be described by the following equation:

$$\Delta E = (IP_A - EA_B) - (IP_B - EA_A) = 2(\chi^0_A - \chi^0_B)$$

As A and B are combined, the electrons pass from B to A provided $\chi^0_A > \chi^0_B$. This process will decrease $\chi^0_A$ and increase $\chi^0_B$, until they become equal: $\chi_A = \chi_B = \chi_{AB}$.

**Absolute hardness of a chemical reagent** is defined as the rate of change in the electron chemical potential:

$$\eta = \tfrac{1}{2}(\partial \mu/\partial N) = \tfrac{1}{2}(\partial^2 E/\partial N^2) = (IP - EA)/2 \tag{3.10}$$

The values of $\chi^0$ and $\eta$ for some molecules are given below:

| | HCl | $BF_3$ | $CH_3Cl$ | $CH_3I$ | $C_2H_4$ |
|---|---|---|---|---|---|
| $\chi^0$ (eV) | 4.7 | 7.8 | 3.8 | 4.9 | 4.4 |
| $\eta$ (eV) | 8.0 | 7.8 | 7.5 | 4.7 | 6.1 |

| | $C_6H_6$ | $N(CH_3)_3$ | $P(CH_3)_3$ | $(CH_3)_2O$ | $(CH_3)_2S$ |
|---|---|---|---|---|---|
| $\chi^0$ (eV) | 4.0 | 1.5 | 2.8 | 2.0 | 2.7 |
| $\eta$ (eV) | 5.2 | 6.3 | 5.9 | 8.0 | 6.0 |

Hard reagents ($BF_3$, HCl, $H_2O$) are characterized by high values of absolute hardness, soft ones ($CH_3I$, $C_6H_6$, $P(CH_3)_3$) by low hardness. It follows from Eqs. (3.9) and (3.10) that both absolute electronegativity $\chi^0$ and absolute hardness $\eta$ of a molecule within the framework of Koopmans' theorem are determined by the energies of its frontier molecular orbitals; since

$$IP_1 = -\varepsilon_{HOMO}, \qquad EA_1 = -\varepsilon_{LUMO}$$

then

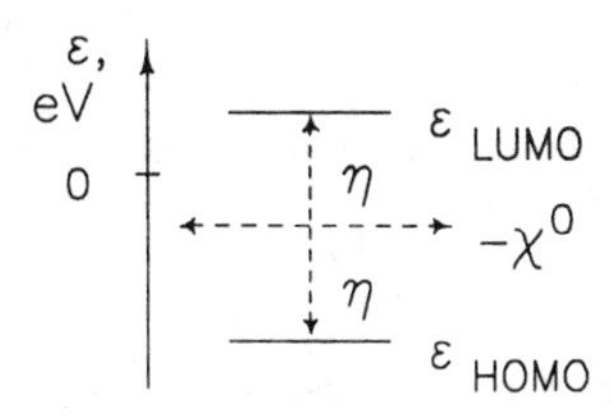


$$\chi^0 = -(\varepsilon_{HOMO} + \varepsilon_{LUMO})/2,$$
$$\eta = -(\varepsilon_{HOMO} - \varepsilon_{LUMO})/2.$$

According to Pearson [90], hardness of a molecule depends on the **energy gap** between the LUMO and HOMO: a **hard molecule** (or ion) has a great gap (which is due to the low energy of the HOMO and high energy of the LUMO), while a **soft molecule** has a small gap (due to the high HOMO energy and low LUMO energy).

The concept of hard and soft molecules (as electron systems) can probably be applied to reacting systems (as supermolecules):

(1) if the gap between the HOMO of one reactant and the LUMO of the other is great, the reacting system is hard; the reaction rate and orientation in this system are determined by the total charges on the reactant atoms;
(2) if the gap between the reactant frontier orbitals is small, the reacting system is soft, while the rate and the direction of the reaction are orbital-controlled.

Qualitative analysis of organic reactions with due consideration of the energy gap between the frontier orbitals of the reacting molecules (Fig. 3.3) is given in [65, 66, 68]. Since not only calculated but also experimentally estimated energies of the HOMO and LUMO of the key organic compounds are now available, the corresponding approaches seem to be advantageous.

## 3.2 ORBITAL STRUCTURE AND BIOLOGICAL ACTIVITY

Comparison of orbital structure of individual molecules and their properties becomes more complicated with transition from reactions in the gas phase or in solution to reactions in living systems. Organic compounds, such as drugs, pesticides, fragrant or toxic substances, produce quite varied effects on a living organism. The mechanisms of their action are not quite understood. Hormones, for example, are present in very small quantities, but they can produce a very strong effect on the growth of the organism and its functions. Vitamins, enzymes, substances involved in transmission of nerve impulses are also very important for normal bodily functions.

The mechanism of biological effects of organic molecules are being studied

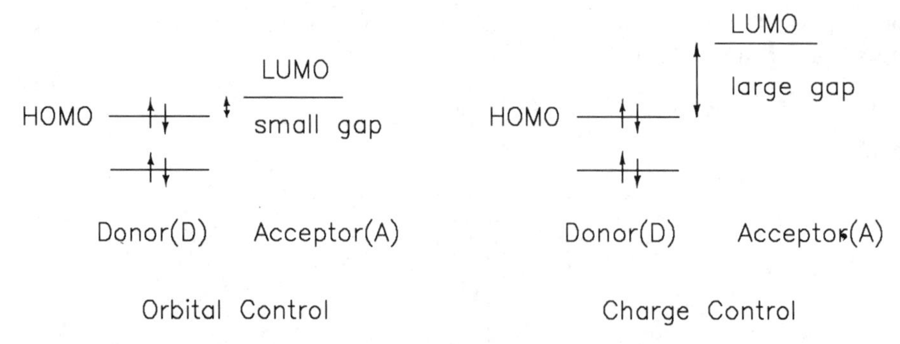


Fig. 3.3. Dependence of reaction control on energy gap between the frontier orbitals of the reagents. (From Klopman [68])

intensively. Molecular biology has explained the structure and the 'operating' principle of macromolecules in living organisms. Quantum biology and molecular pharmacology are younger branches of science. Their objectives are to establish the properties of molecular structures that account for the various biological effects of smaller organic molecules [90–4].

The mechanism of action of small molecules (drugs, vitamins, enzymes, etc.) in biological systems includes the interaction of these molecules with specific active macromolecules known as **receptors**. Synthetic compounds can produce effects similar to those induced by natural substances. Such molecules are called a **agonists**. Other compounds can bind to a cell receptor (the same receptor that is susceptible to natural active substances) and thus nullify the receptor function. Such blockers are called **antagonists**. They can prevent the action of other organic molecules. In molecular pharmacology, drugs can be regarded as both agonists and antagonists, depending on the particular mechanism of their action.

The universal method of molecular pharmacology is the study of the response of a living body to a given dose of a particular drug [92]. Doses are estimated with reference to the dose that kills 50% of experimental animals ($LD_{50}$), or to the median effective dose that produces the desired therapeutic effect in 50% of experimental animals ($ED_{50}$).

The pharmacological effect is usually simulated by absorption since the dose–response curves are quite similar to the adsorption isotherms plotted for heterogeneous catalytic processes. Drug molecules are reversibly bound to the receptor. The greater the number of such occupied receptor centres, the higher the medicative effect:

$$\text{drug} + \text{receptor} \leftrightarrows \text{complex} \rightarrow \text{response}.$$

If this interpretation is correct, the desired pharmacological effect of a particular drug on a biological system premedicated with the appropriate antagonist can be attained only by increasing the drug dose.

Numerous examples confirm a specific interaction of a biologically active substance with the receptor. Even slightest changes in the molecular structure can produce a very strong effect on biological properties [95]. Thus, only one of the four geometric isomers of 10,12-hexadecadiene-1-ol, namely the (10E-12Z)isomer, is a sex attractant synthesized by a female silkworm. Its effect is a billion times stronger than that of the other isomers.

It is easy to understand that if the chemical structure or electron density of the active centre of the receptor were known, the search for active molecules might be much more effective. It turned out to be very difficult to determine experimental characteristics of said centres. Even studies on an isolated receptor failed to reveal its working mechanisms. The modern approach to intermolecular interactions in a living system, known as the 'key–lock' model, can only be effective on the condition that the 'key' and 'lock' are sufficiently labile and are altered on appropriate contacts [96,97].

The available information on the electron structure of molecules gave rise to new hypotheses. Some of them try to estimate the effect of various properties of electron structure on the intensity of biological effects [33, 98]. But in general, a detailed statistical study shows that none of the known properties of a molecule can explain its action in a living system. Among other factors, special importance is attached to at least two properties of molecules: the property responsible for molecular transport to the receptor and the ability of the molecule to interact with the receptor. The role of the electron structure can be decisive in determining either of these properties. The effect of electron structure of organic molecules on their pharmacological properties is illustrated by the following few examples.

The mechanisms of action of organic substances regulating the activity of the nervous system were studied [92, 99–101]. The ability of molecules to form complexes with receptors (or with their models) is of the greatest interest. Frontier electron levels of drug molecules seem to be decisive in these interactions. Thus, correlations of psychotropic activity of drugs with the calculated energies of their HOMOs were established.

Many psychotropic drugs with specific molecular structure and hence specific pharmacological effects were studied [100]. The list of these drugs included dopamine (V, neurotransmitter), mescaline (VI, hallucinogen), 2,5-dimethoxy-4-methylamphetamine (VII, antidepressant), N,N-dimethyltryptamine (VIII, hallucinogen), serotonin (IX, neurotransmitter), chloropromazine (X, neuroleptic), lysergic acid diethylamide, known as LSD (hallucinogen, antagonist to serotonin). It turned out that in estimation of neurotropic activity of a drug not only the first but subsequent ionization potentials of the drug molecules should be considered. The dependence of the drug activity on the average value of the first and second ionization potentials is shown below:

| | V | VI | VI | VII | X | LSD |
|---|---|---|---|---|---|---|
| $(IP_1 + IP_2)/2$, eV | 8.54 | 8.16 | 8.15 | 7.90 | 7.71 | 7.65 |
| $\log ED_{50}$ | 3.52 | 4.40 | 5.30 | 6.52 | 7.00 | 8.22 |

The ability of a drug molecule to replace a [$^3$H]-LSD specimen bound specifically on the corresponding receptor in experiments *in vitro* was estimated with reference to $\log ED_{50}$.

The correlation of biological activity with ionization potentials corresponds to the charge-transfer mechanism, according to which the electron-donating molecule of a psychotropic drug interacts with the electron-withdrawing centre of the receptor.

This interaction can be realized in different ways: by complexation, electrophilic substitution at the aromatic ring of the drug molecule, formation of the hydrogen bond between the receptor and the aromatic ring of the drug, etc. But irrespective of a particular mechanism, the importance of the frontier orbitals in the psychotropic effect is quite convincing.

The dependence of the pharmacological effect of a drug on its electron structure, calculated by quantum chemical methods, has also been studied in detail [91]. Using quantum chemical calculations one can decide which of the two forms of the molecule (determined crystallographically or by measurements in solution) is more probable in a living system, and also obtain a complete set of alternate structures within a certain range of total energies of the molecule. In addition to total energy of the molecule in various conformations, various other properties can also be calculated. Electron distribution and the character of the potential field which is generated by electrons and nuclei are among properties of special importance.

Electrostatic potentials are calculated on the basis of MO eigencoefficients. The electrostatic potential $U(r)$ is calculated from the electron density distribution $\rho(r)$ in the molecule:

$$U(r) = \sum_{\substack{\text{on all} \\ \text{nucleus } \mu}} \frac{Z_\mu}{R_\mu - r} - \int \frac{\rho(r')}{r' - r} \, \mathrm{d}\tau \qquad (3.11)$$

and may be shown graphically in the form of a map. Drawing a line through the points of equal interaction energies of an unperturbed molecule with a single positive charge (proton) gives potential contours. The objective importance of electrostatic potentials is based on their low sensitivity to the quality of wave functions. It can be added that electrostatic potentials around aromatic fragments are usually determined by higher occupied molecular orbitals [92]. Until recently, the main limitation to the use of this fundamental information was its difficult presentation in a compact form. Creation of sophisticated graphic computerized systems has solved this problem.

The interaction of LSD and serotonin IX (and its analogues, isomeric hydroxytryptamines) with the corresponding receptor is an example of correlation of the biological activity with the distribution of the electrostatic potential in a molecule:

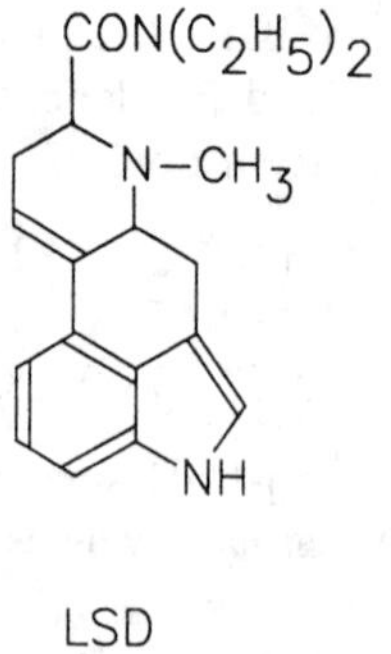

$CON(C_2H_5)_2$ N$-CH_3$ NH $NH_2$ HO NH

LSD 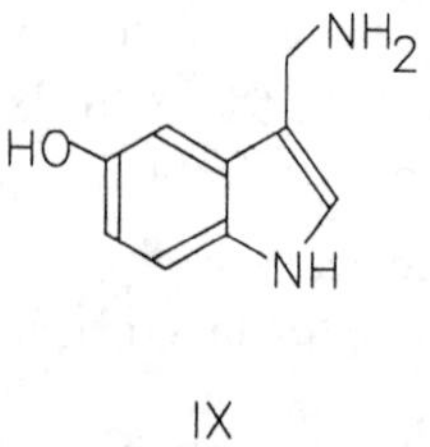 IX

Contours of molecular electrostatic potentials showing the vector of preferable orientation of the molecule relative to the field generated by the positive charge were

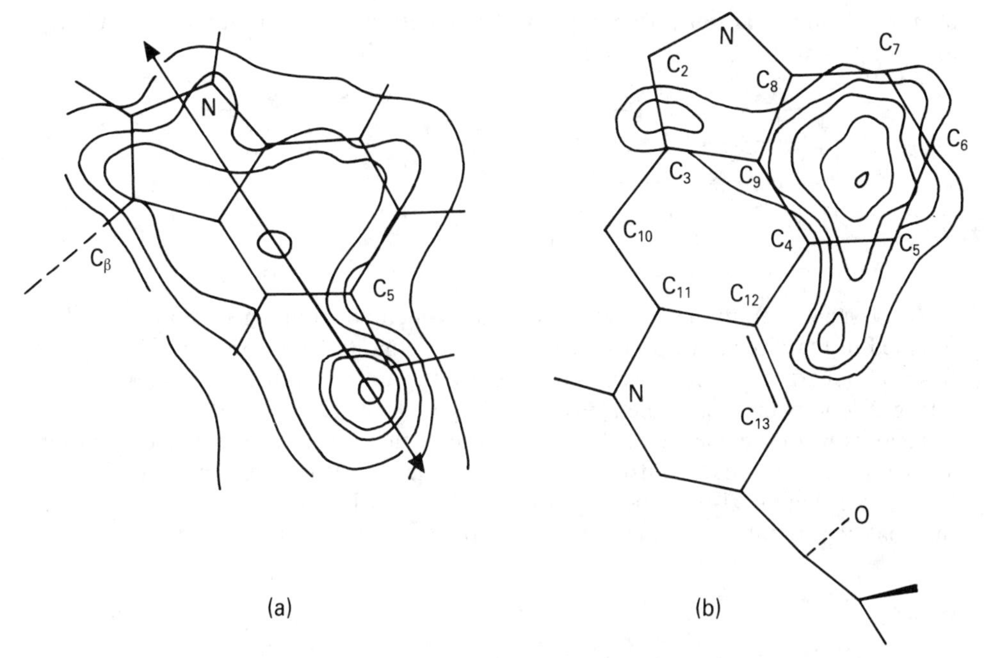


Fig. 3.4. Molecular electrostatic potentials of (a) serotonin and (b) LSD in planes lying above the indole fragment plane (the arrow indicates the orientation vector connecting the minimum points in the regions of maximum potential changes and estimates preferable orientation of the molecule relative to the positive charge placed above the molecule plane). (From Richards [92])

calculated (Fig. 3.4). It turned out, for example, that the directions of these vectors differ substantially for various hydroxytryptamines: the vector of the 6-hydroxytryptamine molecule is almost perpendicular to the vector of the 5-hydroxytryptamine molecule (IX). The different conditions of the interaction with the receptor can be the cause of different biological activity of serotonin and its analogues. At the same time, Fig. 3.4 shows that the specific configuration of the electrostatic potential map of serotonin (the direction of its vector in the first instance) is to a considerable extent repeated in the LSD molecule. This fact, probably, explains their tropism to the same receptor: it is believed that the $C_{12}\leftrightharpoons C_{13}$ bond in the LSD molecule accounts for the effect of the hydroxy group in serotonin. Both compounds are known to produce a strong effect on the function of the central nervous system.

Other examples of correlation between the orbital structure and the effect of molecules on biological systems can be given. The stereochemical conditions for the interaction of morphinans and their analogues (as agonists) with the analgesic receptors were studied [102, 103]. It turned out that the stereoelectronic effects due to nitrogen lone pair interaction with the surroundings compete with the stereoisomerism related to chiral carbon atoms to account for mutual contacts of the morphinan molecule with the receptor.

As the six-membered piperidine ring of the morphinan XI is replaced by the five-membered ring in D-normorphinan XII,the activity of the compound is completely lost:

XI XII

X-ray structural studies showed that the N-methyl group in compound XII is oriented in the direction opposite to the aromatic moity. The absence of activity of this compound is, in this case, explained by the unfavourable orientation of the nitrogen lone pair responsible for the bonding with the receptor [103].

There is further evidence of the importance of the nitrogen lone pair orientation as the parameter that controls bonding to the receptor. The compound 16-$\alpha$-butanomorphinan exists in the conformation XIII in which the piperidine ring A has the boat conformation. The lone pair of nitrogen is strictly oriented toward phenyl.

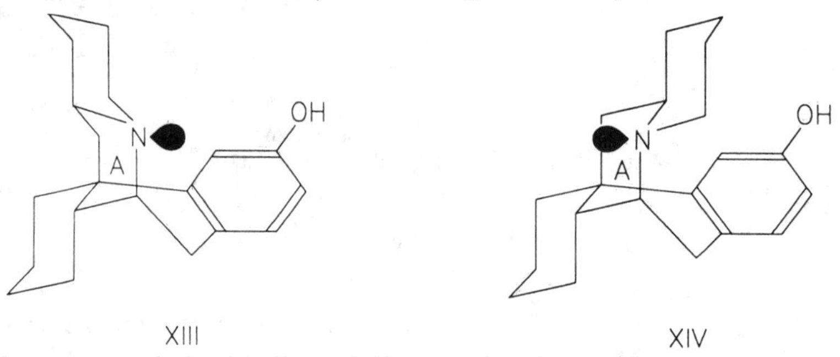


This orientation excludes binding of this morphinan with the receptor. Conversely, the 16-$\beta$-isomer exists in the conformation XIV, which permits coordination on the receptor and acts as an analgesic. It can be added that the absence of activity in compounds XII and XIII cannot be explained by their abnormal basicity (the basicity of N-methylpyrrolidine is 0.38 p$K_a$ units higher than that of N-methylpiperidine) or by steric deformation of the cyclohexane rings. In order to rule out completely the effect of changes in basicity, the activities of N-$CH_3$ and N-$CD_3$ morphinans toward one receptor were compared: the activity of the deuterium morphinan was inferior, although its basicity was 0.12 p$K_a$ units higher.

Based on these data, one can suggest that the interaction of the nitrogen lone pair in morphinan with the electrophilic centre of the receptor results in the electron transfer to the receptor and formation of interneuron bonds responsible for the biological activity of the compound.

## 3.3 ORBITAL STRUCTURE AND CONFORMATIONAL EFFECTS

The dependence of pharmacological properties of organic compounds on their orbital structure suggests a discussion of the correlation between orbital parameters of an organic molecule and its conformation.

The steric structure of an organic molecule, is, on the whole, extremely important for objective estimation of its reactivity and other properties. The so-called conformational effects are especially important in this respect [102,104–107].

In the general case, when explaining the preferred conformation of a molecule, the classical theory of structure describes only those steric interactions that depend on the size of atoms or groups of atoms (substituents). Equation (3.12) sums up the contributions to the energy difference of two conformers ($\Delta E_{ster}$) that are considered in the classical theory of structure:

$$\Delta E_{ster} = \Delta E_r + \Delta E_\theta + \Delta E_{tors} + \Delta E_{non-b} \qquad (3.12)$$

where $\Delta E_r$, $\Delta E_\theta$, $\Delta E_{tors}$ and $\Delta E_{non-b}$ are the energy differences of two conformers arising from their different bond lengths, valence angles, torsional strain and non-bonded atom interactions respectively.

The development of electron concepts resulted in the discovery of an additional energy factor that determines the preference of a definite conformation, $\Delta E\mu$—the difference of electrostatic interactions of the dipoles or charges in the compared conformers of molecules containing polar bonds and charged atoms.

The steric structure of most organic molecules becomes explainable within the framework of the classical concepts: repulsion of sterically close substituents, angular and torsional strains, electrostatic interaction of dipoles, etc. At the same time, theoretical organic chemistry, which systematizes not only rules but also exclusions, describes numerous deviations from 'normal' conformational behaviour of molecules. Such deviations are known as **conformational** effects. Their contribution to the difference of energies of two conformers is described by the term $\Delta E_{eff}$. The difference of energies of two conformers $\Delta E_{conf}$ is thus more completely described by the following formula:

$$\Delta E_{conf} = \Delta E_{ster} + \Delta E_\mu + \Delta E_{eff} \qquad (3.13)$$

Conformational effects have been formulated as a result of generalizations of much experimental material. To emphasize their specific character, they were sometimes given their proper names: the anomeric effect, gauche effect, cis effect, rabbit-ear effect, hockey-stick effect, etc. [106].

Organic compounds in which the **anomeric** effect is present have been studied in detail [107]. This term was first used to designate the tendency of the alkoxy group at the first carbon atom of the pyranose XV ring to assume the axial (a) rather than the equatorial (b) orientation despite unfavourable steric interactions. Later, the anomeric effect was studied in other systems with a similar structure (XV c and d) and was given the name of a generalized anomeric effect: preference of conformation c in the equilibrium ($X = F, NO_2$; $Y = CH_2, O, S$) [106]:

The **gauche** effect is quite common. It is defined as the tendency of the ethane fragment of organic molecules to assume the conformation in which there are

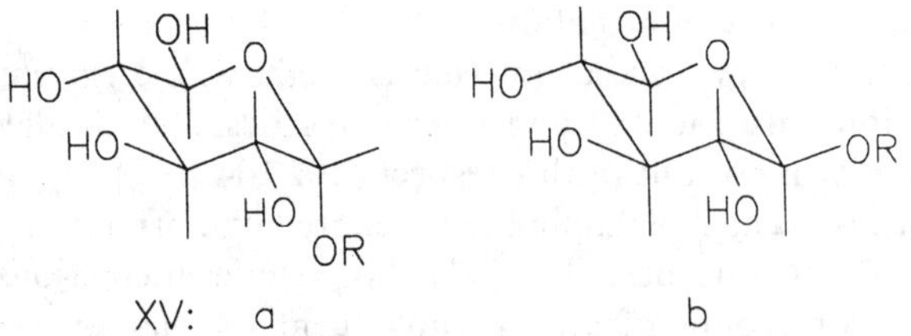

XV: a b

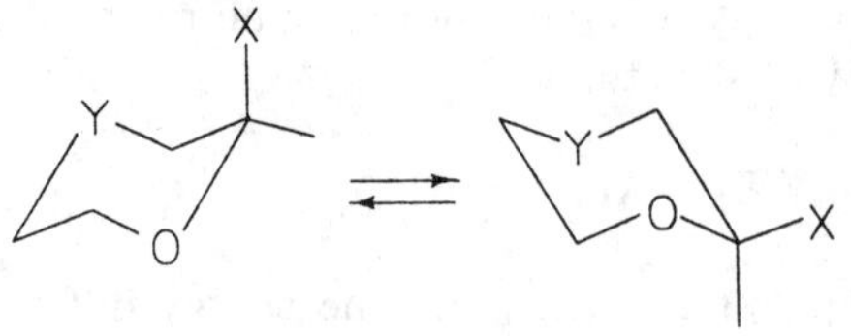

XV: b c

maximum gauche orientations between the adjacent lone pairs and polar bonds [105]. In general, for example, 1,2-dihalogenoethanes can exist in two conformations: anti and gauche. The anti form is more stable in chloro-, bromo- and iodoethane, although both forms are detected by electron diffraction.

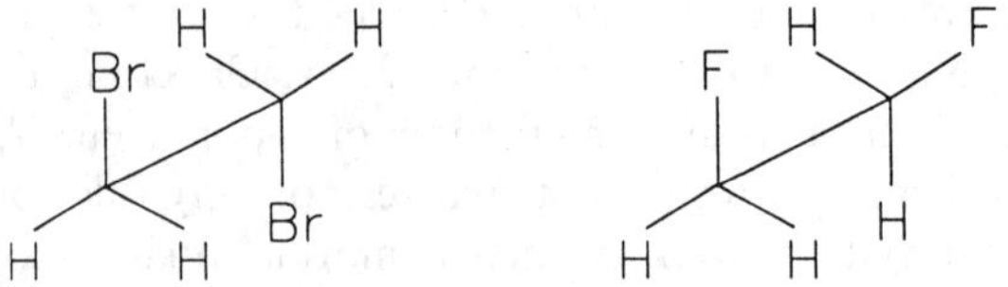

anti- gauche-

The molecule of 1,2-difluoroethane is more stable in the gauche form: at room temperature its content is from 85 to 96%, which practically excludes electron-diffraction detection of the anti form. The difference of energies of anti- and gauche-1,2-difluoroethanes was estimated as 7.37 kJ/mol [108]. The **cis** effect is probably similar to the gauche effect in 1,2-disubstituted ethylenes: cis-1,2-difluoroethylene is $4.5 \pm 0.5$ kJ/mol more stable than the trans isomer.

A peculiar conformational behaviour of substituted cyclohexanes, trans-1,2-disubstituted cyclohexane and trans-1,4-disubstituted six-membered cyclic compounds was studied [106, 109, 110]. The value of the equilibrium constant of 1,2-difluorocyclohexane XVI was determined from comparison of the integral intensities of conformer signals in the low-temperature NMR spectrum of $^{19}F$:

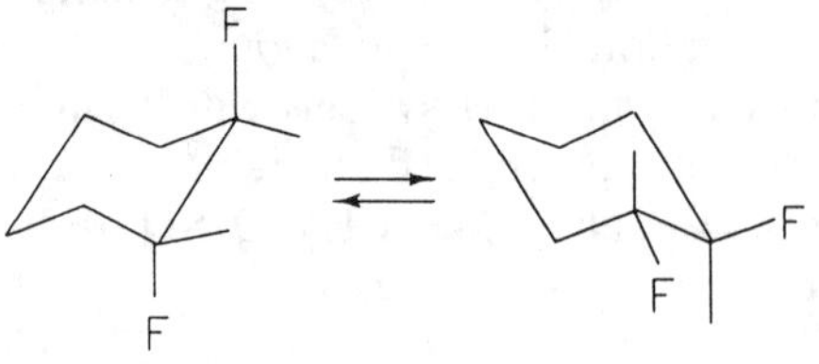

XVI: a b

The di-equatorial conformer XVI (b) dominates in an equilibrated mixture. The proportion of this conformer increases with polarity of the solvent: the equilibrium constant in carbon disuphide is 14.1 ± 1.8 (which corresponds to the value of the conformation effect $\Delta E_{eff}$ equal to 4.3 ± 0.34 kJ/mol), while in deutero-acetone this constant increases to 68 ± 10 [109].

The di-equatorial conformation in trans-1,2-disubstituted cyclohexanes prevails in general. The equations that determine experimental value of the conformational equilibrium $\Delta E_{eq}$ are of the general form:

$$\Delta E_{eq} = \Delta E_X + \Delta E_Y + \Delta E_{X/Y} \tag{3.14}$$

where $\Delta E_X$ and $\Delta E_Y$ are the energies of conformational equilibria of the corresponding monosubstituted cyclohexanes, and $\Delta E_{X/Y}$ is the energy of the gauche interaction of substituents in the di-equatorial conformation. They are widely used for quantitative estimation of conformational effects [106,107].

According to the definition of the conformational effect, the classical concept sometimes fails to explain the most advantageous conformation of the molecule. Zefirov proposed a test for the conformational effect: if the molecular mechanics method does not explain the observed conformation of the molecule, the presence of a specific stabilizing intramolecular interaction can be suggested [106].

A more exacting approach to the analysis of conformational effects is the quantum mechanical approach. It requires solution of the quantum mechanical problem for each conformer of the molecule. It is, however, easy to suggest that if a sufficiently sophisticated approximation is used, the result of the quantum mechanical calculation should agree with the experimental findings, while the concept 'conformational effect' loses its former meaning. In this situation, it is very important to have an opportunity for qualitative estimation of the results of quantum chemical calculations in order to formulate the rules that would explain and predict stereochemical behaviour of organic molecules on the basis of common concepts of organic chemistry. The orbital interaction rules seem to be most effective in estimation of conformational effects (see Section 2.4.2).

Within the framework of the orbital approach, the interaction of doubly occupied molecular orbitals usually destabilizes the system (or its energy remains unchanged), while the interaction of an occupied and unoccupied molecular orbitals stabilizes the corresponding conformation (for more detail, see Section 2.4.2, Fig. 2.10).

The **stabilization energy** $E_{st}$ is proportional to the square of the overlap integrals and inversely proportional to the difference of energies $\Delta E$ of the interacting orbitals [111]:

$$E_{st} \sim S^2/\Delta E$$

The interactions of orbitals that were analysed during the study of conformational effects ($\pi$–$\pi^*$, n–$\pi^*$, $\pi$–$\sigma^*$, n-$\sigma^*$, $\sigma$–$\sigma^*$) obey this relationship [107].

One can suggest that the $\sigma$–$\sigma^*$ interaction with involvement of only CH— and C—C bonds should contribute only insignificantly to conformational effects, mostly because of the high energy difference $\Delta E$ of the corresponding $\sigma$ and $\sigma^*$ orbitals. The lone pair of carbanion $n_C$ must be the most effective donor in the stabilizing interactions, while the donor properties of other orbitals should decrease in the series $n_N > n_O > \sigma_{CC}, \sigma_{CH} > \sigma_{CX}$ (where X = O, N, S, Hal).

Low-energy antibonding $\sigma^*$ orbitals should be the most effective acceptors (next to the unoccupied $\pi$ orbitals in the carbonium ion):

$$\sigma^*_{C-Hal} > \sigma^*_{CO} > \sigma^*_{CN} > \sigma^*_{CC}, \sigma^*_{CH}$$

The strongest stabilizing interaction can be expected in favourable orbital overlapping of the most effective donors and acceptors. This type of interaction prevails in systems where a powerful donor is in the vicinity of a relatively low-energy antibonding orbital.

The effectiveness of the orbital approach to the estimation of conformational effects is evidenced by numerous experimental data. Shown below is the dependence of the percentage of the axial conformer on the nature of the alkoxy group in 2-alkoxytetrahydropyranes [113]:

| Alkoxy group | OH → | $CH_3O$ → | $C_2H_5O$ → | $(CH_3)_2CHO$ → |
|---|---|---|---|---|
| Axial conformer, % | 77 | 72 | 68 | 66 |

| Alkoxy group | $(CH_3)_3CO$ | $ClCH_2CH_2O$ | $Cl_2CHCH_2$ | $Cl_3CCH_2O$ |
|---|---|---|---|---|
| Axial conformer, % | 62 | 77 | 88 | 95 |

It is quite clear that prevalence of the axial conformer in the case with $Cl_3CCH_2O$ depends on the increasing acceptor properties (lowering energy) of the $\sigma^*$(C—O) orbital rather than on the steric dimensions of the alkoxy group.

Various estimations of the nature and role of the anomeric effect are given within the framework of the orbital approach [102]. This effect was first regarded as a destabilizing factor due to the dipole–dipole or UEP–UEP (rabbit-ear effect) repulsion interaction designated by the double headed arrow in structure a:

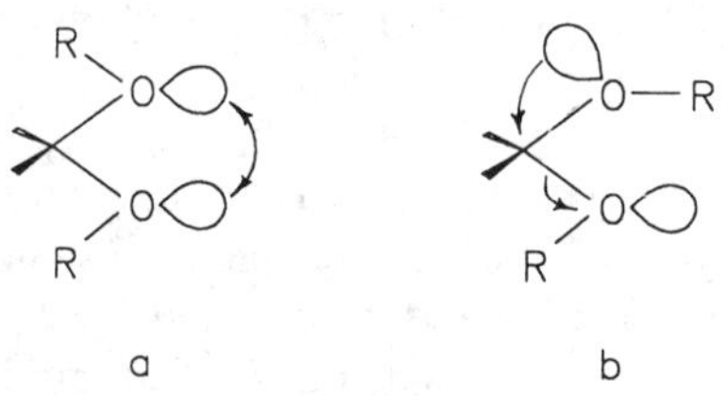


The anomeric effect is regarded now as a stabilizing factor arising from the antiperiplanar orientation of the oxygen atom lone pair (UEP) relative to the polar bond at the $\beta$ position: the stabilizing effect can in this case be attained by partial transfer of a lone pair of one atom to another, more electronegative atom. This process is shown by the curved arrows in structure b. It is interesting to compare the orbital interactions with, for example, the anomeric effect in acid-catalysed isomerization of cis and trans bicyclic acetals. At 80°C, 57% of cis and 43% of trans forms are found in an equilibrium; the stability of the cis isomer is 0.71 kJ/mol higher than that of the trans isomer. If steric interactions in cis acetal are estimated by the value of 6.91 kJ/mol, and if we subtract the entropy factor (1.76 kJ/mol at 80°C) which is due to the existence of cis acetal in the form of a mixture of two conformers, the anomeric effect can be estimated as 5.86 kJ/mol and regarded as the stabilizing $n_O$–$\sigma^*$ interaction of the oxygen lone pair with the $\sigma^*$(C—O) orbital. This interaction can be realized (by stereoelectronic conditions) only in the cis form:

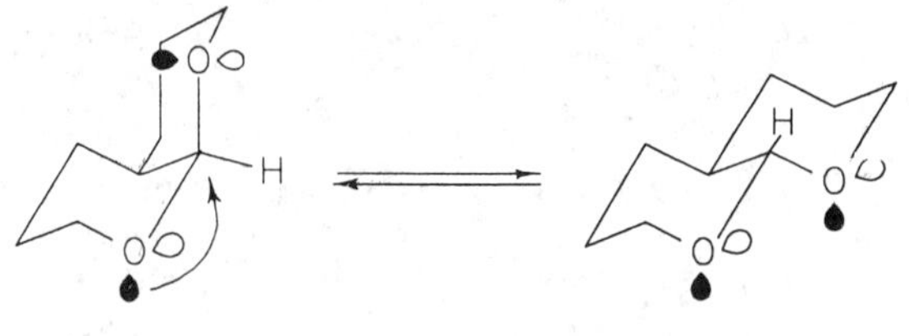

cis– trans–

Conformational effects have been studied in many classes of organic compounds [102, 107]. Only the orbital concepts can explain, for example, why 90% of tert-butylformate is found in the Z form despite the considerable repulsion between the carbonyl and tert-butyl groups:

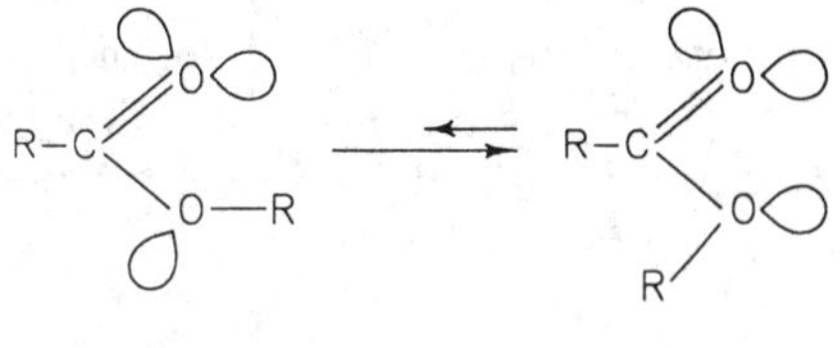

Z–form E–form

Stereoelectronic effects in the ester group are similar to the anomeric effect in acetals; they should be regarded as n–$\sigma^*$ interactions except that the central carbon atom in esters is trigonal, while in acetals it is tetrahedral. The carbonyl oxygen atom of the ester in the Z and E forms (XVIIa and XVIIb respectively) has an electron pair with an antiperiplanar orientation to the C—OR bond. This lone pair and the antibonding ($\sigma^*$) orbital can interact thus to give rise to triple bonding in the carbonyl group.

Another stereoelectronic effect is possible in the Z form due to the antiperiplanar orientation of the lone pair of the ester oxygen and $\sigma$(C—O) bond in the carbonyl group. The esters found in the Z form thus include two anomeric effects (designated a and c), while the esters in the E form have only one anomeric effect (structures b and d).

Additional stabilizing anomeric effect in the Z form of the ester is, probably, higher than in an acetal: the carbonyl group of the ester molecule is polarized much stronger than the C—OR bond of the acetal molecule. As a result, the antibonding $\sigma^*$(C=O) orbital should have lower energy (than the $\sigma^*$(C—O) orbital) and should be more effective in the overlap with the donor lone pair of the ester oxygen. It is believed that the anomeric effect in the Z form of an ester is as high as 12.56 kJ/mol (against 5.86 kJ/mol in the acetal molecule).

The relative stability of the Z form is now the only experimental evidence of importance of the anomeric effects in esters. Relative stability of various form of alkylated esters (dialkoxycarbonium ions) can be regarded as indirect evidence.

According to X-ray structural analysis and calculations, dialkoxycarbonium ions

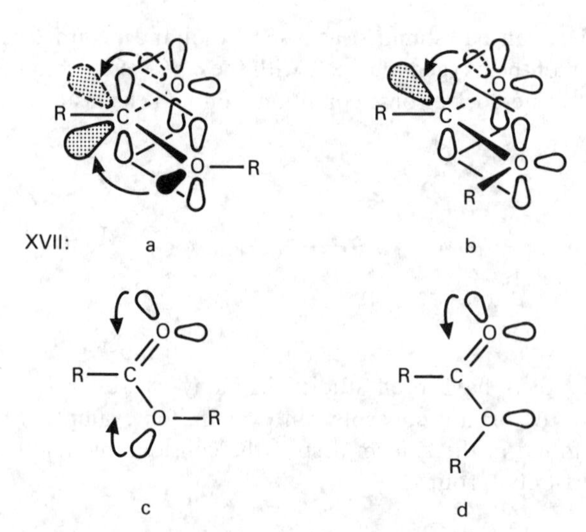


XVIII are planar and can theoretically exist in three various forms: ZZ(a), EZ(b) and EE(c). In the (a) form, both oxygens have one lone pair which is antiperiplanar to the polar C—O bond; in the (b) form, only one oxygen atom has such a lone pair; such atoms are absent in the (c) form:

XVIII: a b c

In agreement with the concept of the anomeric effect, stability of the a–c forms should decrease in the same direction. But since repulsion between substituents R is significant in the ZZ form (a), form (b) is preferable: according to NMR spectra, the methylated molecule of methyl acetate, i.e. dimethoxy(methyl)carbonium salt, only exists at $-30°$ to $-80°C$ in the EZ form (b). Other examples will be discussed in detail in subsequent chapters. Here the role of orbital interactions in the formation of transition states in organic reactions should not be emphasized [65, 102, 113]. Much attention has been given to hydrolysis of carboxylic acid derivatives, i.e. to the specific transitions of acetals, esters and amides. The role of stereoelectronic effects in transformations of carboxylic acid amides XIX has been studied in detail [102]. The amide fragment is planar, and the orbital effects similar to those in the E form of the ester group can therefore be realized in it, viz.: $\pi$ overlap of nitrogen lone pair with the $\pi^*$ orbital of the carbonyl group, and electron delocalization due to oxygen lone pair overlap with the antibonding $\sigma^*$ orbital of the C—N bond.

The nucleophile $Y^-$ approaches almost perpendicularly (at about 109°) the amide fragment plane: the addition product with a tetrahedral central carbon atom is formed. Two n–$\sigma^*$ interactions occur in structure b; oxygen and nitrogen lone pair orbitals are oriented antiperiplanar toward the C—Y bond, while the R′—N bond remains antiperiplanar toward the C—R bond. The R″ group takes the gauche orientation (structure b) instead of the syn-orientation (structure a).

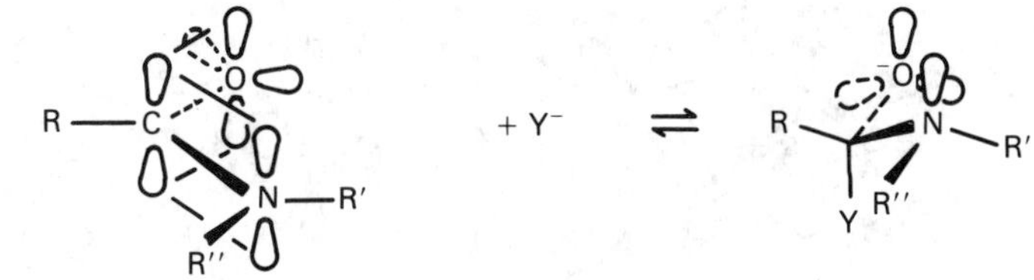

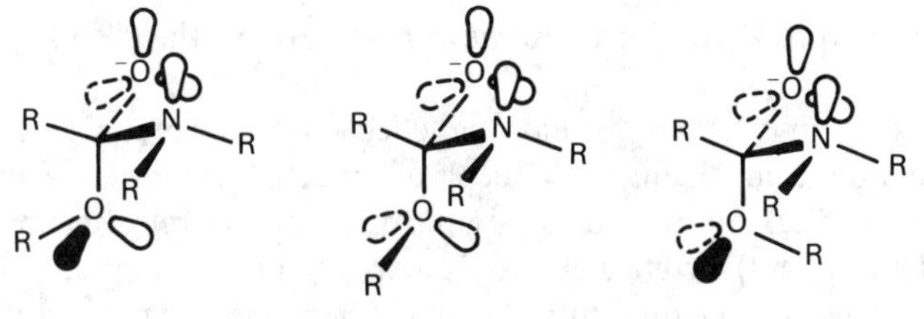

According to the microscopic reversibility principle, the reverse process should occur in the analogous sequence: ejection of the nucleophile $Y^-$ on transition from b to a transforms two mentioned n–$\sigma^*$ interactions (in structure b) into the only n–$\pi^*$ interaction (in structure a). One n–$\sigma^*$ interaction is present in both b and a.

In the case where $Y^-$ is the alkoxy group, both ester and amide can be transformed: the ratio of the products is determined by the orbital interactions in the tetrahedral intermediate. Deslongchamps discusses in detail all possible schemes of stereoelectronic effects; structures b′, b″ and b‴ illustrate some of them [102].

Cleavage, e.g., of structure XIXb’ can give tertiary amide alone, since the n–$\pi^*$ interaction cannot promote ejection of the amino group. Structure b″ can give both tertiary amide and E ester. During degradation of structure XIXb″, in addition to the n–$\pi^*$ interaction, ejection of both the amino- and OR group is attended by the appearance of one n–$\sigma^*$ interaction. At the same time, ejection of the amino group from structure b″ is accompanied, in general, by two n–$\sigma^*$ interactions, while the ejection of OR is accompanied by only one such interaction. Assuming the stereoelectronic factor, both degradations of structure XIXb″ are almost equally probable, but since the OR group is more preferable as the leaving group, the transformation gives tertiary amide. Ejection of the amino group from structure XIXb‴ is preferable as regards the stereoelectronic factor. This process is realized despite the fact that the OR group is preferable as a leaving group.

The results of *ab initio* calculations of aminodihydroxymethane XX, $CH(OH)_2NH_2$ reliably confirms the rule [114]: if two lone pairs are antiperiplanar toward the polar C—Y bond, this bond weakens and its elongation is more pronounced if the Y atom has no lone pair with antiperiplanar orientation toward the polar bond.

The C—N bond, for example, is short (strong) in structure XXa, but it is longer

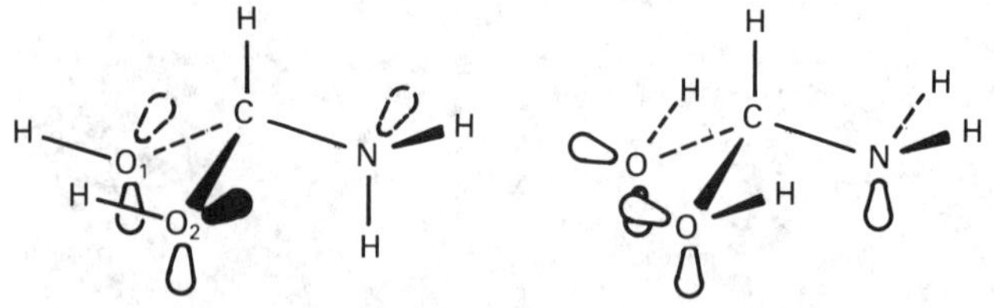

XX: a b

(weaker) in structure XXb. For the same considerations, the C—$O_{(2)}$ bond is weaker in structure a.

At the present time, the conformational effects are interpreted convincingly on the basis of the quantum chemical concept of structure of organic molecules. Even estimations of orbital interactions by simple quantum chemical methods on the basis of localized (fragment) orbitals not only explain but also predict conformational effects. In other words, the quantum chemical interpretation of conformational effects explains the exception to the rules determining the steric structure of organic molecules, and provides an objective basis for the understanding of these rules.

## 3.4 ORBITAL STRUCTURE, COLOUR, AND PHOTOTROPY

Absorption by organic molecules of quanta of electromagnetic radiation with energy from 100–1200 kJ/mol is the subject matter of electronic absorption spectroscopy (see Section 2.3.5). The ability of an organic compound to absorb electromagnetic radiation with energy of 150–300 kJ/mol is sensed by the human eye and is perceived as colour. The parameters of the orbital structure explain these important properties of organic molecules [115,116].

$\pi$ Electron levels of conjugated polyenes (as calculated by the HMO method) are shown in Fig. 3.5. As the number of overlapping $\pi$ orbitals in the molecule increases, the occupied and vacant orbitals get much closer to one another. The one-configurational character of the long-wave electron transition in $\pi$ conjugated hydrocarbons is demonstrated by the linear dependence of the position of the absorption band in the spectrum on the difference of energies of the frontier orbitals [117]:

$$\nu(cm^{-1}) = 14\,800 + 25\,400\Delta\varepsilon, \ \Delta\varepsilon = \varepsilon_{LUMO} - \varepsilon_{HOMO}$$

The electron absorption spectra of aromatic hydrocarbons with condensed cycles has already been discussed in Section 2.3.5. They can be interpreted objectively within the framework of SCF quantum chemical calculations with consideration of the configurational interactions.

The same regularities characterize absorption of electromagnetic radiation by heteroaromatic molecules as well. Table 3.1 gives the comparison of energies (eV) of excited states of pyridine (calculated by the PPP and HMO method) and the energies measured from electron spectra.

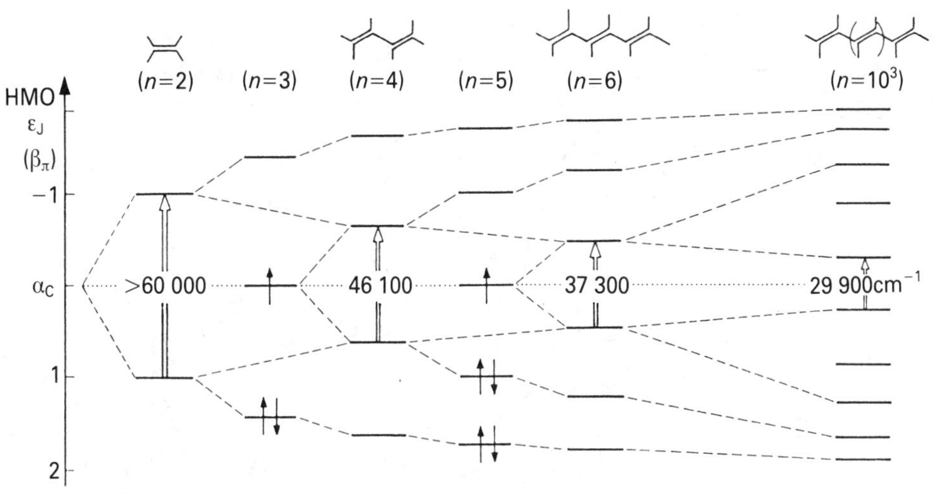


Fig. 3.5. Electron transitions on light absorption in conjugated hydrocarbons containing $n$ $\pi$-electrons in the molecules. (From Bock [117])

**Table 3.1.** Assignment of bands of pyridine electron absorption spectrum

| Band | Absorption band energy (eV) | | |
|---|---|---|---|
| | SCF calculation | HMQ calculation | Experimental |
| 1 | 5.010 | 5.018 | 4.9 |
| 2 | 5.828 | 5.842 | 6.2 |
| 3 | 7.146 | 7.085 | 7.0 |
| | 7.177 | 7.217 | |

(From Peacock [30])

As in the case of pyridine and benzene, the electron absorption spectra of quinoline and naphthalene molecules are practically the same as each other:

| | | | | |
|---|---|---|---|---|
| Quinoline (water) | 226 (4.36) | 275 (3.51) | 299 (3.46) | 312 (3.52) |
| Naphthalene (hexane) | 221 (5.07) | 275 (3.75) | 297 (2.47) | 311 (2.40) |

PPP CI calculations are also used effectively to interpret electron absorption spectra of substituted planar heteroatomic conjugated systems. Figure 3.6 gives the comparison of the spectra and results of calculations (with standard values of valence angles, bond distances, Coulomb and resonance integrals) of pyridines containing methyl, phenyl, styryl and nitrile groups [118].

Electron absorption spectra of organic dyes having even more developed $\pi$ systems are well interpreted within the framework of quantum chemical methods [31, 32, 115, 116]. Calculations of symmetrical cyanines using the MO theory show that these dyes have highly permitted HOMO → LUMO transition which is polarized in the direction of the long axis of the molecule (shorter wavelength transitions are polarized both parallel and perpendicular to the main axis). The corresponding absorption

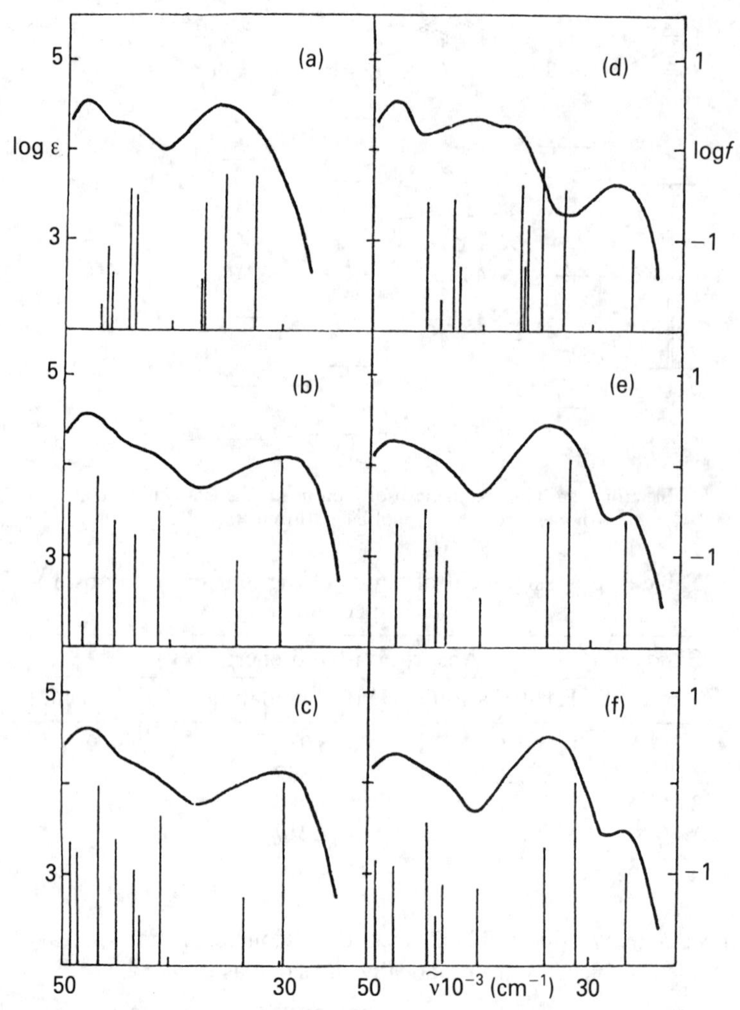


Fig. 3.6. Spectral curves (in log ε − ν coordinates) and results of PPP calculations (vertical lines in log *f* − ν coordinates) of electron absorption of some pyridine derivatives (a) 2,6-diphenyl-3,5-dicyanopyridine; (b) 2-methyl-3,5-dicyano-6-*syn*-styrylpyridine; (c) 2-methyl-3,5-dicyano-6-*anti*-styrylpyridine; (d) 2,6-diphenyl-3,5-dicyanodihydropropyridine; (e) 2-methyl-3,5-dicyano-6-*syn*-styryldihydropyridine; (f) 2-methyl-3,5-dicyano-6-*anti*-styryldihydropyridine. (From Skala *et al.* [118])

$$R_2\overset{+}{N}{=}\diagdown(=)_n\diagdown\ddot{N}R_2 \longleftrightarrow R_2\ddot{N}\diagdown(=)_n\diagdown{=}\overset{+}{N}R_2$$

bands are therefore very narrow (half-width is about 25 nm against 100 nm for azo- and anthraquinone dyes) while the cyanine dyes give extraordinarily bright and pure colours. The character of these electron absorptions shows also that the geometry of the first excited state is very much like the geometry of the ground state [119].

Examples of adequate interpretation of the absorption spectra accounting, first of all, for the frontier orbitals, are also available for other classes of dyes. Thus, a reliable correlation between the calculated and experimental wavelengths in electron absorption spectra has been established for azo dyes XXI [31]:

X, X' = H, $NO_2$, CN, $CH_3O$, $CH_3CO$, OH

Y = H, $CH_3O$, OH

R, R' = H, $CH_3$, $C_2H_5$

XXI

Figure 3.7 shows that the correlation is unreliable only for dyes (XXIIa-c) containing the OH group at the ortho position toward the azo group. These dyes are characterized by strong **intramolecular hydrogen bonding** which causes a bathochromic shift in the long-wave absorption band: ortho methoxyazo dyes XXIII of similar structure obey the dependence shown in Fig. 3.7. By changing the parameters, in order to account for the intramolecular hydrogen bonds, long-wave electron absorption bands of dyes XXII can also be predicted to high accuracy

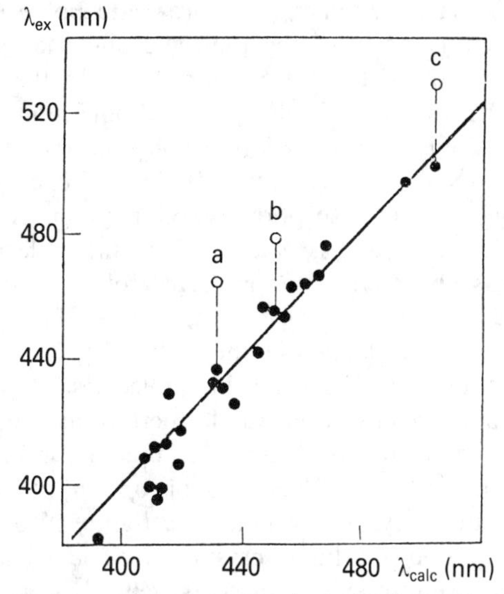


Fig. 3.7. Relationship between experimental and calculated transition energies in electron absorption spectra of some azobenzenes (PPP calculations). (Data by Griffith [31])

Calculations in the $\pi$ approximation also predict reliably the well-known difference in the absorption spectra between the azo- (XXIV) and quinone hydrazone (XXV) tautomeric forms [31]

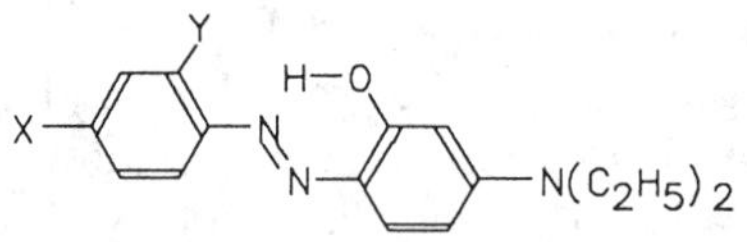

XXII: a: X= Ac; Y= H
b: X= $NO_2$; Y= H
c: X= Y= $NO_2$

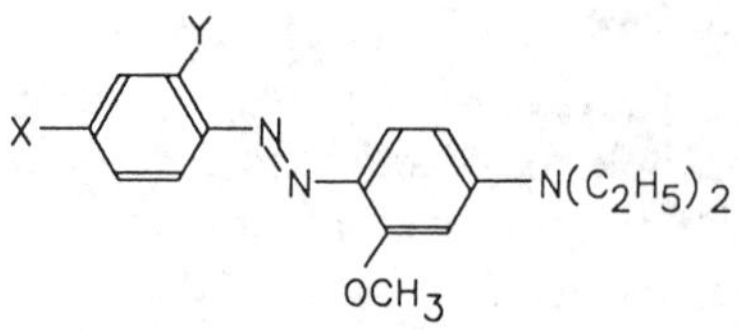

XXIII

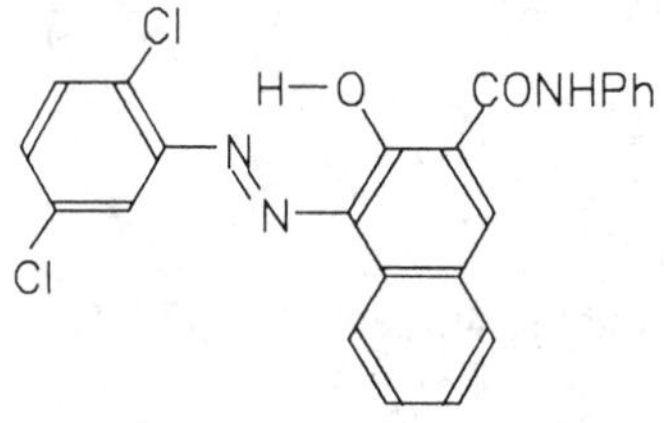

XXIV: $\lambda_{max}(C_6H_{12})$ = 425 nm
$\lambda_{max}$(calc) = 409 nm

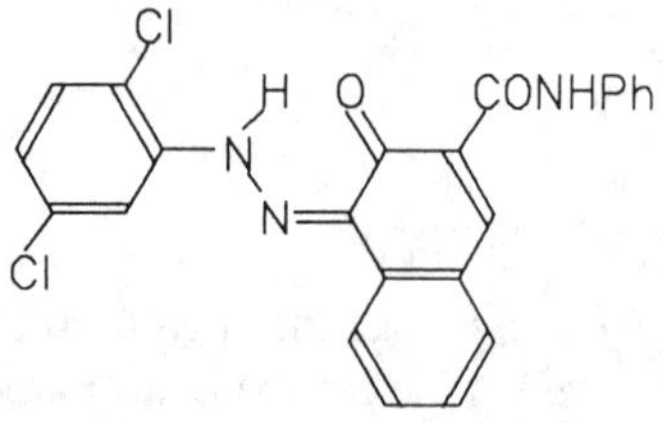

XXV: $\lambda_{max}(C_6H_{12})$ = 503 nm
$\lambda_{max}$(calc) = 506 nm

The important feature of the interaction of organic molecules with electromagnetic radiation includes not only the position of the absorption band in the electron spectrum, but also **intensity** and **polarization** of **absorption**, because each molecule in a given electron state is characterized by the appropriate dipole moment, which could be calculated from the geometry of the molecule and the wave function of the molecular orbitals (for more detail see Section 2.3.5). In the transition from one electron state to another, a molecule changes its dipole moment. The greater this change, the more selective is the absorption of polarized radiation by the molecule. The intensity and the direction of polarization of absorption are especially important for the estimation of electron absorption of oriented molecules. Under normal conditions, when the incident light is polarized and the molecules are arranged at random, light is always absorbed. If the molecules of a dye or any other organic substance are oriented (as in a crystal or polymer film) and a polarized light is used, absorption intensity distinctly changes as the plane of polarization rotates relative to the direction of orientation of the absorbing molecules.

The effect of absorption polarization was formerly only theoretically important, but the use of dye liquid crystals in displays (computers, watches) added practical interest to this phenomenon [31, 120, 121]. In appropriate systems, the molecules of a dye dissolved in a liquid crystal substance are tightly incorporated between its molecules. As the orientation of a liquid crystal molecule is changed by an electric field, orientation of the dye molecule changes as well. Thus, a cell may be coloured under normal illumination, but as voltage is applied, the cell can turn colourless. In this particular case, as voltage is applied, the dye molecules are rearranged so that their long axes are parallel to the direction of the light wave but perpendicular to the electric vector of the wave.

Dyes for liquid crystal display devices are largely selected empirically but quantum chemical calculations (the PPP method included) suggest that selection can be predictable. Here is an example. Red 5-amino-1,4-naphthoquinone has no apparent signs of a liquid crystal dye: its molecule is compact, while the donor (amino) group is perpendicular to the long axis of the molecule. It has been found that this dye becomes easily oriented in liquid crystal media. The dichroic property of 5-amino-1,4-naphthoquinone is explained by the fact that the $\pi$ electron dipole moment (whose value is much higher than that of the $\sigma$ electron moment) is close to the long axis of the molecule and to the direction of the transition moment.

The study of the absorption of electromagnetic radiation by an organic molecule is important not only because it accounts for the colour of a given substance but also because appropriate excitation brings the molecule into a state in which it becomes capable of new chemical (photochemical) transformations. It has been established that while absorbing a UV-light photon with the wavelength of 300 nm, a medium-size molecule is excited as by heating to a temperature of 1500°C. The mechanism of photo-activation of organic molecules to chemical reactions is interpreted in terms of molecular orbitals and states as follows. Absorption of a quantum of light with a given energy by a molecule causes transition of one (or more) electron of the molecule from its HOMO to one of the unoccupied MOs. The molecule thus becomes excited. Its electron population, bond orders, and charges on the atoms become different from those of the original molecule in the ground state, and this makes the molecule capable of new reactions.

**Photochemical transformations** have not been so extensively used in organic synthesis as, say, thermal ones, but some photochemical processes are now realized on a commercial scale. Transformation of ergosterol to vitamin $D_2$(XXVI) in diethyl ether can be considered as an example:

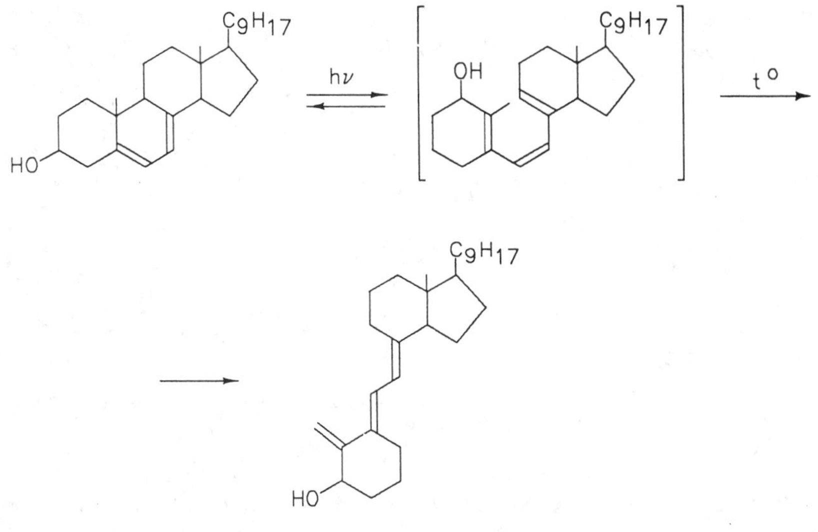

XXVI

The photochromic properties of organic molecules (the ability to reversibly change their colour on absorption of electromagnetic radiation) are now intensively studied. Benzospiropyrans of the indoline series have such properties:

hν
t°

no colour coloured

Photochromic materials can be used to manufacture light-sensitive glass, display devices of computers, and microelectronic materials with high resolution power [122].

The ability of molecules to undergo photochemical transformations stimulates research in the field of sun energy storage by organic substances [123]. Isomerization of norbornadiene to quadricyclane on exposure to light is a good example (Fig. 3.8).

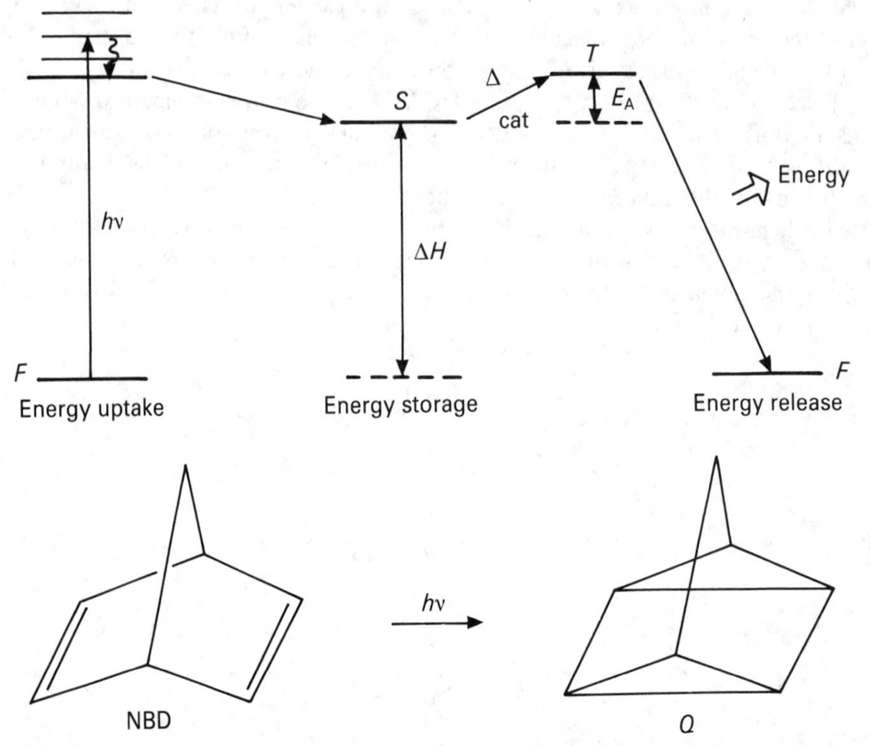


Fig. 3.8. Sun energy storage in photochemical transformation of norbornadiene (NBD) into quadricyclane (Q). (From Wells [123])

In the first step of the process, radiation is absorbed by the molecule $S_0$ with formation of an electron-excited particle $S_1$, having excess vibrational energy. This energy is quickly dissipated while the excited molecule $S_1$ moves to the lower vibrational level and is isomerized to an $S_m$ molecule that is capable of storing excess energy $\Delta H$. If $S_m$ is a stable particle, the energy $\Delta H$ is conserved till the moment when $S_m$ is heated or until the reverse reaction $S_m \rightarrow S_0$ is induced. The formerly absorbed energy $\Delta H$ is released in the course of this transformation.

Photochemical reactions of organic molecules are also important in biochemical processes. As a rule, the frontier orbitals of organic molecules are, first of all, involved in these reactions. As light enters the eye, cis–trans isomerization of 11-cis-retinal (XXVII) occurs [124]:

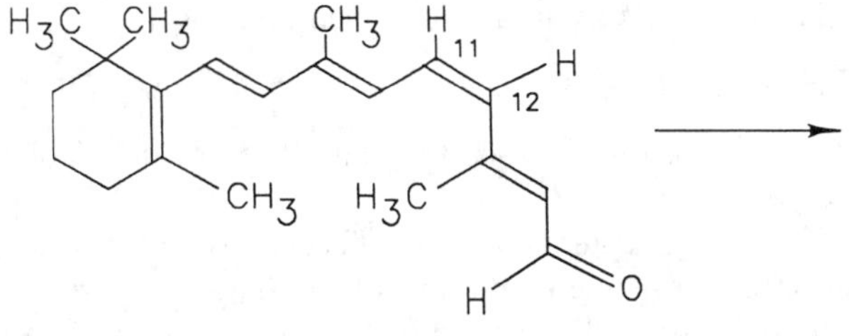

XXVII: cis–

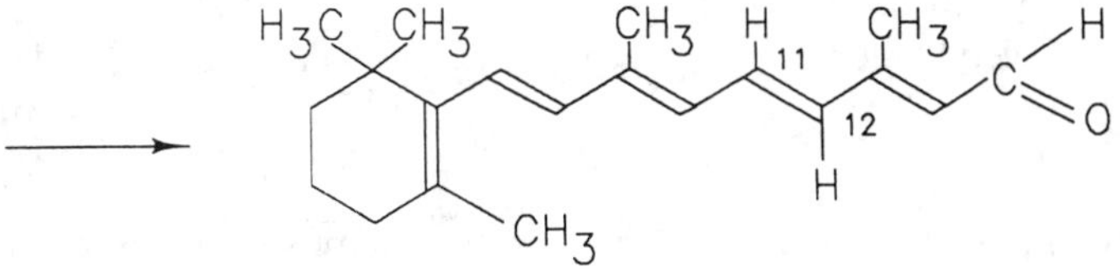

trans–

As a result of the $\pi$–$\pi^*$ transition, the $C_{(11)}$—$C_{(12)}$ double bond in cis⁻ retinal breaks, the polyene molecule straightens and transforms into the trans⁻ configuration. This isomerization changes the shape of the rhodopsin molecule, which, in turn, alters permeability of the appropriate membrane. The $Ca^{2+}$ ions that are involved in the transmission of nerve pulses (which after all accounts for the visual perception in the brain) pass through the membrane. The trans⁻ retinal is converted back to the cis⁻ isomer by the appropriate enzyme-catalysed reactions, and the eye becomes prepared to receive new images.

## 3.5 EFFECT OF SELF-ASSOCIATION AND MEDIUM ON ORBITAL STRUCTURE OF MOLECULES

It is quite evident that the effect of the medium on the electron structure of a molecule should be taken into consideration. The overwhelming majority of reactions occur in conditions under which molecules of organic substances are either solvated by the solvent molecules or associated. This holds for common chemical reactions in solutions, for interactions in biological media, and for solid-phase transformations. The medium produces a substantial effect on the results of spectral measurements [16]. Consider the following examples.

In the transition from the gas phase to solution, the reaction of alkaline hydrolysis of methyl bromide can slow down its rate by almost 20 orders. The nature of the solvent is also very important: the solvolysis rate constant of tert-butyl chloride decreases by 6 orders as ethyl acetate is substituted for water [11]. The reaction rate depends not only on the macroscopic effect of the medium of the reacting molecules. A typical example illustrating the role of specific solvation is the effect of crown ethers. In the presence of 18-crown-6, potassium fluoride is soluble in benzene or acetonitrile and reacts as an active nucleophilic agent:

$$CH_3(CH_2)_7Br + KF \xrightarrow{\text{18-crown-6}} CH_3(CH_2)_7F + KBr$$

In the absence of 18-crown-6, potassium fluoride does not react with alkyl halides in said solvents.

The self-association and solvation effects cause marked changes in frequencies of the IR spectra of organic compounds [16]. Thus, as a substance is converted from vapour to the condensed phase, vibrational frequencies of the carbonyl group always decrease by the value of $\Delta\nu$. This is probably due to the dipole interactions occurring between the carbonyl groups, although dielectric effects can also be important:

| | Acetyl chloride | Phosgene | Acetone | Acetaldehyde | Dimethyl formamide |
|---|---|---|---|---|---|
| $\Delta\nu C{=}O$, $cm^{-1}$ | 15 | 13 | 21 | 23 | 50 |

As a rule, the dipole moments measured in solvents are somewhat low compared with the measurements conducted in the gas phase [17]:

| | $\mu$,D | |
|---|---|---|
| | Gas phase | Benzene (solvent) |
| $C_6H_5NO_2$ | 4.23 | 3.98 |
| $CH_3NO_2$ | 3.54 | 3.13 |
| $(CH_3)_2CO$ | 2.85 | 2.76 |

The interaction of dipole moments of the solvent and solute molecules is sometimes estimated by the value of effective moment $\mu_i$ according to this equation:

$$\mu_S = \mu_O + U_i$$

where $\mu_S$ is the dipole moment of the solute molecule in solvent, and $\mu_O$ is that in the gas phase.

Solvent can also cause chemical shifts in NMR spectra [16] since the solute molecule induces a dipole moment in the solvent which, in turn, creates an appropriate change inside the molecule. As a result, the magnetically active nucleus (e.g. a proton) can be screened additionally by the solvent field.

The conformational transformations are characterized by low activation barriers and small energy differences between the conformers (about 8–9 kJ/mol). The difference in conformer solvation energies can then produce a substantial effect on the parameters of the conformational equilibrium [125]. Thus, trans-1,2-dichloro-cyclohexane in pentane is in the diaxial conformation, while in acetone its conformation is diequatorial.

The **energies of association and solvation processes** are estimated by various methods [11, 82], among which the macroscopic and supermolecular approaches should be noted. According to the **macroscopic (continual) approach**, the energy of a molecule in solvent is regarded as a sum of two major contributions, viz., total internal energy $E_{in}$ of an individual molecule and the solvation energy $E_{solv}$:

$$E = E_{in} + E_{solv}$$

The **solvation energy**, in turn, includes several components: energy of electrostatic interaction of the solute molecule with the solvent ($E_{elst}$); cavitation energy, i.e. the energy of formation of a cavity ($E_{cav}$) where the solute molecule is located; dispersion interaction energy ($E_{disp}$), and the energy of repulsion of valence-unbound atoms ($E_{repl}$):

$$E_{solv} + E_{elst} + E_{cav} + E_{disp} + E_{repl}$$

Each component can be determined from the corresponding model. Electrostatic energy, for example, depends on dielectric constant $\varepsilon$ of the solvent. According to the classical theory proposed by Born and Kirkwood, the energy of a charged solute particle should change proportionally to the value of $1/\varepsilon$, and the energy of the molecule having its own dipole moment, proportionally to the value of $(\varepsilon - 1/(2\varepsilon + 1)$. The other components of the solvation energy can also be estimated. The continual approaches, however, have serious disadvantages: they account only for macroscopic effects of the solvent while the effects of specific solvation are neglected; the effects of the surrounding solvent molecules on the inner energy (the electron energy included) of the dissolved molecule are disregarded.

The environmental effects on the electron structure of a molecule can be estimated within the framework of the orbital concept. The supermolecular (discrete) approaches make use of quantum chemical calculations for molecular aggregations that are regarded as a whole system (supermolecule). The interaction of one solute molecule and one solvent molecule is first studied in detail by changing their relative orientation and distance between them. Next additional molecules of the solvent are added, and their interaction with the solute molecule and with each other are studied.

Solvation effects are now widely estimated at the supermolecular level. Reliability of these estimations is determined by the reliability of steric parameters of the supermolecule and by the quality of the quantum chemical approximation. The study of self-association of organic substances is especially noteworthy. The corresponding data for many organic compounds are available from experimental studies on the molecular crystals and also from optimization of the geometry by the *ab initio* and semi-empirical quantum chemical calculations. The changes in the electron structure occurring during association of simple molecules have been considered in detail. Thus, experimental findings and the results of optimization well agree in predicting steric characteristics of the dimer of water [126] (the point group $C_S$).

The photoelectron spectrum of the dimer in the gas phase was also recorded: dimerization decreases considerably the ionization potentials of electrons of the higher occupied MOs:

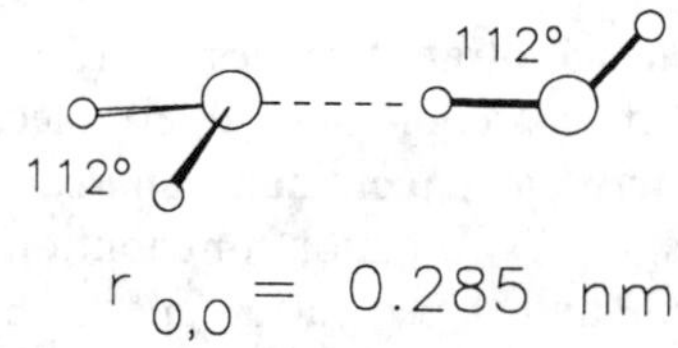

112°

112°

$r_{O,O} = 0.285$ nm

| Monomer $H_2O$ | | Dimer $(H_2O)_2$ | |
|---|---|---|---|
| $IP_i$ (eV) | MO($C_{2v}$) | $IP_i$ (eV) | MO($C_s$) |
| 12.62 | $1b_1$ | 12.10 | $2a''$ |
| 14.74 | $3a_1$ | 14.08 | $8a'$ |

According to the calculations, the HOMO of the dimer is formed mostly by the orbital of the out-of-plane lone pair of the proton-donating molecule, while the next MO is a combination of the orbitals of the in-plane lone pair of the proton-donor molecule and the out-of-plane lone pair of the proton-acceptor molecule.

Ionization potentials of an ammonia dimer were also calculated by the *ab initio* method [127]. Since experimental data on the geometry of the ammonia dimer $(NH_3)_2$ are absent, the calculations were based on the geometric characteristics obtained by optimization of the geometry by the ab initio calculations with the Slater type orbitals in the ST04-31GF basis set. It was established that the dimer $(NH_3)_2$

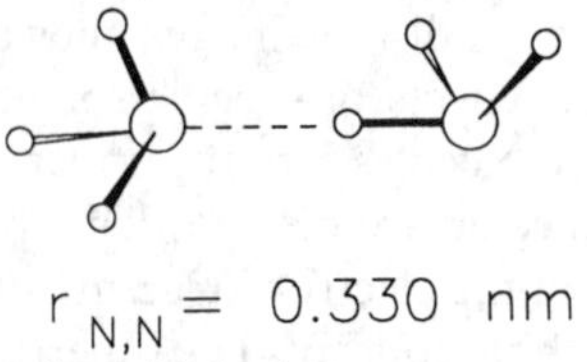

$r_{N,N} = 0.330$ nm

has a staggered conformation in which the monomer molecules do not practically change their conformations and have a linear hydrogen bond N—H ··· N, equal to 3.30 Å. The calculated ionization potentials of the ammonia monomer and dimer are the following:

| Monomer $NH_3$ | | Dimer $(NH_3)_2$ | |
|---|---|---|---|
| $IP_i$ (eV) | MO($C_{3v}$) | $IP_i$ (eV) | MO($C_S$) |
| 11.26 | $3a_1$ | 10.3 | $8a'$ |
| 16.95 | $1e$ | 11.8 | $7a'$ |
| | | 16.0 | $6a'$ |
| | | 16.0 | $2a''$ |
| | | 17.4 | $1a''$ |
| | | 17.4 | $5a''$ |

According to calculations, the HOMO of the dimer $(NH_3)_2$ is formed by the in-plane lone pair orbital, localized over the proton donor molecule $NH_3$. During dimerization, the first ionization potential of the ammonia molecule decreases by 0.96 eV (the calculated energy of dimerization has proved to be 17.17 kJ/mol, which is slightly different from the experimental findings: 18.42–19.26 kJ/mol).

Other results were also obtained on optimization of the ammonia dimer geometry by quantum chemical calculations [128]. The comparison of the linear and cyclic dimers of ammonia belonging to the point groups $C_s$ and $C_{2h}$ respectively, showed that H-bonding between the $NH_3$ molecules in both dimers does not practically cause any changes in the $C_{3v}$ symmetry of the monomer molecule. The cyclic dimer turned out to be about 0.8 kJ/mol more stable than the linear one.

Electron densities in solid ammonia were calculated by the *ab initio* method [129]. A 16-atoms cluster $(NH_3)_4$ was taken as a model. The parameters of its structure were compared with that of the monomer. The comparison showed that the main changes that occurred during condensation of the gas to the solid were the following: (a) polarity of the N—H bond increased; (b) the charge was shifted from the $3a_1$ lone pair orbital toward the region between the N—H bonds, which resulted in a more spherical distribution of electron density around the nitrogen nuclei. At the same time, relative energies of the occupied MOs and the contributions of different atoms to the MOs did not change substantially [130]: the ionization energies and their differences, widths of absorption bands assigned to ionization of the higher occupied electron levels of solid ammonia differed only insignificantly (according to photoelectron spectroscopy) from the energy characteristics of the $3a_1$ and $1e$ molecular orbitals of ammonia gas.

Even strong association of organic molecules does not change the distribution of electron density in the occupied MOs to a great extent. This is confirmed, e.g., by *ab initio* calculations in the STO4-31GF basis set and experimental (PES) studies of the monomer and dimer forms of acetic acid (the geometry of the dimer was determined by electron diffraction) [43, 131]. Monomer and dimer HOMO structures are given below (given are the numbers of MOs, their assignment, and ionization potentials as determined by PES in the gas phase) [43]:

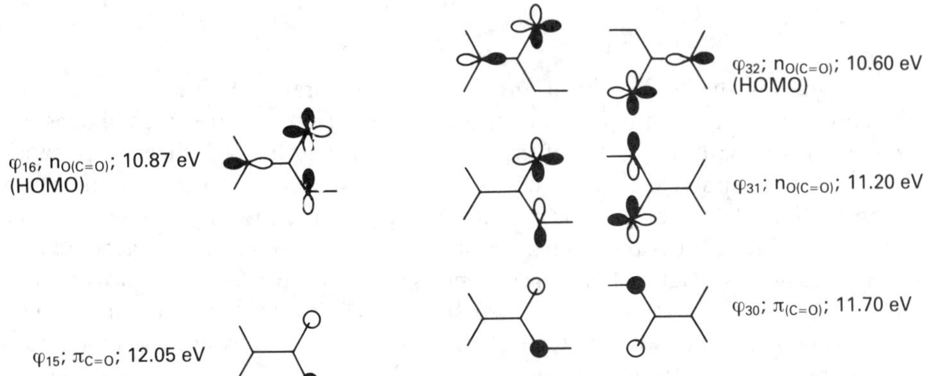

$\varphi_{32}$; $n_{O(C=O)}$; 10.60 eV (HOMO)

$\varphi_{16}$; $n_{O(C=O)}$; 10.87 eV (HOMO)

$\varphi_{31}$; $n_{O(C=O)}$; 11.20 eV

$\varphi_{30}$; $\pi_{(C=O)}$; 11.70 eV

$\varphi_{15}$; $\pi_{C=O}$; 12.05 eV

$\varphi_{26}$; $\pi_{(C=O)}$; (12.10 eV)

MO of monomer

MO of dimer

The changes in the electron structure are far less significant on association of nonpolar and low-polar organic molecules. It was thus found by *ab initio* calculations of a benzene dimer that the changes in the ionization potentials during transition from the monomer to the dimer are small (less than 0.1 eV) and their trustworthy interpretation is therefore impossible [132]. The calculated and experimental ionization potentials (eV) of the benzene monomer and dimer are given below for the purpose of comparison:

| | $IP_1$ | $IP_2$ | $IP_3$ |
|---|---|---|---|
| $IP_i$ of the monomer (eV) | 11.402 | 15.046 | 15.832 |
| Changes in $IP_i$ in the dimer (eV) | +0.054 | +0.062 | +0.061 |
| | −0.037 | −0.050 | −0.031 |

Changes in the energies of the occupied electron shells of polycyclic aromatic hydrocarbons occurring during their transition from the gas to the solid phase, were estimated experimentally (see Section 6.3.4).

As compared with self-association, reliability of estimation of the interaction of organic molecules with the solvent molecules is probably lower. There are no experimental quantum chemical methods to study electron structures of neutral organic molecules surrounded by solvent molecules. Only indirect estimations are possible, by, for example, ESR and absorption spectra of the corresponding ion radicals (see Section 2.3.4) [40, 133]. Numerous examples will be discussed in subsequent chapters. Here we shall only consider the data found for naphthalene and anthracene:

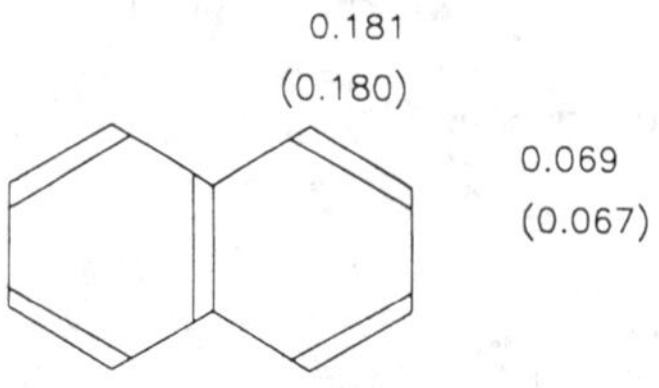

0.181 (0.180) 0.069 (0.067)

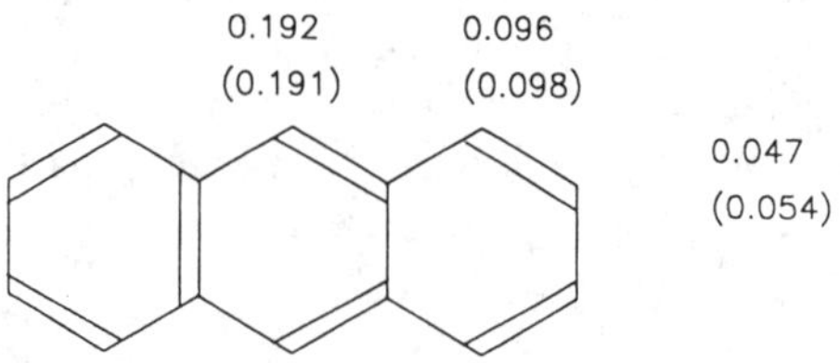

0.192 (0.191) 0.096 (0.098) 0.047 (0.054)

Spin densities in the anion radicals given in the structural formulae were calculated for neutral molecules by the HMO method; parenthesized are experimental densities estimated from the ESR spectra of the anion radicals (solvent, 1,2-dimethoxyethane).

It has been mentioned that the development of the quantum chemical approach for estimation of the solvent effect on the electron structure of a solute molecule is mostly restricted by the failure to measure experimentally the geometry of supermolecules, associates of organic molecules with solvent molecules, since optimization of the steric parameters of these supermolecules on the basis of *ab initio* calculations takes much computer time. In this situation, various approximations and simplified approaches are widely used. Below the parameters of geometric structures of an unsolvated anion radical of 9-(nitromethylene)-fluorene (XXVIII) are compared with the same anion radical solvated with two molecules of water [134]:

| XXVIII: | $XXVIII^{\dot{-}}$ | $XXVIII^{\dot{-}} \cdot 2H_2O$ |
|---|---|---|
| $r_{C-N}$ | 0.1437 nm | 0.1430 nm |
| $r_{C(9)-C(\alpha)}$ | 0.1401 nm | 0.1403 nm |
| $r_{N-O}$ | 0.1223 nm | 0.1226 nm |

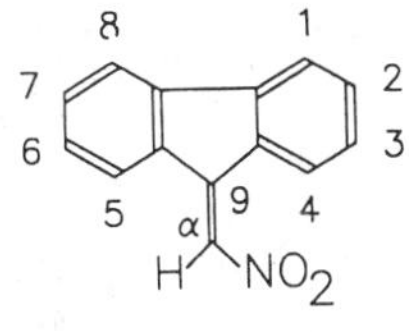

XXVIII

Optimization of the geometry of both structures (the molecules of water were not optimized, while the NO ··· HOH distance was taken to be 0.2 nm) shows that solvation with water molecules does not substantially alter bond lengths in the anion radical. Attachment of two water molecules to the nitro group, however, causes redistribution of spin density in the anion radical. The results of calculations (by the modified NDO and *ab initio* method in the STO-3GF basis) of spin densities in the corresponding particles are given below:

| | XXVIII | | $XXVIII \cdot 2H_2O$ | |
|---|---|---|---|---|
| | MNDO | STO-3GF | MNDO | STO-3GF |
| $C_{(9)}$ | 0.1714 | 0.2108 | 0.1918 | 0.2391 |
| $C_{(\alpha)}$ | 0.1962 | 0.3969 | 0.1815 | 0.3272 |

These calculations do not give true values of electron densities but probably reliably show the shift of the electron density from the $C_{(\alpha)}$ to the $C_{(9)}$ atom on addition of water molecules. It is believed that these changes account for the results of the interaction between 9-(nitromethylene)-fluorene and nucleophilic agents at the $C_{(\alpha)}$ atom in the low-polar dimethyl sulphoxide and at the $C_{(9)}$ atom in water. Within the framework of the frontier orbitals concept (see Section 3.1), the anion radical of the substrate, formed as a result of the electron transfer from a nucleophile (as in the scheme discussed for the electrophilic aromatic substituion [87–9]), should be considered for the reactions with nucleophilic agents. In this case, selectivity of the reaction will depend on the location of an unpaired electron in the substrate anion radical: in order to complete bond formation, the radical-like nucleophile should be attached at the site of highest spin density.

# 4

# Alkanes

## 4.1 ACYCLIC ALKANES

### 4.1.1 Methane

Methane is an extremely important organic substance. It is the first representative of a huge class of hydrocarbons and is the starting compound for the industrial synthesis of acetylene, hydrocyanic acid, methyl alcohol, acetaldehyde, acetic acid, etc. Methane is also unique because it is found in the atmosphere of Saturn and Jupiter, and also in interstellar space. According to some hypotheses, the reactions of methane underlie abiogenic formation of biologically important molecules [135].

The molecule of methane is a regular tetrahedron; according to electron diffraction measurements, the C—H bond is 0.1085 nm long, while the HCH angle is 109.5° [136]. The high spatial symmetry of the methane molecule suggests full equivalence of its four (C—H) bonds and hence full equivalence of the electrons in the valence shell. Such an equivalence of the electrons in the methane molecule is also suggested by the Lewis' octet rule. Later electron concepts, including the concept of the $sp^3$ hybrid orbitals of the C—H bonds, confirm the electron equivalence hypothesis: the eight electrons in the valence shell of methane are localized in pairs in four equivalent C—H bonds. The results of determinations by most physical methods fully agree with the suggestion of equivalence of all C—H bond electrons in methane. Neither NMR nor IR spectra have signals that might be assigned to different C—H bonds. Nevertheless, the states of the electrons in the valence shell of the methane molecule are not equivalent. Photoionization measurements show that these electrons have different energies and two different ionization potentials. The photoelectron spectrum of methane is shown in Fig. 4.1.

The non-equivalence of the eight valence electrons of the methane molecule can logically be explained within the framework of the MO theory [21, 137]. First of all, it can be seen that 1*s* orbitals of the four hydrogens $\chi_1$, $\chi_2$, $\chi_3$, and $\chi_4$, can form four group (fragment) orbitals: $\varphi_1$, $\varphi_2$, $\varphi_3$ and $\varphi_4$:

$$\varphi_1 = 1/\sqrt{4}(\chi_1 + \chi_2 + \chi_3 + \chi_4),$$

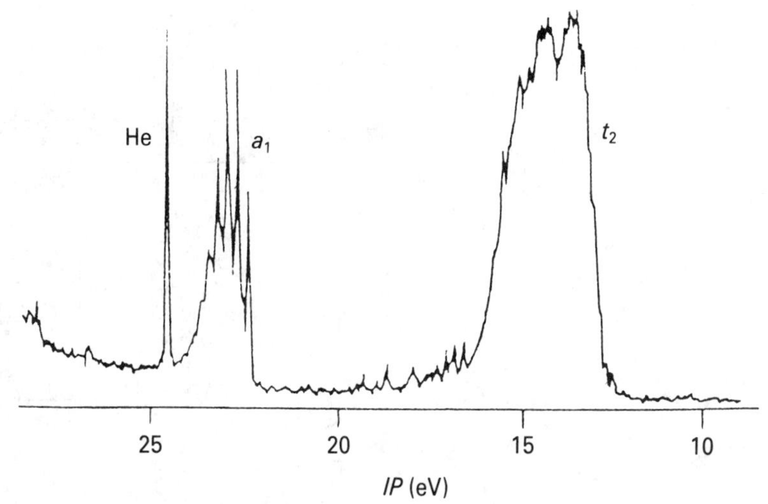


Fig. 4.1. He(II) PE spectrum of methane (peak at 24 eV is assigned to helium). (From Jolly [21])

$$\varphi_2 = 1/\sqrt{4}(\chi_1 + \chi_2 - \chi_3 - \chi_4),$$
$$\varphi_3 = 1/\sqrt{4}(-\chi_1 + \chi_2 - \chi_3 + \chi_4),$$
$$\varphi_4 = 1/\sqrt{4}(-\chi_1 + \chi_2 + \chi_3 - \chi_4).$$

The $\varphi_1$ group orbital has the same symmetry as the 2s orbital of the carbon atom. These two basis orbitals form, in turn, two molecular orbitals of methane:

$$\varphi_{1,2} = 1/\sqrt{2}\chi(2s) \pm 1/2\sqrt{2}(\chi_1 + \chi_2 + \chi_3 + \chi_4).$$

Figure 4.2 shows that the $2p_z$ orbital of the carbon atom and the $\varphi_2$ orbital of hydrogen atoms also have the same symmetry. These orbitals can form two other molecular orbitals in methane:

$$\varphi_{3,4} = 1/\sqrt{2}\chi(2p_z) \pm 1/2\sqrt{2}(\chi_1 + \chi_2 - \chi_3 - \chi_4).$$

One of these orbitals is bonding and the other antibonding. Similar molecular orbitals are formed by the mixing of the $2p_x$ and $2p_y$ orbitals of carbon with the $\varphi_3$ and $\varphi_4$ orbitals of the hydrogen atoms, respectively:

$$\varphi_{5,6} = 1/\sqrt{2}\chi(2p_x) \pm 1/2\sqrt{2}(-\chi_1 + \chi_2 - \chi_3 + \chi_4),$$
$$\varphi_{7,8} = 1/\sqrt{2}\chi(2p_y) \pm 1/2\sqrt{2}(-\chi_1 + \chi_2 + \chi_3 - \chi_4).$$

Thus, the three bonding MOs $\Psi_3$, $\Psi_5$ and $\Psi_7$ have the same symmetry $t_2$ (within the $T_d$ point group, to which the methane molecule belongs), similar energy, and are therefore degenerate. The $\Psi_1$ bonding orbital, formed with the involvement of the $s$ orbital alone, has type $a_1$ symmetry. In agreement with the higher ionization potential of 2$s$ electrons (compared with the 2$p$ electron), the energy level of the $\Psi_1$ orbital is much lower than those of the degenerate $\Psi_3$, $\Psi_5$ and $\Psi_7$ orbitals.

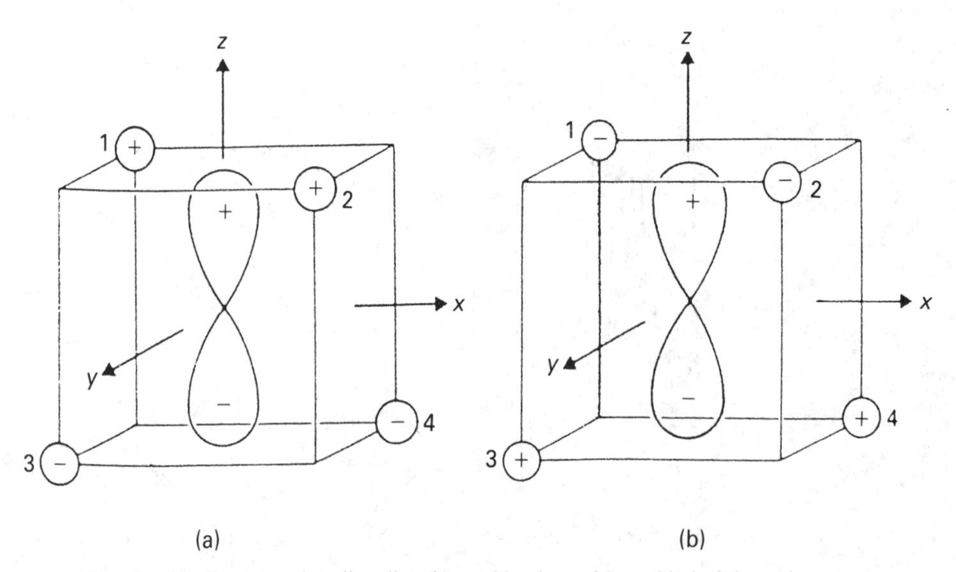


Fig. 4.2. Bonding (a) and antibonding (b) combinations of $2p_z$ orbital of the carbon atom and $\varphi_2$ group orbital of hydrogen atoms (similar combinations are formed by $2p_x$ and $2p_y$ orbitals). (From Jolly [21])

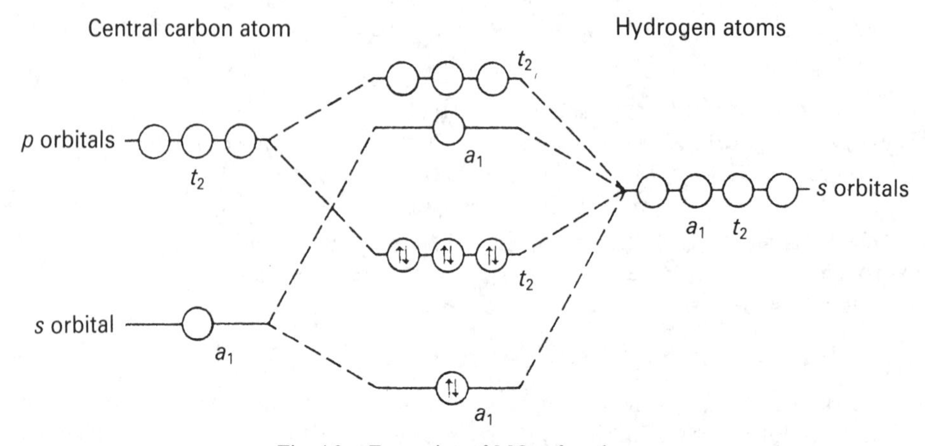


Fig. 4.3. Formation of MOs of methane.

The correlation diagram of the methane molecule is shown in Fig. 4.3. Two valence electrons found in the MO of $a_1$ symmetry thus form the first electron shell and are ionized at 23 eV, while the other six electrons occupy three degenerate MOs of $t_2$ symmetry to form the uppermost electron shell, and start to ionize at 13 eV. The ratio of the intensities of the first and second ionization bands, and splitting of the first absorption band due to the Jahn–Teller effect (see Fig. 4.1) confirm the objective character of the correlation diagram shown in Fig. 4.3.

The photoelectron spectrum of methane has been interpreted not only on the basis of the above analysis in terms of group orbitals. Assignment of the first band to ionization of the degenerate orbitals is confirmed also by *ab initio* quantum chemical calculations, e.g. in the 6-31GF basis set [43] and using the Green functions [138].

Non-equivalence of the eight valence shell electrons in the methane molecule, predicted theoretically by quantum chemical calculations and confirmed experimentally by PE spectroscopy, does not contradict chemical experience; nor does it mean non-equivalence of its four (C—H) bonds. It has already been said (Section 2.2.1) that within the framework of the MO theory, the electron density between atoms is summed up by all occupied MOs. By summing up the densities in the $\varphi_1$, $\varphi_3$, $\varphi_5$ and $\varphi_7$ methane orbitals, one finds that the electron density is the same for all four C—H bonds.

It is worth while to compare PES findings and the MO theory parameters related to the electron structure of methane with the behaviour of this hydrocarbon in chemical reactions. First of all, notice that the main cause of methane's chemical inertness is probably the very low electron-donating properties of its molecule: the ionization potential of methane is the highest among hydrocarbons. If methane is passed through common concentrated acids, no changes occur. It turns out that only effective electron acceptors can remove an electron from the methane molecule. Super-acids, such as $SbF_5 + SO_2$, $HSO_3F + SbF_5 + SO_2$ and $SbF_5 + HF$ were used for the purpose [139].

The reaction begins with acceptance of electron density from the methane molecule by the attacking proton. The intermediate methyl cation reacts then with the molecule of the initial hydrocarbon to form ethane. Subsequent conversions lead to formation of tert-butyl cation and other products of oligomerization (up to hydrocarbons with the molecular weight of 100–700):

$$H_3C{-}H + H^+ \longrightarrow \left[H_3C{\cdots}\langle^{H}_{H}\right]^+ \longrightarrow H_3C^+ + H_2$$

$$H_3C^+ + H{-}CH_3 \longrightarrow \left[H_3C{\cdots}\langle^{H}_{CH_3}\right]^+ \longrightarrow H_3C{-}CH_3 + H^+$$

The donor properties of methane were also discovered in reactions with transition metal complexes. Thus, in the study of the H/D exchange catalysed in alkanes by transition-metal complexes, it turned out that the electron transfer from the methane molecule to $Pt^{II}$ is the limiting step [140]. It was also reported that methane can attach carbon atoms, react reversibly with vapour of some metals on photo-excitation, and interact with organometal compounds [141–3].

Methane is also the most inert aliphatic hydrocarbon in radical reactions that are most characteristic of this class of organic compounds. Relative halogenation rates of various $C(sp^3)$—H bonds of some hydrocarbons are given by way of illustration [5]:

| | $CH_3$—H | $C_2H_5$—H | $(CH_3)_2CH$—H | $(CH_3)_3$—H |
|---|---|---|---|---|
| $k_{rel}$ of reaction: | | | | |
| with $Br_2$ | $7 \times 10^{-4}$ | 1 | 220 | 19400 |
| with $Cl_2$ | $4 \times 10^{-4}$ | 1 | 4.3 | 6.0 |

| | $PhCH_2$—H | $Ph_2CH$—H | $Ph_3$—H |
|---|---|---|---|
| $k_{rel}$ of reaction: | | | |
| with $Br_2$ | 64 000 | $1.1 \times 10^6$ | $6.4 \times 10^6$ |
| with $Cl_2$ | 1.3 | 2.6 | 9.5 |

The chemical inertness of methane is explained not only by its high first ionization potential (low energy level of the HOMO) but also by its extremely low electron affinity (high LUMO energy level). The highly sensitive electron transmission spectroscopy did not discover any resonance lines for methane: a non-resonance character of electron scattering is observed even in the region of high energies of 'irradiating' electrons (up to 9–17 eV).

High energies of the vacant orbitals of methane are confirmed by its UV spectrum. As distinct from most organic compounds, methane does not absorb in the visible or the near UV region (up to 150 nm) of the spectrum. The longest-wave absorption in the methane electron spectrum occurs at 128 nm ($\log \varepsilon = 3.8$) and at 119 nm ($\log \varepsilon = 3.8$). According to calculations by the CNDO and MINDO methods (with accounting for the configurational interactions), this absorption is assigned to the singlet–singlet transitions and is completely localized in the valence shell. In other words, the $\sigma \rightarrow \sigma^*$ transitions of the C—H bond electrons are responsible for the absorption [144]. The C—H bonds are broken in the methane molecule on exposure to a sufficiently high-energy UV radiation. Photochemical reactions of methane occur, for example, on exposure to short-wave UV radiation (xenone resonance lines at 147 nm and 129.5 nm) with formation of hydrogen, ethylene, small amounts of ethane and acetylene. Irradiation of a methane–oxygen mixture causes oxidation of methane with formation of $CO_2$ at 160°C.

To conclude the discussion of the behaviour of methane it is necessary to emphasize the high reliability of the quantum chemical estimations of its properties [24, 38, 145, 146]. The relatively small basis set makes it possible to use perfect *ab initio* and semi-empirical quantum chemical methods. Data on optimization of the geometry of methane, estimation of its reactivity, and interpretation of various spectral properties are available. Some experimentally found properties of methane are compared below with the calculated parameters [146]:

| | Calculated | | Experimental |
|---|---|---|---|
| | STO-3G | 4-31G | |
| $r_{C-H}$ (nm) | 0.1083 | 0.1081 | 0.1085 |
| $E_{diss}$ (kJ/mol) $CH_4 \rightarrow CH_3' + H'$ | 407.4 | 358.8 | 431.2 |
| Proton affinity (kJ/mol) | 504.5 | 479.0 | 494 |
| $IP_1$ (eV) | | | |
| adiab. | 11.78 | 12.14 | 12.66 |
| vert. | 13.29 | 13.50 | 13.60 |

### 4.1.2 Ethane

Ethane is the nearest homologue of methane. Its geometric parameters related to the C—H bond are practically the same as those of methane (Table 4.1).

**Table 4.1.** Structure of ethane and deuterated ethanes as determined by the electron diffraction method

| Bond distance (nm)<br>Angle (degrees) | $CH_3—CH_3$ | $CD_3—CH_3$ | $CD_3—CD_3$ |
|---|---|---|---|
| C—C | 0.1532 | 0.1531 | 0.1530 |
| C—H | 0.1102 | 0.1102 | |
| C—D | | 0.0990 | 0.1099 |
| ∠HCH | 107.51 | 107.30 | |
| ∠DCD | | 107.35 | 107.51 |

(From Iijima [147])

Most noteworthy in the geometry of ethane is that the rotation of methyl groups about the C—C bond is hindered, although the C—C distance in the ethane molecule corresponds to the ordinary bond. The entropy of ethane is lower than one might expect for a molecule with free rotation.

It turned out that ethane has two conformations, staggered and eclipsed, differing in energy by about 12.6 kJ/mol. The staggered conformer is more stable.

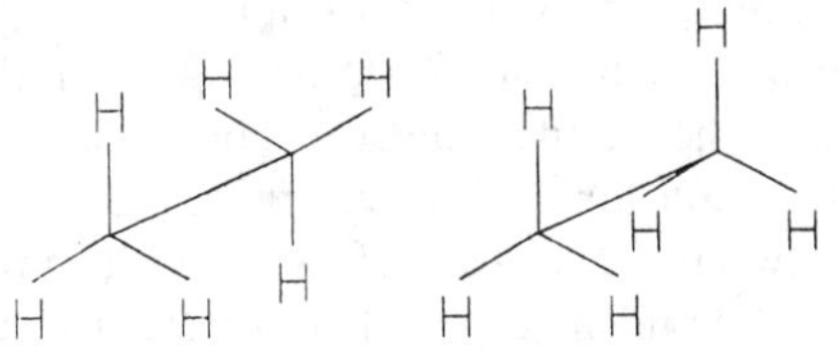

staggered eclipsed

The presence of the C—C bond leads to formation of a new (compared with methane) electron level in the ethane molecule. The analysis of the electron structure becomes more complicated since ionization of the electron of the C—H and C—C orbitals occurs at the same potentials, and the corresponding bands in the PE spectrum of ethane overlap.

Qualitative observations (as well as the results of *ab initio* calculations [38]) suggest that the first band in the PE spectrum of ethane can be assigned to ionization of the C—H orbital [22]:

(1) the frequency of the vibrational structure of the first band is 1170 $cm^{-1}$ (0.145 eV) which is characteristic of symmetric deformational C—H vibrations;
(2) the broad and structureless band at 13.5 eV corresponds to removal of the electron from the $3a_{1g}$(C—C) orbital: the ethane molecule dissociates with the rupture of the C—C bond at 13.6 eV.

The following ionization potentials, types of symmetry of the corresponding MOs, their number, and localization* were determined by PES and *ab initio* quantum chemical calculations in the 6-31G basis set [38]:

* Here and further in the book, localization of MO is determined by the fragment whose contribution to a given MO is maximum.

| $IP_i$ (eV) | No. and type of symmetry ($D_{3h}$ group) | MO localization |
|---|---|---|
| 11.99–12.70 | $1e_g$ | pseudo $\pi(CH_3)^-$ |
| (13.5) | $3a_{1g}$ | $\sigma$(CC) |
| 15.5–(15.9) | $1e_u$ | pseudo $\pi(CH_3)^+$ |
| 20.10 | $2a_{2u}$ | C(2*s*) |

Drawings of some ethane MOs are given below:

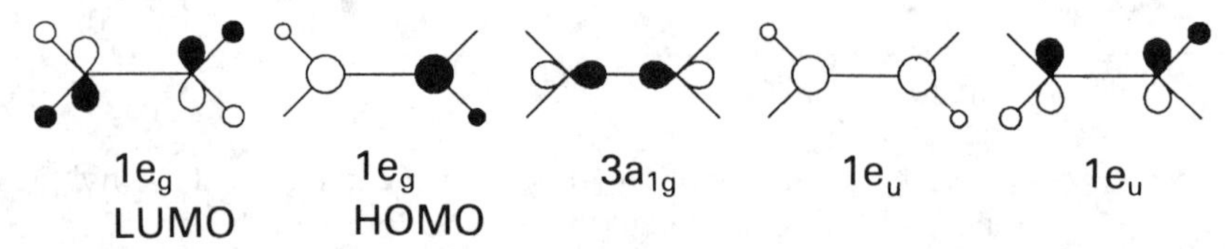


According to calculations, about two thirds of the energy barrier to free rotation of methyl groups in ethane are due to the effects of degenerate $1e_g$ orbitals [148]. It should be noted that degenerate HOMOs of the ethane molecule have pseudo $\pi$ symmetry. Note that free rotation of methylene groups relative to the C—C bond in the ethylene molecule is hindered by the $\pi$ orbital which is also a HOMO.

Other data should also be mentioned in connection with the analysis of the nature of the rotational barrier in the ethane molecule. Thus, according to [149], this barrier is formed largely by the Pauli force of repulsion between the occupied C—H bond orbitals: the distance between the protons of two $CH_3$ groups in the stable staggered conformer is 0.255 nm, while in the eclipsed conformer it is 0.225 nm. The energy of repulsion in the staggered conformer is thus 4.2 kJ/mol per each pair of hydrogen atoms.

The fact that the electrons of ethane C—H and C—C orbitals are ionized at the same energy agrees with the results of experimental studies on reactivity of ethane. Many reactions of methane and ethane with electrophiles (nitration, halogenation, H/D exchange in the presence of super-acids) were studied. Studied also was the H/D exchange in the presence of transition-metal complexes [140, 141, 145]. The D/H exchange in methyl groups and breakdown of the C—C bond compete in ethane reactions with DF + $SbF_5$ at $-78$°C. The C—C bond rupture prevails over the C—H bond rupture. The methane to hydrogen ratio in the reaction products is 8:1. During nitration of ethane with stable salts of nitronium ($NO_2^+PF_6^-$, $NO_2^+2SbF_6^-$, $NO_2^+BF_6$) in a $CH_2Cl_2$–sulpholane mixture, the prevalence of the reaction at the C—C bond is not striking: nitromethane to nitroethane ratio is only 2.9:1. Reactions at C—C bonds are even less frequent in chlorination of ethane in the presence of $SbF_5$: dimethyl chloronium and ethyl chloronium ions are formed in the ratio of 7:3 through the transition states XXIXa and XXIXb respectively.

The decrease in the acid strength causes a further shift in favour of the C—H bond: in the presence of aluminium chloride, chlorine reacts with ethane to give only chloroethane. The catalytic properties of $AgSbF_6$ are rather weak and the corresponding electrophile does not react with ethane at all (but it reacts with strained cycloalkanes). While discussing preference of C—C or C—H MO in the reactions of

$$CH_3-CH_3 + Cl^+ \longrightarrow \left[CH_3\cdots Cl \cdots CH_3\right]^+ \ (\text{XXIXa}) \longrightarrow CH_3Cl + CH_3^+$$

$$CH_3-CH_3 + Cl^+ \longrightarrow \left[CH_3CH_2\cdots (Cl)(H)\right]^+ \ (\text{XXIXb}) \longrightarrow CH_3CH_2Cl + H^+$$

ethane with electrophiles, it is necessary to consider the results of calculation of the steric structure of a protonated ethane molecule: the protonated C—C bond structure (similar to XXIXa) turned out to be 46.05 kJ/mol more stable than the structure with a protonated methyl group (similar to XXIXb) [140].

As in methane, the vacant electron levels of the ethane molecule have rather high energy and cannot be detected by ET spectroscopy.This is confirmed also by the high energy of the long-wave electron transition in the UV spectrum: the corresponding band can only be seen at 121 nm. According to quantum chemical calculations [144], this band can be assigned to the singlet–singlet transition in the valence shells formed mostly by the (C—H) orbitals.

The *ab initio* quantum chemical calculations also explain many other properties of ethane. As estimated by the *ab initio* calculations, the rotational barrier relative to the C—C bond is 12.85 kJ/mol. The experimentally found value is 12.6 kJ/mol. The calculated and measured geometric characteristics of the ethane molecule are given below [146]:

| | Calculated | | Experimental |
|---|---|---|---|
| | STO-3GF | 4-31GF | |
| $r_{C-C}$, nm | 0.1538 | 0.1529 | 0.1532 |
| $R_{C-C}$, nm | 0.1086 | 0.1083 | 0.1102 |
| HCH angle, degree | 108.2 | 107.7 | 107.5 |

### 4.1.3 Medium acyclic alkanes

Higher homologues of methane are also common in nature. Medium and higher aliphatic hydrocarbons, which are the main components of oil distillates, are important intermediates used in organic synthesis. Fragments of their molecules are reproduced in large numbers of living organisms: long aliphatic radicals are contained in many biologically important structures. The steric and electron structure of the hydrocarbon chain largely determines the lipid function in living organisms. Moreover, long aliphatic molecules have marked biological activity even in the absence of functional groups: methane or ethane are odourless while medium and higher alkanes can act on the olfactory organs. At least one alkane, 2-methyl heptadecane, is known as a pheromone (moth attractant).

All linear alkanes, irrespective of their state of aggregation, have tetrahedral angles at the carbon atoms and standard C—C and C—H bond distances [150]. The results of electron diffraction measurements of n-butane, n-pentane, n-hexane and n-heptane are compared below with the findings of x-ray structural analysis of the n-$C_{36}H_{74}$ hydrocarbon:

| | Electronegativity [150] | X-ray structural analysis [151] |
|---|---|---|
| $r_{C-H}$ (nm) | 0.1108 ± 0.0004 | |
| $r_{C-H}$ (nm) | 0.1532 ± 0.0002 | 0.1534 ± 0.0006 |
| CCC angle (degree) | 112.7 ± 0.1 | 112.2 ± 0.3 |

In linear alkanes, the rotation of the molecular fragments about the C—C bond is hindered, and the rotational barriers slightly increase in the transition from ethane to propane and n-butane. Shown below are four conformers of n-butane (XXX):

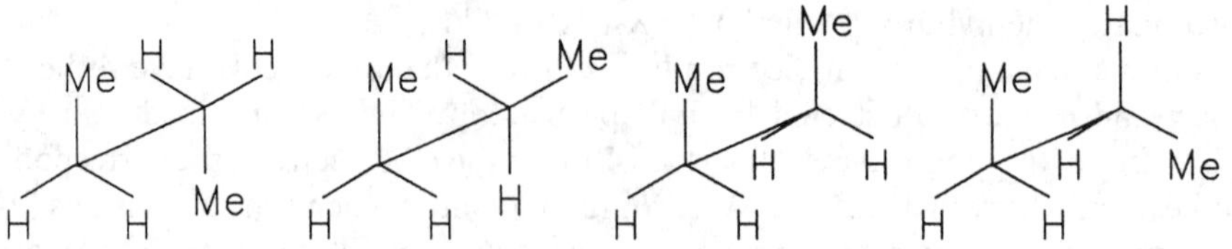

XXX: a b c d

A completely staggered conformer (anti-) XXXa has the highest stability. The stability is slightly lower (by 3.35 kJ/mol) in a partially staggered gauche conformer (XXXb). The perfectly eclipsed isomer XXXc is characterized by high energy (17–25 kJ/mol); the energy is slightly lower (14.24 kJ/mol) in the partly eclipsed isomer XXXd. The barriers were first estimated by Pitzer from the heat capacity of hydrocarbons [152]. Subsequent measurements verified the values of the barriers. The difference between the energies of gauche and anti conformers did not exceed 2.65 kJ/mol. The contribution of the more stable anti conformer was not less than 60% [150].

Owing to the insignificant difference in the energies of different conformers, the molecule of each alkane remains labile as regards its conformation, but its form is mostly zig-zag. The multitude of the conformational forms of alkanes is confirmed by PE spectra. It has been shown that longer hydrocarbon chains of aliphatic hydrocarbons, during their photoionization, are characterized by a greater number of broad overlapping bands corresponding to ionization of various conformers. Figure 4.4 shows the PE spectrum of n-pentane. The bands assigned to $\sigma$(C—C)- and $\sigma$(C—H) orbitals overlap over the entire region of the ionization potentials (for assignment of the bands see Table 4.2).

As one can see, the HOMO of the ethane molecule is formed by $\sigma$(C—H) orbitals, while beginning with propane, and in longer alkanes, the $\sigma$(C—C) orbitals become the HOMOs; these orbitals are mostly localized over the C—C bond region. In accordance with this conclusion, both spectral curves and *ab initio* quantum chemical calculations show that the ionization energies of the higher (C—C) and (C—H) orbitals in linear alkanes differ by not more than 0.5 eV.

Extensive overlapping of the ionization regions of these orbitals probably explain the absence of reports of observation of different conformers of alkanes by their PE

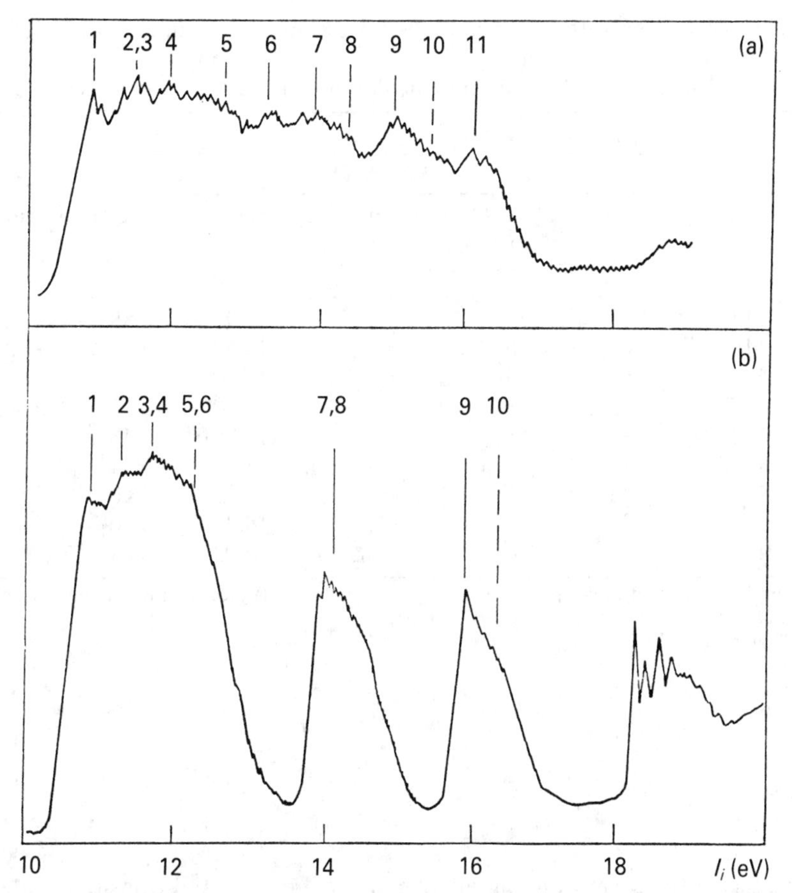


Fig. 4.4. He(I) PE spectra of (a) *n*-pentane and (b) cyclopentane band numbers in the spectra correspond to the numbers of ionization potentials in Tables 4.2 and 4.5. (From Hohlneicher [43])

spectra. In the spectra of polysilanes, the Si analogues of aliphatic hydrocarbons, the ionization region of the $\sigma$(Si—Si) orbitals is distinctly separated from the ionization region of the $\sigma$(Si—H) orbitals. The content of the conformers could thus be evaluated from PES in the series of $Si_nH_{2n+2}$ compounds [153].

The data given in Table 4.2 show that *ab initio* calculations reproduce reliably the values of electron energies in alkanes. The calculations support also the experimental data on low polarity of the C—H and C—C bonds. The full charges on the atoms ($Z \times 10^3$ values in electron charge units) in the molecules of n-butane and n-pentane are given below [144, 154].

The reactions of n-alkanes in superacids and in the presence of transition metal complexes have now been studied extensively [139, 140]. It is unlikely that the differences in the total charges might determine the direction of the attack on the aliphatic hydrocarbon molecule by electrophilic agents. Their reactivity is probably more dependent on the shapes of the highest occupied molecular orbitals, and on

**Table 4.2.** Ionization potentials* and the higher occupied MOs of n-butane and n-pentane (PES-data and calculations in the 4-31G basis set)

| | *n*-Butane | | | | *n*-Pentane | | |
|---|---|---|---|---|---|---|---|
| | | | MO | | | | |
| Nos | $IP_i$ (eV) | $-\varepsilon_i$ (eV) | No and symmetry ($C_{2h}$ group) | Localization | $IP_i$ (eV) | $-\varepsilon_i$ (eV) | Localization |
| 1 | 11.09 | 12.29 | $7a_g$ | $\sigma_{CC}$ | 10.93 | 11.88 | $\pi_{CH_3}, \sigma_{CC}$ |
| 2 | 11.66 | 12.43 | $2b_g$ | $\pi_{CH_2}$ | 11.50 | 12.11 | $\sigma_{CC}, \pi_{CH_3}$ |
| 3 | (12.30) | 12.49 | $2a_u$ | $\sigma_{CC}$ | 11.50 | 12.42 | $\sigma_{CC}$ |
| 4 | 12.74 | 13.33 | $6a_g$ | $\pi_{CH_3}$ | 12.10 | 12.86 | $\sigma_{CC}, \pi_{CH_3}$ |
| 5 | (13.20) | 13.80 | $6b_u$ | $\sigma_{CC}$ | (12.80) | 13.48 | $\sigma_{CC}$ |
| 6 | (14.20) | 15.41 | $5b_u$ | $\pi_{CH_3}$ | 13.30 | 13.94 | |
| 7 | 14.59 | 15.41 | $1b_g$ | $\pi_{CH_3}$ | 13.90 | 14.40 | |
| 8 | (15.00) | 15.90 | $5a_g$ | $\sigma_{CC}$ | (14.40) | 15.30 | $\pi_{CH_3}$ |
| 9 | 15.59 | 17.38 | $1a_u$ | $\pi_{CH_3}$ | 15.03 | 16.04 | |

*Here and in subsequent tables parenthesized are the values of $IP_i$ measured by the position of poorly resolved bands.
(From Kimura *et al.* [38])

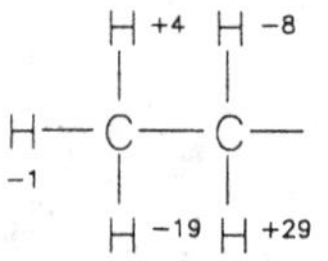

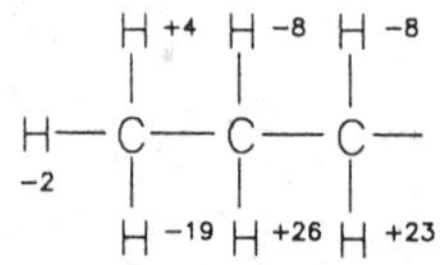

```
    H +4  H -8              H +4  H -8  H -8
    |     |                 |     |     |
H — C  —  C —           H — C  —  C  —  C —
-1  |     |             -2  |     |     |
    H -19 H +29             H -19 H +26 H +23
```

the energies of the corresponding intermediate complexes and transition states [66]. The (C—C) and (C—H) bonds act as donors in these reactions; their electrons form two-electron three-centre bonds with the electrophile. Thus, $NO_2^+PF_6^+$ in the mixture of methylene chloride–sulpholane reacts with propane and n-butane mostly with the rupture of the C—C bonds:

products of propane nitration
$MeNO_2$:$EtNO_2$:iso-$PrNO_2$:n-$PrNO_2$ = 2.8:1:0.5:0.1
products of n-butane nitration
$MeNO_2$:$EtNO_2$:n-$BuNO_2$:tert-$BuNO_2$ = 5:4:1.5:1.

Preferable attack of the electrophile E at the C—C bond suggests formation of a positively charged ion, as in formation of dimethyl chloronium ion in chlorination of ethane.

Among other reactions that require explanation on the basis of the electron structure of alkanes, are the reactions of alkanes catalysed by transition-metal complexes. Although some details of the H/D exchange mechanisms are not clear, it has been established that the rate of exchange strongly depends on the ionization potential of a particular alkane. It is believed that the transition of electron density

$$\left[ CH_3CH_2 \cdots \overset{E}{\vdots} \cdots CH_2CH_3 \right]^+$$

from alkane to the metal atom (e.g., to $Pt^{II}$) is the rate-limiting step for the entire process.

The discovery of electrophilic activation of alkanes by transition-metal complexes has opened new prospects in chemistry of aliphatic hydrocarbons. Beside the H/D exchange, other reactions were studied in detail, e.g. the oxidation and chlorination reactions. The $C_1$—$C_4$ hydrocarbons are, for example, oxidized to alcohols (in acetonitrile) while butane is oxidized by air to acetic acid in the presence of $Co^{III}$ at room temperature [140].

As the number of the C—C bonds increases, the energy of the HOMO of the alkane increases too (the first ionization potential decreases). Similar effects of the interaction of vacant orbitals gradually reduce the energy of the LUMO: according to the electron transmission spectra of a thin film of n-hexane, its electron affinity is 0.9 eV [155].

As in methane and ethane, the first singlet-singlet transitions in longer alkanes are also localized in their valence shells and can be seen only in the short-wave region of the electron absorption spectrum [144]. According to *ab initio* calculations, the first electron transition in propane is associated with excitation of the electrons of the C—H orbitals, while in butane and higher alkanes, the first electron transition occurs mostly due to electron excitation of the C—C orbitals (Table 4.3).

**Table 4.3.** Experimental and calculated energies and intensities of absorption bands in electron absorption spectra of n-alkanes

| Alkane | Experimental | | Calculated (CNDO) | |
|---|---|---|---|---|
| | *E* (eV) | *f* | *E* (eV) | *f* |
| Propane | 8.89 | 0.33 | 8.50 | 0.09 |
| | 9.65 | — | 8.89 | 0.00 |
| | 10.33 | — | 8.92 | 0.08 |
| n-Butane | 8.79 | 0.5 | 8.52 | 0.11 |
| | 9.29 | — | 8.56 | 0.12 |
| | 10.57 | — | 9.05 | 0.05 |
| n-Pentane | 8.76 | 0.6 | 8.27 | 0.12 |
| | 9.22 | — | 8.62 | 0.14 |

(From Salanub, and Sandorfy [144])

In view of the low polarity of the C—C and C—H bonds, the intermolecular forces are relatively weak in alkanes. Attraction effects are called dispersion forces. They are mostly due to electron correlation. These forces are effective only with small distances (if the distance is very small, the repulsion forces become stronger than the attraction forces), because they depend on the area of contact: the greater the contact area in the molecule, the greater the attraction force. Theoretical estimations show that each new methylene group strengthens the attraction force between the adjacent molecules of an aliphatic hydrocarbon by about 4–6 kJ/mol [154].

## 4.2 CYCLOALKANES

Compared with acyclic compounds, new structural effects can be seen in cycloalkanes. These are, first of all, the strain effects: angular, torsional, and transannular strain. Their sum is usually evaluated by the heat of formation, because for compounds $(CH_2)_n$ that differ from one another by one $CH_2$ structural element, the formation heat of each methylene group must be constant [4, 156]. Strain effects are minimal in cyclohexane. Cycle strains of all other cycloalkanes are therefore estimated with reference to cyclohexane. The corresponding data for some cycloalkanes are given in Table 4.4.

**Table 4.4.** Heats of formation and strain energies of cycloalkanes $(CH_2)_n$

| $n$ | Cycloalkane | $\Delta H^0$ (kJ/mol) | $\Delta H^0/n$ (kJ/mol) | Total strain energy |
|---|---|---|---|---|
| 2 | (Ethylene) | +52.34 | +25.96 | 92.11 |
| 3 | Cyclopropane | +53.17 | +17.58 | 113.04 |
| 4 | Cyclobutane | +28.47 | +7.12 | 108.86 |
| 5 | Cyclopentane | −77.00 | −15.49 | 25.12 |
| 6 | Cyclohexane | −123.50 | −20.52 | 0 |
| 7 | Cycloheptane | −118.07 | −16.75 | 25.12 |
| 8 | Cyclooctane | −124.35 | −15.49 | 41.87 |
| 9 | Cyclononane | −132.72 | −14.65 | 54.43 |
| 10 | Cyclodecane | −154.49 | −15.49 | 50.24 |
| 11 | Cycloundecane | −179.61 | −16.33 | 46.05 |
| 12 | Cyclododecane | −230.27 | −19.26 | 16.75 |
| 15 | Cyclopentadecane | −301.45 | −20.10 | 8.37 |
| 16 | Cyclohexadecane | −321.96 | −20.10 | 8.37 |

(From Streitwieser, and Heathcock, [4])

The stable conformer of cyclohexane is not planar: all adjacent $CH_2$ groups would be eclipsed in the planar molecule of any cycloalkane. The chair conformer of cyclohexane is most favourable as regards energy. All bond angles in this conformer are close to tetrahedral, while all hydrogen pairs are completely skewed relative to one another (accurate determinations show a certain tendency to planeness of the chair and enlargement of the angles in the ring to 111.5°). The cyclohexane molecule is rather flexible: one chair conformer is consecutively replaced by another conformer thus overcoming the energy barrier of 45.22 kJ/mol. It is quite natural that both chair conformers are equivalent as regards their energy.

The molecule of cyclopentane is not planar either and has a flexible structure of an envelope; a correlated up and down movement of each methylene group gives a series of structures as if the molecule turns through 360° at a 72° pace (the molecular motion known as pseudo-rotation) [156].

In the transition from cyclohexane to larger rings, repulsion of the hydrogen atoms in the corresponding conformers becomes inevitable. Cycloheptane, for example, has the structure in which the hydrogen atoms are skewed only partly and are partly eclipsed. The strain energy of its molecule is 25.12 kJ/mol. It is also inevitable that the hydrogen atoms in cyclooctane are partly eclipsed, resulting in a further increase in strain energy. Beginning with $C_{15}$ rings, the strain energy becomes equal to 8.4 kJ/mol and it no longer changes: separate segments of the corresponding molecules behave as aliphatic chains and assume conformations in which the hydrogen atoms cannot be eclipsed.

On the whole, the energy barriers between various conformers in medium cycloalkanes are much higher than in their linear analogues. Therefore, although the electron levels of the molecules in both series of the compounds are formed in a similar way, a better resolution of the bands could be expected in PE spectra of cyclic hydrocarbons. This suggestion agrees with experimental findings: in Fig. 4.4 are compared the spectra of n-pentane and cyclopentane. Assignment of the bands in PE spectra of cyclopentane is given in Table 4.5.

**Table 4.5.** Ionization potentials and higher occupied MOs of cyclopentane (PES data and the 4-31G calculations)

| $n$ | $IP_i$ (eV) | Molecular orbitals | | |
|---|---|---|---|---|
| | | $-\varepsilon_i$ (eV) | No. and symmetry ($D_{5h}$ group) | Localization |
| 1 | 11.01 | 11.68 | $3e'_2$ | $\pi_{CH_2}$ |
| 2 | 11.39 | 11.68 | | |
| 3 | 11.82 | 12.84 | $3e'_2$ | $\sigma_{CC}$ |
| 4 | 11.82 | 12.84 | | |
| 5 | (12.00) | 12.87 | $1e''_2$ | $\sigma_{CC}$ |
| 6 | (12.00) | 12.87 | | |
| 7 | 14.21 | 15.06 | $1e''_1$ | $\pi_{CH_2}$ |
| 8 | 14.21 | 15.06 | | |
| 9 | 15.96 | 17.38 | $3a'_1$ | $\pi_{CH_2}$ |
| 10 | (16.5) | 17.41 | $1a''_2$ | $\sigma_{CC}$ |

(From Kimura *et al.* [38])

Table 4.4 shows that strain effects increase considerably with the diminishing size of the cycloalkane ring: strain energies for cyclobutane and cyclopropane are 109 and 113 kJ/mol, i.e. much higher than one might expect in view of the strain due to eclipsed C—H bonds. The corresponding differences are attributed to the angle strain. At first sight, the strain in cyclobutane and cyclopropane does not cause considerable changes in the energies of their electron levels: the first ionization potentials of these compounds differ but slightly from the first potentials of cyclohexane and cyclopentane.

The analysis of PE spectra by *ab initio* calculations (e.g. in the 4-31GF basis set) shows, however, that the energy of the highest C—C orbital in the transition from cyclopentane to cyclobutane and cyclopropane increases by about 1.3 eV, while the energy of the highest C—H orbital in the same series of compounds decreases by 2 eV [26, 38]:

| | Cyclopentane | Cyclobutane | Cyclopropane |
|---|---|---|---|
| *IP* (eV) | | | |
| highest occupied (C—C)MO | 11.82 | 10.7 | 10.5 |
| highest occupied (C—H)MO | 11.01 | 11.7 | 13.2 |

Relative changes in the energies of these orbitals are so significant that their sequence is reversed: the $\sigma$(C—H) orbitals are the highest in medium cycloalkanes, while the $\sigma$(C—C) orbitals are the highest in small rings. It should also be noted that the energy difference between two higher electron levels in cyclopropane is almost 3 eV, and in the cyclopentane and cyclohexane molecules only 0.8 eV.

Without going into the detail of the quantum chemical calculations of cyclobutane and cyclopropane molecules (Table 4.6), and by considering the hybrid orbitals alone, one can conclude the following [6]. Carbon utilizes largely 2*s* orbitals for its bonds with hydrogen atoms, while for its bonds with carbon atoms, it uses the orbitals of the prevalent *p* character ($sp^4$–$sp^5$) (the corresponding HOMOs of cyclopropane are called Walsh orbitals). As a result, the C—C bonds in small rings are curved: the $\theta$ angle in the cyclopropane molecule is 21° (the arrows are directed toward the maximum electron density):

**Table 4.6.** Ionization potentials and higher occupied MOs of cyclopropane and cyclobutane (PES data and *ab initio* calculations)

| $IP_i$ (eV) | Molecular orbitals | | |
|---|---|---|---|
| | $-\varepsilon_i$ (eV) | No. and symmetry | Localization |
| | | Cyclopropane ($D_{3h}$) | |
| 10.5, 11.3 | 10.5 | $3e'$ | CC(2*p*) |
| 13.2 | 12.8 | $1e''$ | CH |
| 15.7 | 15.6 | $3a_1'$ | CC, CH |
| 16.5 | 16.8 | $1a_2''$ | CC, CH |
| — | 20.4 | $2e'$ | CH |
| — | 28.5 | $2a_1'$ | CC(2*s*) |
| | | Cyclobutane $D_{2h}$ ($D_{4h}$) | |
| 10.7, 11.3 | 10.8(11.1) | $4e(3e_u)$ | CC(2*p*) |
| 11.7 | 11.3(11.6) | $4a_1(1b_{1u})$ | $CH_2(\pi)$ |
| 12.5 | 12.0(12.7) | $1b_1(1b_{1g})$ | CC |
| 13.4, 13.6 | 13.4(14.1) | $3e(1e_g)$ | $CH_2(\pi)$ |
| | 16.0(16.7) | $3a_1(3a_{1g})$ | $CH_2(\sigma)$ |
| 15.9 | 16.3(17.3) | $3b_2(1a_{2u})$ | $CH_2(\pi)$ |
| | 18.8(19.5) | $2b_2(2b_{2g})$ | CC, CH |
| | 22.3(23.7) | $2e(2e_u)$ | |
| | 28.2(29.5) | $2a_1(2a_{1g})$ | CC(2*s*) |

(From Schäfer [26])

The prevalent $p$ character of the C—C bonds agrees with the chemical properties of small cycloalkanes:

(1) the behaviour of the cyclopropane fragment in the corresponding reactions is very much like that of the double bond; e.g. in the presence of concentrated acids, cyclopropane reacts with water (with opening of the ring) at a faster rate than olefins attach water at the double bond;
(2) the attack of an electrophile occurs mainly at the C—C orbitals (which are the highest occupied MOs in the cyclopropane molecule) and the reactions are attended by opening of the ring;
(3) the orbitals of the cyclopropane fragment interact effectively with the orbitals of the adjacent reaction centre or substituent [52, 157, 158]; this is manifested by a marked acceleration of the solvolysis of cyclopropylmethyl halides as compared with isopropyl derivatives:

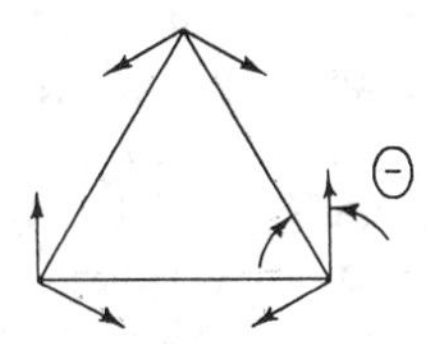

Other specific properties of the cyclopropane molecule can also be noted. Thus, the HCH angle (115°) is greater than in linear alkanes and unstrained cycloalkanes, while the C—H bond length is about the same as in ethylene. Cyclopropane derivatives have stronger C—H and other exocyclic bonds. The diminishing size of cycloalkanone $(CH_2)_nC{=}O$ increases significantly the frequency of C=O valence variations [150]:

| $n$ | 3 | 4 | 5 | 6 | 7 | 8 | 9 | 10 |
|---|---|---|---|---|---|---|---|---|
| $\nu_{CO}$ (cm$^{-1}$) | 1815 | 1780 | 1746 | 1715 | 1705 | 1692 | 1698 | 1649 |

The UV spectra reveal high ability of strained $\sigma$(C—C) orbitals to interact with $\pi$ orbitals. This can be concluded from comparison of the position of long-wave bands in the absorption spectra of some spiroketones. The cycloalkyl fragment is found in these compounds in the plane parallel to $\pi$ orbitals, i.e. its orientation is favourable for the $\sigma$(C—C)–$\pi$ interaction [156].

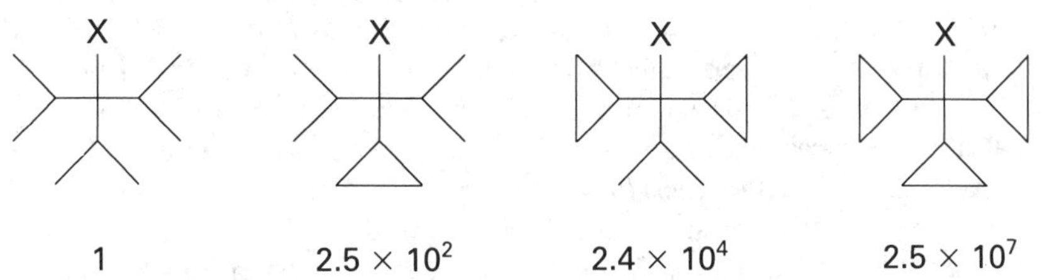


## 4.3 SOME FEATURES OF THE ELECTRON STRUCTURE OF ALKANES AND THEIR ORGANOELEMENT ANALOGUES

Various quantum chemical approaches to the electron structure of saturated hydrocarbons and their analogues have been proposed [159–163]. The Sandorfy

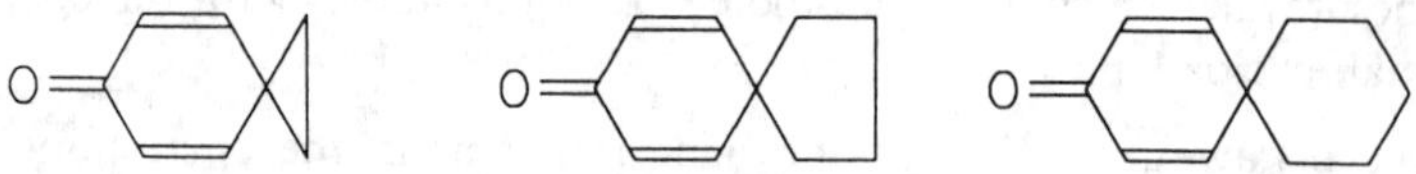

| | | | |
|---|---|---|---|
| $\lambda_{max}$, nm | 274 | 242 | 242.5 |
| $\varepsilon$ | 21900 | 15900 | 14800 |

approximation is among the first noteworthy models. Only the carbon skeletons of the molecules were calculated, while the neglect of the hydrogen atoms was compensated by appropriate parametrization [159]. Hydrocarbon MOs were described mainly as a linear combination of fragment orbitals; the $sp^3$ hybrid orbitals of the C—C and C—H bonds were regarded as the fragment orbitals. If the overlap is neglected, the corresponding calculations can be done within the framework of the simple method (see Section 2.4). Resonance integrals for the overlapping orbitals of the adjacent carbon atoms are designated $\beta$, while the integrals estimating the overlap of the $sp^3$ orbitals of one atom are designated $k\beta$. The integral $\beta$ can be estimated from spectroscopic and thermodynamic data. Thus, using the value of 162.72 kJ for $\beta$, and 0.355 for $k$, Fukui obtained good correlation between experimental and calculated ionization potentials [164] (see Table 4.7).

**Table 4.7.** Calculated HOMO energies and experimental first ionization potentials of alkanes

| n-Alkane | $-\varepsilon_{HOMO}^{calc}$ (eV) | $IP_1^{exp}$ (eV) | n-Alkane | $-\varepsilon_{HOMO}^{calc}$ (eV) | $IP_1^{exp}$ (eV) |
|---|---|---|---|---|---|
| Ethane | 12.21 | 11.76 | Hexane | 10.41 | 10.43 |
| Propane | 11.22 | 11.21 | Heptane | 10.33 | 10.35 |
| Butane | 10.77 | 10.80 | Octane | 10.25 | 10.24 |
| Pentane | 10.55 | 10.55 | Decane | 10.19 | 10.19 |

(From Fukui *et al.* [164])

The successful use of the simple MO method for the analysis of the electron structure of aliphatic hydrocarbons is not limited to reliable determination of their first ionization potentials. PE spectra of saturated hydrocarbons recorded with the use of a He(II) source of ionization have been studied in detail [54]. Two ionization regions are distinctly separated from one another in the spectra of both linear and cyclic alkanes: one in the region of 12–18 eV and the other 18–30 eV. Figure 4.5 illustrates the successive changes in the electron levels with increasing complexity of the alkane molecule (for the $IP_1$ values see Table 4.7).

Ionization bands are sufficiently well resolved in the region of high ionization potentials. This makes it possible to parametrize calculations of energies of the internal electron levels of alkanes within the framework of the simple method. The energy, for example, of the $1a_1$ orbital which, in a $CH_4$ molecule, has predominantly the character of the $2s$ orbital of the carbon atom mixed with the $1s$ orbitals of the hydrogen atoms, can be regarded as the starting electron level that splits in linear, branched, and cyclic hydrocarbons in accordance with the number of adjacent overlapping C—H orbitals. The interaction of these orbitals in all alkanes is

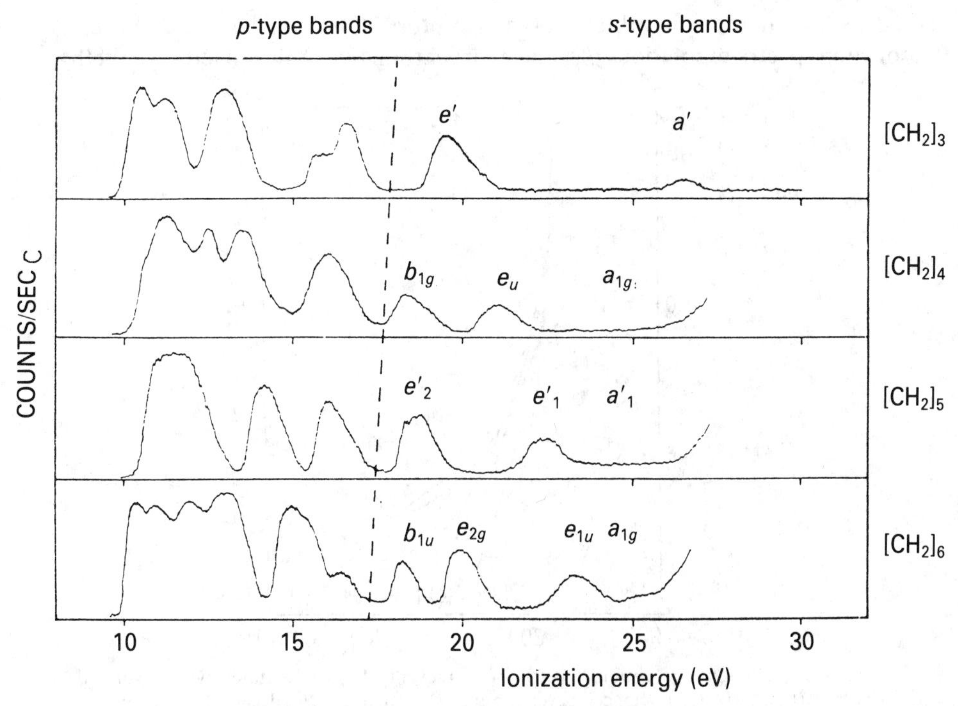


Fig. 4.5. He(I) PE spectra of some cycloalkanes. (From Heilbronner and Maier [54])

determined by the same values of the diagonal $H_{\mu\mu}$ and non-diagonal $H_{\mu v}$ matrix elements. Within the framework of the simple MO method, eigenenergies of saturated hydrocarbons could be determined either by solving the corresponding determinant (see Section 2.2.1) or by similar Coulson equations for conjugated $\pi$ systems [18]:

$$E_j = H_{\mu\mu} + 2\cos[j\pi/(n+1)]H_{\mu\nu} \qquad \text{(for linear alkanes)}$$

and

$$E_j = H_{\mu\mu} + 2\cos(2j\pi/n)H_{\mu\nu} \qquad \text{(for cyclic alkanes)}$$

where $n$ is the number of carbon atoms in the molecule and $j$ varies from 1 to $n$.

While assuming the reliability of this calculation method, it is necessary to note that, in this region of the spectrum (18–30 eV), recorded are ionizations of both bonding and antibonding combinations of $s$ orbitals (as basis orbitals). By contrast, the levels of only bonding combinations of $p$ orbitals are estimated in the region of lower potentials. The results of the simple quantum chemical calculations of the electron structure of alkanes are also fully confirmed by *ab initio* calculations [26, 54].

The orbital structure of alkane molecules, as determined by the combined use of PES and quantum chemical calculations, explains the electron-donating properties of alkanes and their analogues. Saturated hydrocarbons having sufficiently low first ionization potentials turned out to be capable to give off (like donors) their valence electrons to unoccupied orbitals of $\pi$ acceptors. One can compare the electron absorption spectra of solutions of some hydrocarbons (XXXI) with tetracyanoethylene

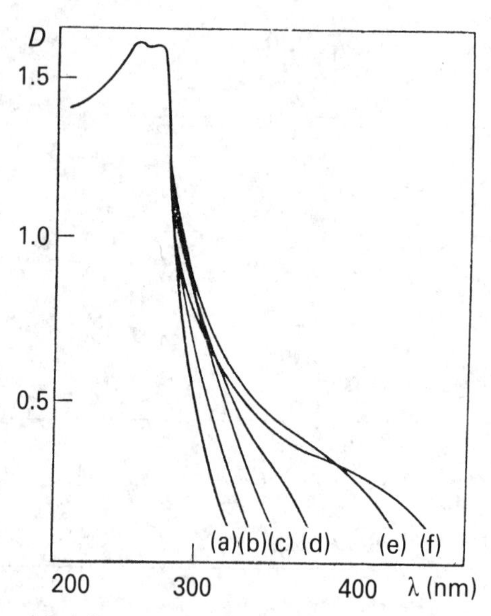


Fig. 4.6. Electron absorption spectra of tetracyanoethylene mixtures with saturated hydrocarbons: with (a) *n*-octane, (b) cyclohexane, (c) bicyclohexyl, (d) decalin, (e) adamantane and (f) 1.1-biadamantyl. (Data by Traven *et al.* [80])

in chloroform (see Fig. 4.6). Compounds XXXI are (a) n-octane, (b), cyclohexane, (c) bicyclohexyl, (d) decalin, (e) adamantane and (f) 1,1-biadamantyl [80]. While absorption of the acceptor in the spectra of TCNE mixtures with compounds XXXI (a through d) at 300 nm shifts only slightly toward longer wavelengths, the bands responsible for the charge transfer can be distinctly recorded in the spectra of those with compounds XXXI e and g. Indicated below are hydrocarbons, their first ionization potentials (eV) and wavelengths of CT bands (nm) of complexes with TCNE: (a) 10.3, 320 (sh); (b) 10.3, 320 (sh); (c) 9.40, 330 (sh); (d) 9.40, 330 (sh); (e) 9.25, 355; (f) —, 380 (sh) (sh stands for shoulder, an insufficiently resolved band).

Strained alicyclic compounds also interact with the $\pi$ acceptors by the CT mechanism. CT complexes of $\pi$ acceptors with tetracyclo($3,2,0,0^{2,7}$, $0^{4,6}$) pentane were reported as well [81].

Polysilanes (silicon analogues of alkanes) have much lower ionization potentials. Besides, the ionization energies of $\sigma$(Si—Si) and $\sigma$(SiC) orbitals differ substantially: the electrons localized over the Si—Si bond are ionized at 7–10 eV, while the electrons occupying the SiC orbitals, at 10.5–12 eV [165]. As a result, as permethylpolysilanes are mixed with TCNE in an inert solvent, well-resolved CT bands can distinctly be seen in the visible region of the spectrum. The number of these bands depends on the number of the occupied electron levels of a polysilane with ionization potentials lower than 9.0 eV (Table 4.8).

A reliable correlation has been established between the CT band energies in

**Table 4.8.** Ionization potentials (eV) of polysilanes (PES-data) and wavelengths (nm) of CT-bands of their TCNE complexes (EAS-data)

| Polysilane | $IP_1$ | $IP_2$ | $IP_3$ | $\lambda_{CT_1}$ | $\lambda_{CT_2}$ | $\lambda_{CT_3}$ |
|---|---|---|---|---|---|---|
| $Si_2(CH_3)_6$ | 8.69 | | | 417 | | |
| $Si_3(CH_3)_8$ | 8.19 | 9.14 | | 480 | 370 | |
| $n$-$Si_4(CH_3)_{10}$ | 7.98 | 8.76 | 9.30 | 520 | 390 | |
| $Si[Si(CH_3)_3]_4$ | 8.24 | | | 458 | | |
| | | 8.11 | | | | |
| $n$-$Si_6(CH_3)_{14}$ | 7.68 | 8.44 | 9.15 | 540 | 435 | |
| cyclo-$Si_6(CH_3)_{12}$ | 7.79 | 8.16 | 9.12 | 555 | 477 | 360 |

(From Traven *et al.* [75])

electron absorption spectra of the corresponding complexes and ionization potentials of polysilanes [75, 166]:

$$h\nu_{CT} = 0.84I_D - 4.25\ (r = 0.981)$$

The parameters of this equation derived for $\sigma$ donors do not differ much from the parameters of the equations describing the interaction between TCNE and $\pi$ donors (see Eqs. (3.2) and (3.3)).

Lower ionization potentials of polysilanes (compared with aliphatic hydrocarbons) explain their behaviour in some chemical transformations. Permethylpolysilanes, for example, actively reduce transition metal chlorides $WCl_6$ and $MoCl_5$. Methyl chlorosilanes and $WCl_4$ (or $MoCl_3$) are thus formed. The high contribution of the Si—Si orbitals to HOMOs of polysilanes explains the prevalent attack at the Si—Si bond in the oxidation–reduction process:

$$Me_3Si - SiMe_3 + WCl_6 \rightarrow 2Me_3SiCl + WCl_4.$$

A linear relationship between the logarithms of relative reaction rates and the first ionization potentials of polysilanes [167] has been established by the competing reaction method.

Other orbital-controlled transformations of polysilanes with electrophilic agents are also known [168, 169]. Lower ionization potentials of the electrons of the Si—Si bonds (compared with the C—C bond), the bathochromic shift of the long-wave absorption in the UV spectra of polysilanes and their higher donor properties compared with alkanes, do not obligatorily mean multiplicity of the SiSi bonds. On the contrary, the transmission of electron effects by Si—Si bonds in polysilanes is at least two times lower than that of carbon-carbon bonds [170].

As the atomic number of the element E increases, the donor properties of the E—E bonds increases too, while the bond probably become weaker. Thus, germanium and tin compounds (analogues of polysilanes) more readily give off their $\sigma$ electrons to $\pi$ acceptors, while the solutions of the corresponding TCNE complexes absorb in the region of longer wavelengths [75]. Cation-radicals of polymetals formed during the CT complexation are, however, less stable than those of polysilanes [75, 79, 171]. Polysilanes form more stable cation radicals than polygermanes or stannanes in the gas phase as well. The stabilities of molecular ions $M^{+}$ of the corresponding compounds of silicon, germanium and tin are as follows:

| Stability of $M^{+}$ ion in mass spectrum at ionizing voltage | $E_2(CH_3)_6$ | | | $E_3(CH_3)_8$ | | $E_4(CH_3)_{10}$ | | $[Si(CH_3)_2]_6$ |
|---|---|---|---|---|---|---|---|---|
| | Si | Ge | Sn | Si | Ge | Si | Ge | |
| 12 eV | 68.7 | 54.7 | 32.2 | 76.3 | 58.6 | — | 57.9 | 100.0 |
| 70 eV | 10.3 | 5.4 | 4.9 | 8.6 | 7.3 | 9.3 | 8.0 | 14.9 |

The mass spectra suggest that the stability of molecular ions of polysilanes is higher than the stability of the corresponding ions of germanium and tin compounds. It indicates higher efficiency of delocalization of the HOMOs of silicon compounds. The cation radical of dodecamethylcyclohexasilane is especially stable. The most favourable conditions for effective mixing of the adjacent Si—Si orbitals are probably realized in the structure of this compound [172].

Looseness of E—E bonds that increases with the atomic number of the element E is confirmed by PES findings. The integral estimating the interaction of the adjacent Ge—Ge orbitals, as found from the PES of permethylpolygermanes, is $-0.39$ eV (against $-0.5$ eV for the interaction of the adjacent Si—Si orbitals) [165, 173].

To conclude the analysis of the electron structure of aliphatic hydrocarbons and some of their analogues, the compounds of the IVA group elements, it should be noted that the discovery of electron donating properties in alkanes launch a new branch in chemistry, where new reactions of alkanes can be expected [140, 141].

Among other important properties of the electron structure of alkanes and their analogues with the IVA group elements, the following should be mentioned: vacant electron levels of alkanes become more available with the increasing number of carbon atoms. Neopentane, for example, is among the simplest aliphatic hydrocarbons

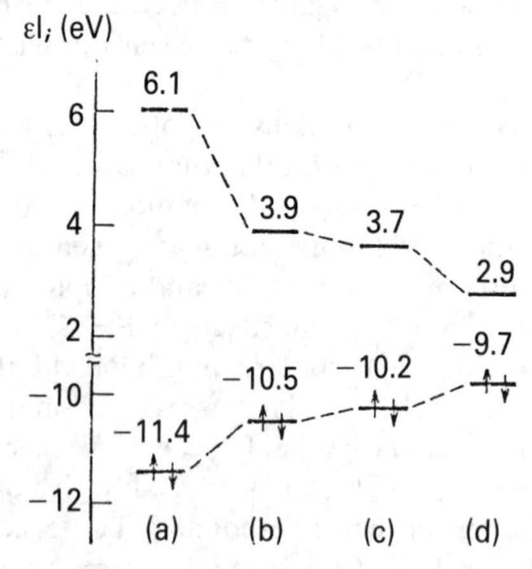


Fig. 4.7. Experimental energies of HOMO (PES) and LUMO (ETS) of tetramethyl derivatives of the IVA group elements $E(CH_3)_4$ (a) E = C; (b) E = Si; (c) E = Ge; (d) E = Sn. (Data by Giordan and Moore [174])

for which the energy of electron capture was measured in the gas phase. The value of electron affinity of this hydrocarbon turned out to be $-6.1$ eV. Compounds of silicon, germanium, and tin behave in a similar way. Most conspicuous is the fact that in the series $CMe_4$, $SiMe_4$, $GeMe_4$ and $SnMe_4$, the LUMO and HOMO energies (the energies of $\sigma$ levels) change almost symmetrically as they approach each other (Fig. 4.7). The same nearing of the occupied and vacant $\pi$ levels is observed for conjugated hydrocarbons characterized by high reactivity with both electrophilic and nucleophilic agents [174].

***Problems***

(1) Why are alkanes inert to electrophiles?
(2) Why are alkanes inert to nucleophiles?
(3) Methane has two *IP*s in the region of ionization of valence electrons: 13.0 and 23.0 eV. Explain in terms of the octet rule and MO theory.
(4) Suggest the relative reactivity of cyclopropane C—C and C—H bonds to electrophiles using the frontier orbital concept.

# 5

# Alkenes and alkynes

## 5.1 ALKENES

### 5.1.1 Ethylene

Ethylene is the key unsaturated hydrocarbon. Its importance in chemical synthesis is extraordinary and its manufacture is therefore immeasurably great compared with any other organic substance. The output of ethylene in the USA in 1980 amounted to 15 million tons, i.e. twice as great as that of propylene, the product that comes next in the list of most common products of chemical manufacture. Ethylene is contained (although in small quantity) in natural gas. It has biological activity and is used to stimulate fruit ripening; it is thus a plant hormone.

As distinct from methane and ethane, ethylene is highly reactive, which is explained by its specific structure.

All atoms of the ethylene molecule lie in one plane (the values of bond distances and angles are given according to [175]) and cannot freely rotate about the CC bond:

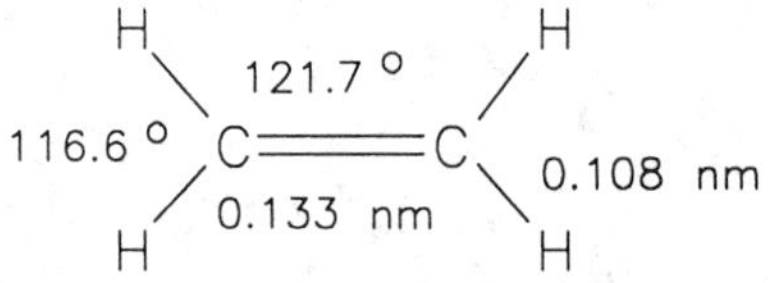

H H
121.7 °
116.6 ° C=C 0.108 nm
0.133 nm
H H

Configurational stability of the ethylene molecule is explained by the fact that its frontier occupied MO is the $\pi$ orbital: the $\sigma$ electron density has a cylindrical symmetry and it practically does not restrict rotation, while the $\pi$ electron is localized over the plane perpendicular to the plane of the molecule and accounts for the existence of the barrier to rotation. 1,2-Dideuteroethylenes, for example, can be interconverted into one another only at 500°C (the activation energy of this conversion is 272 kJ/mol).

PE spectra of ethylene were recorded with various sources [38, 176–178]; their analysis proves the $\pi$ character of ethylene's HOMOs. Table 5.1 gives assignment of the bands in the He(I) spectrum.

**Table 5.1.** Ionization potentials and higher molecular orbitals of ethylene (PES-data and the 6-31G calculations)

| $IP_i$ (eV) | MO | | |
|---|---|---|---|
| | $-\varepsilon_i$ (eV) | No. and symmetry ($D_{2h}$ group) | Localization |
| 10.51 | 12.12 | $1b_{2u}$ | $\pi_{C=C}$ |
| 12.85 | 13.50 | $1b_{2g}$ | $\pi^{-}_{CH_2(pseudo)}$ |
| 14.66 | 15.68 | $3a_g$ | $\sigma_{CC}$ |
| 15.87 | 17.35 | $1b_{3u}$ | $\pi^{+}_{CH_2(pseudo)}$ |

(From Kimura *et. al.* [38])

Some of the ethylene MOs are shown below.

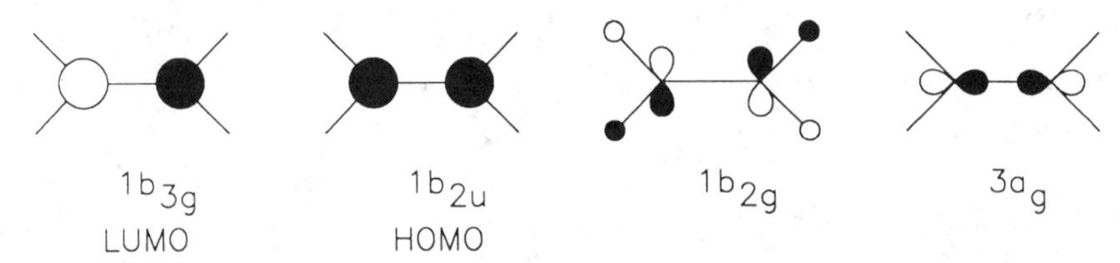


The form of the first band in the PE spectrum of ethylene is the main argument for assigning this band to ionization of its $\pi$ orbital: the band is narrow and has a vibrational structure. Frequency that is determined from the vibrational structure agrees well with the data for vibrational frequency of the double bond in its cation radical.

The regions of $\sigma$ and $\pi$ orbital ionizations are distinctly separated in the ethylene spectrum and it would be reasonable, using the example of this simplest alkene, to get acquainted with the known methods of identification of $\pi$ orbital bands. Among the most convincing of methods is the analysis of the so-called perfluoro-effect [54, 179]: replacement of all hydrogen atoms in the planar $\pi$ system by the fluorine atoms produces no appreciable changes in the energies of the $\pi$ orbitals but decreases considerably the energies of the $\sigma$ orbitals. The energies of the $\pi$ orbitals remain unchanged probably because of the compensation of the acceptor inductive effect of the fluorine atoms by the destabilizing effect of mixing of their $2p\pi$ orbitals with the $\pi$ orbitals of carbon.

Another method of identification of $\pi$ orbital bands consists in comparison of the spectra recorded with different sources of ionization [176, 178]. Figure 5.1 compares the photoelectron [Ne(I), $h\nu = 16.85$ eV] and Penning (Ne*, $h\nu = 16.62$ eV) spectra of ethylene. The intensity of the first band assigned to ionization of the $\pi$ orbital increases significantly in the Penning spectrum. Similar intensification of the first band occurs also with excitation by the helium atoms He* ($h\nu = 19.82$ eV). It is explained by the specific mechanism of Penning ionization: as the molecules are bombarded by excited atoms, the $\pi$ orbitals, which are more sterically available, are preferably ionized.

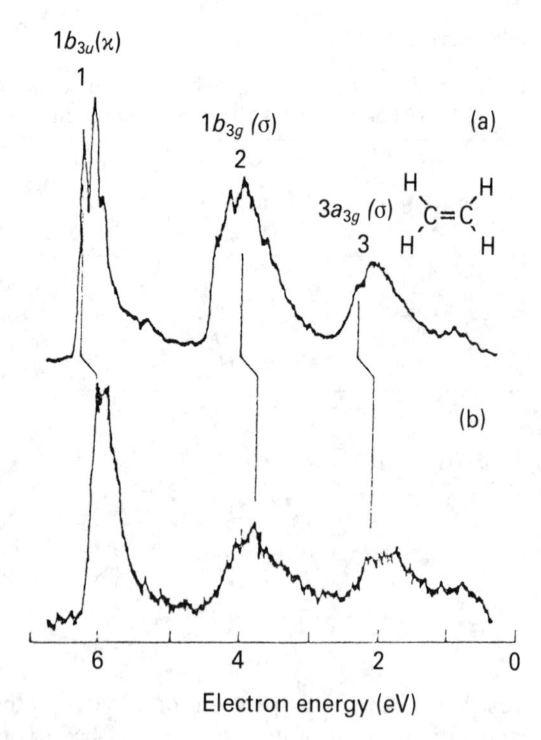


Fig. 5.1. Ne(I) photoelectron and Ne* Penning spectra of ethylene (energies of ionized electrons are plotted against the axis of abscissas; the scale shift is due to the different energies of excitation of Ne(I) (16.85 eV) and of Ne* ($3p_2$) (16.62 eV). (From Munakata *et al.* [178])

And finally, the results of *ab initio* calculations (see Table 5.1) also confirm assignment of the first band in the PE spectrum of ethylene to ionization of $\pi$ electrons. These calculations give objective estimation not only of the ethylene ionization potential; the steric structure of ethylene is also predicted adequately by these calculations [140, 146].

The occupied frontier MO of ethylene is higher than in alkanes, while the unoccupied frontier MO is, on the contrary, lower. According to ETS, electron affinity of the ethylene molecule is $-1.78$ eV [50]. Quasi-vibrational structure was observed in electron capture by $\pi^*$ orbital $1b_{3g}$ of ethylene. Firstly, it helped measurement of the adiabatic value of electron affinity ($-1.55 \pm 0.1$ eV) which satisfactorily agrees with the calculated one ($-1.69$ and $-1.81$ eV). Secondly, the comparison of the vibrational structure of the anion and neutral molecule made it possible to assign vibrations of the anion to symmetric C—C vibrations.

The frontier molecular $\pi$ orbitals account for the properties of ethylene: although the polarity of the C—H bonds in the ethylene molecule is higher (the charge on the carbon atom is $-0.154$ while on the hydrogen atom $+0.077$ electron) than in alkanes [154], it is characterized by orbital-controlled addition (combination) reactions rather than by substitution reactions. Thus, the presence of the higher occupied $\pi$ level, separated in the ethylene molecule from the $\sigma$ orbitals by almost 2.5 eV, accounts for its specific properties and high reactivity in electrophilic addition. Hydrogen halides, the halogens, oxygen and other electrophilic reagents are readily attached at the double bond:

$$+ \ E^+ \rightleftharpoons \ (A) \longrightarrow (B) \xrightarrow{Nu^-}$$

$\pi$ Complexes of ethylene have been calculated by various methods [140, 180]. The obtained results well agree with experimental findings and chemical intuition. The mode of attachment depends largely on the energy that is necessary to convert a symmetric onium ion A into a non-symmetric form B [5, 140, 180, 181].

The properties of ethylene also depend on the frontier unoccupied MO. Thus, the reactions of addition of nucleophiles to ethylene are also known. The importance of unoccupied MOs of ethylene molecule can be demonstrated, e.g. by its complexation with metal ions [182, 183]. Appropriate information is important for understanding the reactivity of olefins on the whole and management of practically important processes such as polymerization of olefins in the presence of Ziegler–Natta catalysts, or catalytic hydrogenation. Nevertheless, according to *ab initio* calculations, $\sigma$ donation prevails over $\pi$ donation in the $C_2H_4 \cdots Ag^+$ complex [184]. The study of ethylene behaviour on active centres of the crystal of titanium and nickel fluorides showed the prevalence of $\pi$ donation from metal to olefin: the electrons pass from the occupied d orbitals of the transition metal to $\pi^*$ or $\sigma$ *MOs of ethylene [185]. This effect activates the C—C bond of ethylene: localization of the electrons in the antibonding orbitals elongates and weakens this bond.

The lowest unoccupied $\pi$ electron level in the ethylene molecule manifests itself in the electron absorption spectrum: as distinct from alkanes, ethylene absorbs due to the $\pi \rightarrow \pi^*$ transition of electrons in the region of much longer wavelengths, i.e. at 185 nm [186]. Ethylene is a vivid example of changes in the geometry of an organic molecule exposed to UV radiation: as an electron passes from the HOMO to the LUMO, rotation of the methylene groups about the one-electron $\pi$ bond becomes possible.

### 5.1.2 Linear and cyclic alkenes

The electron structure of ethylene homologues illustrates the character of the interaction between saturated and unsaturated fragments in an organic molecule. In the orbital approach, we deal with the interaction of the $\sigma$ orbitals of the alkyl groups with the adjacent $\pi$ orbitals of double bonds. PES data show that the effects of $\sigma$–$\pi$ interaction can be quite significant.

From Table 5.2 it follows that as the alkyl group increases in size, the energies of all occupied electron levels of the alkene molecule increase too. The changes in the first ionization potentials are most illustrative: each methyl group attached at the double bond lowers the ionization potential of the $\pi$ orbital electrons by 0.6–0.7 eV. This effect is too great to be attributed to the inductive effect of the methyl groups alone [53].

**Table 5.2.** Ionization potentials and higher occupied MOs of some alkenes (PES-data and quantum chemical calculations)

| Alkene | $IP_i$ (eV) | | | | | | | |
|---|---|---|---|---|---|---|---|---|
| Propene ($C_s$) | 9.73 ($2a''$) | 12.2 ($10a'$) | 13.1 ($9a'$) | 14.4 ($1a''$) | 14.4 ($8a'$) | 15.9 ($7a'$) | | |
| cis-Butene ($C_{2v}$) | 9.12 ($2b_1$) | 11.7 ($6b_2$) | 12.7 ($7a_1$) | 13.5 ($5b_2$) | 14.1 ($1a_2$) | 14.1 ($6a_1$) | 14.5 ($1b_1$) | 16.1 ($5a_1$) |
| Cyclopropene ($C_{2v}$) | 9.70 ($2b_1$) | 11.0 ($3b_2$) | 12.7 ($6a_1$) | 15.1 ($1b_1$) | 16.7 ($5a_1$) | 18.3 ($2b_2$) | 19.6 ($4a_1$) | |
| Cyclobutene ($C_{2v}$) | 9.43 ($2b_1$) | 11.0 ($5b_2$) | 11.8 ($7a_1$) | 12.8 ($1a_2$) | 13.4 ($6a_1$) | 15.5 ($1b_1$) | 16.4 ($5a_1$) | |
| Cyclohexene ($C_2$) | 8.94 | 10.7 | 11.3 | 11.7 | 12.8 | 13.2 | 13.2 | |

(From Bieri *et al.* [187]; Masclet *et al.* [188])

The role of the interaction of the $\pi$(C=C) orbitals with the adjacent C—H and C—C bonds in alkenes becomes overt in the following series of unsaturated hydrocarbons [189]:

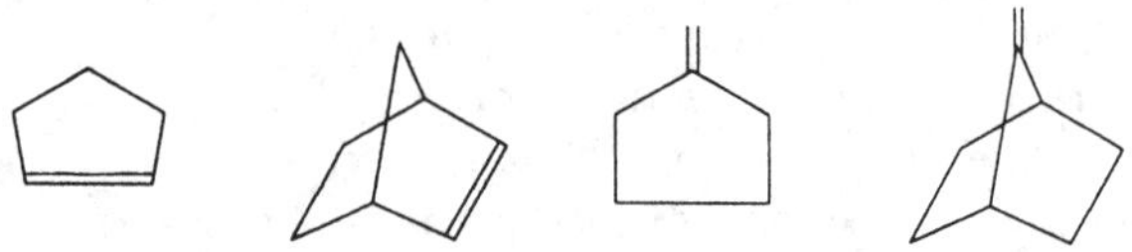

XXXII ($IP_1$, eV): a(9.20) b(8.95) c(9.15) d(9.40)

The rigidity of the norbornane framework rules out appreciable $\pi$ mixing between the adjacent $\sigma$ bonds and exocyclic $\pi$(C=C) orbital in XXXIId: this orbital is only subject to the inductive effect of the alkyl fragments.

At the same time, according to calculations in the 4-31GF basis, Kimura *et al.* assign HOMO of propene to a purely $\pi$(C=C) orbital since its interaction with the methyl group orbitals is insignificant (Table 5.3).

**Table 5.3.** Ionization potentials and higher occupied MOs of propene (PES-data and calculations in the 4-31G basis set)

| $IP_i$ (eV) | MO | | |
|---|---|---|---|
| | $-\varepsilon_i$ (eV) | No. and symmetry ($C_s$ group) | Localization |
| 10.03 | 9.59 | $2a''$ | $\pi_{C=C}$ |
| 12.31 | 13.18 | $10a'$ | $\sigma_{C-C}$ |
| 13.23 | 14.15 | $9a'$ | $\pi_{CH_3}$, $\sigma_{C=C}$ |
| 14.48 | 15.30 | $1a''$ | $\pi_{CH_3}$ |
| (14.80) | 15.73 | $8a'$ | $\sigma_{C=C}$ |
| 15.90 | 17.35 | $7a'$ | $\pi_{CH_2}$ |

(From Kimura *et al.* [38])

Other estimations of alkene properties on the basis of *ab initio* calculations should also be noted. Thus, rotational barriers of alkyl groups attached at the double bond were calculated. The most stable conformation of the methyl group adjacent to the double bond in the molecules of propene, isobutene, cis- and trans-butenes is the conformation in which the C—H bonds are skewed toward the double bond [24]. The stability of the skewed conformation of cis-2-butene was found to be by 44.38 kJ/mol greater than the eclipsed conformation (accounting for the standard values of the angles and lengths of bonds). The results of calculations depend largely

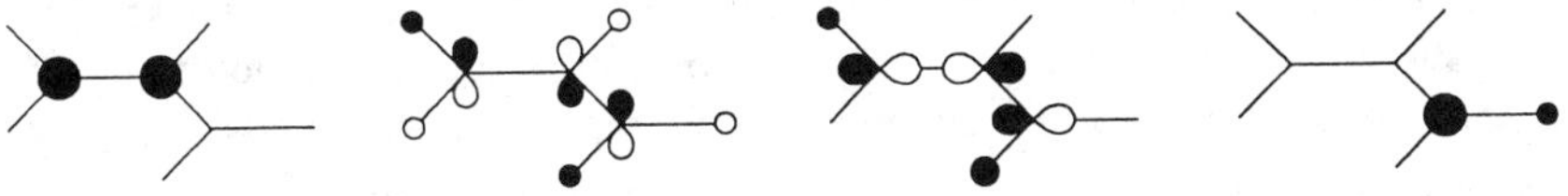

2a'', HOMO 10a' 9a' 1a''

on the geometry of the molecule. When calculating with optimization of the geometry, the eclipsed conformer is however 5.02 kJ/mol more stable, which agrees with experimental findings [190].

Heats of hydrogenation are compared with the parameters of the electron structure of alkenes. It is believed that the following data can be regarded as confirmation of the hyperconjugation effect [9, 187]:

| | $CH_2{=}CH_2$ | $CH_3CH{=}CH_2$ | $CH_3CH{=}CHCH_3$ |
|---|---|---|---|
| $\Delta H$ (kJ/mol) | 137.3 | 126.0 | 119.7 |
| $IP_1$ (eV) | 10.51 | 10.03 | 9.12 |

| | $(CH)_2C{=}CH_2$ | $(CH_3)_2C{=}C(CH_3)_2$ |
|---|---|---|
| $\Delta H$ (kJ/mol) | 118.9 | 111.4 |
| $IP_1$ (eV) | 9.24 | 8.27 |

It can be seen that the heat of hydrogenation decreases monotonously with increasing energy of the $\pi$ orbital.

The comparison of the electron structure with heats of hydrogenation of cycloalkenes is complicated by the strain effects. Small rings are characterized by high angular strain in their molecules and hence by extremely high heats of hydrogenation. Large rings are characterized by the effects of strain due to the eclipsed new C—H bonds in the hydrogenation product molecules and hence due to lower heats of hydrogenation [156].

Substituents at the double bond produce an appreciable effect on the anion states as well. Below are the first values of electron affinity (eV) for some alkenes: ethylene, − 1.78; propene, − 1.99; cis-butene, − 2.22; trans-butene, − 2.10; tetramethylethylene, − 2.24 [46, 191]. It can be seen that substitution of the methyl group for the hydrogen atom in the ethylene molecule increases the energy of LUMO by about 0.2 eV (cf.: the increase in the energy of HOMO is almost 0.7 eV). As the methyl group is introduced, a weak vibrational structure characteristic of the ethylene ET spectrum disappears, indicating the lower stability of the corresponding anion radical.

## 5.2 ALKYNES

Acetylene is the simplest hydrocarbon with a triple bond. It is a very important intermediate and its output in the USA is as great as 1 million tons a year. Acetylene does not occur in nature, but there are many natural compounds with triple bonds. Tridecapentainene $CH_3—(C{\equiv}C)_5—CH{=}CH_2$ has been, for example, isolated from sunflower oil. Acetylene is a biologically active substance with a weak narcotic action.

The steric structure of acetylene is simple: all its atoms form a straight line ($D_{\infty h}$ symmetry): $r_{CH} = 0.1061$ nm; $r_{CC} = 0.1203$ nm. The PE spectrum of acetylene is also simple: the first ionization band is separated from subsequent ones by 5 eV; it has fine vibrational structure and is assigned to ionization of the degenerate $\pi$ level [38, 42, 192]. Table 5.4 illustrates assignment of the other spectral bands.

**Table 5.4.** Acetylene higher occupied MOs (PES data and calculations in the 6-31G basis set)

| $IP_i$ (eV) | MOs | | |
|---|---|---|---|
| | $-\varepsilon_i$ (eV) | No. and symmetry | Localization |
| 11.40 | 11.02 | $1\pi_u$ | $\pi_{C=C}$ |
| 11.40 | 11.02 | | |
| 16.36 | 18.36 | $3\sigma_g$ | $\sigma_{CC}$ |
| 18.69 | 20.62 | $2\sigma_u$ | $\sigma_{2s}$ |

(From Kimura *et al.* [38])

Quantum chemical calculations of the acetylene electron structure accounting for the reactivity of its triple bond are noteworthy. It is known that the rate of bromine addition to olefins increases with electron density in the region of the double bond. It may appear at first sight that the concentration of electron density in alkynes in the region of the triple bond is even higher and can lead to their higher reactivity in electrophilic addition. But the triple bond is less active in the reactions of electrophiles than the double bond is. This can probably be explained by the fact that $\pi$ electron density is localized in the acetylene molecule in the orthogonal planes rather than in one plane, as is the case with the alkene molecule. Such localization does not create more favourable conditions for addition of an electrophile since, according to calculations [145], the electrophilic agents attack the acetylene molecule at an exactly right angle toward the molecule axis at the centre of the C—C bond in the plane of one of the orbitals:

H—C—C—H + $E^+$ ⇌ H—C—C—H ($E^+$) ⟶ (E)(H)C—C(+)—H

The lower energy of the occupied frontier electron level cannot increase the tendency to reactions with electrophiles either: $IP_1$ of acetylene is almost 1 eV higher than $IP_1$ of ethylene.

Calculations also confirm the complexity of the acetylene electron structure: the $\sigma$ electron charge on the acetylene molecule in the region of the C—C bond is equal to about 2, while the $\pi$ charge is only 1.2 electron per one $\pi$ bond. Thus, about 40% of the $\pi$ charge of the triple bond is outside the region confined between two carbon atoms.

The results of calculations explain the somewhat shorter C—H bond in acetylene compared with that in ethane: 0.1060 against 0.1093 nm. Electron saturation of the triple bond correlates with the higher acidity ($pK_a$) of acetylene [4, 193]: methane, 50, ethylene, 44, and acetylene, 25.

The high polarity of the C—H bonds in the acetylene molecule is quite important for its reactivity [154]: the charge on the carbon atom is $-0.182$, and on the hydrogen atom $+0.182$ electron.

According to ET spectroscopy, the first value of electron affinity of acetylene is $-2.6$ eV [46]. The occupied and vacant electron levels of acetylene are thus symmetrically shifted in opposite directions relative to the ethylene levels: HOMO of acetylene is 0.9 eV lower while its LUMO is higher by 0.82 eV compared with similar orbitals of ethylene.

While estimating the electron structure of acetylene homologues one can note that substitution of the methyl group for the hydrogen in the acetylene molecule appreciably lowers the potential of degenerate highest occupied $\pi$ orbitals: the first ionization potential of propyne is 10.37 eV. Although, according to *ab initio* calculations [38], the methyl group orbitals do not contribute to the highest electron level of propyne, Bock gives a different estimation using the simple method of calculation: the integral describing the interaction between $\sigma$(C—H) and $\pi$(C≡C) orbitals is 2.0 eV [53].

The influence of steric structure on the degree of splitting of the degenerate $\pi$ orbital level of the triple bond was studied in cycloalkynes. Fig. 5.2 shows PE spectra of cyclooctyne (XXXIII) and 3,3,7,7-tetramethylcycloheptyne (XXXIV). Both

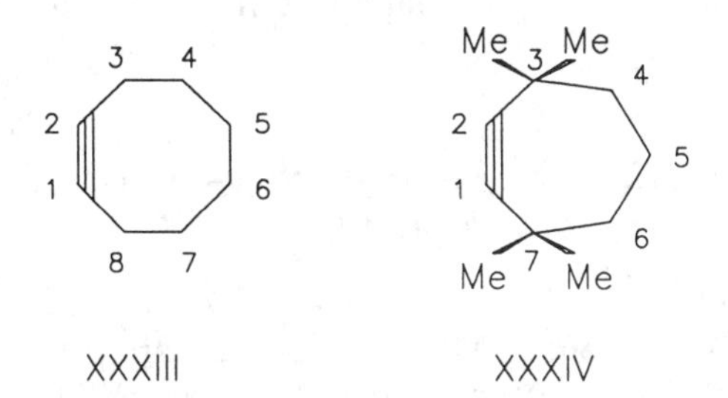


hydrocarbons have a considerable strain in their rings. One could expect that the 21.5° deviation of the simple C—C bonds 2–3 and 8–1 from the axis of the triple bond would remove degeneracy of the $\pi$ orbitals arranged in mutually perpendicular planes. But there is no such evidence in the spectrum of cyclooctyne.

Only much greater deviation of the bonds associated with the addition of four methyl groups at positions 3 and 7 of cycloheptyne can, as suggested by the shape of the first ionization band in the PE spectrum, lead to a distinct removal of degeneracy of $\pi$ orbitals [54].

Inclusion of the triple bond in the ring appreciably decreases the energy of the

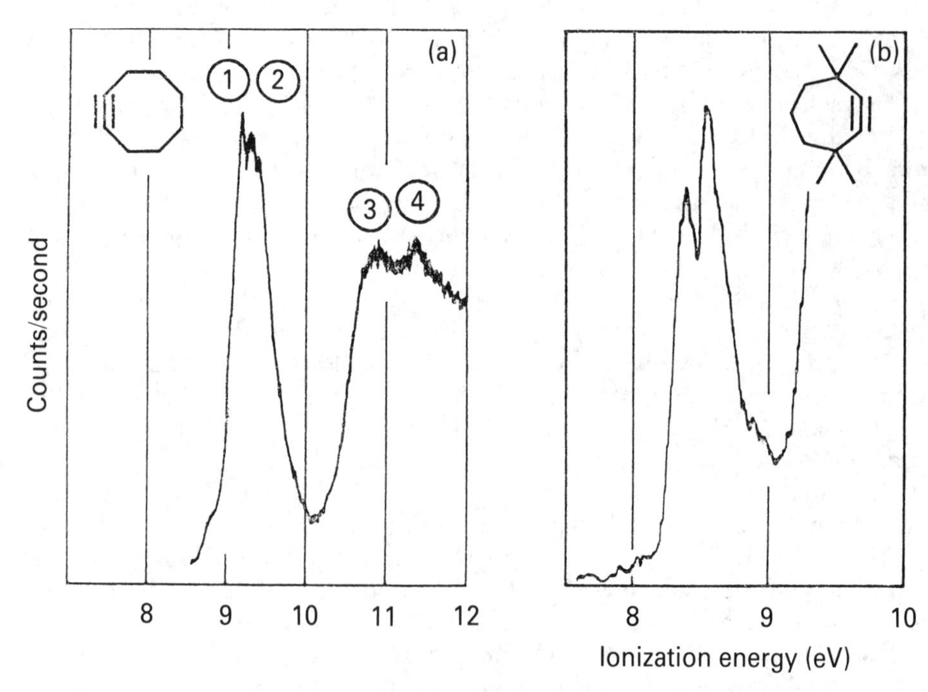


Fig. 5.2. He(I) photoelectron spectra of cyclooctyne XXXIII and 3,3,7,7-tetramethylcycloheptyne XXXIV. (From Heilbronner and Maier [54])

lower unoccupied $\pi^*$ orbital and hence increases the electron affinity [194]:

| | But-2-yne | Di(tert-butyl)-acetylene | Cyclooctyne |
|---|---|---|---|
| $EA_1$ (eV) | −3.43 | −3.10 | −2.18 |

## 5.3 ELECTRON STRUCTURE OF POLYENES AND WOODWARD–HOFFMANN RULES

The properties of alkanes, alkenes, and alkynes demonstrate the importance of symmetry ($\sigma$ or $\pi$) of the frontier MOs: if $\pi$ orbitals are present in a hydrocarbon molecule, they are the first to be attacked by both electrophiles and nucleophiles.

The properties of alkadienes and polyenes illustrate the importance of the signs of frontier $\pi$ orbitals on different atoms of the molecule. These signs account, for example, for the results of the electrocyclic reactions of alkadienes and polyenes that are common in organic synthesis and especially in the synthesis of natural compounds [195].

An **electrocyclic reaction** is one in which ions or radicals are not formed and the electrons move from the HOMO of one molecule to the LUMO of the other in the cyclic transition state. Electrocyclic reactions include, for example, reactions in which the $\sigma$ bond between the terminal atoms of a linear system is formed at the expense of $\pi$ orbitals localized on these atoms and the inverse transformations.

Since formation of $\sigma$ bonds in the a → b and c → d reactions is possible only if the interacting p orbitals overlap by their lobes with the same sign, the a ⇌ b transformations can occur only on the condition that the terminal orbitals rotate in opposite directions, i.e. undergoing contrarotatory motion, while the c ⇌ d transformations are possible with the orbitals rotating in the same direction, i.e. undergoing conrotatory motion. Depending on this condition, the cyclization products have different configurations:

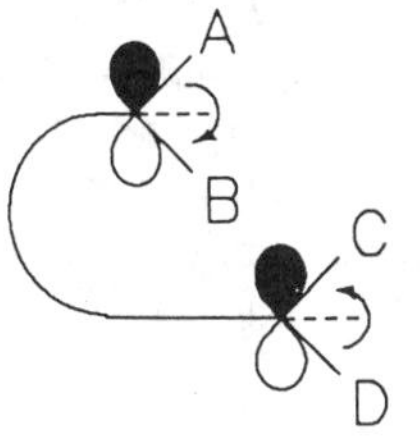

A 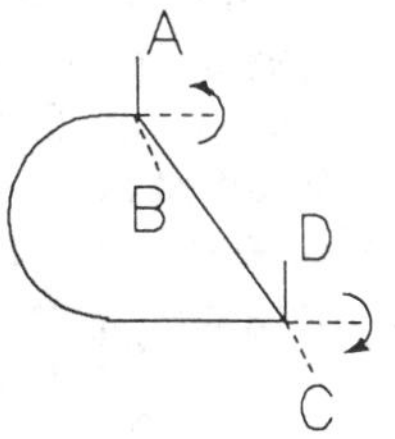 B C D a A B D C b

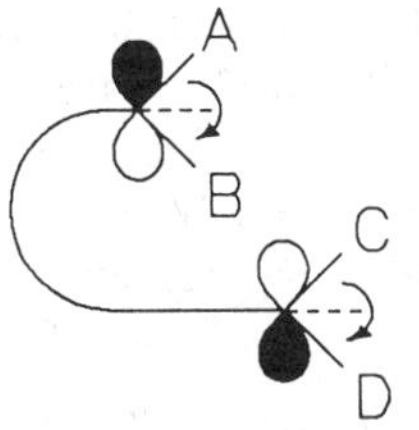

A B C D c A B C D d

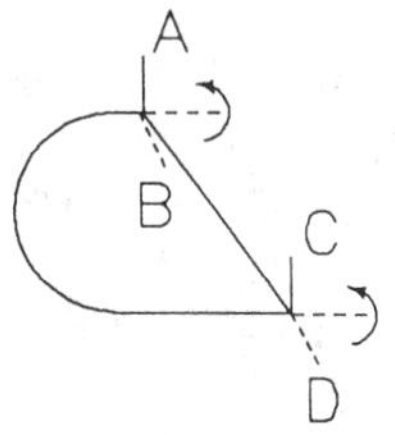

This is confirmed by experiment: both photochemically and thermally induced transformations are stereospecific. The results of electrocyclic reactions are explained by the Woodward–Hoffmann rules [76–8], although the former formulations were later corrected by more perfect calculations [82, 196, 197]. According to the Woodward–Hoffmann rules, the structure of the product formed in thermal processes in these systems is determined by the HOMO symmetry, and in photochemical reactions by the symmetry of the LUMO:

(1) linear systems containing $4n\pi$ electrons can in thermal reactions undergo cyclization by the conrotatory mechanism only, while in photochemical reactions, by the contrarotatory mechanism;
(2) linear systems containing $(4n + 2)\pi$ electrons can undergo cyclization in thermal reaction by the contrarotatory mechanism only and in photochemical reactions by the conrotatory mechanism.

The Woodward–Hoffmann rules were first formulated as empirical ones [76]. In the study of the electron structure of molecules by experimental methods of quantum chemistry these rules appear in the form of eigenvalues and eigencoefficients of frontier molecular orbitals of the appropriate unsaturated compounds.

### 5.3.1 $4n\pi$ electron systems: buta-1,3-diene

Consider, for example, the electron structure of buta-1,3-diene, the hydrocarbon with 4 $\pi$ electrons. At room temperature, it exists in the transoid form (95–7%), in which the ordinary and double bonds are 0.1465 and 0.1245 nm respectively [198].

Various sources of ionization were used and various model compounds were studied for assignment of the bands of butadiene PE spectrum (Table 5.5) [38, 54, 176, 199]. Ionization spectra of butadiene, 1,1,4,4-tetrafluorobutadiene and perfluorobutadiene are compared in Fig. 5.3. It can be seen that the first two bands in the butadiene spectrum change their positions to the minimum extent on the introduction of four or six fluorine atoms into its molecule; these two bands are assigned to ionization of $\pi$ orbitals.

The relationship between the intensities of ionization bands with He(I) and He(II) sources of ionization also agrees with the assignment of the first two bands to the $\pi$ orbitals [176]:

| Band No. (MO localization | $1(\pi^{-}_{C=C})$ | $2(\pi^{+}_{C=C})$ | $3(\sigma_{C-C})$ | $4.5(\pi^{\mp}_{CH_2})$ | $6.7(\sigma^{\mp}_{C=C})$ |
|---|---|---|---|---|---|
| He(I)/He(II) | 0.58 | 0.54 | 1.0 | 1.2 | 1.2 |

As He(II) source is used, the intensities of 'genuine' $\pi$ orbital bands only increase; the ionization of the electrons of pseudo-$\pi$ orbitals of the $CH_2$ groups, as well as those of $\sigma$ orbitals, gives bands of equal intensity in both spectra. Finally, in the Penning spectrum, these are only the first two bands whose intensities increase according to the requirement of $\pi$ electron ionization [199].

Unoccupied $\pi$ levels of butadiene were also studied experimentally: two resonances were found in its ET spectrum: $EA_1 = -0.62$ eV and $EA_2 = -2.80$ eV [46]. The electron levels of ethylene and buta-1,3-diene molecules are compared in Fig. 5.4.

The symmetry of the occupied frontier MO of buta-1,3-diene unambiguously indicates that the bonding interaction of the terminal carbon atoms is possible only with conrotatory motion. Only this mechanism can underlie the thermal cyclization of butadiene, which is confirmed by the exclusive formation of conformer a:

Conversely, the symmetry of the unoccupied frontier MO of butadiene-1,3 indicates that when excited photochemically, butadiene undergoes cyclization only by the contrarotatory mechanism; conformer b is thus formed.

**Table 5.5.** Ionization potentials and higher occupied MOs butadiene-1,3 (HeI PES data and calculations in the 6-31G basis set)

| $IP_i$ (eV) | MO | | | $-\varepsilon_i$ (eV) (6-31G, CI) |
|---|---|---|---|---|
| | $-\varepsilon_i$ (eV) | No. and symmetry ($C_{2h}$ group) | Localization | |
| 9.09 | 8.63 | $1b_g$ | $\pi_{C=C^-}$ | 8.72 |
| 11.55 | 11.99 | $1a_u$ | $\pi_{C=C^+}$ | 11.54 |
| 12.35 | 13.08 | $7a_g$ | $\sigma_{C-C}$ | 12.24 |
| 13.70 | 14.54 | $6b_u$ | $\pi_{CH_2^+}$ | 13.84 |
| (14.0) | 15.09 | $6a_g$ | $\pi_{CH_2^-}$ | 14.22 |

(From Kimura *et al.* [38])

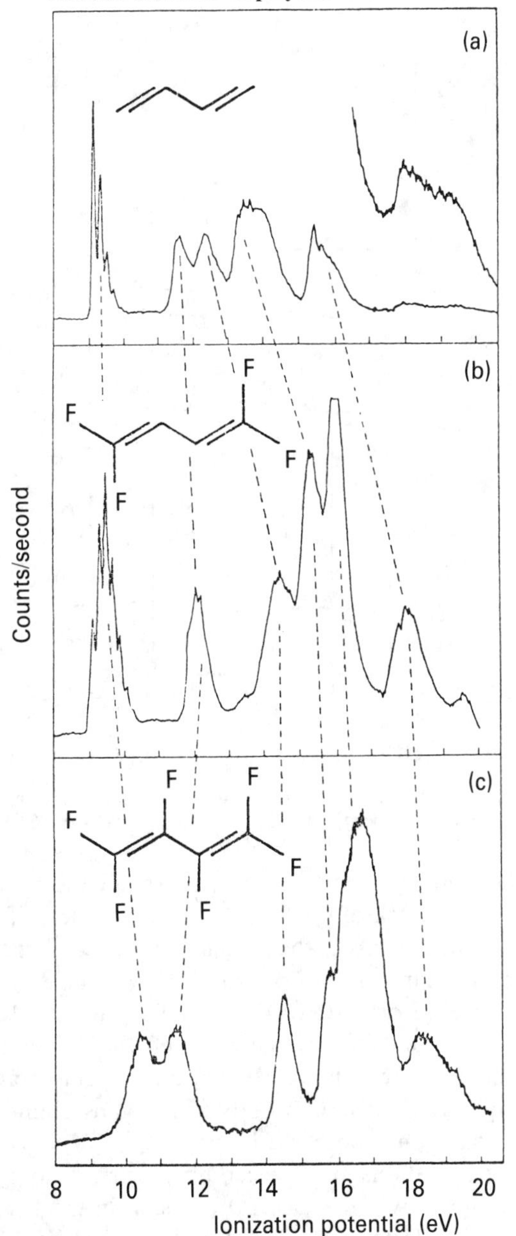


Fig. 5.3. He(I) photoelectron spectra of (a) butadiene, (b) 1,1,4,4-tetrafluorobutadiene and (c) perfluorobutadiene. (From Heilbronner and Maier [54]).

The reactions with opening of cyclobutene rings are highly stereospecific as well [200]. Pyrolysis of XXXVa at 320–340°C with intermediate formation of cyclobutene gives 4-deutero-1-acetoxybutadienes XXXVIa and b. The structures of the isomers formed agree only with conrotatory opening of the cyclobutene ring:

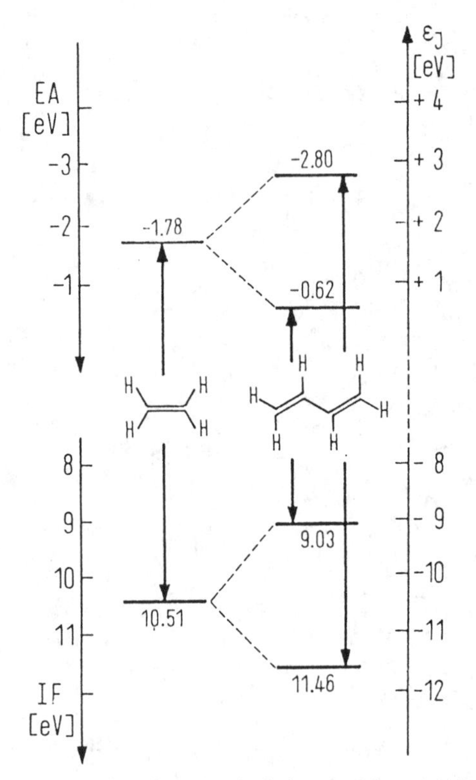


Fig. 5.4. Experimental energies of $\pi$MO of ethylene and buta-1,3-diene (occupied MOs according to PES and unoccupied MOs according to ETS). (From Bock [117])

The shapes of the frontier orbitals of dienes explain the results of Diels–Alder reactions within the framework of the frontier orbitals concept [201]. According to the symmetry of the frontier MOs of the reagents in the reaction between butadiene-1,3 and ethylene, both charge transfers (from HOMO of ethylene to LUMO of butadiene, and conversely from HOMO of butadiene to LUMO of ethylene) will lower the order of the $\pi$ bond in ethylene and shift the double bond in butadiene. Energy diagrams for the changes in the MOs of the reagents were calculated taking account of the changed bond distance: both energy gaps diminish substantially as the the reagents approach one another (Fig. 5.5).

Extended Hückel calculations show a certain preference for the charge transfer from the HOMO of diene over the LUMO of olefin in the transition state of the Diels–Alder reaction [77]. The importance of the butadiene HOMO interaction with the LUMO of ethylene is suggested by the abrupt increase in the cycloaddition rate

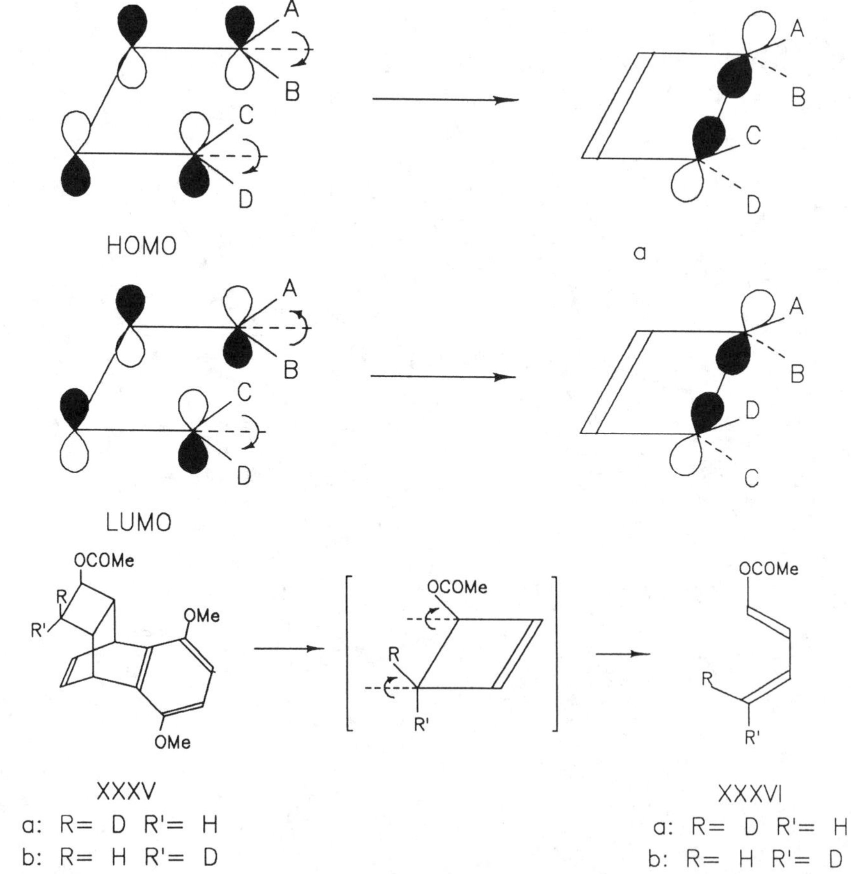


with decreasing energy of the alkene LUMO. Below are half-wave potentials of polarographic reduction of some cyanoethylenes and also rate constants of their addition to cyclopentadiene ($k_a$; $IP_1$ of this diene is 8.58 eV) and to 9,10-dimethylanthracene ($k_b$; $IP_1$ of this diene is 7.02 eV) [202, 203]:

| $E_{1/2}$ (*V*) | $-1.84$ | $-1.28$ | $-1.29$ | $-0.15$ |
|---|---|---|---|---|
| $k_a \times 10^6$ | 0.104 | 8.1 | 9.1 | $4.3 \times 10^6$ |
| $k_b \times 10^6$ | 0.89 | $1.39 \times 10^2$ | $1.31 \times 10^2$ | $1.3 \times 10^6$ |

Studies on the Diels–Alder reaction mechanism continue. Although selectivity is usually regarded as an argument for the synchronous process, other mechanisms can be considered as well. According to Dewar *et al.* [197], stereoselectivity can be explained within the framework of the two-steps mechanism if the biradical or ionic type intermediate product is converted into the end product at a faster rate than it is isomerized at the expense of the inner rotation. New hypotheses concerning the mechanism of cycloaddition of alkenes to dienes do not argue the objective parameters of the electron structure of individual molecules, determined by quantum chemical calculations and electron spectroscopy, or the importance of these parameters in separate stages of the Diels–Alder reactions.

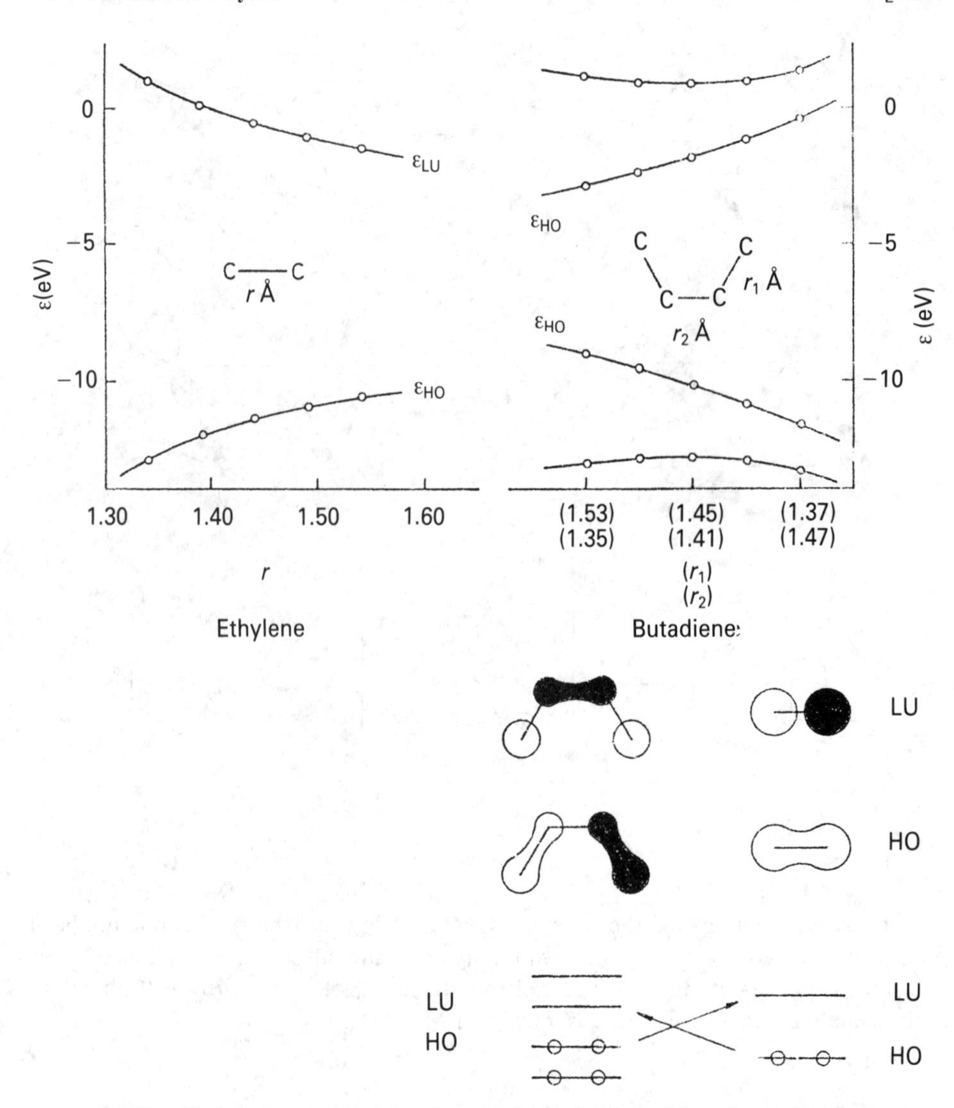


Fig. 5.5. Changes in energies of frontier orbitals of ethylene and butadiene molecules as they approach one another in the Diels–Alder reaction. (From Fukui and Fujimoto [201])

To conclude the discussion of the electron structure of buta-1,3-diene, it is reasonable to estimate the effectiveness of the interaction of two ethylene orbitals in its molecule. A reliable quantitative estimation of this interaction can be derived from the analysis of the PE spectrum of this hydrocarbon since the ionization of two $\pi$ levels is reliably separated in it from the region of ionization of $\sigma$ orbitals [53, 54].

By assigning the first ionization potential (9.09 eV) to the (−) combination of the $\pi$ orbital of ethylene fragments:

$$\pi^- = 1/\sqrt{2}(\pi_a - \pi_b); \qquad \varepsilon_\pi^- = -IP_1 = -9.09 \text{ eV}$$

**Table 5.6.** Ionization potentials and higher occupied MOs of cis- and trans-hexa-1,3,5-trienes (PES data and SPINDO calculations)

| cis isomer | | | trans isomer | | |
|---|---|---|---|---|---|
| $IP_i$ (eV) | MO | | $IP_i$ (eV) | MO | |
| | $-\varepsilon_i$ (eV) | No. and symmetry ($C_{2v}$ group) | | $-\varepsilon_i$ (eV) | No. and symmetry ($C_{2h}$ group) |
| 8.32 | 8.88 | $2b_1(\pi)$ | 8.29 | 8.89 | $2a_u(\pi)$ |
| 10.27 | 10.51 | $1a_2(\pi)$ | 10.26 | 10.55 | $1b_g(\pi)$ |
| 11.5 | 11.43 | $9b_2$ | 11.6 | 11.86 | $10a_g$ |
| 11.9 | 11.83 | $1b_1(\pi)$ | 11.9 | 11.81 | $1a_u(\pi)$ |
| 12.6 | 12.61 | $10a_1$ | 12.6 | 12.23 | $9b_u$ |

(From Beez *et al.* [204])

and the second potentials (11.55 eV) to the (+) combination of the $\pi$ orbitals:

$$\pi^+ = 1\sqrt{2}(\pi_a + \pi_b); \qquad \varepsilon_\pi^+ = -IP_2 = -11.55 \text{ eV},$$

it is easy to find the corresponding diagonal matrix elements $H_{aa}$ and $H_{bb}$ from the equation:

$$H_{aa} = H_{bb} = -\frac{I_2 - I_1}{2} = -\frac{11.55 - 9.09}{2} = -10.32 \text{ eV}$$

The non-diagonal element $H_{ab}$, estimating the effect of interaction of two ethylene $\pi$ orbitals in the butadiene molecule, is determined by solving the secular determinant relative to this element:

$$\begin{vmatrix} (H_{aa} - 9.09) & H_{ab} \\ H_{ba} & (H_{bb} - 9.09) \end{vmatrix} = 0$$

$$H_{ab} = H_{ba} = H_{\pi\pi} = \sqrt{[(H_{aa} - 9.09)(H_{bb} - 9.09)]} = -1.23 \text{ eV}$$

The higher value of the $H_{aa}$ and $H_{bb}$ integrals (−10.32 eV) as compared to the value predicted from the first ionization potential of ethylene (−10.51 eV) is due to the inductive effect of the adjacent unsaturated fragment.

### 5.3.2 (4*n* + 2)$\pi$ electron systems: hexa-1,3,5-trienes

The Woodward–Hoffmann rules also determine the behaviour of linear systems containing $(4n + 2)\pi$ electrons. Consider the parameters of the orbital structure of hexa-1,3,5-trienes and its cyclization reactions.

The molecule of this triene has $\pi$MOs formed by $p_\pi$ orbitals of the six carbon atoms. According to the PE spectrum of hexatriene, the degree of mixing of atomic $p_\pi$ orbitals does not decrease with their increasing number in the molecule. Assignment of the ionization potentials of cis- and trans-hexa-1,3,5-trienes is given in Table 5.6. Excellent agreement between the calculated and experimental energies of the occupied MOs suggests high reliability of MO configurations available from the quantum chemical calculations. It can also be seen that the occupied frontier MOs in both isomers differ by almost 2 eV from their closest occupied $\pi$ MOs.

The configurations of the frontier MOs of the cis-isomer (substituted hexatrienes) are shown below. These configurations show that thermal cyclization can occur only as a contrarotatory process with formation of the isomer a, while photochemical reaction occurs in the conrotatory manner with formation of the isomer b:

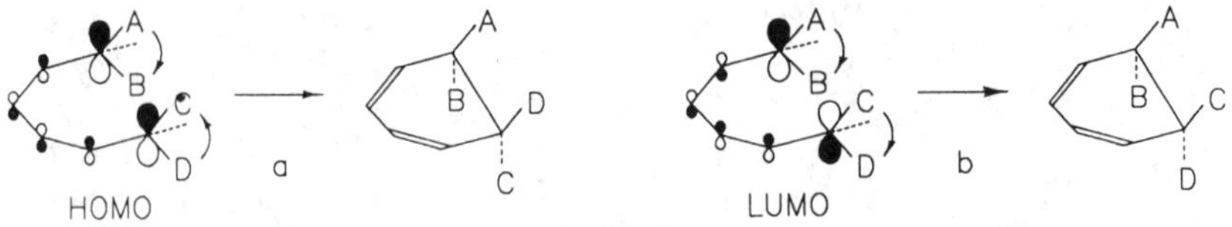

### 5.3.3 Cyclohexa-1,3-diene and cyclohexa-1,4-diene

Inclusion of double bonds at position 1, 3 in the ring does not decrease the degree of mixing of ethylene fragment $\pi$ orbitals. On the whole, the sequence of MOs found for buta-1,3-diene holds also for cyclohexa-1,3-diene: $\pi^-$, $\pi^+$. This is confirmed by the experimentally found ionization potentials of cyclohexa-1,3-diene and their assignment (Table 5.7). It can be seen that the degree of splitting of the occupied $\pi$ orbital levels in the molecule of this hydrocarbon is not lower than in the butadiene molecule, while the ionization bands are shifted toward the region of lower potentials, probably due to the electron-donating inductive effect of two saturated carbon atoms:

The formation of $\pi$ orbitals of a cyclohexa-1,4-diene molecule is not so conspicuous. Dienes with isolated double bonds do not practically differ from mono-olefins by their reactivity: electrophilic addition with such alkadienes occurs independently at each double bond. One might expect that double bonds in cyclohexa-1,4-diene are also isolated and that their ionization potentials do not differ from that of cyclohexene.

This suggestion has not been confirmed: the PE spectrum indicates unambiguously that the ethylene fragment $\pi$ orbitals interact in the cyclohexa-1,4-diene molecule:

**Table 5.7.** Ionization potentials and higher occupied MOs of cyclohexa-1,3-dienes and -1,4 (PES data and calculations in 4-31G basis set)

| Cyclohexa-1,3-diene | | | | Cyclohexa-1,4-diene | | | |
|---|---|---|---|---|---|---|---|
| | | MO | | | | MO | |
| $IP_i$ (eV) | $-\varepsilon_i$ (eV) | Number | Localization | $IP_i$ (eV) | $-\varepsilon_i$ (eV) | Number | Localization |
| 8.25 | 8.00 | 11*a* | $\pi_{C=C^-}$ | 8.82 | 8.68 | 11*b* | $\pi_{C=C+}$ |
| 10.7 | 11.24 | 11*b* | $\pi_{C=C^+}$ | 9.88 | 9.71 | 10*b* | $\pi_{C=C^-}$ |
| 11.3 | 12.09 | 10*a* | $\sigma_{CC}$ | 11.0 | 12.02 | 9*b* | $\sigma_{CC^+}$, $\sigma_{CH}$ |
| 11.8 | 12.80 | 10*b* | $\sigma_{CC}$ | 12.0 | 13.42 | 11*a* | $\sigma_{CC^+}$ |
| 12.7 | 13.74 | 9*a* | $\pi_{CH_2}$ | 13.3 | 14.54 | 8*b* | $\sigma_{CC^-}$ |
| 13.5 | 14.88 | 9*b* | $\sigma_{CC}$ | 13.7 | 14.85 | 10*a* | $\pi_{CH_2}$ |
| 14.0 | 15.55 | 8*b* | $\sigma_{CH^-}$ | 13.7 | 15.26 | 9*a* | $\sigma_{CH}$ |
| 14.0 | 15.70 | 8*a* | $\sigma_{CC}$ | 14.7 | 15.87 | 7*b* | $\pi_{CH_2^+}$ |

(From Kimura *et al.* [38])

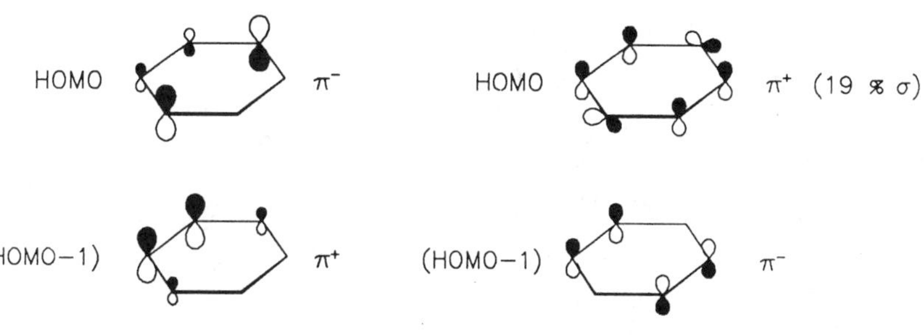


the first and the second bands assigned to $\pi$ orbitals differ by more than 1 eV (see Table 5.7).

Noteworthy is the altered sequence of $\pi$ orbitals: the sequence established formerly for buta-1,3-diene and cyclohexa-1,3-diene on the one hand, and the sequence discovered in cyclohexa-1,4-diene, on the other hand, are different. The shapes of the occupied $\pi$ orbitals in both cyclohexadienes were shown above: the higher occupied $\pi$ orbital of cyclohexa-1,4-diene is the $\pi^+$(C=C) orbital (the bonding combination of ethylene $\pi$ orbitals) as distinct from the antibonding combination of ethylene $\pi$ orbitals in cyclohexa-1,3-diene. The formation of $\pi$ levels in cyclohexadiene molecules is given in Fig. 5.6. The reversal of $\pi$ orbitals in cyclohexa-1,4-diene is explained by the fact that $\sigma$ orbitals of methylene groups are also involved in the interaction of ethylene $\pi$ orbitals: the so-called 'through-bonds' effect is thus realized.

It can be seen from the scheme that in the absence of the through-bonds effect, one might expect splitting of the $\pi$ orbitals in the left part of the diagram owing to their through-space interaction ('through-space' effect). This effect might preserve the normal sequence of the $\pi$ levels: $\pi^-$, $\pi^+$. But due to the through-bonds effect, according to the symmetry demands, pseudo-$\pi$ orbitals of methylene fragments can only interact with the (+) combination of $\pi$ orbitals. This interaction raises the (+) combination to the highest level.

Unoccupied $\pi$ orbitals in cyclic dienes are formed in the same way, i.e. by the interaction of ethylene orbitals. It can be seen that hydrocarbons with conjugated double bonds, and also dienes with isolated double bonds, have two resonance signals in their ET spectra. Thus, two resonance signals are seen in the spectrum of cyclohexa-1,4-diene. This fact indicates unambiguously that, in addition to the occupied $\pi$ orbitals, unoccupied ethylene $\pi^*$ orbitals also interact in this hydrocarbon. It is noteworthy that the normal sequence of $\pi^*$ orbitals can be suggested for the anion states of cyclohexa-1,4-diene: $(-)\pi^*$ and $(+)\pi^*$ (in the order of their lowering energy).

The scheme of formation of unoccupied $\pi^*$ levels of cyclohexa-1,4-diene is also given in Fig. 5.6. One can see that its $\pi^*$ levels are destabilized by mixing with the $\sigma$ orbitals to a smaller degree than the occupied $\pi$ levels. It has already been mentioned that mixing with the methyl group orbitals destabilizes the cation-radical state of ethylene by 0.78 eV, and the anion-radical state only by 0.21 eV.

### 5.3.4 Norbornadiene and related compounds

Since the symmetry of frontier $\pi$ orbitals is very important for estimation of the direction and the result of the reaction, it is useful to consider other examples of

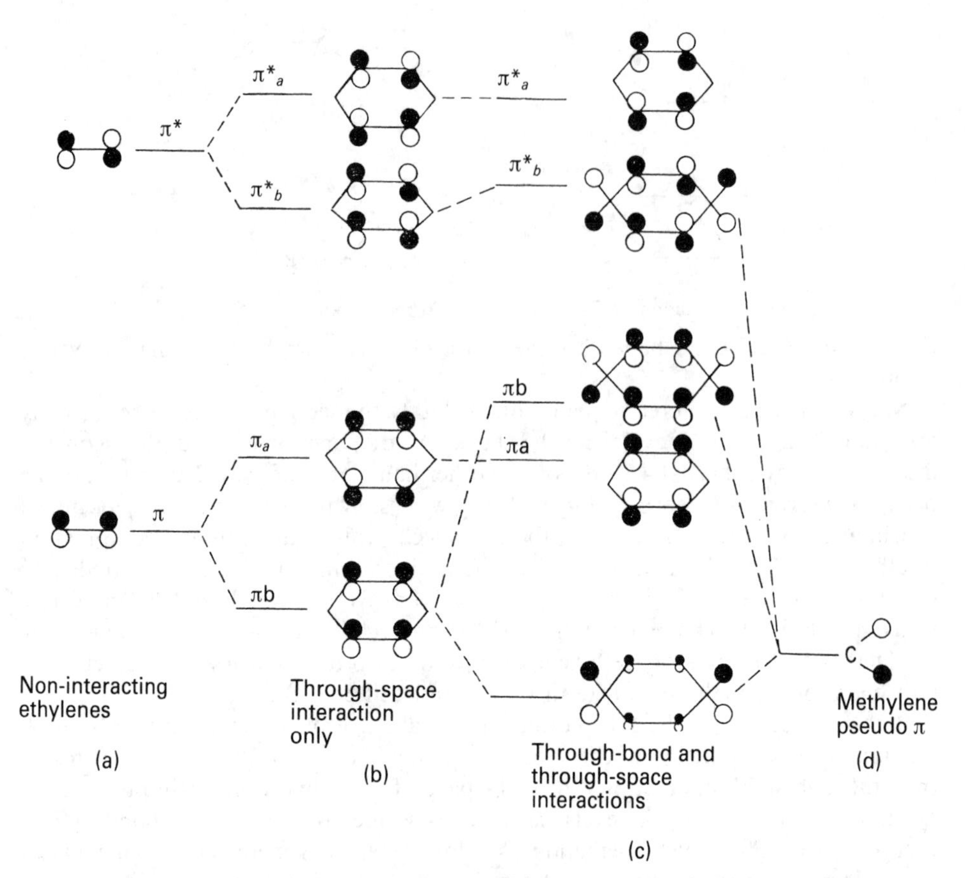


Fig. 5.6. Interaction of orbitals of a separate fragment in a cyclohexa-1,4-diene molecule: (a) ethylene molecule before mixing; (b) effect of $\pi$ orbital mixing through space; (c) effect of $\pi$ orbital mixing through space and bonds; (d) orbitals of methylene groups before mixing. (From Jordan and Burrow [46])

establishment of $\pi$ orbital sequences. In addition to various quantum chemical calculations, the comparison of spectral characteristics of compounds with similar structure can also give objective information.

The molecular conformation is altered substantially in the transition from cyclohexa-1,4-diene to norbornadiene. Diminution of the dihedral angle in norbornadiene to 110° decreases (as distinct from cyclohexadiene) the importance of the interaction of both occupied and vacant ethylene $\pi$ orbitals through the methylene fragments in the norbornadiene molecule: the $\pi$ and $\pi^*$ orbitals of the norbornadiene molecule have normal sequences [53, 54].

The sequence of occupied $\pi$ orbitals in norbornadiene was established by comparison of PE spectra and estimation of orbital interactions within the framework of the HMO calculations of the following bicyclic hydrocarbons (XXXVII): norbornene (a), norbornadiene (b), isopropylidene norbornadiene (c), and isopropylidene norbornane (d).

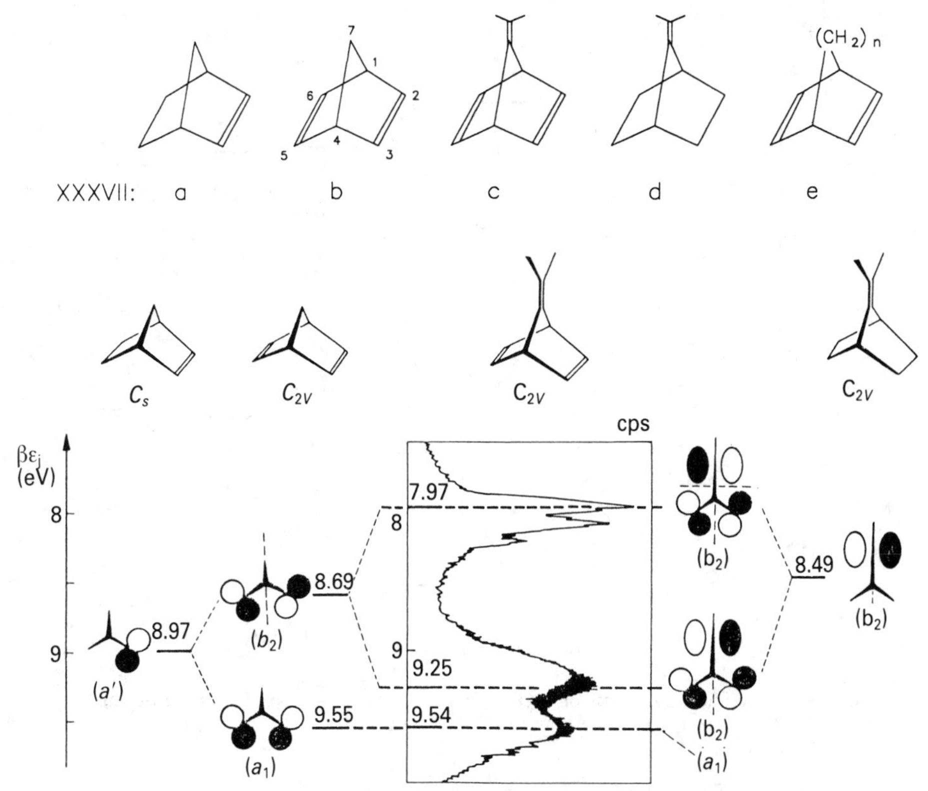


Fig. 5.7. Orbital interactions in the molecules of norbonadiene and isopropylidene norbornadiene XXXVIIc. He(I)-PE spectrum of compound XXXIIe is given in the centre. (From Bock [117])

In accordance with the diagram of $\pi$ orbital mixing in these compounds (Fig. 5.7), the type of symmetry of the highest occupied molecular $\pi$ orbital of norbornadiene was determined by testing its ability to interact with the isopropylidene $\pi$ orbital. This orbital is antisymmetric relative to the plane passing through atoms 1, 4, and 7, and can only interact with the $(-)$ combination of ethylene $\pi$ orbitals of the norbornadiene fragment in the molecule of compound XXXVIIc. It can be seen from the PE spectrum that this interaction takes place, and that it is characterized by a sufficiently high absolute value of the resonance integral $H_{\pi\pi}$.

The interaction with the $\sigma$(C—H) orbitals is insignificant and it does not alter the sequence of $\pi$ or $\pi^*$ orbitals in the norbornadiene molecule. Comparison of its ET spectrum with that of cyclohexa-1,4-diene shows that the ground anion state of norbornadiene is lower (Fig. 5.8), in accordance with the $(+)\pi^*$ symmetry of the LUMO. The second resonance signals in the ET spectra of both compounds appear at about the same energies of the bombarding electrons and correspond to the $(-)$ combinations of the $\pi^*$ orbitals of ethylene.

Noteworthy regularities of the electron structure were established during studies on some bicycloalkadienes with the general formula XXXVIIe: norbornadiene, $n = 1$;

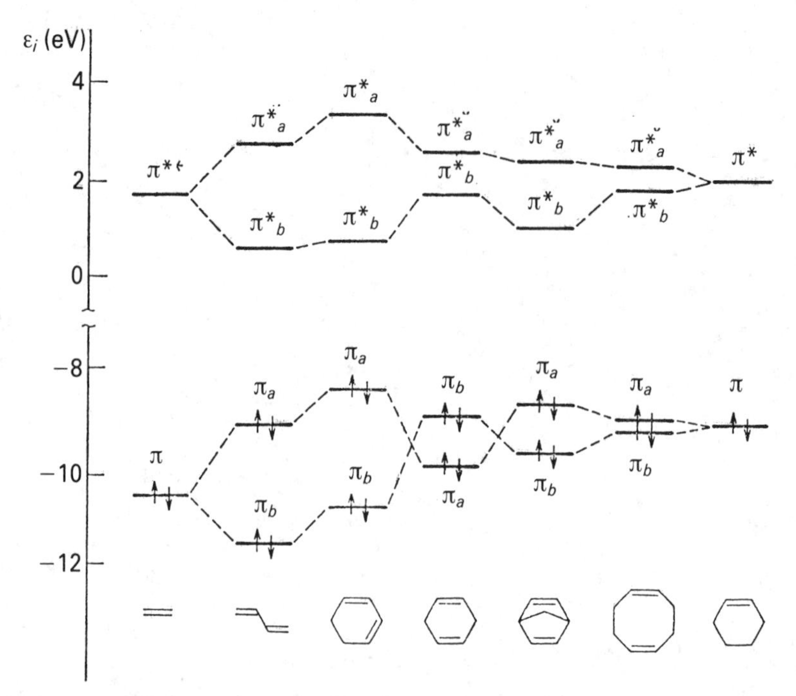


Fig. 5.8. Experimental energies of MOs of some alkadienes (occupied MOs according to PES, vacant MOs according to ETS). (From Jordan and Burrow [46])

bicyclo[2,2,2]octadiene, n = 2; bicyclo[3,2,2]nona-6,8-diene, n = 3; and bicyclo [4,2,2]deca-7,9-diene, n = 4.

With preservation of constant energy of the $\pi$ orbitals of ethylene fragments, the value of the dihedral angle $\theta$ (the angle between the plane of 1, 2, 3, 4 atoms and the plane of 4, 5, 6, 1 atoms) increases regularly with the value of $n$ in these compounds. This, in turn, systematically changes the conditions of interaction of the ethylene $\pi$ orbitals, the through-bonds and through-space effects included.

The sequence of the $\pi$ orbitals in bicyclo[2,2,2]octadiene was determined experimentally as well. To that end, the contours of the $\pi$ ionization bands in its PE spectrum were compared with the spectrum of barrelene in which the 'normal' sequence of $\pi$ orbitals was proved by the Jahn–Teller splitting of the corresponding photoionization bands.

One could reasonably suggest that the integral of ethylene $\pi$ orbital overlapping should increase in the series of bicycloalkadienes XXXVII with the angle $\theta$. Splitting of the $\pi$ orbitals in nona- and decadienes XXXVIIe ($n = 3$ and $n = 4$) should decrease accordingly. It can be concluded for both compounds as regards their $\pi$ orbitals: the (+) combinations of $\pi$ orbitals of ethylene fragments become their HOMOs, as in cyclohexa-1,4-diene. Interestingly, attaining its minimum in nonadiene ($n = 3$), the splitting degree of $\pi$ levels in decadiene ($n = 4$) increases again to about 1 eV at $n = \infty$ (this value was obtained by extrapolation, taking into account the splitting of the $\pi$ levels in cyclohexa-1,4-diene [54]). The ionization potentials of $\pi$ orbitals of compounds XXXVIIe are given below:

| | $n = 1$ | $n = 2$ | $n = 3$ | $n = 4$ | $n = \infty$ | Cyclohexa-1,4-diene |
|---|---|---|---|---|---|---|
| $IP_1$ (eV) | 8.70 | 8.82 | 9.0 | 8.95 | 8.60 | 8.82 |
| $IP_2$ (eV) | 9.55 | 9.45 | 9.20 | 9.30 | 9.60 | 9.88 |
| $\Delta IP_{1,2}$ (eV) | 0.85 | 0.63 | 0.20 | 0.35 | 1.00 | 1.06 |

The dependence of the highest $\pi$ MO sequence on $n$ is shown in Fig. 5.9.

The data on the electron structure of bicycloalkadienes best of all agree with competition of the through-bonds and through-space effects of ethylene orbital mixing. Through-space interactions between ethylene $\pi$ orbitals prevail in norbornadiene and bicyclooctadiene molecules having lower dihedral angles $\theta$. In bicyclononadiene and bicyclodecadiene, whose molecules have greater dihedral angles $\theta$, the through-bonds interactions prevail between the $\pi$ orbitals.

Quantitative analysis of the through-bonds and through-space effects in bicycloalkadienes was also conducted on the basis of quantum chemical calculations [205]. The regularities established for the interaction of the occupied and vacant $\pi$ orbitals in cyclic dienes are shown in Fig. 5.8.

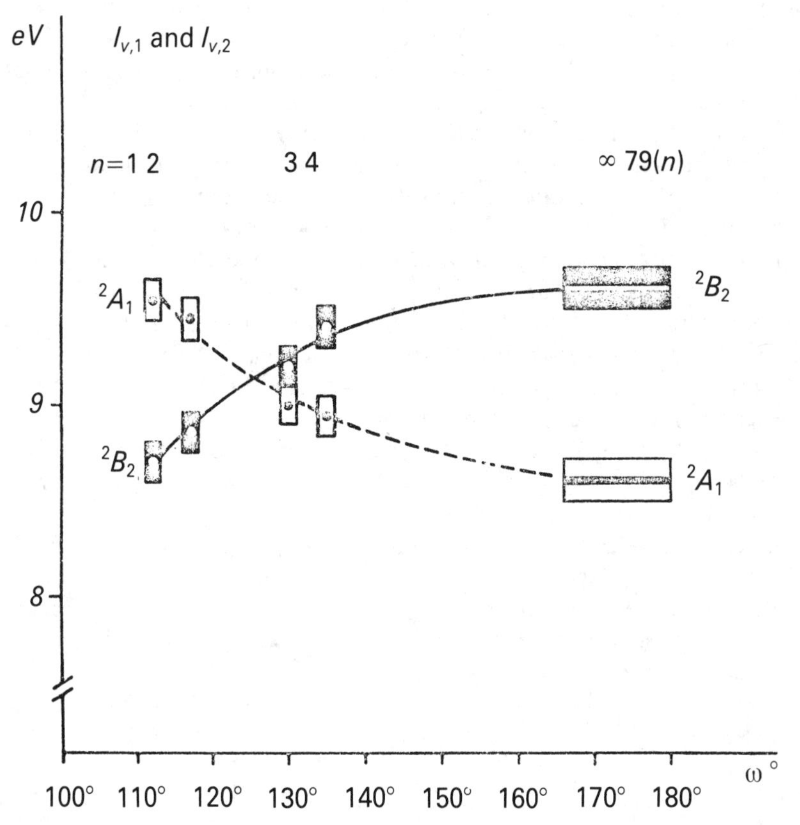


Fig. 5.9. Dependence of ionization potentials of higher occupied MOs $a_1$ and $b_2$ of bicycloalkadienes XXXVIIe on dihedral angle $\theta$. (From Heilbronner and Maier [54])

***Problems****

(1) Compare the ability of ethane and ethylene to release an electron. Evaluate the relative reactivity of ethane and ethylene to electrophiles.

(2) Compare the ability of ethane and ethylene to attach an electron. Evaluate the relative reactivity of ethane and ethylene to nucleophiles.

(3) Suggest what products are formed in the following $A_E$ reactions of alkenes:

(a) $CH_3 - CH = CH_2 + H_2SO_4 \rightarrow$

(b) $Cl - CH = CH_2 + HCl \rightarrow$

(c) $CH_3O - CH = CH_2 + ROH \xrightarrow{H^+}$

(d) $CH_2 = CH - NO_2 + H_2O \xrightarrow{\text{weak acid}}$

(4) Why does ethylene absorb light in a longer wavelength region of the U-V spectrum (at 185 nm) than ethane (at 121 nm)?

(5) Why are alkynes only about $10^{-2}$ to $10^{-3}$ as reactive in $A_E$ reactions as the corresponding alkenes?

(6) Compare "hardness" (after Pearson) of ethane, ethylene and acetylene molecules.

(7) Why does buta-1,3-diene more easily give release an electron than ethylene? Suggest the relative reactivity of ethylene and butadiene to electrophiles.

(8) Why does buta-1,3-diene more easily accept an electron than ethylene? Suggest the relative reactivity of ethylene and butadiene to nucleophiles.

(9) Explain why buta-1,3-diene has two one-electron $\pi$ ionizations although both its double $\pi$ bonds are equivalent in terms of the octet rule.

(10) Butadiene accepts readily an electron with the formation of an anion-radical. Suggest the distribution of an unpaired electron on the carbon atoms.

(11) Why do electron-withdrawing substituents in the alkene and electron-donating substituents in a diene increase the rate of Diels–Alder addition?

(12) Why do electron-donating substituents in an alkene and electron-withdrawing substituents in a diene decrease the rate of Diels–Alder addition?

(13) Suggest which of the possible isomers is formed from (2E, 4E)-hexa-2,4-diene: (a) on thermal cyclization; (b) on photocyclization.

(14) Suggest which of the possible isomers is formed from (2E,4Z,6E)-octa-2,4,6-triene: (a) on thermal cyclization; (b) on photocyclization.

(15) Suggest a stereochemical result of cycloaddition of (Z)-but-2-ene and ethylene. Why is this reaction thermally forbidden but takes place on exposure to U-V radiation?

(16) Why does buta-1,3-diene absorb light in a longer wavelength region of the U-V spectrum (at 217 nm) than ethylene (at 185 nm)?

(17) Explain why the Diels–Alder cycloaddition of buta-1,3-diene is stereospecific:

(a) cis-4,5-di(carboethoxy)cyclohexene is formed with diethyl maleate as a dienophile;

(b) trans-4,5-di(carboethoxy)cyclohexene is formed with diethyl fumarate as a dienophile.

---

* In Solving these problems use results of HMO calculations of the reactants FMO given in the Appendix.

# 6

# Aromatic hydrocarbons

Aromatic hydrocarbons are an important class of organic substances. The chemistry of these compounds underlies the manufacture of organic dyes, monomers, and many other substances. Benzene, toluene, naphthalene and xylene are consumed in millions of tons. The study of various physical and chemical properties of aromatic hydrocarbons has resulted in formulation of the concepts and theories that are essential for organic chemistry in general. These include the concept of mesomerism, the theory of aromaticity, orientation rules, etc. The analysis of the reactivity of aromatic hydrocarbons in the terms of the frontier orbitals concept and the concept of orbital and charge control of their reactions is also of substantial importance [206–211]. A new step in the study of aromatic hydrocarbons is connected with environmental problems.

## 6.1 ANNULENES

Annulenes are monocyclic polyenes with conjugated double bonds. Although the compounds of this class are not widely used and do not show any biological activity, their electron structure was studied in detail during the development of the concepts of electron delocalization and aromaticity.

$\pi$ Electron structure of annulenes is, within the framework of the MO theory, similar to that of linear polyenes except that the interaction of $2p\pi$ orbitals localized over the 1st and $n$th carbon atom of an annulene is accounted in the appropriate $H$ matrix. The general formula of the secular determinant, which enables calculation of eigenvalues of any cyclic hydrocarbon $C_nH_n$ within the HMO method, is shown below:

$$\begin{vmatrix} (H_{11}-E) & H_{12} & 0 & \cdots & H_{1n} \\ H_{21} & (H_{22}-E) & H_{23} & \cdots & 0 \\ 0 & H_{32} & (H_{33}-E) & \cdots & 0 \\ \cdots & \cdots & \cdots & \cdots & \cdots \\ H_{n1} & 0 & 0 & \cdots & (H_{nn}-E) \end{vmatrix} = 0$$

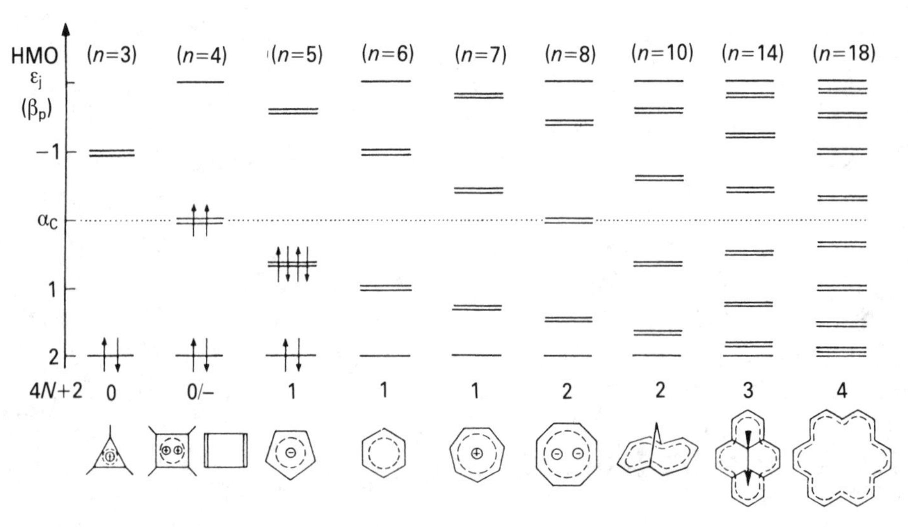


Fig. 6.1. Eigenenergies of annulenes $C_nH_n$ according to HMO calculations. (From Bock [117])

The calculated eigenvalues of annulenes $C_nH_n$ are given in Fig. 6.1.

### 6.1.1 [4]Annulene (cyclobutadiene)

In full agreement with experiment, even quantum chemical calculations by the HMO method predict instability of electron structure of annulenes that have open electron shells (e.g. of the biradical type) and the total number of $\pi$ electrons other than $(4n + 2)$. Thus, planar molecules of cyclobutadiene and cyclooctatetraene turned out to be unstable: according to the calculations, their HOMOs are non-bonding (their energies are equal to $\alpha$) and there is one unpaired electron in each of the two degenerate orbitals.

According to the MO theory, a non-planar (or planar but not square) structure should stabilize [4]annulene, cyclobutadiene. Distortion of the square conformation leads to splitting of the degenerate singly occupied MOs. As a result, the molecule is stabilized by insertion of both electrons into the lower level, which is formed after removal of degeneracy. The specificity of the electron structure mentioned explains the difficulties of the isolation of cyclobutadiene. This compound could be obtained only at very low temperatures. Experiments show that a cyclobutadiene molecule has, under these conditions, a non-planar conformation with almost localized double bonds [212]. Cyclobutadiene complexes with transition metals turned out to be more stable. In the complex $C_4H_4Fe(CO)_3$, electron density is shifted from the ring to the metal. The state of the aromatic $\pi$ electron doublet, occupying the stable lowermost $\pi$ orbital, is thus attained in the cyclobutadiene fragment. The cyclobutadiene ring turns out to be planar and square [213] and becomes capable of entering electrophilic aromatic substitution reactions [214].

### 6.1.2 [6]Annulene (benzene)

[6]Annulene, or benzene, is the simplest aromatic hydrocarbon. This accounts for its importance not only in the chemistry of aromatic compounds but also in organic

chemistry in general. Although benzene proper is not contained in living organisms, the benzene structure very often occurs in biological material; there are a great number of biologically active compounds that are actually benzene derivatives [124, 125]. Aromatic amines are used as neurotropics; benzaldehyde derivatives are known as pheromones; benzoquinone derivatives have the properties of insect repellants.

More than a hundred years ago, Kekule proposed the structural formula of benzene — a regular hexagon with alternating single and double bonds. At first sight, it may appear that the 'mystery' of the benzene structure was disclosed very rapidly: benzene was discovered by Faraday in 1825 and as soon as 1865 Kekule suggested the formula that is used by chemists until the present time.

But the conventional character of Kekule's formula was already evident at the time when he suggested it: two isomers for the ortho-substituted benzenes, as predicted by the Kekule structure, were not known:

It was soon shown that alternation of three single and three double bonds could not explain the unique physico-chemical properties of benzene, such as equal length of each of the six carbon–carbon bonds (0.139 nm); absence of reactions with bromine or potassium permanganate — the reactions that are so characteristic of unsaturated compounds; substantial differences in heat effects of consecutive addition of three moles of hydrogen, viz., +23.45, −111.80, and −119.74 kJ per mole. In order to reconcile his formula with experimental findings, Kekule proposed in 1872 that, owing to vibrational motion of carbon atoms about the middle position, the double bonds moved so that two formulae (rather than one) might describe the actual state of the benzene molecule. Many other formulae were later proposed for benzene:

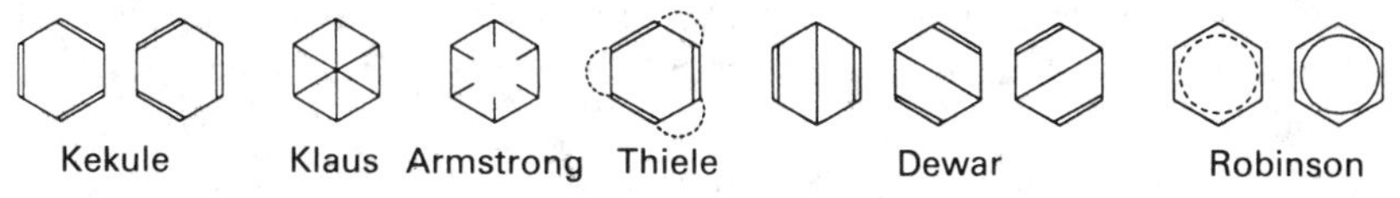


After discovery of electrons and creation of the Lewis' theory, chemists tried to describe completely the uniform distribution of electron density in the benzene molecule by numerous approaches. According to the valence-bond method, the benzene molecule was presented by two Kekule formulae and three Dewar formulae. All other possible formulae (the formulae mentioned included) are combinations of these five independent canonical structures. The valence-bond method objectively described the mesomeric state of the benzene molecule and estimated its advantages over the state described by the Kekule structure as the resonance energy (154.9 kJ/mol). The valence-bond method estimated the contribution of each canonical structure: 78% (total) were the Kekule structures and 22% the Dewar structures. The equivalence of all carbon–carbon bonds in benzene and its specific properties were proposed finally by the Robinson formulae.

According to the MO method, the unique properties of benzene account for the ideal steric conditions for the most effective overlap of the $2p$ orbitals of the adjacent carbon atoms: electron diffraction in the gas state [215] and x-ray studies of the molecular crystal at $-3°$ [216] show that all valence angles in the planar molecule of benzene are 120°, and all (C—C) bond distances are 0.139 nm.

At the present time, the electron structure of benzene has been studied by various quantum chemical calculations. Using the benzene molecule, Hückel demonstrated effectiveness of his simple method of the molecular orbital theory in the $\pi$ approximation (see Section 2.2.1). Later, many new calculation methods were proposed within the framework of the MO theory; they were also used to calculate the benzene molecule [20, 23, 51, 217, 218, 219]. All these later calculations suggest that the benzene HOMO is formed by two degenerate $\pi$ orbitals: $\pi_2$ and $\pi_3$. But the relative arrangement of subsequent $\pi_1$ and highest $\sigma$ orbitals was argued for a long time.

The first PE spectra of benzene demonstrated considerable overlap of the regions of $\pi$ and $\sigma$ level ionizations, although it was believed (before development of PE spectroscopy) that the $\pi$ levels of all organic molecules occupied a relatively narrow region and had markedly lower energy than the $\sigma$ levels. It was only the first ionization potential in the PE spectrum of benzene (9.24 eV) that was assigned unambiguously by the first investigators to ionization of the electrons of the degenerate highest occupied $\pi$MOs. The validity of this assignment was confirmed not only by quantum chemical calculations but also by the specific character of the first ionization band [220]: it has vibrational structure which is common for the symmetry $e_{1g}$ vibrations, although Jahn–Teller splitting, which is common in photoionization of the bonding $(1e_{1g})^4$ shell, was not recorded. This was explained by a weak bonding character of the electron in the HOMO of benzene. Figure 6.2 shows the occupied molecular orbitals of benzene.

Assignment of the second ionization potential seemed to be not so obvious. First reports suggested that the second potential of 11.44 eV could be attributed to ionization of the electrons in the lower occupied $\pi$ orbital. The $\pi$ and $\sigma$ levels were 'identified' by measuring the perfluoro-effect and by the study of photoionization of benzene using (in addition to the He(I) radiation) other sources such as Ne(I), He(II), He*(I) and Ne*(I).

Thus, the effect of fluorine substitution for hydrogen on the energies of MOs of different symmetry was estimated by comparing the spectra of benzene and hexafluorobenzene [179, 221, 222]. As in other unsaturated compounds, the transition from benzene to hexafluorobenzene only slightly changes the energies of the $\pi$ orbitals and markedly lowers the energy of the $\sigma$ orbitals. Changes in the energies of the corresponding orbitals were compared on the basis of the $X_\alpha$ calculations [210, 223]. Assignment of ionization potentials in the benzene spectrum is shown in Table 6.1.

Assignment of the second band in the PE spectrum to ionization of the degenerate $\sigma$ level is confirmed by the comparison of intensities of ionization bands in the spectra recorded with Ne(I) (16.85 eV) and Ne* (16.62 eV) sources. The intensities of the first and the third bands only increase significantly in the Penning spectrum in accordance with their assignment to ionization of $\pi$ orbitals (Fig. 6.3) [224, 225].

Graphic representation of all 15 occupied molecular orbitals of benzene with their assignment to the corresponding bands in the PE spectrum is given in Fig. 6.2 [20,

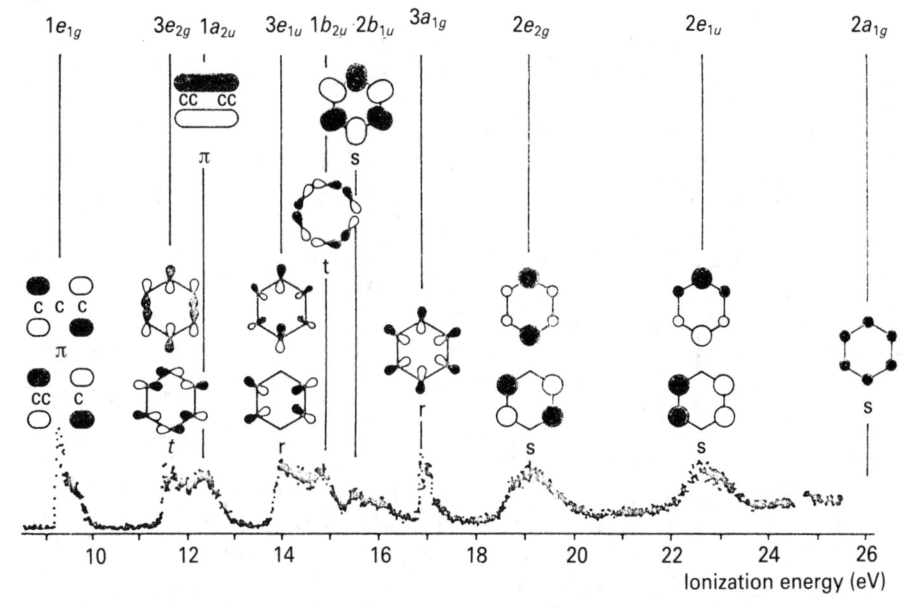


Fig. 6.2. Occupied benzene MOs (MO energies according to He(I)-PES; MO configurations according to *ab initio* calculations). The benzene ring plane coincides with the plane of the $\sigma$ orbital drawing and it is perpendicular to the plane of $\pi$ orbital ($1e_{1g}$ and $1a_{2u}$) drawing. (From Lindholm [218])

**Table 6.1.** Ionization potentials and occupied molecular orbitals of benzene (PES data and $X_\alpha$-calculations)

| $IP_{\text{exp}}$ (eV) | No. and symmetry ($D_{6h}$ group) | MO number | $IP_{\text{exp}}$ (eV) | No. and symmetry ($D_{6h}$ group) | MO number |
|---|---|---|---|---|---|
| 9.24 | $1e_{1g}(\pi)$ | 15 | 15.40 | $2b_{1u}(\sigma)$ | 7 |
| | | 14 | 16.84 | $3a_{1g}(\sigma)$ | 6 |
| 11.49 | $3e_{2g}(\sigma)$ | 13 | 19.20 | $2e_{2g}(\sigma)$ | 5 |
| | | 12 | | | 4 |
| 12.30 | $1a_{2u}(\pi)$ | 11 | 22.60 | $2e_{1u}(\sigma)$ | 3 |
| 13.90 | $3e_{1u}(\sigma)$ | 10 | | | 2 |
| | | 9 | 25.90 | $2a_{1g}(\sigma)$ | 1 |
| 14.70 | $1b_{2u}(\sigma)$ | 8 | | | |

(From Bloor and Sherrod [223])

33, 218]. The configurations of MOs show that the $\sigma$ orbitals of benzene are formed by the same scheme that has already been discussed for alkanes and cycloalkanes (see Sec. 4.3). Thus, orbitals with the ionization energies at 18–25 eV in the benzene molecule are linear combinations of carbon 2*s* orbitals: their symmetries and coefficients are similar to the corresponding parameters of the $\pi$ orbitals.

Thus, the $\pi$- and $\sigma(2s)$ levels can be described independently using the same simple Hückel method when discussing the electron structure of aromatic hydrocarbons as

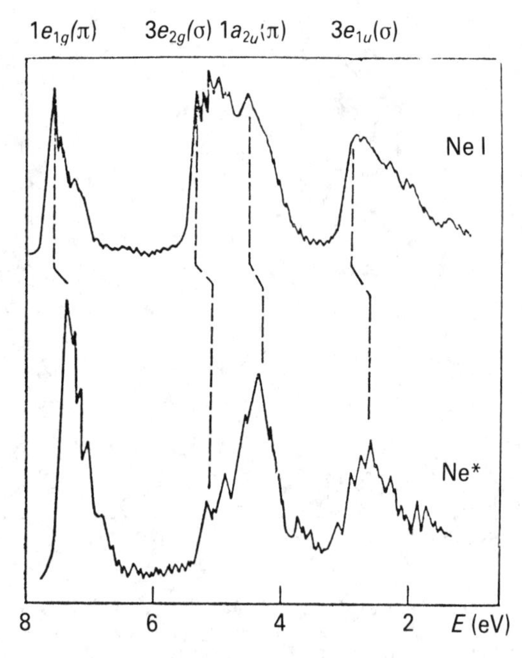


Fig. 6.3. Relative intensities of spectral bands in Ne(I)-photoionization and Ne*-Penning ionization of benzene. (From Munakata *et al.* [178].

well. Parametrization of orbital interaction for both $\pi$ and $\sigma$ levels is possible using PES findings [54].

Anion-radical states of the benzene molecule were also studied in detail. Two anion states were discovered by electron transmission spectroscopy: the lower state estimating the first value of electron affinity $EA_1 = -1.15$ eV and the state with higher energy ($EA_2 = -4.85$ eV).

The first anion state has the vibrational structure of the resonance peak (Fig. 6.4) which suggests its stability. A detailed analysis of the vibrational structure of the ETS signals for benzene is given in [46]. The vacant electron levels of the benzene molecule were also studied in [45, 174, 226].

It is noteworthy that as in the acetylene series, the incorporation of the triple bond into the cyclic molecule with conjugated double bonds increases markedly its electron affinity. It is believed that this explains the marked electrophilic properties of dehydrobenzene, the intermediate formation of which is postulated in many reactions of nucleophilic aromatic substitution. Electron affinity $EA_1$ of dehydrobenzene is 0.1–0.6 eV against the $-1.15$ eV for the benzene molecule [194].

On the whole, as already mentioned in Section 2.2, experimental data (PES and ETS) confirm the symmetrical arrangement of vacant and occupied MOs of benzene as an alternant hydrocarbon (as predicted by the MO theory). One can compare the energies of benzene $\pi$ orbitals according to the PE- and ET spectra in Koopmans' approximation (Fig. 6.6).

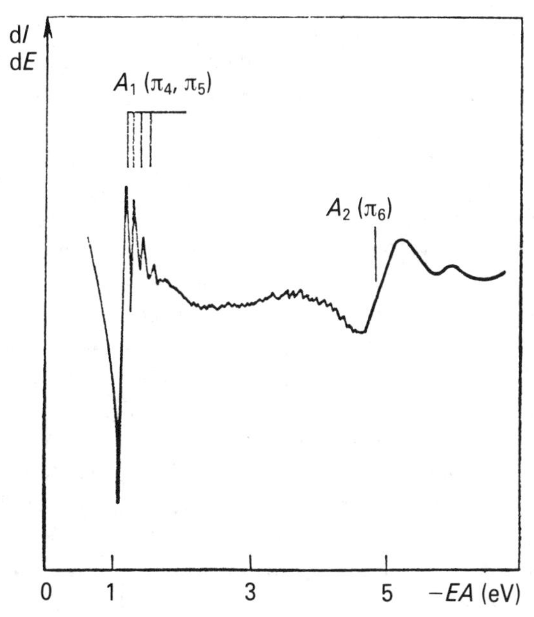


Fig. 6.4. Electron transmission spectrum of benzene. (From Jordan and Burrow [46])

Both vacant and occupied frontier $\pi$ levels of benzene are formed by degenerate orbitals which however differ in their coefficients (see Section 2.2.1 for details).

At first sight, these parameters indicate non-equivalence of different positions in the benzene molecule. Calculations of spin densities at each of the six carbons in the cation radical and anion radical of benzene show, however, complete equivalence of all positions of its molecules:

$$\rho_{1,4}^{\text{theor}} = \frac{(1\sqrt{3})^2 + 0}{2} = 0.167$$

$$\rho_{2,3,5,6}^{\text{theor}} = \frac{(1\sqrt{12})^2 + (1/2)^2}{2} = 0.167$$

The ESR spectrum of the benzene anion radical agrees well with the prediction of the theory: the interaction of an unpaired electron of the anion radical in the degenerate $\pi_4$ and $\pi_5$ orbitals with six protons gives seven equidistant lines with the intensity ratio of 1:6:15:20:15:6:1 and the splitting constant $a_{\text{H}} = 3.75$ Oe [50].

In accordance with the spectrum and the known equation [50]:

$$a_\mu = Q_{\text{C—H}} \cdot \rho_\mu$$

the spin density in the electron shell formed by the degenerate $\pi_4$ and $\pi_5$ orbitals is the same in all carbons:

$$\rho_{1\text{–}6}^{\text{exp}} = 3.75/22.5 = 0.167 \qquad \text{(assuming that } Q_{\text{C—H}} = 22.5 \text{ Oe).}$$

Coincidence of spin densities $\rho^{\text{theor}}$ and $\rho^{\text{exp}}$ should be noted. It suggests similarity of structures of the corresponding electron level in various phases: quantum chemical calculations refer to an isolated benzene molecule while the ESR spectrum findings refer to the anion radical of benzene generated in the liquid phase (dimethoxyethane).

In this connection, the parameters of the electron structure of benzene in the gas and solid phases should be compared as well. The corresponding data were obtained by PE and ET spectroscopy. Measurements of the ET spectrum of solid benzene films confirmed the structure of its vacant electron levels in the solid phase: resonance signals were discovered in two energy regions of the bombarding electrons, viz. at 1.5–3.0 eV and at 3.9–4.6 eV [155].

Changes in the structure of the benzene electron levels in transition from a molecule to the molecular crystal were the subject of a special quantum chemical analysis [132]. To that end, the molecules of benzene and its dimer (the geometry of the dimer was given by x-ray structural analysis [216]) were calculated in the same approximation. Compared with an individual molecule, only infinitesimal changes were discovered in the intermolecular structure of the dimer (see Section 3.5).

Changes in the electron structure of the benzene molecule induced by external contact are estimated by the PE spectra of benzene complexes with transition metals [227]. In the spectrum of bis($\pi$ benzene) chromium, the first bands at 5.4 and 6.4 eV are assigned to ionization of the occupied d orbitals of chromium with the $a_{1g}$ and $e_{2g}$ geometry. Striking similarity of the PE spectra of bis($\pi$ benzene)chromium and benzene was discovered in the region of ionization potentials above 8 eV; given below are vertical values of $IP_1$ (parenthesized are the intervals of ionization bands):

| $C_6H_6$ | $(\pi\text{-}C_6H_6)_2Cr$ |
|---|---|
| 9.24 (9.2–9.8) | 9.6 (9–10) |
| 11.48 (11.4–13.0) | 11.5 (10.8–13.0) |
| 13.9 (13.8–15.5) | 13.8 (13.5–15.0) |

The spectral bands of the complexes are assigned as in the spectrum of benzene proper. Slight splitting of the third band at 9–10 eV of the complex is assigned to ionization of the MO which are mostly $e_{1g}$ orbitals of the ligand mixed with the metal orbitals. The formation of the complex with the transition metals is mainly associated with splitting and stabilization of the frontier occupied $\pi$ electron level of benzene. The absence of appreciable splitting of any other band of the ligand suggests that the other orbitals of the benzene molecule are not involved in the formation of the complex by the ligand-metal or ligand-ligand interaction.

The complexes of benzene with $\pi$ acceptors, metal cation, and electrophiles have been studied in detail [88]. It turned out, e.g., that in complexes of benzene with $Ag^+$, $Cu^+$ and their analogues (XXXVIII), the metal ion is localized over the molecular plane between two adjacent carbon atoms (structure a):

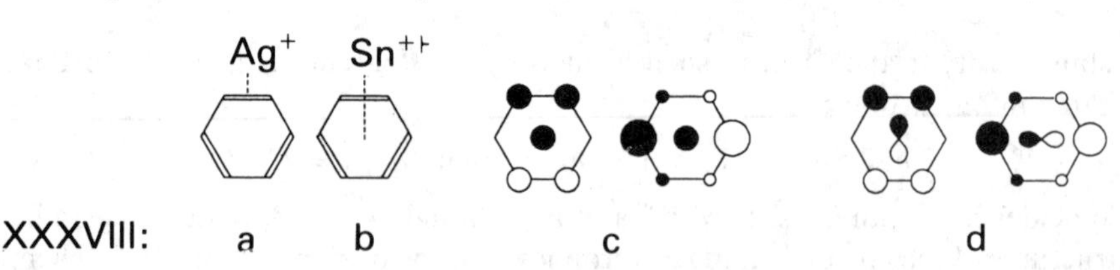

In benzene complexes with the $Sn^{2+}$, $Pb^{2+}$, $Ga^{+}$, and $Th^{+}$ ions, the metal is located on the axis of the 6th order, i.e. symmetrically relative to each of the six carbons (structure b).

The specific structure of the complexes is explained by the symmetry of the orbitals that are involved in complexation. Thus, LUMOs of the $Ag^{+}$ and $Cu^{+}$ ions are 5*s* and 4*s* orbitals. The differences between their symmetry and the symmetry of benzene $e_{1g}\pi$ orbitals rule out any interaction within the framework of structure c: overlapping of the benzene HOMO and ns orbitals of the metal is only possible in structure a.

LUMOs of the $Sn^{2+}$, $Pb^{2+}$, $Ga^{+}$ and $Th^{+}$ ions are p orbitals; they have symmetry $e_1$ which allows the overlap with the benzene HOMO, as shown in structure d.

Although the highest occupied $\sigma$ level of the benzene molecule is higher than the lowest $\pi$ level, the electron absorption spectrum is adequately described in the $\pi$ approximation. All quantum chemical calculations show that three absorption bands in the region of the long wavelengths of the spectrum at 256 nm ($\log \varepsilon = 2.2$), 203 nm ($\log \varepsilon = 3.9$), and 180 nm ($\log \varepsilon = 4.8$) are due to the configurational interaction of the electron transitions between the frontier occupied ($\pi_2$ and $\pi_3$) and vacant ($\pi_4$ and $\pi_5$) orbitals of the benzene molecule [30].

### 6.1.3 [10]Annulenes

In accordance with the diagram shown in Fig. 6.1, stable systems with closed $\pi$ electron shells are formed only when $(4n + 2)$ $\pi$ electrons are located in the occupied MOs: the MO theory predicts the aromatic properties in [10], [14], [18], [22], and [26]annulenes. The data available on the reactivity of these annulenes, the analysis of their structure by x-ray and NMR spectroscopy confirm this prediction [5, 209, 211]. Consider the electron structure of [10]annulenes in more detail.

[10]Annulenes (XXXIX) can have at least two appreciably different conformations: *a*-completely cis, and *b*-cis–trans–cis–trans:

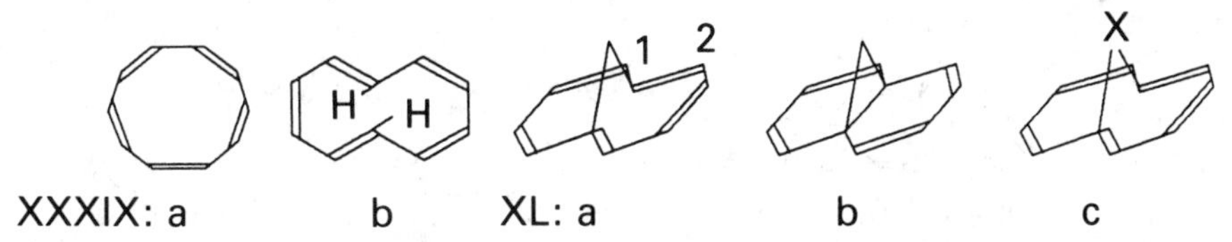


In order to assume a planar conformation, the molecules of cyclodecapentaene XXXIX should overcome a considerable strain: in the conformation *a* the angles should be 144° instead of 120°, while in the conformation *b* (with preservation of the angles for the $sp^2$ state), considerable steric interactions between the hydrogen atoms are inevitable. The crystals of cyclodecapentaene in the conformation *a* were prepared at a temperature of −80°C: all signals in its PMR spectrum were observed in the region of olefin protons [209].

[10]Annulene in the conformation *b* has not been obtained. During numerous attempts to synthesize this substance, the investigators observed formation of naphthalene derivatives (in whose molecules two interfering hydrogens were replaced by the bond between the 1st and 6th carbon atoms). The interfering hydrogens of

cyclodecapentaene XXXIXb are replaced in 1,6-methane[10]annulene by the methylene bridge group XLa*.

1,6-Methane[10]annulene cannot be in a completely planar conformation, but overlapping of the $2p_\pi$ orbitals of the carbon atoms at the periphery is sufficiently effective to give the aromatic character to this compound. This conclusion is also confirmed by the length of the peripheral C—C bonds [209–211]: the molecule of 1,6-methane[10]annulene-2-carboxylic acid has a planar perimeter with 'benzene' bonds (0.138–0.142 nm). The spectra and reactions of various derivatives of bridged [10]annulenes with the general structure XLc [$X = (CH_2)_n$, $C(CH_3)_2$, $C{=}CH_2$, $C{=}C(CH_3)_2$, CO, O, NH, NR] were studied in detail. Since the bridge group at atoms 1 and 6 does not interfere seriously with the interaction of the $\pi$ orbitals in the peripheral ring, the chemical behaviour of these annulenes also obeys the rule of aromaticity ($4n + 2$). These compounds actively react with electrophiles by the general scheme of electrophilic aromatic substitution: they are brominated with bromine, acylated in the presence of $SnCl_4$, and nitrated by $Cu(NO_3)_2$ in acetic anhydride [211, 228–230].

Annulenones of the general structure XLI are a convincing example of delocalization of electron density within the [$10\pi$] perimeter (bridged compounds are only known for high $n$ values). Delocalization of $\pi$ electrons in their molecules depends on the contribution of the mesomer structure XLIb to the ground state:

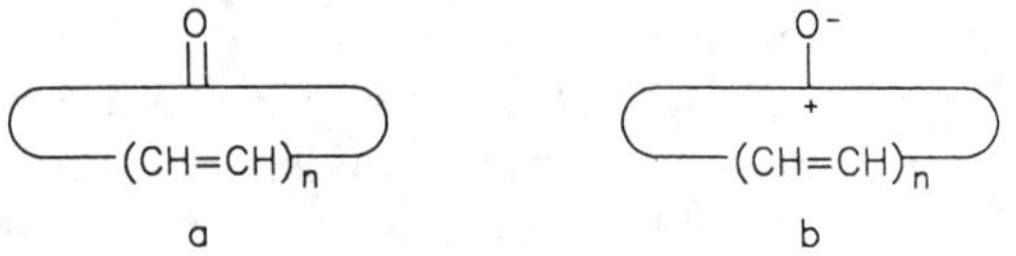

XLI: a $(CH{=}CH)_n$ (C=O); b $(CH{=}CH)_n$ ($C^+$–$O^-$)

Although the intensity of the ring current in these annulenones is not high, it was established that protonation of oxygen increases appreciably $\pi$ electron delocalization. The corresponding effects are illustrated by $pK_a$ values: while aliphatic ketones have low basic properties (in the form of conjugated acids, their $pK_a$ values range from

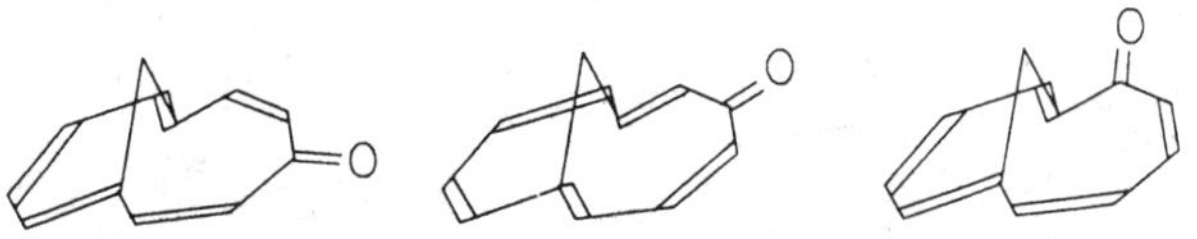

$pK_a = -0.6$ $+0.5$ $-0.4$

$-6$ to-7), bridged annulenones are much stronger bases [231]. (The structural formulae are subscribed by the values of $pK_a$ of conjugated acids).

The first PE spectrum of 1,6-methane[10]annulene was reported by Schmidt *et al.* in 1972 [232]. According to calculations of the planar cyclodecapentaene molecule (Fig. 6.5), three bands might be expected to be assigned to ionization of $\pi$ orbitals. If fact, they turned out to be five. This unambiguously indicated that degeneracy of two higher $\pi$ electron levels, characteristic of the [10]annulene molecule, is removed

* All attempts to discover 1,6-methane[10]annulene in solution in the form of the tricyclic norcaradiene XLb were a failure for a long time although quantum chemical calculations predicted its presence at room temperature in at least 0.05% amount. The form XLb has recently been detected in $\pi$ complexes of cobalt [211].

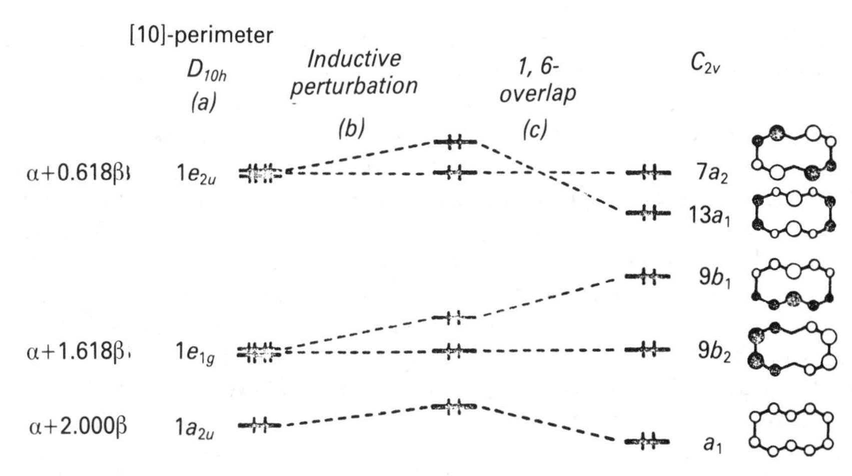


Fig. 6.5. Transformation of $\pi$MO on transition from cyclodecapentaene to 1,6-methane[10]annulene: (a) $\pi$MO of cyclopentaene; (b) and (c) $\pi$MO of 1,6-methane[10]annulene ((b) the inductive effect of the $CH_2$ group alone is taken into account). (From Heilbronner and Maier [54])

in the molecule of 1,6-methane[10]annulene. Schmidt *et al.* proposed that the inductive effect of the $CH_2$ bridge group and conjugation between $2p$ orbitals of the 1st and 6th carbon atoms are responsible for the removal of degeneracy of $1e_{2u}$ and $1e_{1g}$ levels. The effect of orbital interaction is, according to some estimations, 40% of the standard value for the adjacent carbon atoms in the benzene molecule [211]. The description of experimentally studied electron levels was thus satisfactory within the framework of the HMO method (Table 6.2).

**Table 6.2.** Ionization potentials of bridged [10] annulenes XL (PES data)

| Band No. | MO | *IP* (eV) | | | | | |
|---|---|---|---|---|---|---|---|
| | | X=$CH_2$ | $CF_2$ | C=$CH_2$ | C=O | NH | O |
| 1 | $7a_2$ | 7.92 | 8.19 | 7.98 | 8.47 | 8.01 | 8.14 |
| 2 | $13a_1$ | 8.39 | 8.79 | 8.23 | 9.14 | 8.01 | 8.45 |
| 3 | $9b_1$ | 9.25 | 8.73 | 9.29 | 10.36 | 9.59 | 10.10 |
| 4 | $9b_2$ | 10.36 | 10.70 | 10.32 | 10.80 | 10.64 | 10.74 |

(From Andrea *et al.* [230])

What is most surprising in the proposed scheme is that it completely disregards the non-planeness of the molecule of 1,6-methane[10]annulene: the deviation of the bonds from the plane to $\theta_{1,2} = 34.0°$; $\theta_{2,3} = 19.7°$ and $\theta_{3,4} = 0°$ was established by x-ray studies [233]. Furthermore, subsequent attempts to account for the twist of the double bonds in 1,6-methane[10]annulene gave a much worse description of the PE spectrum. It is believed that the molecular structures of bridged [10] and [14]annulenes in the gas and crystal states are different. In any case, the deviations of double bonds from the plane of conjugation in these compounds produce an insignificant effect on the energies of the corresponding electron levels.

It should be added in conclusion that the sequence of molecular orbitals of 1,6-methane[10]annulene, first proposed in [232], has been confirmed by CNDO/S, MNDO [230], and also by *ab initio* calculations [234].

PE spectra of 1,6-methane[10]annulene derivatives with various substituents (Me, MeO, F, etc.) in the peripheral ring and in the bridge were studied. The result may appear unexpected at first sight: the introduction of the substituent into the peripheral ring destabilizes all $\pi$ orbitals to about the same extent. The methyl group, e.g. at position 2, decreases the corresponding ionization potentials by about 0.2 eV. As shown below, this effect differs substantially from the effect of the same substituents in the benzene molecule and is explained by non-zero coefficients at position 2 in all occupied $\pi$ orbitals of the 1,6-methane[10]annulene molecule (Fig. 6.5).

Transformation of the unoccupied frontier $\pi$ orbitals of cyclodecapentaene caused by the introduction of the methylene bridge at positions 1 and 6 has also been studied:

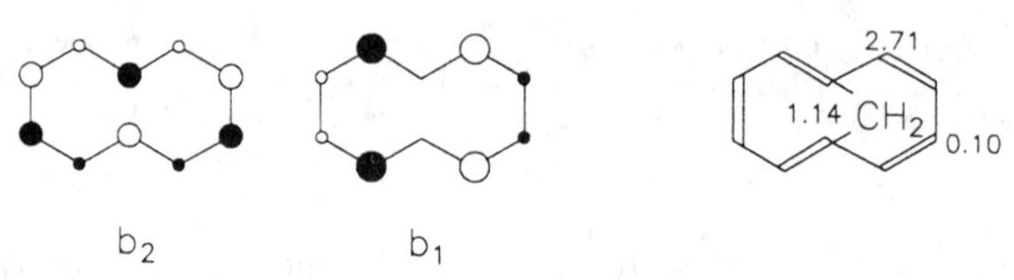


Of the two degenerate unoccupied frontier MOs of cyclodecapentaene, only the $b_2$ orbital can be influenced by the methylene group, and it therefore increases its energy. The $b_1$ orbital has zero coefficients at the site of substitution and it remains uninvolved. The values of HFI constants $a_H$ measured in the ESR spectrum of the anion-radical of 1,6-methane[10]-annulene show that the $b_1$ orbital is the LUMO of this hydrocarbon: the unshared electron is located in this orbital [235].

## 6.2 ALKYL BENZENES

### 6.2.1 Toluene

The studies on the properties of aromatic hydrocarbons with alkyl groups as substituents have resulted in formulation of the concept of hyperconjugation in organic chemistry.

The field effect of the alkyl groups alone should explain the decrease in their donor properties in the corresponding reaction series (including $R—C_6H_4—X$): $Me_3C > Me_2CH > MeCH_2 > Me$. This sequence is observed in many cases, e.g. in measuring dipole moments of alkyl benzenes in the gas phase. But there are examples where this sequence does not hold. Below are values for R in benzyl bromides $R—C_6H_4—CH_2Br$, their dipole moments, relative rate constants of the nucleophilic substitution for the bromine atom:

| R | $CH_3$ | $CH_3CH_2$ | $(CH_3)_2CH$ | $(CH_3)_3C$ |
|---|---|---|---|---|
| $k_{rel}$ | 1.00 | 0.90 | 0.81 | 0.82 |
| $\mu, D$ | 0.37 | 0.58 | 0.65 | 0.70 |

This sequence was explained by the increased ability of the C—H bonds to conjugation, i.e. by the hyperconjugation effect [137]. Controversial estimations of this fact were later proposed.

The results of studies on the steric structure of a toluene molecule, for example, do not prove the preference of the orthogonal conformation, which should be the most favourable for the interaction of $\sigma$(C—H) and $\pi$ orbitals [236]:

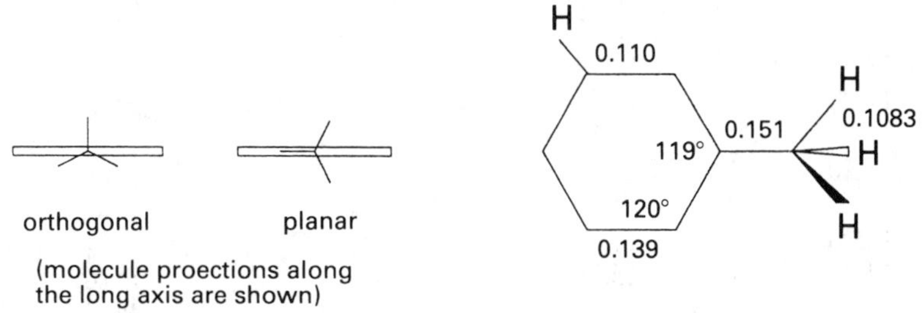

orthogonal planar

(molecule proections along the long axis are shown)

According to microwave spectroscopy, the rotational barrier of the methyl group in the toluene molecule is not higher than 58.6 kJ/mol, while the valence angle and bond distance differ only insignificantly from the standard values. No rotational barrier of the methyl group was found in the toluene molecule by *ab initio* quantum chemical calculations in the STO-3G basis set: 'planar' and 'orthogonal' conformations differ by not more than 25.12 kJ/mol, which is beyond the limits of calculation accuracy [237].

The PE spectrum of toluene was interpreted in various approximations [38, 238, 239]. Splitting of the highest occupied benzene orbitals (of both $\pi$ type, $e_{1g}$ and $\sigma$ type, $e_{2g}$) is unambiguously confirmed by *ab initio* calculations in both the minimal basis and DZ basis. Palmer *et al.* used the tabulated values of valence angles and bond distances for the C—$CH_3$ fragments in these calculations [239]. Strict *ab initio* calculations of toluene in the STO 4-31G basis set were also performed with interpretation of the PE spectrum by Kimura *et al.*; the experimental data on the geometry of the toluene molecule were used in these calculations [38]. The measured ionization potentials and the results of calculations are given in Table 6.3 for the purpose of comparison. These data also indicate splitting of the degenerate benzene orbitals with the introduction of the methyl group into the benzene molecule, but they however do not reveal appreciable mixing of the benzene $e_{1g}\pi$ orbital with the substituent orbitals.

By contrast, the calculations performed by Palmer [239] confirm unambiguously the interaction of the $\sigma$(C—H) orbitals of the methyl group with a deeper benzene $\pi$ orbital ($\pi_1$, $a_{2u}$ orbital). The conditions for the interaction of this pair of orbitals in the toluene molecule are most favourable. Thus, in addition to the possible steric overlap, said orbitals have about the same values of energies: the ionization potentials of their electrons, measured from the PE spectra of methane and benzene, are 13.0 and 12.3 eV respectively.

The energy effect of this $\sigma$–$\pi$ interaction is, however, small, because the ionization potential of the occupied $2b_1$ MO of toluene is only 0.3–0.4 eV lower than the ionization potential of the $a_{2u}\pi$ orbital of benzene. At the same time, this is probably the $\sigma$(C—H)–$\pi(a_{2u})$ interaction that explains the diagram of $\pi$ densities in the toluene

**Table 6.3.** Ionization potentials and occupied orbitals of toluene (PES data and 4-31G calculations)

| $IP_{i,\text{exp}}$ (eV) | MO | | |
|---|---|---|---|
| | $-\varepsilon_i$ (eV) | No. and symmetry | Localization* |
| 8.83 | 8.68 | $2b_1$ | $\pi_3$ |
| 9.36 | 8.99 | $1a_2$ | $\pi_2$ |
| 11.43 | 12.74 | $12a_1$ | $B(19)$ |
| 11.98 | 13.01 | $10b_2$ | $B(18)$ |
| 11.98 | 13.15 | $11a_1$ | $\pi_1$ |
| (13.20) | 14.51 | $9b_2$ | $\pi_{CH_3}$ |
| 13.70 | 15.25 | $10a_1$ | $B(15)$ |
| 14.00 | 15.67 | $1b_1$ | $\pi_{CH_3}$ |
| 14.00 | 15.75 | $8b_2$ | $B(16)$ |
| 15.10 | 16.89 | $9a_1$ | $B(13)$ |
| (15.30) | 17.14 | $7b_2$ | $B(14)$ |
| 16.45 | 18.67 | $8a_1$ | $B(12)$ |

*The letter B is used to designate MOs localized in the benzene ring; parenthesized are the numbers of these MOs in the benzene's orbital set.
(From Kimura *et al.* [38])

molecule as determined by the STO-3G calculations [237]; the net value of $\pi$ electron density transfer into the ring is 0.009.

Although according to PE spectroscopy and quantum chemical calculations, the electron effects due to the attachment of the methyl group to the benzene ring are not great, it should be noted that degeneracy of the benzene highest occupied $\pi$ orbitals is removed in toluene: according to Kimura [38] these orbitals are separated by an energy of 0.53 eV. The parameters of the electron spectroscopy and quantum chemical calculations are important for estimation of toluene reactivity and the spectral properties. Some examples are given below.

Depending on 'hardness' of an electrophile, the toluene molecule is subject to prevalent attack at the para- or ortho position. Reactions with 'soft' electrophiles (e.g. brominating agents) are characterized by early transition states. The orientation is determined by the orbital interactions in these reactions: 'soft' electrophiles attack at the para position in accordance with the distribution of electron density in the frontier occupied MO of toluene. The corresponding data were obtained not only by quantum chemical calculations. Constants of interaction of the unpaired electron with the proton in the cation radical of toluene were measured by ESR spectroscopy: $a_H$ (para) = 11.8 Oe, $a_H$ (meta) = O, $a_H$ (ortho) = 2.1 Oe [240].

Reactions with 'hard' electrophiles (e.g. the chloronium cation) are characterized by late transition states and charge control. The ortho position of toluene is mostly attacked in these reactions. These regularities are confirmed by the quantum chemical calculations. Their results show that the toluene molecule must react at the para position with electrophiles having low vacant orbitals (i.e. 'soft' orbitals), and at the ortho position with electrophiles having high vacant orbitals ('hard' orbitals) [68, 241].

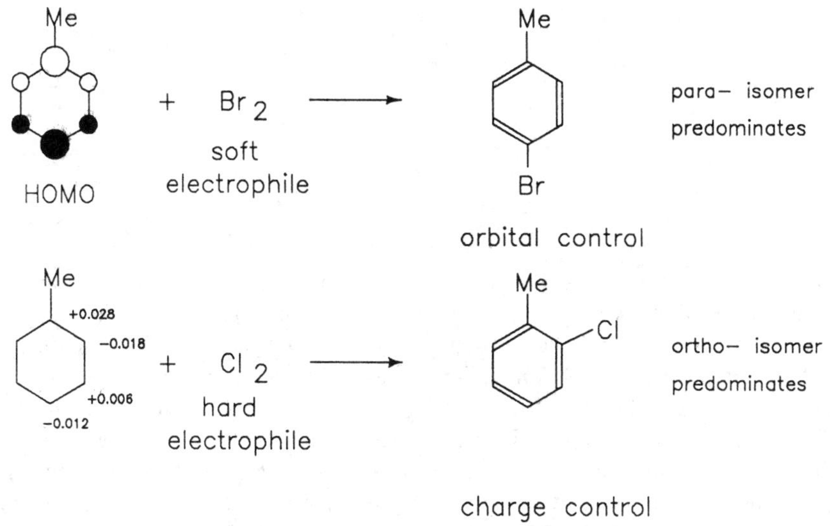

Me

HOMO + $Br_2$ soft electrophile → para– isomer predominates

orbital control

Me +0.028 −0.018 +0.006 −0.012

+ $Cl_2$ hard electrophile → ortho– isomer predominates

charge control

Weak splitting is characteristic of the unoccupied $\pi$ orbitals of toluene as well [242]. The appropriate data are discussed in detail in a subsequent section, while here we shall note that the ETS method gives only two (not three, as one might expect with distinct LUMO splitting) values for electron affinity: $EA_1 = -1.10$ eV and $EA_2 = -4.80$ eV. Nevertheless, the unoccupied frontier MO of toluene is not degenerate. According to the ESR spectrum of the anion radical of toluene, splitting of the unpaired electron is the same on the ortho- and meta protons, but it is an order lower on the proton at the para position to the methyl group. The measured values of HFI constants and spin densities calculated in the toluene anion radical are shown below [243]:

| | ortho | meta | para |
|---|---|---|---|
| HFI constant, Oe | 5.12 | 5.45 | 0.59 |
| Spin density | | | |
| HMO calculations | 0.25 | 0.25 | 0.00 |
| exp., ESR | 0.23 | 0.20 | 0.02 |

According to these data, the frontier unoccupied MO of toluene is the $\pi$(asym)orbital of the benzene ring. This orbital has changed its energy only insignificantly due to the effect of the methyl group.

As a result of the weak effect of the methyl group on the energy of the unoccupied MO, the electron affinity of toluene increases only insignificantly compared with that of benzene, but the lifetime of the toluene anion radical is almost halved [46].

As several methyl groups are attached to the benzene ring, the character of their interaction with the aromatic $\pi$ system does not change. The valence angles and bond distances in the mesitylene molecule are the same as in toluene:

$r$ (nm)

| | |
|---|---|
| $C_{(1)} - C_{(2)}$ | 1.401 |
| $C_{(1)} - C_{(7)}$ | 1.509 |
| $C_{(7)} - H_{(13)}$ | 1.111 |
| $C_{(2)} - H_{(8)}$ | 1.098 |
| $\angle C_{(2)}C_{(1)}C_{(6)}$ | $118.2 \pm 0.2°$ |
| $\angle C_{(1)}C_{(7)}C_{(13)}$ | $111.0 \pm 0.8°$ |

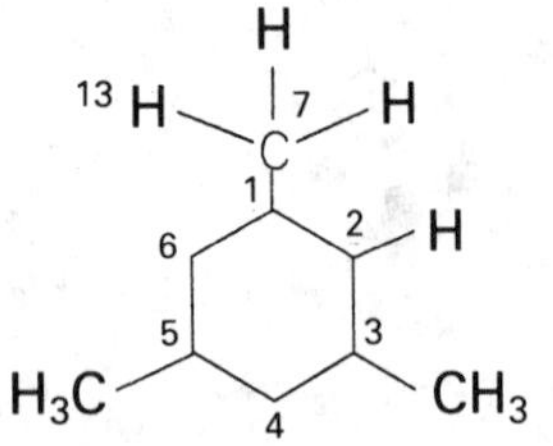

The comparison of structural parameters of methylbenzenes and unsubstituted benzene shows a regular decrease in the endocyclic angle at the carbon atom to which the methyl group is attached. This occurs also in toluene (118.9°) and mesitylene (118.2°). For the purpose of comparison, it can also be noted that this angle is even smaller in p-xylene (117.1°). (The steric structure of various substituted benzenes is discussed in detail in [244]). As in toluene, the rotational barrier of the methyl group in the mesitylene molecule is very small. According to calculations, its conformation with 'planar' arrangement of the methyl group is the most stable, but the rotational barrier is only 83.7 kJ/mol [244].

**Table 6.4.** Ionization potentials of some methylbenzenes (PES data and PMO-calculations)

| Me position | Symmetry group | $IP_1$ (eV) | $IP_2$ (eV) |
|---|---|---|---|
| 1 | $C_{2v}$ | $9.00(b_1)$ | $9.30(a_2)$ |
| 1, 2 | $C_{2v}$ | $8.75(b_1)$ | $9.00(a_2)$ |
| 1, 3 | $C_{2v}$ | $8.75(b_2)$ | $9.05(b_1)$ |
| 1, 4 | $D_{2h}$ | $8.60(b_1)$ | $9.15(b_{3g})$ |
| 1, 2, 3 | $C_{2v}$ | $8.60(a_2)$ | $8.60(b_1)$ |
| 1, 2, 4 | $C_{1h}$ | $8.50(a'')$ | $8.95(a'')$ |
| 1, 3, 5 | $D_{3h}$ | $8.65(e'')$ | $8.65(e'')$ |
| 1, 2, 3, 5 | $C_{2v}$ | $8.30(b_1)$ | $8.60(a_2)$ |

(From Klessinger [245])

According to PES data, the effect of the methyl groups in the benzene ring on the energy of its electron levels is additive [245], the total effect depending on their mutual location. The values of the first and second ionization potentials of various methylbenzenes are given in Table 6.4 by way of illustration.

The results of the analysis of the electron absorption spectra on the basis of the PPP CI calculations for toluene, para-xylene and mesitylene are given in Table 6.5.

### 6.2.2 tert-Butylbenzene, neopentylbenzene and their organoelement analogues

Detailed information on the interaction between $\sigma$ and $\pi$ orbitals has been obtained for the series of tert-butylbenzene, neopentylbenzene and their analogues, phenyl and benzyl derivatives of the IVA group elements.

It is believed that in the molecules of tert-butylbenzene analogues XLII, the interaction of $\sigma$(E—C) orbitals of the $EMe_3$ substituent with $\pi$ orbitals of the benzene ring is responsible for the changes in the energy levels. Efficiency of this interaction

**Table 6.5.** Experimental and calculated energies and intensities of EAS bands of toluene, para-xylene, and mesitylene

| Compound | Band No. | $\Delta E_{exp}$ (eV) | $f_{exp}$ | Symmetry of electron transition | $\Delta E_{calc}$ PPP CI (eV) | $f_{calc}$ | Polarization* |
|---|---|---|---|---|---|---|---|
| Toluene | 1 | 4.61–4.72 | 0.004 | $^1B_2$ | 4.698 | 0.007 | $x$ |
| | 2 | 5.83–5.96 | — | $^1A_1$ | 5.879 | 0.021 | $y$ |
| | 3 | 6.57;6.65 | 0.713 | $^1B_2$ | 6.525 | 1.092 | $x$ |
| | | | | $^1A_1$ | 6.527 | 1.158 | $y$ |
| *p*-Xylene | 1 | 4.51–4.61 | 0.008 | $^1B_{1u}$ | 4.624 | 0.025 | $x$ |
| | 2 | 5.60–5.83 | — | $^1B_{3u}$ | 5.762 | 0.061 | $y$ |
| | 3 | 6.38;6.46 | 0.815 | $^1B_{3u}$ | 6.500 | 1.166 | $y$ |
| | | | | $^1B_{1u}$ | 6.506 | 1.058 | $x$ |
| Mesitylene | 1 | 4.55–4.67 | 0.002 | $^1A_{2'}$ | 4.628 | 0 | |
| | 2 | 5.51–5.83 | — | $^1A_{1'}$ | 5.718 | 0 | |
| | 3 | 6.21;6.31 | 0.698 | $^1E'$ | 6.413 | 1.143 | $y$ |
| | | | | $^1E'$ | 6.413 | 1.143 | $x$ |

*$Y$ is the long axis of the molecule.
(From Marschner, and Goetz, [246])

was estimated within the framework of the simple method by solving the second-order determinant:

$$\begin{vmatrix} (H_\pi - \varepsilon) & H_{\pi\sigma} \\ H_{\sigma\pi} & (H_{\sigma(\mathrm{EC})} - \varepsilon) \end{vmatrix} = 0$$

The found $H_{\sigma\pi}$ integral values show that the ability of the $\sigma$(CE) and $\pi$ orbitals to mix in XLII compounds decreases in the series C > Si > Ge > Sn. The corresponding data and also the results of calculations of the frontier MO energies in PhE $(Me)_3$ compounds in the $\pi$ approximation (PES parametrization) are given in Table 6.6. Ref. [247], [248].

Phenylsilanes absorb in the region of the longest wavelengths of the UV spectrum: e.g. para-bis(trimethylsilyl)benzene $Me_3Si—C_6H_4—SiMe_3$ absorbs at 226 and 270 nm, against 214 and 263 nm in the UV spectrum of para-bis(tert-butyl)benzene. This is probably due to involvement of the vacant 3*d* orbitals of silicon in the unoccupied MOs of phenylsilanes and the appropriate decrease in their energies. The contribution of the 3*d*(Si) orbitals is estimated quantitatively by the CNDO/2 calculations with and without incorporation of the 3*d* orbitals in the basis [249, 250].

Since the vacant $d_{xz}$ orbitals have $b_1$ symmetry (within the framework of the $C_{2v}$ symmetry group), they can probably mix with the occupied $\sigma$ and $\pi$ orbitals of the same symmetry thus decreasing the energy level of the corresponding HOMOs, and the absolute value of the $H_{\sigma\pi}$ integral estimating the efficiency of the $\sigma$–$\pi$ interaction in $C_6H_5EMe_3$ compounds (XLII). Besides, rather weak interaction between the $\pi$ and $\sigma$(E—C) orbitals in these compounds is probably explained by the fact that the

**Table 6.6.** Ionization potentials, bands positions in UV spectra, HOMO energies and effects of $\sigma$ and $\pi$ corbital interactions in $PhEMe_3$ (XLII) and $PhCH_2EMe_3$ (XLIII) compounds

| Compound (E) | $IP^*_{E-C}$ (eV) [247] | PES Ionization potentials of XLII and XLIII, eV [248] | | | | | | | | $H_{\sigma(EC)\pi}$, eV[†] | $\varepsilon_{HOMO}$, eV | UV spectrum [249] $\lambda(1g\varepsilon)$ |
|---|---|---|---|---|---|---|---|---|---|---|---|---|
| XLII: | | | | | | | | | | | | |
| $a$(E=C) | 11.40 | 8.83 | 9.31 | 10.92 | 11.39 | 12.11 | 12.72 | 12.93 | 14.93 | −1.03 | −8.85 | 208(3.93), |
| | | ($\pi_2$) | ($\pi_3$) | | $\sigma_{CC}$ | ($\pi_1$) | | | | | | 258(2.30) |
| $b$(E=Si) | 10.50 | 9.00 | 9.26 | 10.48 | 10.60 | 11.56 | 11.91 | | | −0.60 | −9.01 | 211(4.03), |
| | | ($\pi_2$) | ($\pi_3$) | $\sigma_{SiC}$ | | | | | | | | 260(2.48) |
| $c$(E=Ge) | 10.20 | 9.00 | 9.25 | 9.98 | 10.30 | 11.63 | 11.97 | 12.80 | 13.91 | −0.54 | −9.01 | 208(4.08), |
| | | ($\pi_2$) | ($\pi_3$) | $\sigma_{GeC}$ | | $\sigma_{CC}$ | ($\pi_1$) | | | | | 258(2.30) |
| $d$(E=Sn) | 9.70 | 8.94 | 9.29 | 9.71 | 10.02 | 11.58 | 12.02 | 12.72 | 13.87 | −0.48 | −8.95 | 209(4.09), |
| | | ($\pi_2$) | ($\pi_3$) | $\sigma_{SnC}$ | | $\sigma_{CC}$ | ($\pi_1$) | | | | | 252(2.78) |
| XLIII: | | | | | | | | | | | | |
| $a$(E=C) | 11.40 | 8.77 | 9.13 | 10.73 | 11.41 | 12.10 | 12.49 | 13.81 | | −1.21 | −8.73 | 211(4.67), |
| | | ($\pi_2$) | ($\pi_3$) | | $\sigma_{CC}$ | ($\pi_1$) | | | | | | 259(2.24) |
| $b$(E=Si) | 10.50 | 8.42 | 9.08 | 10.53 | 11.40 | 11.98 | 12.78 | 13.55 | | −1.33 | −8.48 | 222(3.94), |
| | | ($\pi_2$) | ($\pi_3$) | $\sigma_{CS}$ | $\sigma_{CC}$ | ($\pi_1$) | | | | | | 267(2.63) |
| $c$(E=Ge) | 10.20 | 8.40 | 9.12 | 10.41 | 11.43 | 12.01 | 12.92 | 13.67 | | −1.39 | −8.36 | 225(3.86), |
| | | ($\pi_2$) | ($\pi_3$) | $\sigma_{CGe}$ | $\sigma_{CC}$ | ($\pi_1$) | | | | | | 269(2.53) |
| $d$(E=Sn) | 9.70 | 8.21 | 9.21 | 9.90 | 10.32 | 11.42 | 11.99 | 12.83 | 13.25 | −1.35 | −8.24 | 236(3.74), |
| | | ($\pi_2$) | ($\pi_3$) | $\sigma_{CSn}$ | | $\sigma_{CC}$ | ($\pi_1$) | | | | | 272(2.69) |

*$I_{E-C}$ denotes ionization potentials of $\sigma_{E-C}$ orbital in $Me_4E$ (PES).
[†]$H_{\sigma(CE)\pi}$ integral and $c_{CE}$ coefficient values are given for XLIII.
(From Traven, and Stepanov, [62])

favourable factor of closeness of the energies of the interacting orbitals on the transition from Si to Sn is, to a certain degree, compensated by the unfavourable factor of the increasing $C_{Ar}$—E(Me) bond distance in the same direction.

The ability of various $\sigma$(C—E) orbitals to interact with $\pi$ orbitals can be estimated more objectively by comparing the appropriate spectral characteristics of benzene derivatives XLIII. The $C_{Ar}$—C(E) bond distance in these compounds does not depend on a particular E, and the possibility of overlapping of the *d* orbitals of the E atom with the benzene $\pi$ orbital is excluded. Ionization potentials measured by PES [248], the $H_{\sigma(CE)\pi}$ integral values, and the position of the long wavelength absorption bands in the UV spectra of the $PhCH_2EMe_3$ compounds are also given in Table 6.6. It can be seen that the interaction of the $\sigma$(C—E) and $\pi$ orbitals in these compounds increases with decreasing difference in the orbital energies to attain its maximum for $\sigma$(C—Ge) and $\sigma$(C—Sn) orbitals.

The data given in Table 6.6 agree with the experimental findings.

Among other types of organic compounds of the IVA group elements, in which the $\sigma$ and $\pi$ orbitals actively interact, arylpolysilanes are noteworthy. The s(Si-Si)—$\pi$ interactions in them were estimated (taking account of the stereoelectronic factor) by PE spectra of arylpolysilanes and absorption spectra of their CT complexes with TCNE [251-255]. The effects of $\sigma$-$\pi$ conjugation were also studied in detail for other organometallic compounds [256, 257].

In the gas phase, the anions of alkylbenzenes are formed at lower energies of the bombarding electrons than the benzene anion. Below compared are the electron affinities of alkylbenzenes (with reference to benzene) in the gas phase and in solution:

| $\Delta EA$ (eV): | Toluene | p-Xylene | m-Xylene |
|---|---|---|---|
| gas phase | $0.04 \pm 0.02$ | $0.08 \pm 0.02$ | $0.09 \pm 0.02$ |
| solution | $-0.022$ | $-0.0333$ | $-0.059$ |

| $\Delta EA$ (eV): | o-Xylene | Isopropyl-benzene | tert-Butyl-benzene |
|---|---|---|---|
| gas phase | $0.03 \pm 0.03$ | $0.07 \pm 0.02$ | $0.09 \pm 0.02$ |
| solution | $-0.100$ | $-0.055$ | $-0.070$ |

In all cases, $\Delta EA$ was measured as the *EA*(alkylbenzene)-*EA*(benzene) difference. Positive $\Delta EA$ indicates that alkylbenzene has greater electron affinity than $C_6H_6$ does. At the same time, ETS findings show that the lifetime of the alkylbenzene anion-radical (like that of toluene) is shorter: the vibrational structure of the resonance signals is absent in their ET spectra [174, 258].

Electron affinity of alkylbenzenes in solvents was estimated in the reaction

$$C_6H_6 + RR'C_6H_4^{\dot{-}} \rightleftharpoons C_6H_6^{\dot{-}} + RR'C_6H_4$$

Solutions of known benzene and alkylbenzene concentration were reduced by the K/Na alloy in a mixture of tetrahydrofuran and dimethoxyethane. $K_{eq}$ was determined by the ESR spectroscopy. It was found that, as distinct from the measurements in the gas phase, the alkyl group in the benzene nucleus hinders one-electron reduction of the corresponding alkylbenzene in solvent (probably due to the increasing solvation energy of the anion containing the bulky alkyl group) [259].

The energies of the unoccupied levels of $Me_3EC_6H_4EMe_3$ compounds (where E = C, Si, Ge, Sn) were also estimated [174, 259]. According to ETS, introduction of either of the $EMe_3$ substituents somewhat increases stability of the anion state but the effects are most pronounced with the elements (Si, Ge, Sn) having unoccupied *d* orbitals [174]. Measured in the gas phase, the energies of the occupied and vacant MOs of alkylbenzenes and their analogues, containing atoms of the group IVA elements, are compared in Fig. 6.6.

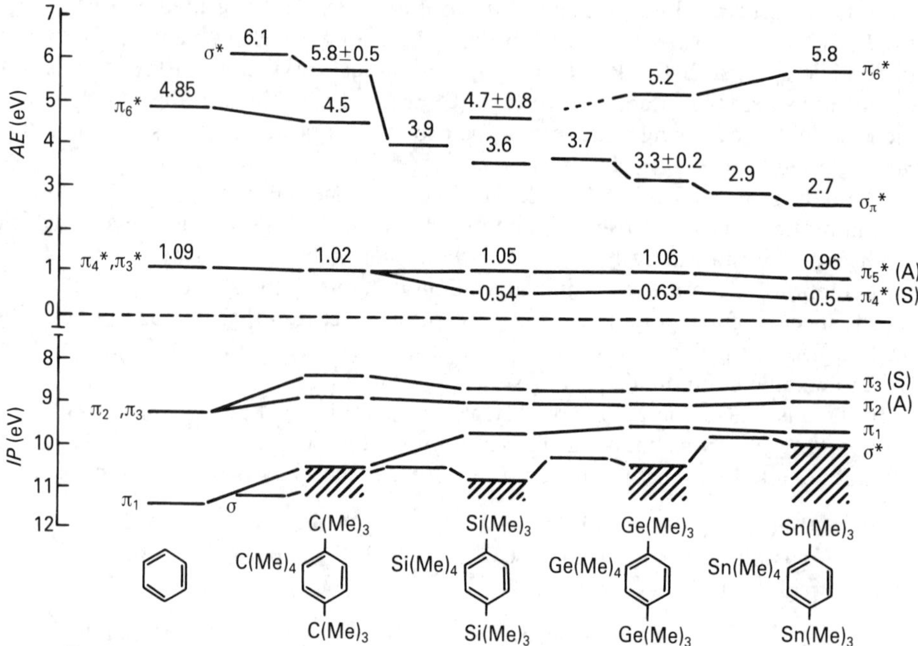


Fig. 6.6. Experimental energies of $\pi$ orbitals of benzene derivatives *p*-$Me_3E$-$C_6H_4$-$EMe_3$ (occupied orbitals according to PES and vacant — according to ETS). (From Giordan and Moore [174])

Splitting of the lowest unoccupied $\pi$ orbital in alkylbenzenes is confirmed by the results of the analysis of the ESR spectra of their anion radicals [50, 229]. Like in the anion radical of toluene, an unpaired electron in the anion radicals of tert-butylbenzene and other alkylbenzenes is found in the antisymmetric $\pi$ orbital of the benzene ring.

By contrast, the splitting constants of unpaired electron on the ring protons of the trimethylsilylbenzene anion radical illustrate a different way to remove of degeneracy of the benzene vacant orbitals: an unpaired electron in these anion radicals is found in the $\pi$(symm)orbital [260]:

| Position | $a_H$ | $\rho_{exp}$ | $\rho_{calc}$ |
|---|---|---|---|
| $CH_3$ | 0.24 | — | 0.118 |
| $C(CH_3)$ | — | — | 0.282 |
| ortho | 2.66 | 0.095 | 0.096 |
| meta | 1.06 | 0.038 | 0.060 |
| para | 8.18 | 0.292 | 0.287 |

### 6.2.3 α(Alkyl)phenylcyclopropanes

The specific example of interaction between the alkyl group and the π system can be observed in phenylcyclopropane and its α-alkyl substituted derivatives (XLIV) [33]

XLIV: a, orthogonal conformation  b, planar conformation

Calculations of phenylcyclopropane performed by various methods [24, 261] showed that the perpendicular (relative to substituent R) conformation *a* is preferable to both phenylcyclopropane and its α-alkyl substituted derivatives. It is in this conformation that the 2*p* orbital of the $C_\alpha$ atom contributes substantially to the occupied frontier MO of phenylcyclopropane. By changing the substituent R one can estimate the ability of the corresponding σ(C—R) orbital to interact with the π system of the benzene ring (Table 6.7).

**Table 6.7.** Ionization potentials (eV) of some cyclopropane derivatives

| Compound | $IP_1$ | $IP_2$ | $IP_3$ | $IP_4$ |
|---|---|---|---|---|
| Toluene | 8.9 | 9.13 | | |
| Cyclopropane | 10.90 | | | |
| Methylcyclopropane | 10.10 | 10.90 | | |
| XLIV R=H | 8.66 | 9.21 | 10.53 | 11.11 |
| R=$CH_3$ | 8.73 | 9.17 | 10.09 | 10.59 |
| R=$C_2H_5$ | 8.70 | 9.17 | 9.95 | 10.50 |
| R=$CH(CH_3)_2$ | 8.63 | 9.12 | 9.74 | 10.38 |
| R=$C(CH_3)_3$ | 8.63 | 9.15 | 9.63 | 10.33 |

(From Prins *et al.* [261])

It is important to note that the first ionization potential of phenylcyclopropane is lower than that of α-methyl- and α-ethyl substituted compounds. This fact probably indicates the hyperconjugation effect: preference of the σ(C—H) orbitals, as compared to the σ(C—C) orbitals, in their interaction with benzene πMOs.

### 6.2.4 Styrene and Stilbene

#### *Styrene*

The ability of styrene to enter Diels–Alder and other cycloaddition reactions, its photochemical activity and problems of stereoregular polymerization account for the significant interest in the study of the electron structure of this hydrocarbon.

The electron levels of styrene depend on the conditions of interaction of the π orbitals of the benzene and ethylene fragments in its molecule. According to the

Raman spectrum [262], the styrene molecule is planar. This conformation is also the most stable according to *ab initio* calculations in the STO-3G basis set [237]. The rotational barrier of the vinyl group of the styrene molecule is, however, low (about 7.95 kJ/mol). This is explained by destabilization of the planar form at the expense of steric repulsion between the $\alpha$-hydrogen in the vinyl group and the hydrogen in the ortho position of the phenyl ring. With optimization of the geometry, the C—C=C angle in the planar conformer increases to 128°, while the rotational barrier increases to 18.50 kJ/mol.

The interaction of the $\pi$ orbitals of the benzene and ethylene fragments in the styrene molecule was estimated quantitatively by PE spectroscopy [263–6]. The PE spectrum of styrene has three bands in the region of low ionization potentials at 8.42, 9.13, and 10.55 eV. The first two bands correspond to splitting of the degenerate benzene $\pi(1e_{1g})$ orbitals. The first band has fine structure with the frequency of 1533 $cm^{-1}$, which corresponds to the C—C vibrations of the ring (the corresponding vibrations in the IR spectrum of benzene have a frequency of 1585 $cm^{-1}$). The second band has no fine structure and is assigned to ionization of the unperturbed $\pi$(asymm) orbital of the benzene ring. The third band could, by its position at 10.55 eV, be attributed to ionization of the unperturbed orbital of the vinyl group (in the ethylene molecule, this orbital is ionized at 10.5 eV). But it should be noted that the third band of the styrene spectrum has vibrational structure with a frequency of 969 $cm^{-1}$, corresponding to the vibrations of the C—C bond between the benzene ring and the vinyl group.

The first band in the PE spectrum of styrene thus corresponds to ionization of electrons of the antibonding combination of benzene $\pi$ and ethylene $\pi$ orbitals, and the third band to ionization of the electrons of the bonding combination of the same orbitals.

The first three and the fifth ionization bands in the Penning spectrum have higher intensities, which suggest ionization of $\pi$ orbitals [178]. It is important to note that the position of the fifth band at 12.2 eV is close to the ionization potential (12.3 eV) of the benzene $1a_{2u}\pi$ orbital. This agrees with CNDO/2 calculations according to which the lowest occupied benzene $\pi$ orbitals are only slightly mixed with the $\pi$ orbitals of the vinyl group [263]. Changes in the $\pi$ electron structure due to the interaction of the $\pi$ orbitals of the adjacent fragments associated with the introduction of the vinyl group into the benzene molecule are shown in Fig. 6.7.

It is interesting that considerable interaction of the orbitals of the substituent and the benzene ring causes only slight changes in the $\pi$ population of the styrene molecule (compared with benzene) [237]:

$CH_2=CH$

0.996

1.002

0.998

1.000

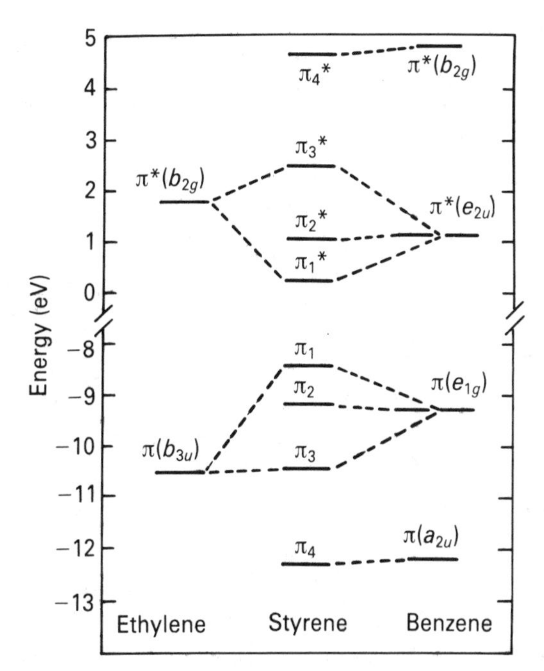


Fig. 6.7. Formation of $\pi$MO of styrene according to PES and ETS findings. (From Burrow *et al.* [268])

The scheme of formation of the highest occupied $\pi$ orbitals in the styrene molecule makes it possible to estimate objectively the steric structure of its various derivatives [264]. The conformational analysis is based on the fact that splitting between the first and the third bands in the PE spectrum is most sensitive to the dihedral angle between the vinyl group and the benzene ring planes. Using the CNDO/S method, Kobayashi *et al.* calculated the differences of energies $\Delta\varepsilon_{1,3}$ of the orbitals (that correspond to the first and third ionization potentials) for all styrene conformations differing in the value of the dihedral angle from 0 to 90° at 15° intervals. The obtained data were compared graphically with the values of the angle $\theta$. By measuring the experimental value of $\Delta IP_{1,3}$ from the PE spectrum, one can determine reliably the value of the dihedral angle. The results of the conformational analysis of various substituted styrenes according to PE and UV spectra are given in Table 6.8.

Indene can be considered as an example of a styrene derivative. All electron levels of indene are higher than those of styrene by 0.3 eV. The fragment of the vinyl group in the indene molecule is fixed rigidly in the benzene ring plane. Since the styrene molecule in the gas phase is also planar, the higher electron levels of indene are explained by the inductive effect of the methylene group [267].

The vacant levels of the styrene molecule are also shown in Fig. 6.7 [268]. In complete agreement with the properties of alternant hydrocarbons, the vacant and occupied electron levels of styrene are symmetrical: according to the ET spectrum, there are four vacant $\pi$ levels in the styrene molecule. The $\pi_1^*$ orbital is the bonding combination of the $\pi^*(b_{2g})$ orbital of ethylene and $\pi^*(e_{2u})$ orbital of benzene; the

**Table 6.8.** Ionization potentials (eV) and steric structure of substituted styrenes (PE- and U-V spectra data)

| Compound | $IP_1$ | $IP_2$ | $IP_3$ | $IP_4$ | $\theta$ (PES) | $\theta$ (UVS) |
|---|---|---|---|---|---|---|
| Styrene | 8.49 | 9.27 | 10.55 | 11.52 | — | — |
| 2-Methylstyrene | 8.53 | 8.99 | 10.37 | 11.22 | 38 ∓ 7 | 31 |
| α-Methylstyrene | 8.52 | 9.18 | 10.12 | 11.26 | 29 ∓ 6 | 38 |
| cis-β-Methylstyrene | 8.48 | 9.18 | 10.26 | 11.30 | 22 ∓ 6 | 35 |
| trans-β-Methylstyrene | 8.34 | 9.09 | 10.25 | 11.46 | 12 ∓ 7 | 0 |
| trans-β-Ethylstyrene | 8.30 | 9.09 | 10.19 | 11.37 | 12 ∓ 4 | — |
| trans-β-tert-Butylstyrene | 8.18 | 9.05 | 10.10 | 11.02 | 8 ∓ 8 | — |
| cis-β-tert-Butylstyrene | 8.85 | 9.27 | 9.5 | 10.31 | 72 | — |
| 2,4-Dimethylstyrene | 8.22 | 8.80 | 10.11 | 11.06 | 41 ∓ 8 | 30.7 |
| 2,6-Dimethylstyrene | 8.48 | 8.62 | 10.04 | 11.04 | 68 ∓ 10 | 54 |

(From Marschner and Goetz [246])

value of $EA_1$ is $-0.25$ eV. The $\pi_3^*$ orbital is the antibonding combination of the same orbitals; the value of $EA_3$ is $-2.48$ eV. The $\pi_2^*$ and $\pi_4^*$ orbitals are vacant benzene orbitals that are not involved in orbital mixing: the $\pi_2^*$ is the benzene $e_{2u}$ orbital ($EA_2 = -1.05$ eV) and the $\pi_4^*$ orbital is the $b_{2g}$ benzene orbital ($EA_4 = -4.67$ eV).

***Stilbene***

Stilbene and its derivatives are able to interact effectively with electromagnetic radiation. On exposure to light, they undergo isomerization, cyclization, etc. They are used as valuable optic bleaching agents, photographic compounds, and intermediates for the manufacture of heterocyclic compounds.

Stilbene and its derivatives can exist in two configurations relative to the double bond:

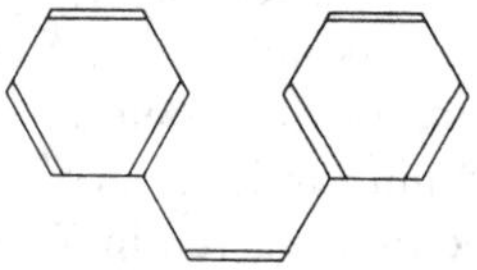

12 13 11 14 10 9 8 7 1 2 6 3 5 4

(Z)−stilbene 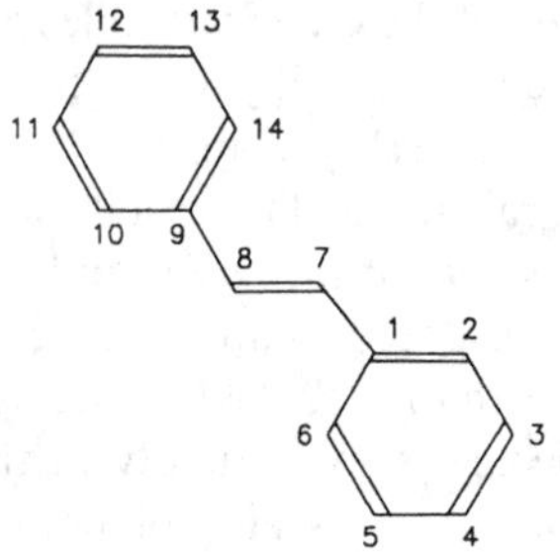 (E)−stilbene

The molecule of Z-stilbene is non-planar according to the measurements in the liquid and gas states. Quantum chemical calculations by various methods predict the values of the dihedral angle $\theta$ of 40–45° [269].

The interactions that might distort the planar conformation of the E-stilbene molecule are absent, but this molecule is not planar either: the dihedral angle between the benzene rings is 30°.

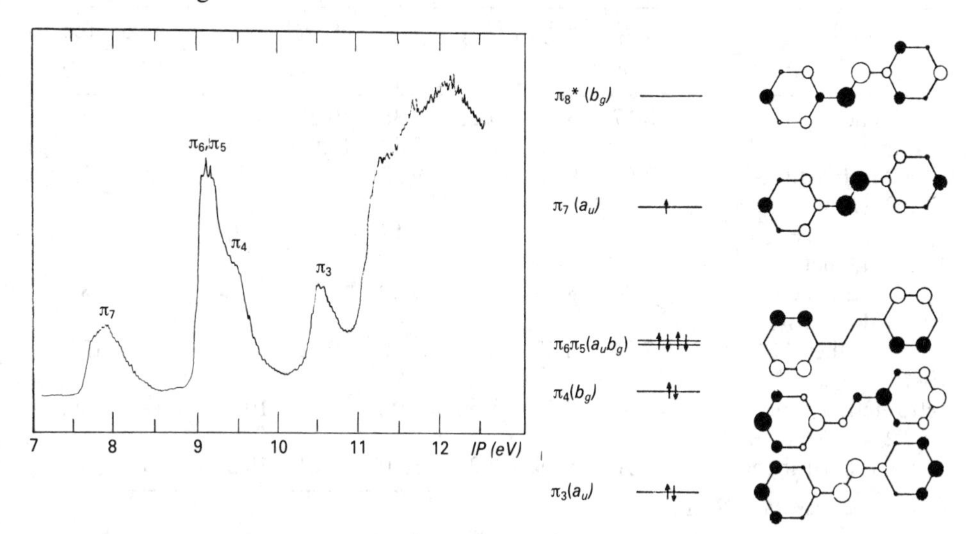


Fig. 6.8. He(I) PE spectrum and $\pi$MOs (according to HMO calculations) of stilbene. (From Haselbach *et al.* [270])

The first bands in the PE spectrum of E-stilbene at 7.87, 9.08, 9.50 and 10.51 eV are unambiguously assigned to ionization of the $\pi_3$–$\pi_7$ orbitals [270] (Fig. 6.8). Ionization of two lower occupied orbitals $\pi_1$ and $\pi_2$ are assigned to the bands at 11.3 and 12.73 eV with a smaller degree of certainty. The assignment was however verified by the Penning spectrum, by the study of the 'perfluoro' efect [271], and shown in Fig. 6.8. The values of ionization potentials of stilbene and its derivatives are given in Table 6.9.

As for the corresponding substituted styrenes, Kobayshi *et al.* used PE spectra and quantum chemical calculations to study the electron structure and conformations of some stilbene derivatives (Table 6.9). To that end, they compared the differences of the first and fifth ionization potentials as measured from the PE spectra and calculated by the CNDO/S method. These potentials correspond to ionization of the $\pi_7$ and $\pi_3$ orbitals, which are delocalized over the entire molecule and therefore depend on conformation of the molecule to a greater degree than the other potentials.

The PES data and quantum chemical calculations were successfully used to interpret the electron absorption spectra of cation radicals of stilbene and its methyl substituted compounds [270].

The values of electron affinity of the E-stilbene molecule in the gas phase were measured by ET spectroscopy. The values found were $-0.97$, $-2.53$, $-4.4$ and $-5.0$ eV. (It should be noted that this method cannot be used to measure positive values of electron affinity [41].) These findings agree with the estimations made on the basis of the quantum chemical calculations within the framework of the MO LCFO model. Four values of electron affinity were also determined for Z-stilbene: $-0.92$, $-1.68$, $-3.4$, and $-4.67$ eV.

**Table 6.9.** Ionization potentials (eV) and steric structure of substituted stilbenes (PES data)

| Compound | $IP_1$ | $IP_2$ | $IP_3$ | $IP_4$ | $IP_5$ | $IP_6$ | $\theta$, degree | |
|---|---|---|---|---|---|---|---|---|
| | | | | | | | PES | UVS |
| Z-Stilbene | 8.17 | 8.99 | 9.22 | 9.36 | 10.27 | 11.3 | 45 | 42.2 |
| E-Stilbene | 7.87 | 9.08 | 9.08 | 9.50 | 10.51 | 11.3 | 33 | (32.2) |
| (E)-4-Methylstilbene | 7.63 | 9.02 | 9.02 | 9.33 | 10.33 | 11.1 | 32 | (32.2) |
| (E)-2,2′-Dimethyl-stilbene | 7.83 | 8.72 | 8.72 | 9.17 | 10.27 | 11.0 | 40 | 37.3 |
| (E)-3,3′-Dimethyl-stilbene | 7.74 | 8.73 | 8.73 | 9.20 | 10.25 | 11.1 | 37 | (32.2) |
| (E)-4,4′-Dimethyl-stilbene | 7.54 | 8.82 | 8.82 | 9.00 | 10.12 | 11.0 | 36 | (32.2) |
| (E)-2,2′,5,5′-Tetramethylstilbene | 7.63 | 8.37 | (8.60) | 9.02 | 10.06 | 10.8 | 40–1 | 39.4 |
| (E)-2,2′,4,4′,6,6′-Hexamethylstilbene | 7.90 | 8.29 | 8.29 | 8.80 | 9.61 | 10.5 | 60 | 59.7 |

(From Kobayashi *et al.* [269])

HFI constants [46] were used to evaluate experimentally the density of the unpaired electron in the stilbene anion radical. As a result, the character of the distribution of electron density in the frontier vacant MO of stilbene was adequately predicted by calculation by the HMO method:

| At. No. | $a_H$ (Oe) | $\rho$(exp) ($Q = 28$) | $\rho$(*HMO calc*) |
|---|---|---|---|
| 2, 6, 10, 14 | 2.48 | 0.089 | 0.075 |
| 3, 5, 11, 13 | 0.56 | 0.020 | 0.0075 |
| 4, 12 | 4.00 | 0.143 | 0.103 |
| 7, 8 | 4.51 | 0.161 | 0.177 |

### 6.2.5 Ethynylbenzenes

The ethynyl fragment is unambiguously oriented by the long axis of ethylnylbenzenes Ph—C≡C—R. This structure is favourable for overlapping of the $\pi$ orbitals of the adjacent fragments.

As expected, the spectra of ethynylbenzene and its derivatives are simple to interpret [263, 272, 273]. Quantum chemical calculations show that in the region of low ionization energies are located four occupied $\pi$ orbitals: bonding and antibonding combinations of the $\pi$ orbital of the ethynyl fragment and the benzene $b_1$ orbital, and two orbitals ($a_2$ orbital of the benzene nucleus and $b_2$ orbital of the ethynyl group) that remain localized in the appropriate fragments. Four well resolved ionization bands in the PE spectra correspond to these orbitals. Fig. 6.9 shows the $\pi$MOs, the results of calculations of their energies by the HAM/3 method [272], and the first ionization potentials of ethynylbenzene.

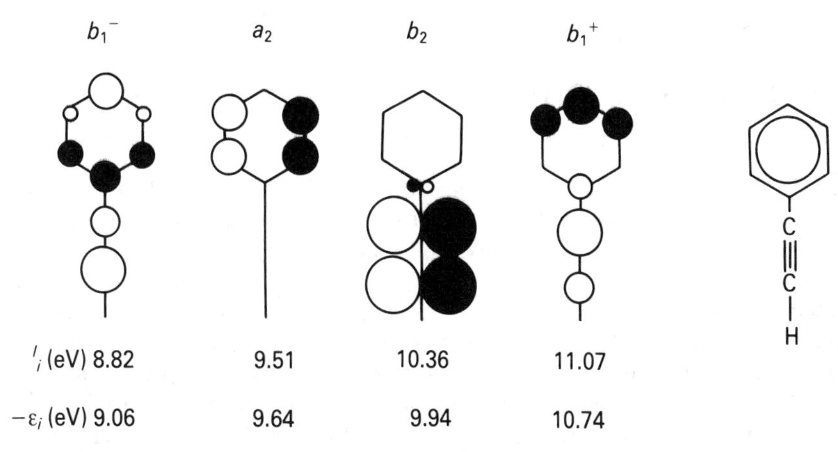


Fig. 6.9. Configurations of higher occupied $\pi$MOs of ethynylbenzene, their energies (HAM/3 calculations) and the corresponding $IP_i$ values (PES findings). (From Elbel [272])

According to calculations and PE spectra, $\pi$ orbitals of the benzene fragment and ethynyl group are mixed effectively in ethynylbenzene. As in the case of styrene, this interaction, however, does not induce substantial changes in $\pi$ populations. Nevertheless, there is an important difference between styrene and ethynylbenzene: the vinyl group slightly increases the $\pi$ population at the ortho and para positions of the benzene, while the ethynyl group slightly decreases it [237]:

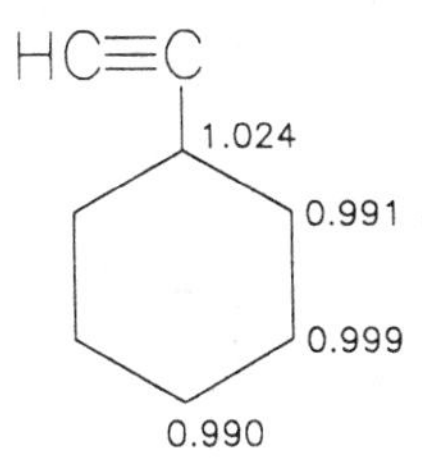


The $\sigma$ and $\pi$ electrons are 'withdrawn' from the benzene ring: the experimental value of the dipole moment is 0.73$D$ (the calculated dipole moment is 0.52$D$).

According to the PE spectra of 1-propynylbenzene, the substitution of the methyl group for the hydrogen of the ethynyl group destabilizes (to different extent) all four $\pi$ orbitals of ethynylbenzene: 8.42, $b_1^-$; 9.26, $a_2$; 9.64, $b_2$; 10.60, $b_1^+$.

Among the PE spectra of other ethynylbenzenes Ph—C≡C—R, the spectrum of tolan (R = Ph) is noteworthy: comparison of the measured intensities of ionization bands in the region of 7–10 eV with those calculated by the SPINDO method permits the calculation of the conformational composition of this compound in the gas phase. According to the PE spectra, tolan in the gas phase is present as a mixture of two conformers, planar $D_{2h}$ ($\theta = 0°$) and non-planar ($\theta = 40°$), in the ratio of 1:0.82 (the values of the dihedral angles are given in parentheses).

## 6.3 POLYCYCLIC AROMATIC HYDROCARBONS

### 6.3.1 Biphenyl, terphenyl, paracyclophanes

***Biphenyl***

Biphenyl derivatives include valuable dyes and biologically active compounds. In recent years, they have been widely used in the manufacture of liquid crystals.

The molecular structure of biphenyl and its derivatives depends on at least two factors: the interaction of $\pi$ orbitals of the adjacent benzene ring 'flattens' the molecule while the steric strain due to repulsion of the hydrogen atoms (or substituents) at the o,o' positions resists this 'flattening'. The steric structure of biphenyl therefore depends on the phase: in the gas phase, it is non-planar [274] but it is planar in the solid phase [275]. According to electron diffraction data, the dihedral angle in the biphenyl molecule is $42 \pm 2°$. For the same reason, the geometry of the biphenyl derivatives depends largely on particular substituents in the phenyl rings. Both biphenyl and its derivatives were studied in detail by electron spectroscopy and quantum chemical methods [33, 276–278]. Their ionization potentials are given in Table 6.10.

$(CH_2)_n$

XLV *a* $(n = 1)$
*b* $(n = 2)$
*c* $(n = 3)$
*d* $(n = 4)$

**Table 6.10.** Ionization potentials (eV) of biphenyl and its analogues

| Compound | $IP_1$ | $IP_2$ | $IP_3$ | $IP_4$ | $IP_5$ | $IP_6$ |
|---|---|---|---|---|---|---|
| Biphenyl | 8.34($\pi_6$) | 9.04($\pi_5$) | 9.20($\pi_4$) | 9.82($\pi_3$) | 11.17($\pi_2$) | 12.7($\pi_1$) |
| XLVa | 7.91 | 8.77 | 9.09 | 9.84 | 11.04 | 13.2 |
| XLVb | (8.1) | (8.9) | (9.0) | (9.8) | | |
| XLVc | 8.3 | 8.8 | 9.1 | 9.76 | | |
| XLVd | 8.5 | 8.8 | 9.0 | 9.4 | | |

(From Gleiter *et al.* [276])

Molecular $\pi$ orbitals of biphenyl arranged in the order of their increasing energies are shown in Fig. 6.10 (according to HMO calculations). It can also be seen from the figure that the energies of $\pi$MO change with the dihedral angle $\theta$. But various MOs respond in a different way to the increase in the angle. The bonding and antibonding combinations of benzene $\pi$(asymm) orbitals ($\pi_4$ and $\pi_5$ orbitals of biphenyl) do not respond to the changes in the angle $\theta$ at all, since they have zero coefficients at the site of attachment. By contrast, the bonding and antibonding combinations of the benzene $\pi$(symm) orbitals with high coefficients at the site of attachment strongly depend on the dihedral angle: as the angle increases, the antibonding combination (the occupied frontier MO($\pi_6$) of biphenyl) is stabilized due to decreasing antibonding interaction, while the bonding combination ($\pi_3$ orbital) is destabilized due to decreasing bonding overlap.

In their study of the orbital structure of biphenyls in the terms of the HMO

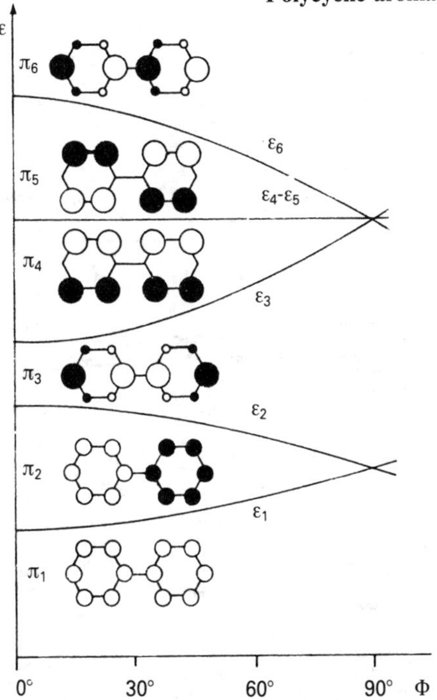


Fig. 6.10. Dependence of energies of occupied $\pi$ orbitals of biphenyl on dihedral angle $\theta$. (From Klessinger and Rademacher [277])

method, Turner *et al.* established the dependence between the degree of splitting of the energy levels of $\pi_3$ and $\pi_6$ orbitals and the value of the dihedral angle $\theta$:

$$\Delta\varepsilon = \varepsilon(\pi_3) - \varepsilon(\pi_6) = 2\beta\cos\theta \tag{6.1}$$

A set of $\Delta IP = IP(\pi_6) - IP(\pi_3)$ values has been found for some biphenyls of the known structure. In accordance with Eq. (6.1), the values obtained within the framework of the Koopmans' theorem turned out to be dependent linearly on the values of $\cos\theta$:

$$\Delta IP = 1.65\cos\theta + 0.35 \text{ eV} \tag{6.2}$$

Equation (6.2) can be used to estimate the conformation of any substituted biphenyl by its PE spectrum. It turns out that $\theta$ values obtained by the proposed scheme agree well with the data obtained by other methods. It is important to note that the straight line (6.2) does not pass through the origin: one cannot disregard the interaction of the $\pi$ orbitals of one benzene ring with the $\sigma$ orbitals of the C—C bonds of the adjacent ring in conformations with high values of the angle $\theta$. This $\sigma$–$\pi$ mixing explains the presence of $\Delta IP$ even at $\theta = 90°$, i.e. in complete absence of $\pi$–$\pi$ overlapping.

The conformational analysis proposed by Turner *et al.* was successfully used by many authors [33, 276–8]. Gleiter *et al.*, for example, used Eq. (6.2) to study the steric structure of the compounds XLV. They found out that the difference in the first and fourth ionization potentials, corresponding to the $\pi_6$ and $\pi_3$ orbitals

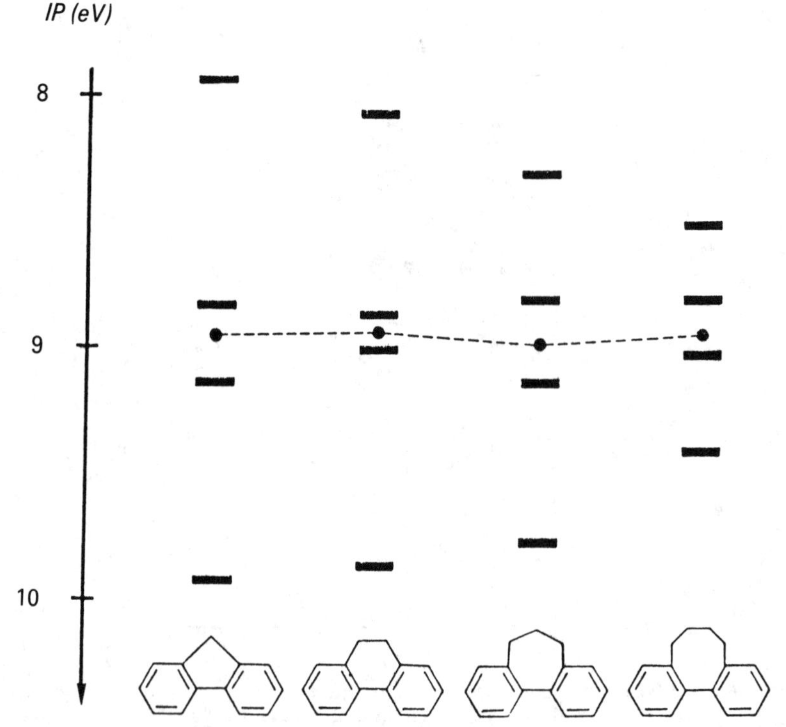


Fig. 6.11. Energies of higher occupied $\pi$ electron levels of some o,o′-disubstituted biphenyls XLV according to PES. (From Gleiter [276])

respectively, decreases unambiguously with increasing $n$ (the number of methylene groups in the molecules of these compounds; see Fig. 6.11). Accordingly, dihedral angles in these biphenyl derivatives change from 0 to 80°; $a - 0°$; $b - 20°$; $c - 40°$; $d - 80°$ [276]. Among the known biphenyls, 2,2′-diethynylbiphenyl is noteworthy. The difference between the first and fourth ionization potentials of this compound being only 0.86 eV allows, according to Eq. (6.2), to determine the angle between the planes of the benzene rings (72°). (According to Gleiter *et al.*, this angle was evaluated as 80–90°).

The orbital structures of neutral molecules of biphenyl and its derivatives, and also of their cation radicals, were compared by their PES and electron absorption spectra. The cation radicals were prepared by exposing hydrocarbons in a solid argon matrix at 20 K to UV radiation (mercury and argon lamps) [279]. The data for various biphenyls and their analogues are given in Table 6.11.

On the whole, the electron absorption spectra of cation radicals of biphenyl and its derivatives recorded in solid argon agree with their absorption spectra in solution (690 and 380 nm) and in solid matrices (700 and 390 nm). Thus, absorption of cation radicals in the long wavelength region is assigned to the $\pi_3 \rightarrow \pi_6$ transition (the $\pi_6$ orbital in the cation radical is singly occupied). Within the framework of one-configurational approach, the energy of this transition should be directly estimated by the difference between the first and the fourth ionization potentials (available from

**Table 6.11.** Energies of electron transitions in cation radicals of biphenyl derivatives and estimations of their steric structure (EAS- and PES data)

| Compound | $\lambda_{max}$ (nm) | $\varepsilon_{\lambda(EAS)}$ (eV) | | $\Delta IP$(PES) $= IP_4 - IP_1$ (eV) | $\theta$ (degrees) |
|---|---|---|---|---|---|
| | | calc. | exp. | | |
| Naphthalene | 675.2 | 1.85 | 1.84 | 1.85 | 0 |
| Phenanthrene | 899.8 | 1.38 | 1.38 | 1.38 | 0 |
| Fluorene | 641.1 | 1.96 | 1.93 | 1.93 | 0 |
| Dihydrophenanthrene | 642 | 1.95 | 1.93 | 1.84 | 20 |
| Biphenyl | 669.4 | 1.82 | 1.85 | 1.48 | $42 \pm 2$ |
| 4-Methylbiphenyl | 674.5 | 1.80 | 1.83 | 1.45 | $42 \pm 2$ |
| 2-Fluorobiphenyl | 675.8 | 1.75 | 1.83 | 1.40 | $49 \pm 5$ |
| 2,2′-Difluorobiphenyl | 705 | 1.65 | 1.76 | 1.15 | $60 \pm 5$ |
| Perfluorobiphenyl | 900 | 1.20 | 1.38 | 0.90 | $70 \pm 2$ |
| 2-Methylbiphenyl | 746 | 1.40 | 1.66 | 0.90 | $70 \pm 5$ |
| 2,2′-Dimethylbiphenyl | 749 | 1.20 | 1.65 | 0.70 | $78 \pm 5$ |

(From Andrews *et al.* [279])

the PE spectra). A considerable hypsochromic shift of this band in the absorption spectrum of the biphenyl cation radical (compared with the predicted value $\Delta IP = IP_4 - IP_1$) is probably explained by flattening of the biphenyl molecule as it is converted to the cation radical state. It has already been mentioned that flattening of the biphenyl molecule must destabilize the $\pi_6$ orbital and stabilize the $\pi_3$ orbital. Strong absorption of the biphenyl cation radical at 382 nm can be assigned to the transition between $\pi_2$ and $\pi_6$ orbitals but this transition is forbidden by the symmetry rule. It is more probable that the said absorption is explained by the $\pi_6 \rightarrow \pi_7$ transition, the first $\pi \rightarrow \pi^*$ type transition in the biphenyl cation radical.

In good agreement with the proposed interpretation of the PE spectra are the electron absorption spectra of the cation radicals of phenanthrene, fluorene, and dihydrophenanthrene, whose molecules are either planar or almost planar. The data given in Table 6.11 show that the energies of the bands in their electron absorption spectra agree well with the corresponding estimations of the band energies from PE spectra.

According to the ET spectrum, the first value of the electron affinity ($EA_1$) of biphenyl in the gas phase is $-0.37$ eV [46]. Below is the distribution of an unpaired electron in the frontier vacant MO of biphenyl, as estimated by ESR of the anion radical [280] and the HMO method. Comparison of the calculated and experimental data indicates that the unpaired electron is located in the bonding combination of $\pi$ orbitals of the $b_1$ symmetry of the two benzene nuclei:

| position | $a_H$ | Electron density | |
|---|---|---|---|
| | | exp. | HMO calc. |
| ortho | 2.73 | 0.097 | 0.090 |
| meta | 0.43 | 0.015 | 0.020 |
| para | 5.46 | 0.195 | 0.158 |

***Terphenyl***

The molecule of terphenyl is planar in the solid phase and almost planar in the gas phase: the dihedral angle between the benzene rings is 0–20° [281, 282].

The first three bands in the PE spectrum in the gas phase are assigned unambiguously to ionization of $\pi$ electrons. This assignment is confirmed by the 1:4:1 ratio of their intensities which corresponds to the degree of degeneracy of $\pi$(asymm) orbitals of the benzene fragments in the terphenyl molecule.

The PE spectrum of a solid terphenyl film was also recorded. As in other polycyclic aromatic hydrocarbons (see Section 6.3), adequate correlation of both spectra is observed (the polarization energy of $-1.6$ eV; Fig. 6.12):

| *IP* (eV) | | | | | |
|---|---|---|---|---|---|
| gas phase | 8.20 | 9.17 | 10.10 | 11.20 | 11.67 |
| solid phase | 6.6 | 7.6 | 8.7 | 9.7 | — |

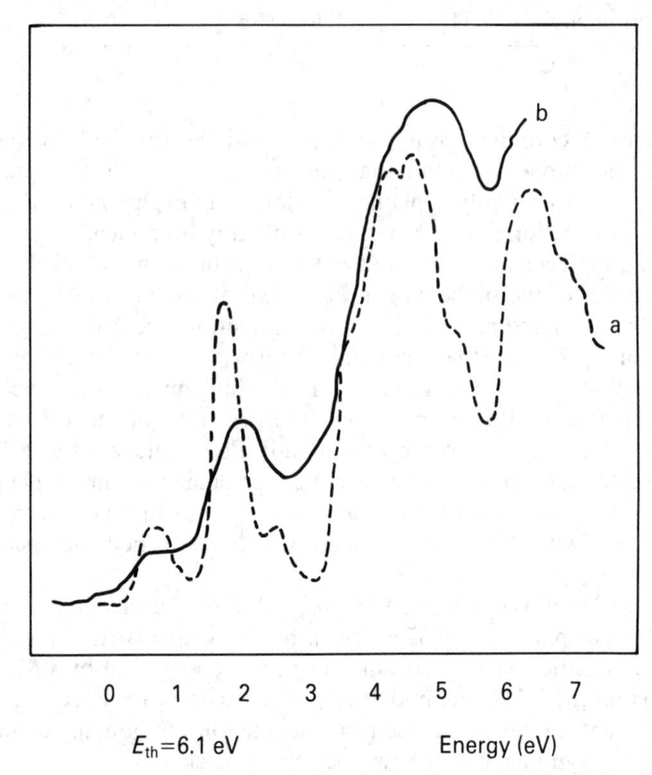


Fig. 6.12. He(I) PE spectra of terphenyl (a) in the gas phase and (b) in the form of a solid film. (From Hino *et al.* [281])

Assuming that the first three bands in the spectrum of the solid film correspond to ionization of benzene $\pi$ orbitals, a certain increase in the splitting between the first and the third bands (2.1 eV against 1.9 eV in the gas phase spectrum) is explained by slight distortion of the planar conformation of the molecule of this hydrocarbon in the gas phase.

***Paracyclophanes***

Cyclophanes were studied in detail because transannular orbital mixing is possible in them. Thus, the structures of the [2,2]cyclophanes were studied, from [2,2](1,4)cyclophane (*a*) to [2,2,2,2,2,2](1,2,3,4,5,6)cyclophane (*b*). The study of their PE spectra and the electron absorption spectra of cation radicals showed quite unexpectedly that cyclophane *b* has not the lowest value of $IP_1$ as might be expected.

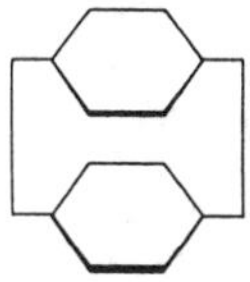

a 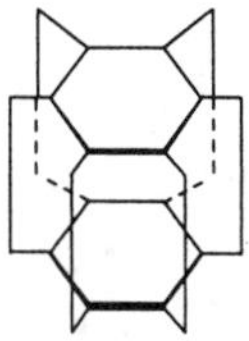 b

Since in the transition from para-xylene to [2,2]cyclophane, the ionization potential decreases by 0.6–0.7 eV, it was expected that a similar effect in the compound *b* might lower the first ionization potential to 6.0 eV. No special 'cyclophane' effect is needed to explain these results; they are explained by competition of orbital interactions in cyclophanes through space (the planes of the benzene rings in these compounds are only separated by a distance of 0.278 nm instead of the 0.340 nm, the sum of the van der Waals' radii) and through $\sigma$ bonds [283, 284]. These effects were evaluated by the analysis of PE spectra of cyclophanes with perfluorinated ethylidene bridges [54].

### 6.3.2 Naphthalene

Naphthalene is widely used in organic synthesis. It is manufactured in considerable amounts in thermal treatment of coal. Naphthalene molecules can also be produced by living organisms. Many naphthalene derivatives have been detected in plants and insects, e.g. 1,4,5-trihydroxynaphthalene was obtained from wall-nut leaves, and 1,4-naphthoquinone derivatives (lapahol, lauson, lomatole) from the foliage of tropical trees.

The molecular structure of naphthalene is very advantageous thermodynamically: steric strains are practically absent in the naphthalene molecule while the conditions for the formation of delocalized $\pi$ orbitals in it are quite favourable. These properties explain why numerous attempts to synthesize [10]annulene ended in formation of naphthalene derivatives.

The traditional electron theory and the valence bond method can be used to draw the structural formula of naphthalene. But the greater number of benzene nuclei in the molecule of the aromatic hydrocarbon complicates the selection of an adequate formula: the number of formulae that should be taken into account in the estimation of the mesomeric state of the system with $2n$ $\pi$ electrons is determined by the equation $N = 2n!/[(n!)(n+1)!]$, i.e. the mesomeric state of the naphthalene molecule should be described by 42 formulae.

Even this set of canonical formulae could not however explain many experimental data concerning the behaviour of polycyclic aromatic hydrocarbons in chemical reactions, orientation rules included. While explaining the results of electrophilic substitution in naphthalene, it was, for example, supposed (in addition to many other hypotheses) that the substitution was attended by the distribution of conjugation not over the entire aromatic system but in one benzene ring only. This agrees more readily with the formulae *a* and *b*, of which *a* is known as a non-symmetrical Erlenmeyer formula:

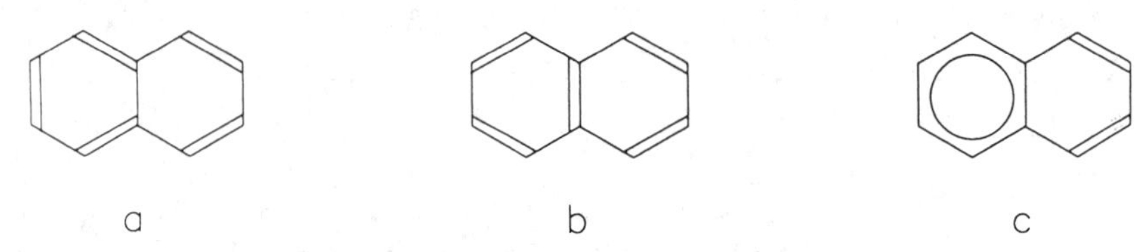


Clar [67] adheres to the same hypothesis within the framework of the sextet concept: he suggested that the molecule of naphthalene might be represented by formula *c* where there is only one aromatic sextet. Formulae *a* and *c* suggest the presence of the butadiene system (that reacts with electrophiles at the 1,4 position) in the right ring of the naphthalene molecule. But there are no structural data proving non-equivalence of two benzene rings in the naphthalene molecule.

Fukui analysed the reactivity of naphthalene in the terms of the MO theory to find good agreement with the experimental data. For more detail see Section 6.3.5, since here we shall limit ourselves only to the discussion of the molecular and electron structure of naphthalene.

The important difference between the molecules of benzene and naphthalene is non-equivalence of the C—C bonds in the naphthalene molecule although both six-membered rings are equivalent [67]:

0.1425
0.1361
0.1421

Within the framework of the MO theory, the $\pi$ orbitals of naphthalene are formed by the interaction of $2p_z$ orbitals of the ten carbon atoms lying in one plane. Calculations give ten molecular $\pi$ orbitals, five of which are doubly occupied while the other five are vacant. All $\pi$ orbitals of naphthalene have different energies and are non-degenerate. The naphthalene molecule thus has five $\pi$ electron shells occupied by ten electrons (the graphic representation of the corresponding molecular $\pi$ orbitals is given in Fig. 6.13). Quantum chemical calculations of the naphthalene molecule were later made in more strict approximations: CNDO/S2 [285], modified INDO/1 [286], INDO/S [225], and *ab initio* [287].

The specific property of the electron structure of naphthalene is that its molecule (as distinct from the benzene molecule) contains not one but at least four $\pi$ electron shells lying above the highest $\sigma$ level.

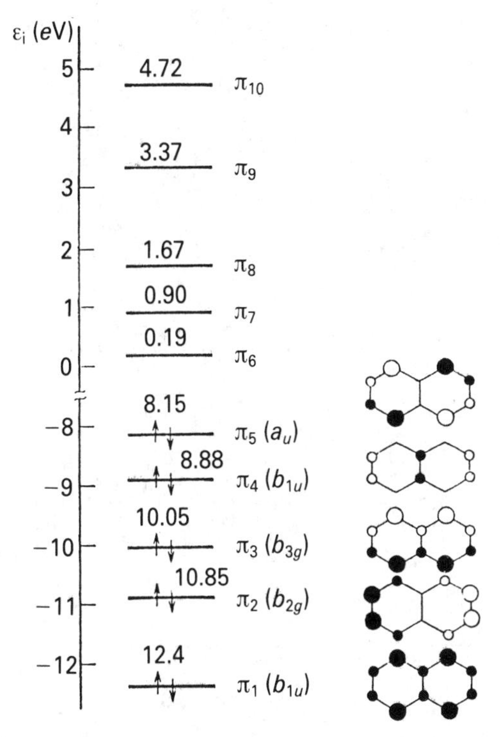


Fig. 6.13. $\pi$ Orbital structure of naphthalene (energies of occupied MOs estimated from PES, and of unoccupied orbitals from ETS findings).

The sequence of the molecular $\sigma$ and $\pi$ orbitals was determined by PE spectra of naphthalene and its derivatives (Fig. 6.14). Identification of PE spectral bands corresponding to the $\pi$ and $\sigma$ levels of naphthalene was promoted by the study of the perfluoro effect and comparison of the spectra using various sources of ionization. Thus, the comparison of the PE spectra of naphthalene and perfluoronaphthalene shows that ionization of electrons of the $\sigma$ level corresponds to potentials exceeding 11 eV [221].

The bands related to all occupied $\pi$ orbitals, the LOMO included, are accurately assigned by Penning spectroscopy. Using excited Ne* atoms as the source of ionization to study naphthalene increases intensities of the bands related to $\pi$ orbitals (compared with the Ne(I) spectrum; see Fig. 6.14). Penning ionization thus demonstrates better steric accessibility of $\pi$ electrons compared with $\sigma$ electrons not only in simple unsaturated compounds but also in quite complicated aromatic hydrocarbons. As the studied molecule interacts with the excited neon (or helium) atoms, the attack on the aromatic hydrocarbon by electrophiles is probably modelled by the Penning ionization [225, 271].

The comparison of band intensities in the spectra recorded with He(I) and He(II) sources also is in favour of assignment of the bands at 8.15, 8.88, 10.10, 10.85 and 12.40 eV in the PE spectra to ionization of the $\pi$ orbitals: the bands related to the $\pi$ orbitals have higher intensity in the He(II) spectra [288].

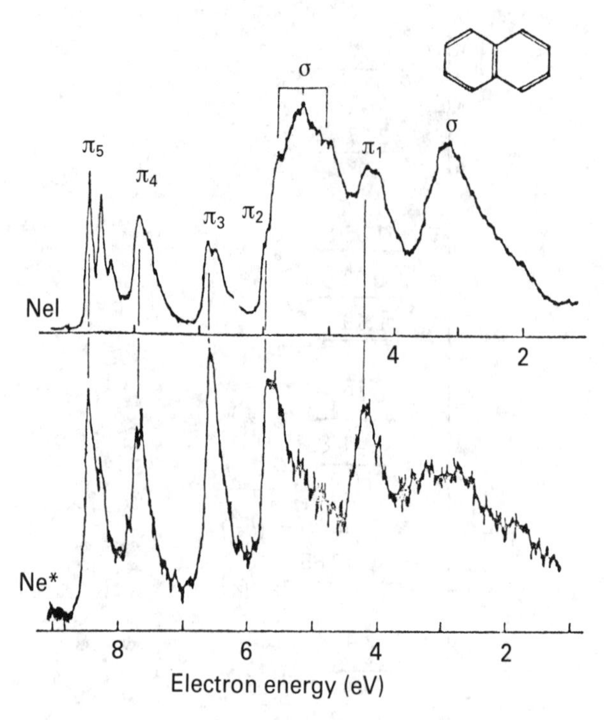


Fig. 6.14. Ne(I) PE spectrum and Ne* Penning spectrum of naphthalene. (From Munakata *et al.* [271])

While discussing the orbital structure of naphthalene, it is interesting to compare the PE spectrum of naphthalene in the gas phase with the electron absorption spectra of its cation radical generated in the solid phase. Direct comparison of these spectra is based on the fact that, according to SCF calculations, the electron absorption bands of aromatic hydrocarbon cation radicals usually correspond to one-configurational electron transition into singly occupied HOMO from one of the lower doubly occupied molecular orbitals [289]. The energies and probabilities of electron transitions in the naphthalene cation radical are given in Fig. 6.15. Energies of $\pi$ MOs are given in compliance with PES and Koopmans' theorem. The symmetry of MOs and electron transition symmetry were evaluated on the basis of the HMO method.

It follows from Fig. 6.15 that of the four $\pi \rightarrow \pi$ transitions, only $A_u \leftarrow B_{3g}$ and $A_u \leftarrow B_{2g}$ transitions are allowed by symmetry. They should be characterized by bands at 653 and 459 nm in the electron absorption spectrum. The experimental absorption spectrum of the naphthalene cation radical, generated by exposure to a helium lamp in a solid argon matrix at 20 K, includes absorption bands at 675 and 461 nm [290].

Vacant electron levels of the naphthalene molecule were studied by two complementary methods of spectroscopy: ESR and ET.

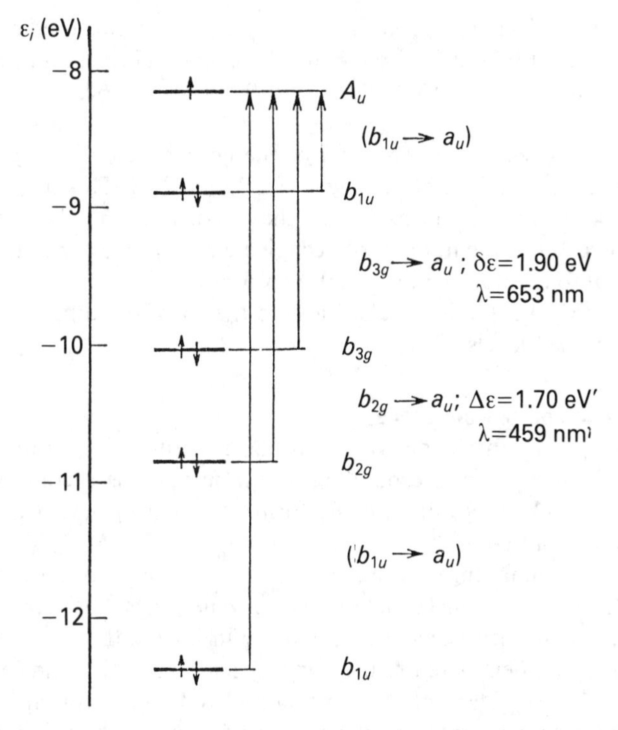


Fig. 6.15. Interpretation of electron absorption spectrum of naphthalene cation radical according to PES of naphthalene (parenthesized are symmetry forbidden transitions. (Data by Andrews *et al.* [290])

Thus, the ESR spectrum of the anion radical contains information about the symmetry of the frontier vacant MO of naphthalene. The spectrum includes 25 lines, which correspond to unpaired electron splitting on two proton groups (four protons in each group):

$$N = \prod(2nZ + 1) = (2 \times \tfrac{1}{2} \times 4 + 1)(2 \times \tfrac{1}{2} \times 4 + 1) = 25$$

with splitting constants $\alpha_H$ of 4.95 and 1.83 Oe respectively [50]. Spin densities on four carbon atoms in the naphthalene anion-radical are 0.180 and on the other four atoms 2.7 times lower, i.e. 0.067.

These results correlate well with the quantum chemical calculations. Thus, according to HMO calculations, spin density on the 1st, 4th, 5th, and the 8th carbon atoms in the naphthalene anion-radical is 0.181, while on the 2nd, 3rd, 6th and 7th carbons spin densities are 0.069.

While the ESR spectrum of the anion-radical gives information only on the symmetry of the lowest vacant level, the ET spectrum can be used to estimate the energy of higher vacant states of the naphthalene molecule. According to quantum chemical calculations (HMO calculations included), the number of anion states in the naphthalene molecule is at least five, in accordance with the number of non-degenerate vacant molecular $\pi$ orbitals. In accordance with this prediction, five anion states were discovered by the ETS method. These states correspond to the following values of electron affinities of naphthalene: $-0.19$, $-0.90$, $-1.67$, $-3.37$ and $-4.72$ eV [46].

Figure 6.13 illustrates the comparison of the energies of occupied and vacant $\pi$ electron levels of naphthalene according to PES and ETS. This diagram is another example of experimental confirmation of the validity of the Pairing theorem for alternant hydrocarbons: the number of occupied electron $\pi$ levels in the naphthalene molecule is equal to the number of vacant $\pi$ levels; moreover, the values of the energy gaps between the vacant levels are also close to the corresponding values of the gaps between the occupied levels [291].

### 6.3.3 Anthracene and phenanthrene

Anthracene and phenanthrene derivatives occur in nature. The anthracene fragment (mostly in the form of the corresponding anthraquinone) is a component of many natural dyes while phenanthrene (in the form of a hydrogenated fragment) is the main component of all steroids.

Anthracene and phenanthrene are the simplest polycyclic aromatic hydrocarbons, differing by the type of annelation of the benzene rings: the benzene nuclei are condensed linearly in anthracene and angularly in phenanthrene. The molecules of these hydrocarbons differ therefore in their physical properties and reactivity.

The geometric parameters of the anthracene and phenanthrene molecules are compared below: although both are planar, bond distances in them change differently from benzene [67, 292]:

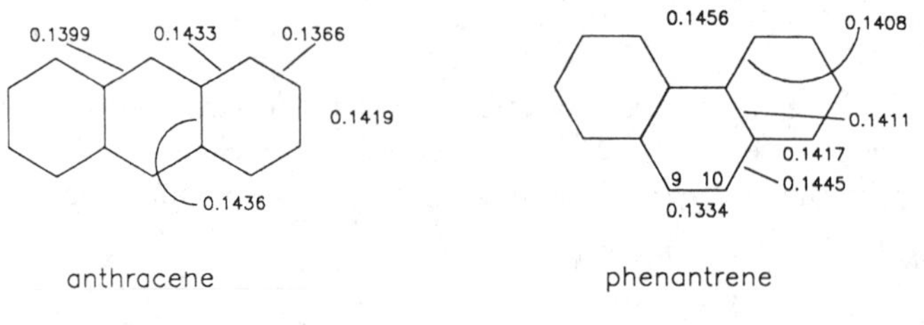

0.1399 0.1433 0.1366
0.1419
0.1436

anthracene

0.1456 0.1408
0.1411
0.1417
9 10 0.1445
0.1334

phenantrene

Non-equivalence of C—C bonds is especially characteristic of the phenanthrene molecule: the bond distance between the 9th and 10th carbons is probably the shortest among the other C—C bonds in aromatic hydrocarbons.

The molecules of anthracene and phenanthrene differ in their electron structure as well. Their ionization potentials are given in Table 6.12 [221, 225]. Assignment of the ionization bands in the anthracene PE spectrum is more complicated than in naphthalene: the ionization of two lower $\pi$ levels is masked by a broad band of ionization of the higher occupied $\sigma$ orbitals at 11–14 eV. Trustworthy conclusions were derived from the spectra recorded with different sources of ionization.

The first four intense bands in the Penning spectrum of anthracene are assigned to ionization of $\pi$ orbitals [271]. This assignment agrees with earlier measurements of PE spectra of anthracene [287, 293], and perfluoroanthracene [225]. But Penning ionization also selects two bands at 11.9 and 12.8 eV from the strongly overlapping

**Table 6.12.** Ionization potentials and occupied MOs of anthracene and phenanthrene (PES data and MO LCAO calculations)

| $IP_i$ (eV) | Anthracene *No. and MO symmetry* | | | | | $IP_i$ (eV) | Phenanthrene *No. and MO symmetry* | | | | |
|---|---|---|---|---|---|---|---|---|---|---|---|
| | CNDO/S | MINDO/2 | HMO | PPP | EHMO | | CNDO/S | MINDO/2 | HMO | PPP | EHMO |
| 7.47 | $b_{3g}$ | $2b_{2g}$ | $2b_{2g}$ | $2b_{2g}$ | $2b_{2g}$ | 7.86 | $b_2$ | $4b_1$ | $4b_1$ | $4b_1$ | $4b_1$ |
| 8.57 | $b_{2g}$ | $2b_{3g}$ | $2b_{3g}$ | $2b_{3g}$ | $2b_{3g}$ | 8.15 | $a_2$ | $3a_2$ | $3a_2$ | $3a_2$ | $3a_2$ |
| 9.23 | $a_u$ | $1a_u$ | $1a_u$ | $1a_u$ | $1a_u$ | 9.28 | $a_2$ | $2a_2$ | $2a_2$ | $2a_2$ | $2a_2$ |
| 10.26 | $b_{3g}$ | $1b_{2g}$ | $2b_{1u}$ | $2b_{1u}$ | $1b_{2g}$ | 9.89 | $b_2$ | $3b_1$ | $3b_1$ | $3b_1$ | $3b_1$ |
| 10.40 | $b_{1u}$ | $2b_{1u}$ | $1b_{2g}$ | $1b_{2g}$ | $1b_{1u}$ | 10.59 | $b_2$ | $2b_1$ | $2b_1$ | $2b_1$ | $2b_1$ |
| 10.80 | $b_{1g}$ | | | | | 10.60 | $a_2$ | | | | |
| 11.20* | $a_g$ | | | | | | | | | | |
| 11.50* | $b_{3u}$ | | | | | | | | | | |
| 11.90 | $b_{2g}$ | $1b_{3g}$ | $1b_{3g}$ | $1b_{3g}$ | $1b_{3g}$ | | | | | | |
| 12.10* | $b_{2u}$ | | | | | | | | | | |
| 12.80* | $b_{3u}$ | | | | | | | | | | |
| 12.80 | $b_{2u}$ | | | | | | | | | | |

*Tentative values
(From Veszpremi [225])

region of potentials at 11–14 eV. In accordance with their intensities, these two bands can also be assigned to ionization of $\pi$ orbitals, viz., lower occupied $\pi_2$ and $\pi_1$ MOs of anthracene. The intensities of other bands in the Penning spectrum are markedly lower and can be attributed to ionization of $\sigma$ orbitals.

An important argument for assignment of the ionization bands are quantum chemical calculations of anthracene. Table 6.12 shows that, except for two closely located $2b_{1u}$ and $1b_{2g}$ $\pi$ orbitals, all calculations of anthracene predict the same sequence of MOs in the region of 6–13 eV.

The enlargement of the hydrocarbon shifts but slightly the ionization threshold of $\sigma$ orbitals to the region of lower potentials (see Section 6.3.4): the electrons of the highest occupied $\sigma$ MO of anthracene are ionized at 10.8 eV.

As the first ionization potential decreases with increasing number of benzene rings in the molecule of an aromatic hydrocarbon, the stability of its cation radical, which is formed under various conditions, increases. Cation radicals of anthracene and its derivatives were, e.g., generated in strong acids [294], by electrochemical reactions [199], and using a laser source [295].

The electron absorption spectra of these cation radicals were also studied. The electron absorption spectra of the cation radicals produced electrochemically from various anthracene derivatives have an intense long-wave band in the region of 700–50 nm (the cation radical of anthracene absorbs, e.g. at 717 nm, $h\nu = 1.73$ eV; Fig. 6.16). It can be suggested that this band corresponds to the $a_u \rightarrow b_{3g}$ transition in the cation radical. This is a symmetry-allowed transition (the symmetry of electron states is given according to [221]).

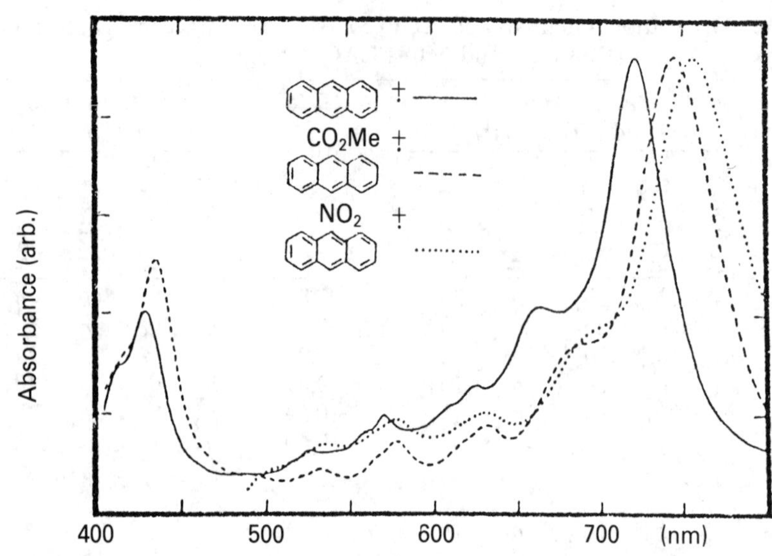


Fig. 6.16. Electron absorption spectra of cation radicals of anthracene and its substituted compounds. (From Masnovi *et al.* [199])

According to PES data, the energy difference between the corresponding states can be estimated as the difference between the first and the third ionization potentials $\Delta E = 1.76$ eV. Another band in the absorption spectrum of the anthracene cation radical and its derivatives is found at 440 nm ($h\nu = 2.88$ eV). It probably corresponds to the symmetry-allowed $b_{1u} \rightarrow b_{3g}$ transition. According to the PES of anthracene, the energy of this transition can be estimated as the difference between the 1st and the 5th ionization potentials: $\Delta E = 2.93$ eV. It is not clear from [199] whether or not the absorption spectra of cation radicals have bands at 1130 nm that would correspond to transitions between HOMO and the $\pi$ orbital that comes next to it. But the $b_{2g} \rightarrow b_{3g}$ transition in the anthracene cation radical is a symmetry-forbidden transition in any case.

Cation radicals of anthracene derivatives generated by picosecond laser pulses from their TCNE complexes have similar electron absorption spectra: e.g. the cation radical of 9-cyanoanthracene absorbs intensively at 750 nm [295].

It can be noted in conclusion that the electron absorption spectra of anthracene and phenanthrene have been studied on the basis of their calculations by the SCF methods in the $\pi$ approximation. All calculations give the same types of symmetry of electron transitions and reproduce adequately the experimental spectra [30].

It is important to note a close correlation between the character of the spectra and the photoelectron properties of hydrocarbons. Thus, being a semiconductor, anthracene becomes a conductor of electricity on exposure to light at the wavelength of 366–400 nm [296].

While discussing the electron structure of anthracene and its derivatives, it is necessary to mention the data reported in [297] that describe still another function of the substituent in the molecule of a polycyclic aromatic hydrocarbon which appears on the interaction of the hydrocarbon with a solid surface. The spectral characteristics of anthracene and some substituted anthracenes in a non-polar organic solvent and in the film adsorbed on $TiO_2$ were compared. It was found that the electron absorption spectrum of the unsubstituted anthracene adsorbed on titanium dioxide differs substantially from that of anthracene dissolved in a non-polar solvent. These differences are practically absent in the spectra of the substituted anthracenes. The positions of bands ($\lambda$, nm) in fluorescence spectra of anthracene and 9-methylanthracene in cyclohexane solution and on a solid surface of $TiO_2$ are given for the purpose of comparison:

| | Cyclohexane | | $TiO_2$ | |
|---|---|---|---|---|
| Anthracene | 380 | 400 | 425 | 445 |
| 9-Methylanthracene | 390 | 412 | 390 | 410 |

Similar results were obtained in the comparison of the spectra of other polycyclic aromatic hydrocarbons and their substituted derivatives recorded under the same conditions. It appears that the electron structure of an aromatic hydrocarbon depends on the mechanism of adsorption on a solid surface. The molecules of unsubstituted hydrocarbons interact with the $TiO_2$ surface over the entire $\pi$ system. As a result, adsorption is attended by significant changes in the spectral characteristics. The molecules of substituted hydrocarbons are bound to the $TiO_2$ surface through the substituent (through the methyl group, e.g. in the case of 9-methylanthracene) in such a way that the $\pi$ electron level of the hydrocarbon fragment remains unchanged.

### 6.3.4 Higher polycyclic hydrocarbons

The molecular structure of polycyclic aromatic hydrocarbons (PAH) has been studied in much detail. The possibilities of manufacture of new dyes have not been exhausted: polycyclic aromatic fragments are components of many natural pigments that have no analogues among synthetic dyes. 4,9-Dihydroxyperylene-3,10-quinone and erythroafin can be mentioned by way of illustration [298, 299]. It is believed that reduction of such pigments in nature explains the presence of some polycyclic aromatic hydrocarbons, e.g. perylene, in high concentration in sea deposits [300, 301]. The aromaticity of PAH has been discussed in detail [207, 302–4]. Moreover, the use of polycyclic aromatic hydrocarbons in the manufacture of micro-electronic materials has been recently reported by many authors [305–8].

The study of PAH at the molecular level is also stimulated by the problems of ecology and medicine. PAHs are contained in large amounts in the exhaust gas of engines. Some of these hydrocarbons are carcinogens [301, 309, 310].

The steric structure of PAH in the crystal state is characterized by non-equivalence of the C—C bonds that increase with the number of benzene nuclei in the molecule [67, 310]. Below are the data of x-ray structural analysis of molecular crystals of coronene and ovalene:

It can be seen that the inner rings lose, to a considerable extent, their aromatic 'benzene' character: each inner bond elongates tending to an ordinary C—C bond distance.

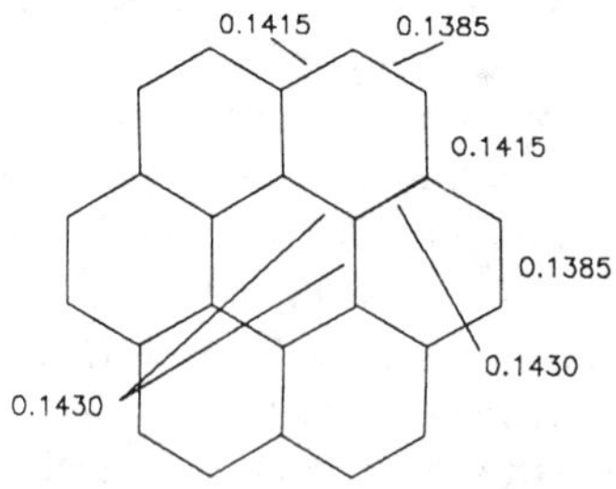

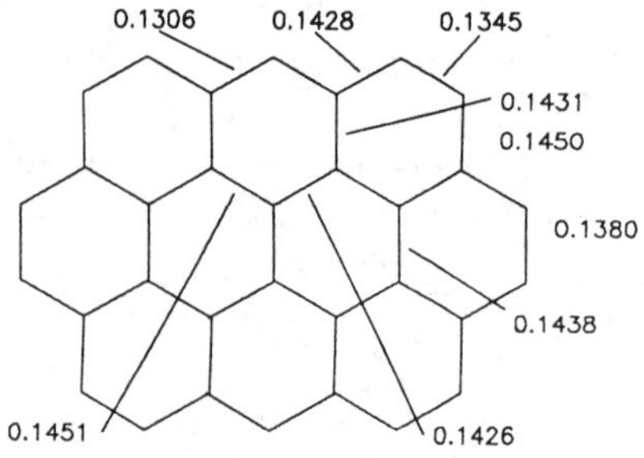

**Table 6.13.** Ionization potentials and higher occupied $\pi$ orbitals of polycyclic aromatic hydrocarbons (PES data and MO LCAO calculations)

| Nos. | Hydrocarbon (symmetry group) | *IP* (eV) and cation-radical state symmetry | | | | | |
|---|---|---|---|---|---|---|---|
| (1) | Benzene ($D_{6h}$) | 9.24 | 9.24 | 12.25 | | | |
| | | $e_{1g}$ | $e_{1g}$ | $a_{2u}$ | | | |
| (2) | Naphthalene ($D_{2h}$) | 8.15 | 8.87 | 10.08 | 10.83 | | |
| | | $a_u$ | $b_{1u}$ | $b_{2g}$ | $b_{3g}$ | | |
| (3) | Anthracene ($D_{2h}$) | 7.41 | 8.54 | 9.19 | 10.18 | 10.28 | |
| | | $b_{2g}$ | $b_{3g}$ | $a_u$ | $b_{2g}$ | $b_{1u}$ | |
| (4) | Phenanthrene ($C_{2v}$) | 7.86 | (8.15) | 9.28 | 9.89 | 10.59 | |
| | | $b_1$ | $b_2$ | $a_2$ | $b_1$ | $b_1$ | |
| (5) | Tetracene ($D_{2h}$) | 6.97 | 8.41 | (8.41) | 9.56 | 9.70 | 10.25 |
| | | $a_u$ | $b_{2g}$ | $b_{1u}$ | $a_u$ | $b_{3g}$ | $b_{2g}$ |
| (6) | Benz[a]anthracene ($C_s$) | 7.41 | 8.04 | 8.86 | 9.38 | 9.91 | 10.36 |
| | | $a''$ | $a''$ | $a''$ | $a''$ | $a''$ | $a''$ |
| (7) | Pyrene ($D_{2h}$) | 7.41 | 8.26 | 9.00 | 9.29 | 9.96 | |
| | | $b_{3g}$ | $b_{2g}$ | $b_{1u}$ | $a_u$ | $b_{1u}$ | |
| (8) | Triphenylene ($D_{3h}$) | 7.88 | 7.88 | 8.65 | 9.68 | 9.68 | 10.60 |
| | | $e''$ | $e''$ | $a_1''$ | $e''$ | $e''$ | $a_2''$ |
| (9) | Benz[a]phenanthrene ($C_{2v}$) | 7.60 | 8.02 | 8.98 | 9.18 | 9.96 | 10.22 |
| | | $b_1$ | $a_2$ | $a_2$ | $b_1$ | $b_1$ | $a_2$ |
| (10) | Chrysene ($C_{2h}$) | 7.59 | 8.10 | 8.68 | 9.43 | 9.72 | 10.52 |
| | | $a_u$ | $a_u$ | $b_g$ | $b_g$ | $a_u$ | $b_g$ |
| (11) | Perylene ($D_{2h}$) | 7.00 | 8.55 | 8.68 | 8.90 | 9.34 | 10.40 |
| | | $a_u$ | $b_{2g}$ | $b_{1u}$ | $b_{3g}$ | $b_{2g}$ | $b_{1u}$ |
| (12) | 1,12-Benzoperylene ($C_{2v}$) | 7.19 | 7.86 | 8.70 | 8.85 | 9.05 | 9.88 |
| | | $a_2$ | $b_1$ | $a_2$ | $b_1$ | $a_2$ | $b_1$ |

(From Heilbronner and Maier [293] and Boschi *et al.* [315])

Most PAHs have been studied by PE spectroscopy [54, 287, 293, 311–16]. Table 6.13 gives their ionization potentials.

On the whole, PE spectra of PAH look like those of naphthalene, anthracene and phenanthrene. A series of sharp peaks with distinct vibrational structure (removal of electrons from $\pi$ orbitals is attended by transition of hydrocarbon molecules to the excited vibrational states) can be seen in the region of potentials below 11 eV. The first ionization bands of PAH have the most pronounced vibrational structures with the vibrational frequency of $1350 \pm 40\ \mathrm{cm}^{-1}$, regardless of the molecular size. The 0–0 vibrational components of the first $\pi$ bands are especially intense, which, in the first approximation, can be regarded as a proof of the absence of appreciable changes in the geometry due to ionization of the HOMOs. The parameters of the vibrational structure of the $\pi$ bands were compared with the frequencies of the corresponding vibrations measured for neutral molecules.

A series of strongly overlapping bands (due to ionization of $\sigma$ orbitals) begins at ionization potentials about 11 eV in the PE spectra of PAH. We have already seen that ionization of the highest $\sigma$ levels is detected in the benzene PE spectrum at 11.44 eV; it shifts to 11 eV in the naphthalene and anthracene PE spectra and does not change with a further increase in the number of benzene rings in a molecule of PAH.

The effect of the increasing number of benzene rings on the electron structure of PAH has been studied by various quantum chemical methods [54, 287, 315, 316]. As a rule, all agree in assignment of at least $\pi$ levels. The assignment of the ionization potentials of perylene is given below by way of illustration [315]:

| Experiment | 7.00 | 8.55 | 8.68 | 8.90 | 9.34 | 10.40 |
|---|---|---|---|---|---|---|
| HMO | $2a_u$ | $3b_{1u}$ | $3b_{3g}$ | $2b_{2g}$ | $2b_{3g}$ | $1a_u$ |
| PPP | $2a_u$ | $3b_{1u}$ | $3b_{3g}$ | $2b_{2g}$ | $2b_{3g}$ | $1a_u$ |
| EHMO | $2a_u$ | $3b_{3g}$ | $3b_{1u}$ | $2b_{2g}$ | $2b_{3g}$ | $1a_u$ |
| MINDO/2 | $2a_u$ | $3b_{3g}$ | $3b_{1u}$ | $2b_{2g}$ | $2b_{3g}$ | $2b_{1u}$ |

One can see that the absolute values of ionization potentials are predicted by various quantum chemical calculations to different degrees of accuracy. Boschi, *et al.* [287] have discussed the following relationships between the calculated and measured ionization potentials:

PPP: $IP_i\,(\mathrm{eV}) = (8.999 \pm 0.014) - (0.893 \pm 0.012)\varepsilon_i$;
EHMO: $IP_i\,(\mathrm{eV}) = (-11.707 \pm 0.233) - (1.615 \pm 0.018)\varepsilon_i$;
MINDO/2: $IP_i\,(\mathrm{eV}) = (0.174 \pm 0.106) - (0.920 \pm 0.011)\varepsilon_i$;

Correlations of the results of the HMO calculations and PES data are especially noteworthy because the derived equations:

$$IP_i(\mathrm{PES}) = \alpha + \lambda_i(\mathrm{HOMO})\beta \qquad (6.3)$$

where:

| $-\alpha$ (eV) | $-\beta$ (eV) | |
|---|---|---|
| 5.94 | 2.94 | [315] |
| $6.55 \pm 0.34$ | $2.73 \pm 0.33$ | [316] |

are examples of parametrization of quantum chemical calculations by the PES data. Thus the found parameters ($\lambda = -6.5$ eV; $\beta = -2.75$ eV) are widely used for calculation of energies of the electron levels. Reliable reproduction of the energies of

**Table 6.14.** Measured and calculated ionization potentials $IP_i$ (eV) of $C_{18}H_{12}$ isomeric hydrocarbons

| Band No. | Chrysene $IP_i$ | | Benz[a]anth-racene $IP_i$ | | Triphenylene $IP_i$ | | Tetracene $IP_i$ | | 3,4,-Benzophen-anthrene $IP_i$ | |
|---|---|---|---|---|---|---|---|---|---|---|
| | exp. | calc. | exp. | calc. | exp. | calc. | exp. | calc. | exp. | calc. |
| 1 | 7.61 | 7.60 | 7.42 | 7.40 | 7.86 | 7.98 | 7.01 | 6.88 | 7.62 | 7.71 |
| 2 | 8.10 | 8.19 | 8.03 | 8.14 | 8.63 | 8.68 | 8.41 | 8.34 | 8.00 | 8.00 |
| 3 | 8.68 | 8.64 | 8.82 | 8.78 | 9.66 | 9.67 | 8.60 | 8.59 | 8.96 | 8.93 |
| 4 | 9.44 | 9.41 | 9.34 | 9.29 | 10.05 | 9.93 | 9.56 | 9.53 | (9.13) | 9.05 |
| 5 | 9.73 | 9.69 | 9.90 | 9.87 | | | (9.70) | 9.78 | 9.95 | 10.05 |
| 6 | 10.52 | 10.58 | 10.40 | 10.38 | | | 10.25 | 10.31 | 10.26 | 10.25 |

(From Ghosh, P.K. [33])

occupied orbitals of PAH by the HMO method was, for example, demonstrated by the data derived for linearly annelated PAH [33, 316].

The best reproduction of experimental ionization potentials within the framework of the HMO method could be obtained with introduction of corrections for bond non-equivalence in a neutral molecule and in the formed cation radical of PAH. Table 6.14 gives the comparison of experimental and calculated ionization potentials for isomeric hydrocarbons $C_{18}H_{12}$ based on the real geometry of the molecules ($\lambda = -5.782$ eV, $\beta = -3.199$ eV are used in the calculations [33]).

PE spectra of solid films were measured and the effect of the phase on the molecular electron structure of PAH estimated [313, 317]. PE spectra of solid films of naphthalene, pentacene, perylene, coronene and p-terphenyl were studied using various ionization sources [30]. It turned out that the bands in the spectrum of a solid film were shifted for each hydrocarbon by 1.7 eV toward the region of lower potentials compared with the spectrum recorded for the gas phase. These spectral properties are explained within the framework of the so-called polarization model. The electron levels of the solid state characterized by the ionization potential $IP_s$ and electron affinity $EA_s$ are mainly formed by the electron levels measured in the gas phase, i.e. by $IP_g$ and $EA_g$, but they are shifted to the magnitude of polarization energy, $P$. Thus the shift, as measured from the spectra of perylene, is 1.7 eV, which is the measure of the polarization energy of this hydrocarbon.

Polarization energies of the overwhelming majority of other hydrocarbons are (according to PES) about the same and equal to 1.6–1.7 eV. But in especially large hydrocarbon molecules (e.g. compounds XLVI(a)–(f), the polarization energies are markedly lower. According to [313], the polarization energy of (a) violanthrene A, (b) isoviolanthrene A, (c) violanthrene B and (d) isoviolanthrene B is 1.5 eV, while those of (e) tetrabenzoperylene and (f) tetrabenzopentacene are 1.24 and 1.15 eV respectively.

It is believed that polarization energy is determined by two parameters: the intermolecular (packing density in a crystal) and intramolecular (polarizability of the hydrocarbon molecule). A decrease in either of these parameters can cause a reduction in the polarization energy. Measurements showed, however, that molecular packing

XLVI: a b

c d

e f

density in (e) and (f) decreases only by 10 per cent, which is insufficient to explain the observed changes in $P$. The molecules of tetrabenzoperylene (e) and tetrabenzopentacene (f) are probably characterized by lower polarizability. This conjecture was confirmed by appropriate calculations.

It has already been shown for naphthalene and anthracene that useful information on the electron structure of aromatic hydrocarbons is derived from the comparison of their PE spectra and the electron absorption spectra of their cation radicals. Exposure to $\gamma$ radiation of more than twenty various polycyclic aromatic hydrocarbons in a solid freon matrix at $-77°C$ resulted in the formation of their cation radicals [289]. Quantum chemical calculations showed that the transitions responsible for the new bands in the electron absorption spectra of cation radicals are mainly one-configurational and can therefore be compared directly with the energies of the corresponding cation-radical states available from PE spectra in the form of their ionization potentials. A good linear dependence between the energies of bands observed in the electron absorption spectra of cation radicals and the differences of the corresponding ionization potentials (PES) has been derived [289]:

$$E = (0.952 \pm 0.047)\Delta IP \pm (0.036 \pm 0.091) \qquad (6.4)$$

The slope of the straight line obtained differs only insignificantly from unity: the matrix effect (due to transition from the gas phase to a solid matrix in freon) does not influence the energy of the intramolecular levels in cation radicals of polycyclic aromatic hydrocarbons.

Positive values of electron affinity of PAH cannot be estimated by the ETS method, but energies of their higher vacant electron levels were determined [226]. First (positive) values of electron affinity of PAH were measured in the gas phase by other methods [18, 291, 318] (see Table 6.15).

**Table 6.15.** The values of electron affinity $EA_1$ of some PAH

| Compound | $EA_1$ (eV) | Compound | $EA_1$ (eV) |
|---|---|---|---|
| Naphthalene | $0.152 \pm 0.016$ | Pyrene | $0.579 \pm 0.064$ |
| Phenanthrene | $0.308 \pm 0.024$ | Dibenz[a,h]anthracene | $0.676 \pm 0.122$ |
| Chrysene | $0.419 \pm 0.036$ | Benz[a]anthracene | $0.696 \pm 0.045$ |
| Benz[e]pyrene | $0.486 \pm 0.155$ | Benz[a]pyrene | $0.829 \pm 0.121$ |
| Benz[e]phenanthrene | $0.542 \pm 0.040$ | Azulene | $0.587 \pm 0.065$ |
| Anthracene | $0.552 \pm 0.061$ | | |

(From Holy [318])

Electron affinity of PAH in solvents is estimated by their polarographic reduction:

$$\mathrm{ArH} + e^- \rightarrow \mathrm{ArH}^-$$

since subsequent protonation of the anion radical formed is about 1000 times slower than addition of the electron. The polarographic half-wave potential $E_{1/2}^{\mathrm{red}}$ is a kind of objective measure of electron affinity of PAH [318]: half-wave potentials of some aromatic hydrocarbons were compared with the energies of unoccupied frontier MOs calculated by the HMO method and a trustworthy linear dependence between the data was obtained (Table 6.16):

$$E_{1/2}^{\mathrm{red}} = a + b\varepsilon_{\mathrm{LUMO}} = (2.41 \pm 0.09)\varepsilon_{\mathrm{LUMO}} - (0.43 \pm 0.06) \qquad (6.5)$$

Examples of such dependences are discussed in detail in [18]. We shall only note here that the properties of the solvent, e.g. in the transition from aqueous dioxane to 2-methoxyethanol, have only insignificant effect on the parameters $a$ and $b$: changes in solvation energies bear but little effect on the energies of occupied and vacant electron levels in anion radical and neutral molecules of aromatic hydrocarbons (cf. data in [291]).

Comparison of the data given in Tables 6.13 and 6.16 shows that the energy of frontier occupied MO gradually increases while the energy of the frontier vacant MO decreases with increasing number of the benzene rings in the PAH molecule. As the molecule of PAH grows in size, the electron levels become closer to one another (cf. the corresponding diagrams of $\pi$ electron levels of benzene and naphthalene in Section 6.2.2 and 6.3.2 [291]).

It has already been said that high stability of the first anion states of PAH makes it impossible to measure the first electron affinities by the ETS method. At the same time, these anion radical states are reliably recorded by ESR spectroscopy. The same method was used to study PAH cation radicals. On the whole, the special character of ion radicals of aromatic hydrocarbons has become a striking experimental confirmation of the main concepts of the MO theory: in accordance with the Pairing theorem, the values of spin densities at definite positions $\mu$ of alternant aromatic hydrocarbon molecules:

$$|c_{i\mu}|^2 = |c_{j\mu}|^2$$

**Table 6.16.** Half-wave potentials $E_{1/2}^{\mathrm{red}}$ of polarographic reduction of some hydrocarbons, their LUMO energies (HMO), and spin densities $c_\mu^2$ in frontier MOs

| Hydrocarbon | $-E_{1/2}^{\mathrm{red}}$ (V) | $-\lambda^{(\mathrm{HMO})}$ ($\beta$) | Position ($\mu$) in molecule* | $c_\mu^2$ |
|---|---|---|---|---|
| Naphthalene | 2.50 | 0.618 | 1 | 0.181 |
| | | | 2 | 0.069 |
| Anthracene | 1.96 | 0.414 | 1 | 0.096 |
| | | | 2 | 0.047 |
| | | | 9 | 0.192 |
| Phenanthrene | 2.46 | 0.605 | 1 | 0.116 |
| | | | 2 | 0.002 |
| | | | 3 | 0.099 |
| | | | 4 | 0.054 |
| | | | 9 | 0.172 |
| Pyrene | 2.12 | 0.445 | 1 | 0.136 |
| | | | 2 | 0.000 |
| | | | 4 | 0.087 |
| Perylene | 1.67 | 0.347 | 1 | 0.083 |
| | | | 2 | 0.013 |
| | | | 3 | 0.108 |
| Tetracene | 1.58 | 0.295 | 1 | 0.056 |
| | | | 2 | 0.034 |
| | | | 5 | 0.148 |

*Enumeration of ArH positions in the molecule is indicated below:
(From Holy [318])

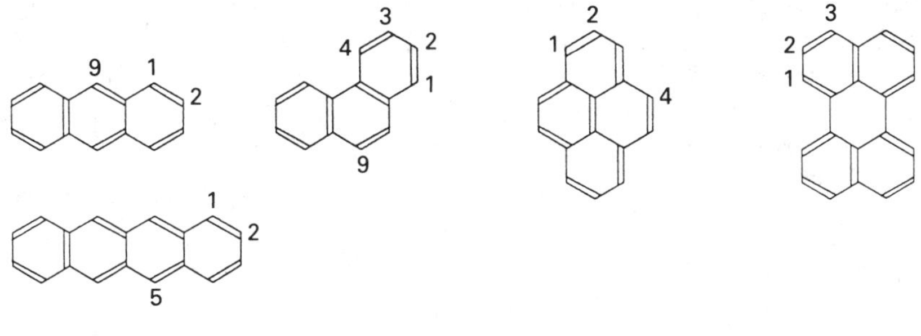

where $i$ is the HOMO index and $j$ is the LUMO index, determined from ESR spectra of anion radicals, turned out to be practically the same as those determined from the ESR spectra of cation radicals (Table 6.17).

Polycyclic aromatic hydrocarbons readily attach more than one electron. Pyrene, for example, is reduced by lithium at low temperature with formation of a doubly negative ion having anti-aromatic properties (this is confirmed by a high-field shift

**Table 6.17.** Splitting constants $a_H^-$ and $a_H^+$ in ESR spectra of ion radicals of anthracene, pyrene, and perylene

| Hydrocarbon | Position* | Splitting constant | |
|---|---|---|---|
| | | Anion radical | Cation radical |
| Anthracene | 1 | 2.74 | 3.06 |
| | 2 | 1.51 | 1.38 |
| | 9 | 5.34 | 6.53 |
| Pyrene | 1 | 4.75 | 5.38 |
| | 2 | 1.09 | 1.18 |
| | 4 | 2.08 | 2.12 |
| Perylene | 1 | 3.08 | 3.10 |
| | 2 | 0.46 | 0.46 |
| | 3 | 3.53 | 4.10 |

*Enumeration of positions in ArH molecules is given in the footnote to Table 6.16.
(From Gerson [50])

of the signals in the PMR spectrum). When reduced by sodium or potassium, pyrene forms a quadruply charged anion having the properties of the aromatic state: it has low-field shift of the signals in the PMR spectra [304]. Similar properties are seen in perylene and also in alkyl- and phenyl derivatives of pyrene.

The analysis of electron absorption spectra of cation and anion radicals of PAH is also rather informative for the study of their electron structure [319]. It appeared that charges on the molecules of aromatic hydrocarbons do not cause significant reconstruction in their electron levels compared with neutral molecules and the corresponding electron transitions remain one-configurational in the first approximation [289].

Estimations of energies of PAH frontier electron levels are available from their electron absorption spectra, since the concepts of the MO theory successfully explain them (Fig. 6.17). Many empirical relationships have been formulated in the course of EAS studies [30, 67, 293, 312]. It was noted, for example, that linear annelation shifts appreciably the absorption in the electron spectrum toward longer wavelengths (thereby increasing the reactivity of the hydrocarbon; cf. [312]).

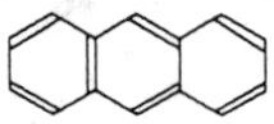

λ, nm 380 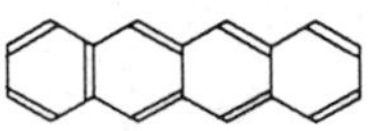 471 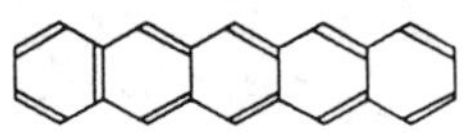 575.5

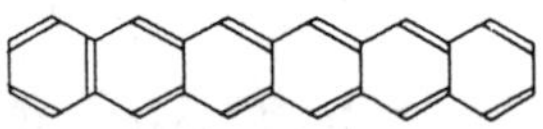

693

By contrast, angular annelation is characteristic of aromatic compounds of low reactivity. Thus, while heptacene is a dark-green, highly reactive and unstable compound, tetrabenzoanthracene, which also has seven benzene rings in the molecule, is a colourless, highly stable compound with low reactivity.

As far back as in 1954, the electron spectra of PAH were explained satisfactorily by the suggestion of one-configurational electron transitions from only two occupied to only two lower vacant electron levels. If each of the mentioned levels is non-degenerate (structure (a)), the transition corresponding to the p band (according to Clar's classification) is described by the wave function $\Psi(A \to A')$. This approximation turned out to be universal. The only exception are hydrocarbons in which MOs, coming next to the frontier MOs ($B$ and $C$ or $B'$ and $C'$ in structure (b) respectively), are degenerate or almost degenerate:

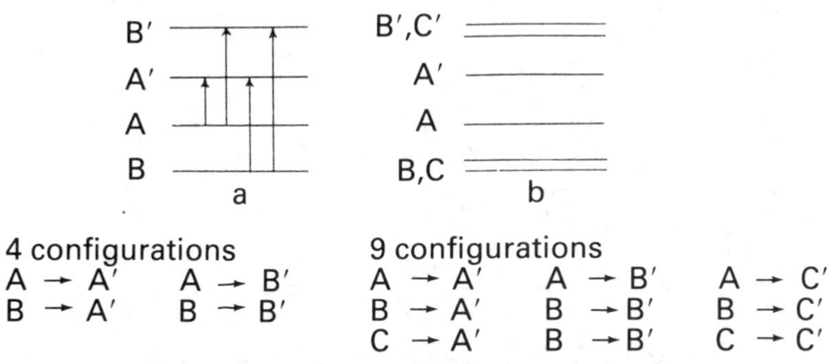


Interpretation of the UV spectra of such hydrocarbons requires account to be taken of the interaction of at least nine configurations (see structure (b)). Thus, the wave function $2^{-1/2}[\Psi(A \to B') + \Psi(B \to A')]$ is assigned to the $\alpha$ band of lower energy. Higher-energy bands (designated $\beta$ and $\beta'$ in Clar's classification) are described by the function $B \to B'$ and $2^{-1/2}[\Psi(B \to A') + \Psi(A \to B')]$ respectively. The results of calculations of electron absorption spectra of some acenes, taking account of configurational interaction, are given in Table 6.18.

Assignment of electron absorption spectra of PAH given by Dewar and Longuet–Higgins [30] was later confirmed by the analysis of the UV and PE spectra of a great number of PAH [293, 311–12]. By analysing the spectra of more than 60 PAH, Schmidt and Clar established a reliable dependence between the energies of long wavelengths in their electron absorption spectra (e.g. $p$ and $a$ bands) and the first and second ionization potentials measured from their PE spectra.

The role of frontier orbitals of PAH in electron transitions recorded by the electron absorption spectra is confirmed by the polarographic findings as well [291]. Absolute values of polarographic redox potentials $E^0$ have been found to correlate with the parameters of the electron spectra of many polycyclic hydrocarbons. It was found that the energy of the intense long-wave transition $h\nu$ in the electron absorption spectrum changes linearly with the value of the oxidation ($E^0_{\mathrm{ox}}$) and reduction potential ($E^0_{\mathrm{red}}$) (with $+1/2$ and $-1/2$ slopes respectively). Graphs in Fig. 6.18 show that the $h\nu$ value decreases in proportion to the growth of the reduction potential $E^0_{\mathrm{red}}$ and the decrease in the oxidation potential $E^0_{\mathrm{ox}}$.

It is important to note that at $h\nu \to 0$, both dependences tend to the same value which is close to the $E_{\mathrm{C}} = 4.39$ eV:

**Table 6.18.** Experimental and calculated absorption bands (wave numbers $\bar{\nu}$ are indicated) in electron spectra of some PAH

| PAH | $\bar{\nu}$ (cm$^{-1}$) (exper) | SCF calculation taking account of 9 configurations | | Wave function | HMO calculation taking account of 25 configurations |
|---|---|---|---|---|---|
| | | symmetry | $\bar{\nu}$ (cm$^{-1}$) | | |
| Benzene | 37 900 | $B_{3u}$ | 39 563 | $1/\sqrt{2}\ [\Psi(B \to A') + (A \to B')]$ | |
| | 48 000 | $B_{2u}$ | 44 704 | $1/\sqrt{2}\ [\Psi(A \to A') - (B \to B')]$ | |
| | 54 500 | $B_{3u}$ | 57 195 | $1/\sqrt{2}\ [\Psi(B \to A') - (A \to B')]$ | |
| | | $B_{2u}$ | 57 197 | $1/\sqrt{2}\ [\Psi(A \to A') + (B \to B')]$ | |
| Naphthalene | 31 800 | $B_{3u}$ | 35 448 | $1/\sqrt{2}\ [\Psi(B \to A') - (A \to B')]$ | 32 397 |
| | 34 600 | $B_{2u}$ | 35 449 | $\Psi(A \to A')$ | 36 227 |
| | 45 500 | $B_{1g}$ | 47 972 | $1/\sqrt{2}\ [\Psi(C \to A') + (A \to C')]$ | 44 403 |
| | 52 600 | $B_{3u}$ | 50 224 | $1/\sqrt{2}\ [\Psi(B \to A') + (A \to B')]$ | 46 193 |
| Anthracene | 26 400 | $B_{2u}$ | 28 738 | $\Psi(A \to A')$ | 29 400 |
| | — | $B_{3u}$ | 33 147 | $1/\sqrt{2}\ [\Psi(A \to B') - (B \to A')]$ | 30 000 |
| | 39 550 | $B_{1g}$ | 38 172 | $1/\sqrt{2}\ [\Psi(C \to A') + (A \to C')]$ | 36 500 |
| | 45 300 | $B_{1g}$ | 43 877 | $1/\sqrt{2}\ [\Psi(C \to A') - (A \to C')]$ | 39 700 |
| Naphthacene | 21 200 | $B_{2u}$ | 24 822 | $\Psi(A \to A')$ | 25 100 |
| | — | $B_{3u}$ | 32 400 | $1/\sqrt{2}\ [\Psi(C \to A') - (A \to C')]$ | 28 700 |
| | 36 800 | $B_{1g}$ | 33 112 | $1/\sqrt{2}\ [\Psi(A \to B') - (B \to A')]$ | 31 500 |
| | 43 500 | $B_{1g}$ | 38 858 | $1/\sqrt{2}\ [\Psi(A \to B') + (B \to A')]$ | 34 400 |

(From Peacock [30]

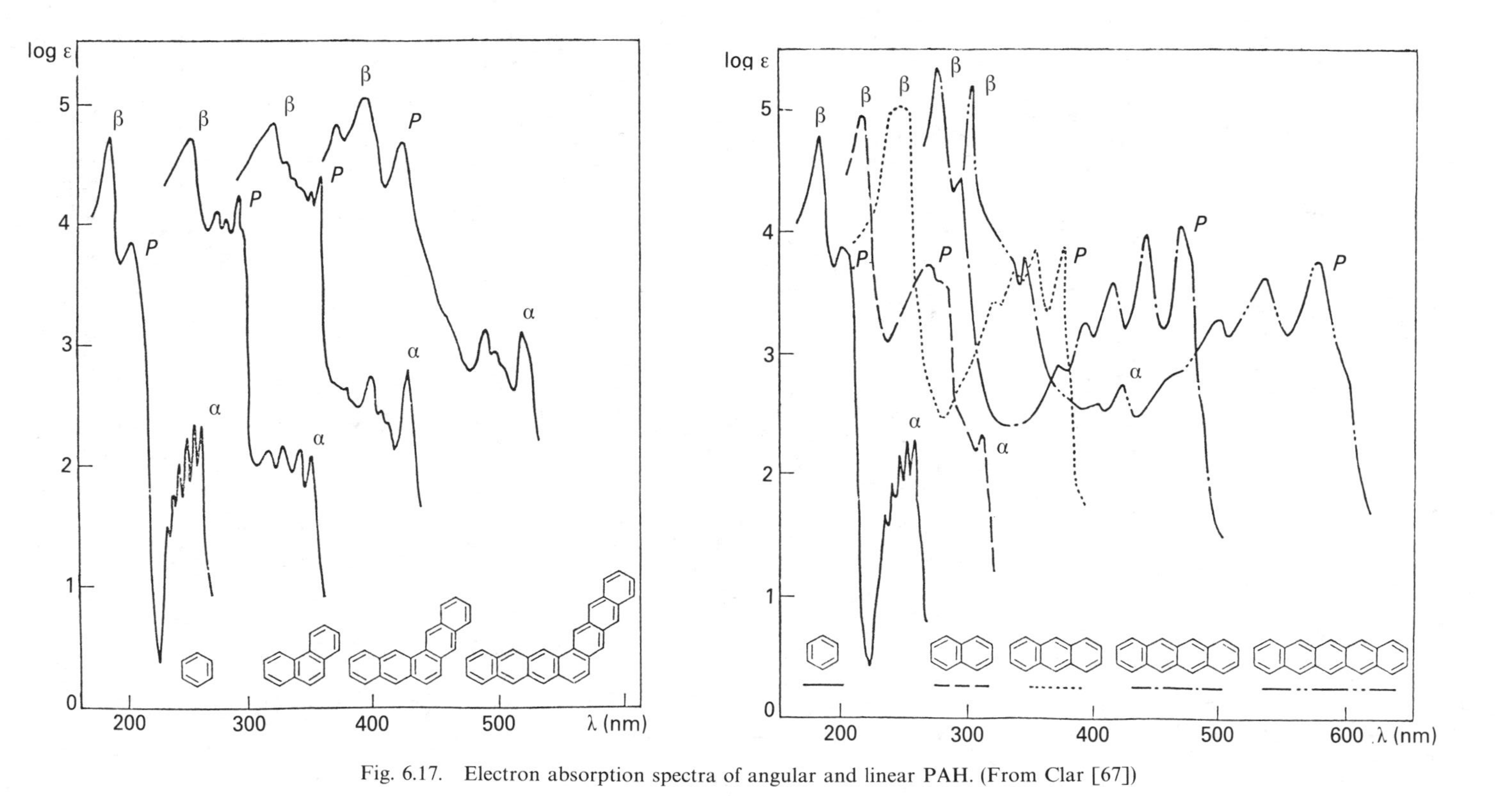


Fig. 6.17. Electron absorption spectra of angular and linear PAH. (From Clar [67])

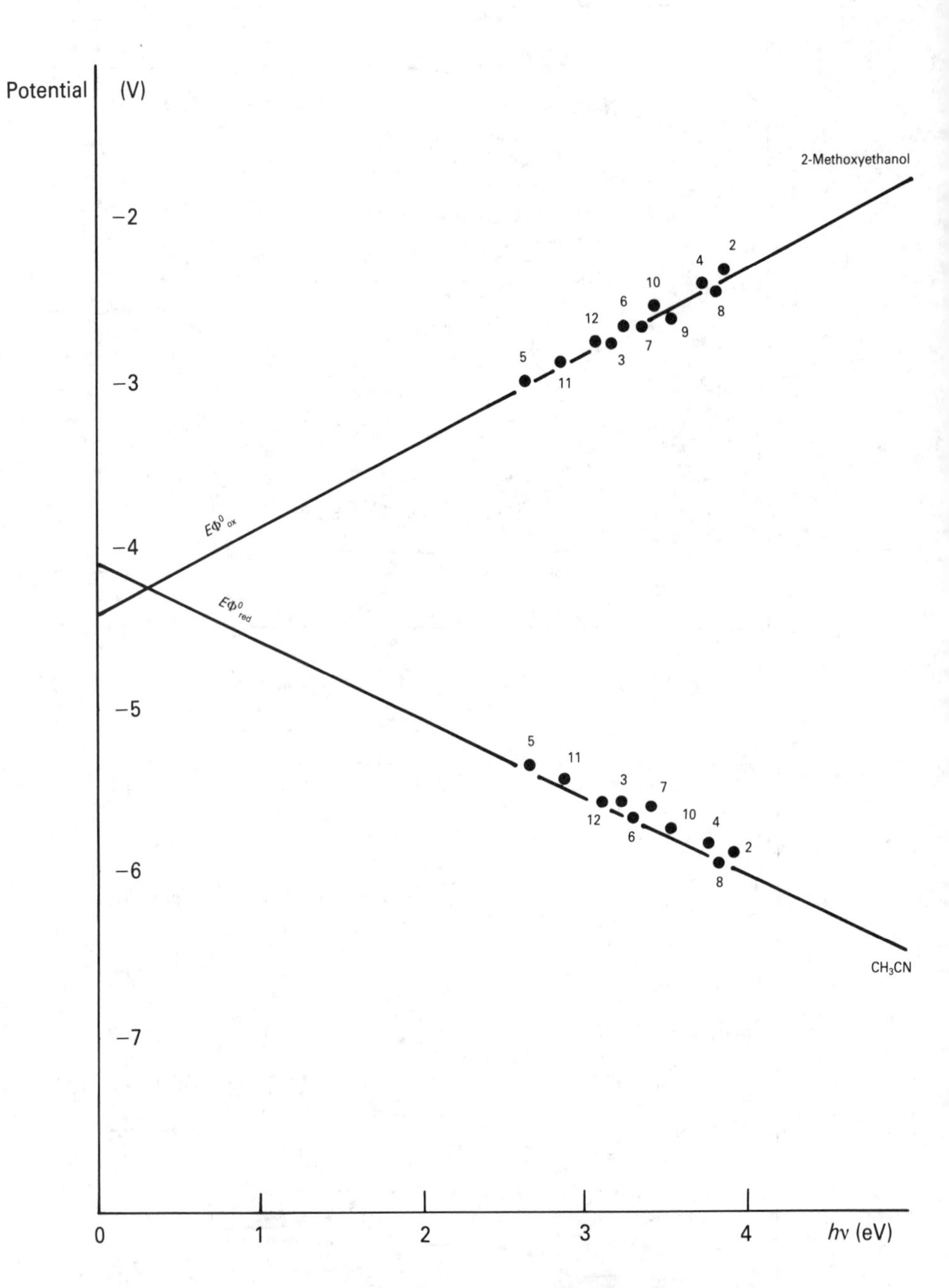


Fig. 6.18. Dependence of *hν* energies of long-wave bands in electron absorption spectra on oxidation and reduction potentials ($\phi^0_{ox}$ and $\phi^0_{red}$ respectively) of PAH (the numerals correspond to the numbers of compounds in Table 6.13). (From Notoya and Matsuda [291])

$$E^0_{ox} = E_C + \tfrac{1}{2}h\nu$$

$$E^0_{red} = E_C - \tfrac{1}{2}h\nu$$

The oxidation and reduction potentials of conjugated aromatic hydrocarbons thus correspond to the energy levels of the frontier molecular orbitals, the levels being symmetric relative to the Fermi level in graphite.

### 6.3.5 Frontier orbitals and reactivity of polycyclic aromatic hydrocarbons

Fukui has explained the behaviour of naphthalene in electrophilic reactions within the framework of the frontier orbitals concept. According to this concept, the difference between electron densities in the highest occupied MO of a naphthalene molecule is sufficiently high for a preferable attack on it by an electrophilic agent at position 1 (for more detail see Section 3.1).

The results of electrophilic substitution with other aromatic hydrocarbons can be explained in a similar way [63, 64, 67]. Thus, anthracene, perylene, phenanthrene, pyrene, and tetracene are nitrated, acylated, and halogenated mostly at positions 9, 3, 9, 1 and 5 respectively, i.e. at the positions characterized by the highest electron density of their HOMOs (Table 6.16). Benz[a]phenanthrene is the only exception among the hydrocarbons listed in Table 6.13: the electrophilic reactions of this hydrocarbon occur mostly at position 2 because of the steric hindrance at position 5 (the $C_{(5)}$ to $C_{(4)}$ distance is 0.3 nm).

The molecules of naphthalene, anthracene and perylene are highly symmetric and they therefore have only two or three non-equivalent positions. It is therefore important to discuss some examples of the orbital-controlled electrophilic reactions of polycyclic hydrocarbons having molecules with lower symmetry.

All the 12 positions of a benz[a]anthracene molecule, which are capable potentially of substitution reactions, are non-equivalent. Below follows a comparison of ionization potentials measured and assigned by HMO calculations. Good agreement between the experimental findings and the theoretically calculated values of ionization potentials indicates adequacy of the description of the electron structure (first of all, of the occupied orbitals) of benz[a]anthracene by the simple MO LCAO method:

| | | | | | | |
|---|---|---|---|---|---|---|
| $IP_i$(PES) (eV) | 7.41 | 8.04 | 8.86 | 9.38 | 9.91 | 10.36 |
| $-\varepsilon_i$(HMO) (eV) | 7.89 | 8.56 | 9.25 | 9.66 | 10.06 | 10.45 |

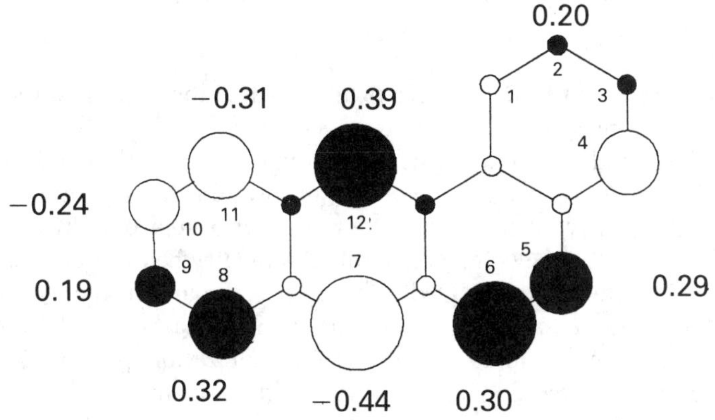

This conclusion is also confirmed by the distribution of electron density of the frontier occupied MO of benz[a]anthracene as determined by the semi-empirical CNDO/S3 method [314].

Estimation of electron density distribution in the HOMO within the framework of the frontier orbitals theory can be used to predict the electrophilic attack on the benz[a]anthracene molecule at positions 7 and 12. Experiments show that the electrophilic reactions (chlorination, nitration, etc.) occur at these sites [67].

The leading role of HOMO of PAH becomes evident in other reactions of these compounds as well. Thus, the properties of their complexes with $\pi$ acceptors depend on the parameters of the highest occupied electron levels of the donating molecules [320–4].

Polycyclic hydrocarbons form coloured complexes with $\pi$ acceptors such as tetracyanoethylene, trinitrofluorenone, tetracyanobenzene, etc. The spectral characteristics of some complexes are given in Table 6.19. It was shown in Section 3.1 that, with the same acceptor, the position of the charge-transfer band in the absorption spectrum of a CT complex depends linearly on the ionization potential of the donor molecule. It can be seen from Table 6.19 that the following equation holds for the charge transfer between PAH and tetracyanoethylene:

$$h\nu_{\mathrm{CT}(1)} = 0.80 IP_{\mathrm{D}(1)} - 4.20 \qquad (6.6)$$

The equation shows that the first CT absorption band arises from the electron transfer from the HOMO of the polycyclic aromatic compound to the LUMO of tetracyanoethylene.

The absorption spectra of CT complexes of polycyclic hydrocarbons usually contain several CT bands. Orgel was the first to study the multiplicity of electron absorption spectral bands of TCNE complexes with substituted benzenes. He explained it by the splitting of the degenerate $e_{1g}$ level of benzene in response to substitution [320]. According to this approach, the appearance of shorter CT bands in the absorption spectra of the complexes is explained by the electron transfer to the LUMO of TCNE from the lower occupied molecular orbitals of the polycyclic hydrocarbon rather than from its HOMO. Below follows the equation calculated for all the CT bands of the spectra of the PAH-TCNE complexes:

$$h\nu_{\mathrm{CT}(i)} = 0.85 IP_{\mathrm{D}(i)} - 4.58 \qquad (6.7)$$

This equation differs but little from Eq. (6.6), only describing the position of the first CT bands (cf. Section 3.1).

Equations (6.6) and (6.7) and the data given in Table 6.19 show that the number of CT bands in the visible region of the absorption spectra of the complexes depends unambiguously on the energies of the upper occupied electron levels of the hydrocarbon molecules: the complexes of high-symmetry benzene and coronene (the hydrocarbons with degenerate HOMO) have one CT band in the visible region of the absorption spectra, while the complexes of hydrocarbons with undegenerate highest occupied electron levels (having several ionization potentials lower than 9.25 eV) have two or three CT bands in their visible spectra.

The donor–acceptor interactions, which are not limited to the charge transfer

**Table 6.19.** CT bands in electron absorption spectra of PAH-TCNE complexes and ionization potentials of donors (PES)

| Donor | CT$_1$ band | | | $IP_1$ (eV) | CT$_2$ band | | | $IP_2$ (eV) | CT$_3$ band | | | $IP_3$ (eV) |
|---|---|---|---|---|---|---|---|---|---|---|---|---|
| | $\bar{\nu}_{max} \times 10^{-3}$ (cm$^{-1}$) | $h\nu$ (eV) | $\lambda$ (nm) | | $\bar{\nu}_{max} \times 10^{-3}$ (cm$^{-1}$) | $h\nu$ (eV) | $\lambda$ (nm) | | $\bar{\nu}_{max} \times 10^{-3}$ (cm$^{-1}$) | $h\nu$ (eV) | $\lambda$ (nm) | |
| Benzene | 26.1 | 3.22 | 385 | 9.24 | — | | | | — | | | |
| Biphenyl | 20.0 | 2.48 | 500 | 8.34 | 25.7* | 3.19 | 389 | 9.04 | — | | | |
| Phenanthrene | 19.0* | 2.35 | 526 | 7.86 | — | | | (8.15) | 28.2 | 3.49 | 355 | 9.28 |
| Naphthalene | 18.2 | 2.26 | 549.5 | 8.15 | 23.4* | 2.90 | 427 | 8.87 | — | | | |
| Triphenylene | 18.0 | 2.23 | 556 | 7.88 | 22.1* | 2.74 | 452 | 8.65 | — | | | |
| Fluorene | 17.7 | 2.19 | 565 | 7.91 | 23.8* | 2.95 | 420 | 8.77 | 24.7 | 3.06 | 405 | 9.09 |
| Stilbene | 16.8 | 2.08 | 595 | 7.87 | 26.0* | 3.22 | 385 | 9.08 | — | | | |
| Chrysene | 15.9 | 1.97 | 629 | 7.59 | 18.8* | 2.33 | 532 | 8.10 | 22.9 | 2.84 | 437 | 8.68 |
| 1,2,5,6-Dibenzanthracene | 14.5 | 1.80 | 690 | 7.41 | 17.1 | 2.12 | 585 | | 21.9 | | | |
| 1,2-Benzanthracene | 14.3 | 1.77 | 699 | 7.41 | 18.5* | 2.29 | 541 | 8.04 | 25.1 | 3.12 | 398 | 8.86 |
| Coronene | 14.2 | 1.76 | 704 | | — | | | | — | | | |
| Pyrene | 14.0 | 1.74 | 714 | 7.41 | 20.5* | 2.54 | 488 | 8.26 | 25.8 | 3.20 | 388 | 9.00 |
| Anthracene | 13.5 | 1.67 | 741 | 7.41 | 21.5* | 2.67 | 465 | 8.54 | — | | | |
| Perylene | 11.2 | 1.39 | 893 | 7.00 | — | | | | | | | |

*Averaged position of the wide overlapped CT band.
(From Briegleb *et al.* [321])

alone, but are followed also by covalent bonding between the donor and acceptor, are of special interest. Thus, many polycyclic aromatic hydrocarbons react as dienes with appropriate dienophiles according to the Diels–Alder reaction [295, 323-5]. These reactions begin with formation of an intermediate CT complex and terminate by addition of TCNE to the aromatic hydrocarbon. Such reactions occur, for example, with anthracene derivatives [326].

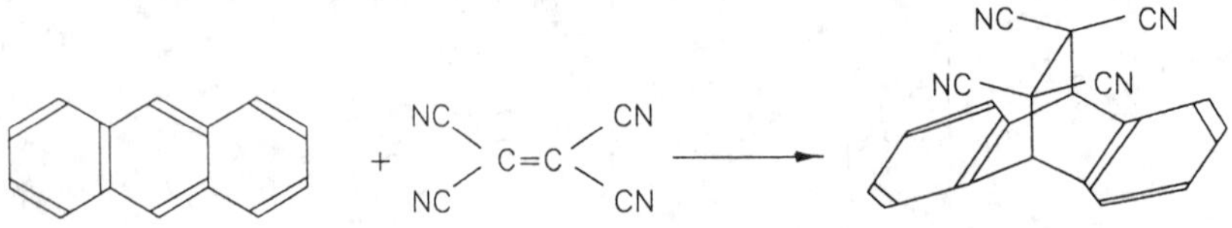

The rate of the cyclo-addition reaction obeys the second-order kinetic equation:

$$w = k_2[\mathrm{ArH}][\mathrm{TCNE}]$$

The logarithms of the second-order rate constants depend linearly on the energies of the corresponding bands of the CT absorption spectra:

$$\log k_2 = -15.4h\nu_{\mathrm{CT}} + 26$$

The slope of the line nears unity when CT band energies are compared with the activation energies of the addition reactions. These conclusions suggest unambiguously that the Diels–Alder reactions between aromatic hydrocarbons and tetracyanoethylene are orbital-controlled and they depend on the electron transfer from the HOMO of the hydrocarbon to the LUMO of the acceptor [326, 327].

Among the orbital-controlled reactions of polycyclic aromatic hydrocarbons, biochemical transformations should also be noted [328, 329]. Anthracene and phenanthrene, for example, have no biological potency but they are converted into 9,10-dihydroxy derivatives by biochemical oxidation. The MO approach explains this selectivity: the electron densities of the occupied frontier MOs of these hydrocarbons have their maximum values at the $C_{(9)}$ and $C_{(10)}$ atoms.

Carcinogenicity of some polycyclic hydrocarbons has been studied intensively [314, 329, 330]. The MO theory seems to adequately explain the results of these studies. It has been found that, for example, the introduction of two methyl groups into the benz[a]anthracene molecule increases markedly the carcinogenic power only if the hydrocarbon is methylated at positions 7 and 12. The total energy of the $\pi$ electron levels in 7,12-dimethylbenz[a]anthracene is almost 2 eV higher than the total energy of the orbitals in the starting benz[a]anthracene. In view of the fact that the electron density in HOMO is the highest at positions 7 and 12, Akiyama and Harvey believe that 'such a perturbation of $\pi$ orbital structure is certainly large enough to alter reactive and binding properties in a manner which could ultimately influence the biochemical activity of 7,12-dimethylbenz[a]anthracene and its metabolites.' [314].

The examples illustrating the role of frontier occupied MOs in the behaviour of polycyclic hydrocarbons during their interaction with electrophilic agents should be supplemented by the findings demonstrating the importance of their unoccupied frontier MOs. It has already been shown that these orbitals are sufficiently low and readily accept electrons. It is easy to suggest that the formed anion radicals play an important role in the reactions of these hydrocarbons with nucleophilic agents. In fact, the readily formed anion radicals of PAH react subsequently depending on the distribution of the electron spin density in the LUMO of the starting polycyclic aromatic hydrocarbons [318]. The addition of the proton, for example, to the anion radical of naphthalene at positions 1 and 4 results in formation of 1,4-dihydronaphthalene. The anion radical of anthracene is protonated at positions 9 and 10 to give 9,10-dihydroanthracene. If the newly formed anion radical reacts with, for example, $CO_2$, the interaction occurs at the site of the highest coefficient of the LUMO as well: the reduction of naphthalene is followed by treatment with $CO_2$ with formation of 1,4-dihydronaphthalene-1,4-dicarboxylic acid, while phenanthrene gives 9,10-dihydrophenanthrene-9,10-dicarboxylic acid.

Other polycyclic hydrocarbons are reduced in the Birch reaction in a similar way. Pyrene, for example, is reduced in three steps, viz.:

(1) the reaction first occurs at positions 4 and 5 characterized by the maximum values of the LUMO coefficient of pyrene;
(2) next, the protons are added at positions 9 and 10 characterized by the maximum values of the LUMO coefficients of phenanthrene;
(3) the reduction is completed by formation of derivative *a* in compliance with the maximum values of the LUMO coefficients in the para-positions of the benzene rings of biphenyl:

10 1 2 3 4 5 6 7 8 9 $\xrightarrow{+2a^-}$ $[\ ]^{2-}$ $\xrightarrow{+2H^+}$ $\xrightarrow{+2e^-}$ $\xrightarrow{+2H^+}$

$\xrightarrow{+2e^-}$ $\xrightarrow{+2H^+}$ a

### *Problems*

(1) In reactions with some electrophiles, benzene releases an electron, thus forming a cation radical. Suggest an unpaired electron distribution in this cation radical.
(2) Explain why benzene has two IPs in the region of ionization of $\pi$ electrons: 9.24 and 12.2 eV. Use the octet rule, the model of hybridized orbitals and the MO theory.
(3) The values of log $k_{rel}$ of electrophilic nitration of polycyclic aromatic hydrocarbons are given below:
naphthalene, 0; chrysene, 0.87; pyrene, 1.56; perylene, 2.22.
Explain.

(4) Suggest at which position pyrene undergoes an electrophilic attack. Make the corresponding suggestions for perylene, anthracene, phenanthrene, and tetracene.
(5) Perylene is easily reduced by alkali metals. Suggest an unpaired electron distribution in the perylene anion radical. Make the corresponding suggestions for anthracene, pyrene, phenanthrene, and tetracene.
(6) Anthracene undergoes Diels–Alder reactions at position 9 and 10. Use the canonical formulae and the frontier orbitals concept to explain the results of the reactions.
(7) Naphthalene forms a cation radical on electrophilic nitration. Suggest an unpaired electron distribution in the naphthalene cation radical. Make the corresponding suggestions for anthracene, pyrene, phenanthrene, and tetracene which are also easily oxidized by electrophiles.
(8) Explain why the longest wavelength absorption band is bathochromically shifted in the UV spectrum of PAH with an increasing number of annelated benzene rings in its molecule: benzene (264 nm), naphthalene (314 nm), anthracene (379 nm), naphthacene (472 nm).

# 7

# Halogenohydrocarbons

## 7.1 HALOALKANES

Haloalkanes are widely used in organic synthesis. Reactions of haloalkanes have proved to be extremely important in the development of the fundamental principles of the theory of organic reactions. The studies on these compounds resulted in development of the concepts of electronic and steric effects of substituents and elaboration of the theory of nucleophilic substitution at aliphatic carbon atoms.

The halogen atoms differ from the carbon atom by their electronegativity and the presence of lone electron pairs in their valence shells. As a result, substitution of the hydrogen atom for the halogens alters substantially the properties of the molecule [4]:

(1) the C—Hal bond distance is greater than the C—H bond ($r_{C-H} = 0.1090$ nm)

| | F | Cl | Br | I |
|---|---|---|---|---|
| $r_{C-Hal}$ (nm) | 0.1385 | 0.1784 | 0.1929 | 0.2139 |

(2) the C—Hal bond is strongly polarized; the dipole moments $\mu$ of halomethanes $CH_3Hal$ measured in the gas phase are given below:

| | F | Cl | Br | I |
|---|---|---|---|---|
| $\mu$ (D) | 1.82 | 1.94 | 1.79 | 1.64 |

The C—Hal bonds are characterized not only by their polarity but also by their high polarizability. The smaller the atom, the lower the polarizability: I > Br > Cl > F. The van der Waals' radius of bromine is similar to that of the methyl group, but its polarizability is much higher.

The unique properties of fluorohydrocarbons are noteworthy. Their boiling points, for example, are closer to the boiling points of the corresponding hydrocarbons than one might expect from the difference in their molecular weights: ethane boils at −89°C, perfluoroethane at −79°C.

The main properties of the electron structure of haloalkanes can be demonstrated in halomethanes. The first PE spectra showed an unexpected behaviour of fluoromethane [331]: the narrow intensive band that was expected from the ionization of electrons of degenerate $2p$ orbitals of the fluorine atom was absent. Such bands are usually associated with photo-ionization of the halogen compounds. The bands can, for example, be seen in the spectra of all hydrogen halides: degenerate lone pairs of the halogen manifest themselves in each spectrum by narrow ionization bands in which spin-orbital splitting increases with the atomic number of the halogen (Fig. 7.1) [22].

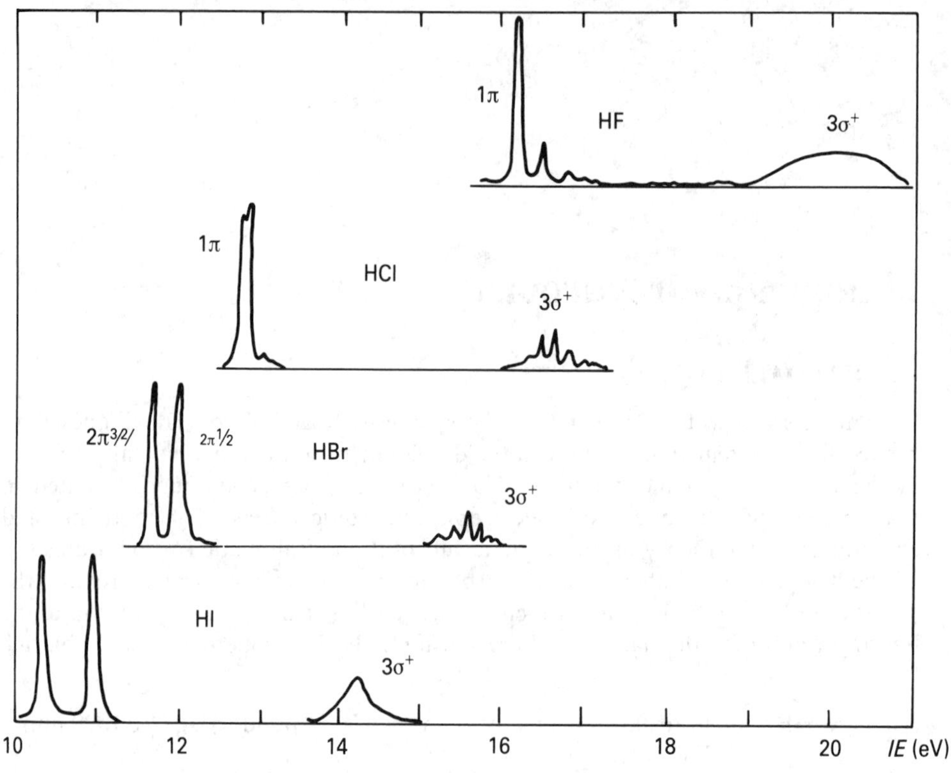


Fig. 7.1. He(I) PE spectra of hydrogen halides. (From Ballard [22])

The PE spectrum of fluoromethane is shown in Fig. 7.2. Other fluorides $EH_3F$ (e.g. E = Si, Ge) have similar spectra: none of the two broad bands can be assigned to ionization of lone pairs of the fluorine atom [332]. The results of *ab initio* quantum chemical calculations of fluoromethane assign these broad bands to strong interaction of the fluorine atom lone pairs with the $\sigma$(C—H) orbitals of the methyl group. According to calculations in the STO-3G basis set, the first occupied electron levels of all fluoromethanes are formed by the interaction of the fluorine lone pairs with the (C—H) orbitals. Furthermore, contributions of the $2p$ orbitals of fluorine are essential to other electron levels of fluoromethane: the antibonding interaction of the $2p$(F) and (C—H) orbitals can be seen in the HOMO, while the band at 17 eV is assigned to the bonding combination of the same orbitals [38].

As distinct from fluoromethane, specific narrow ionization bands correspond to non-bonding $p$ orbitals of the valence shell in the spectra of other halomethanes (the spectrum of chloromethane is shown in Fig. 7.2). This fact agrees poorly with the

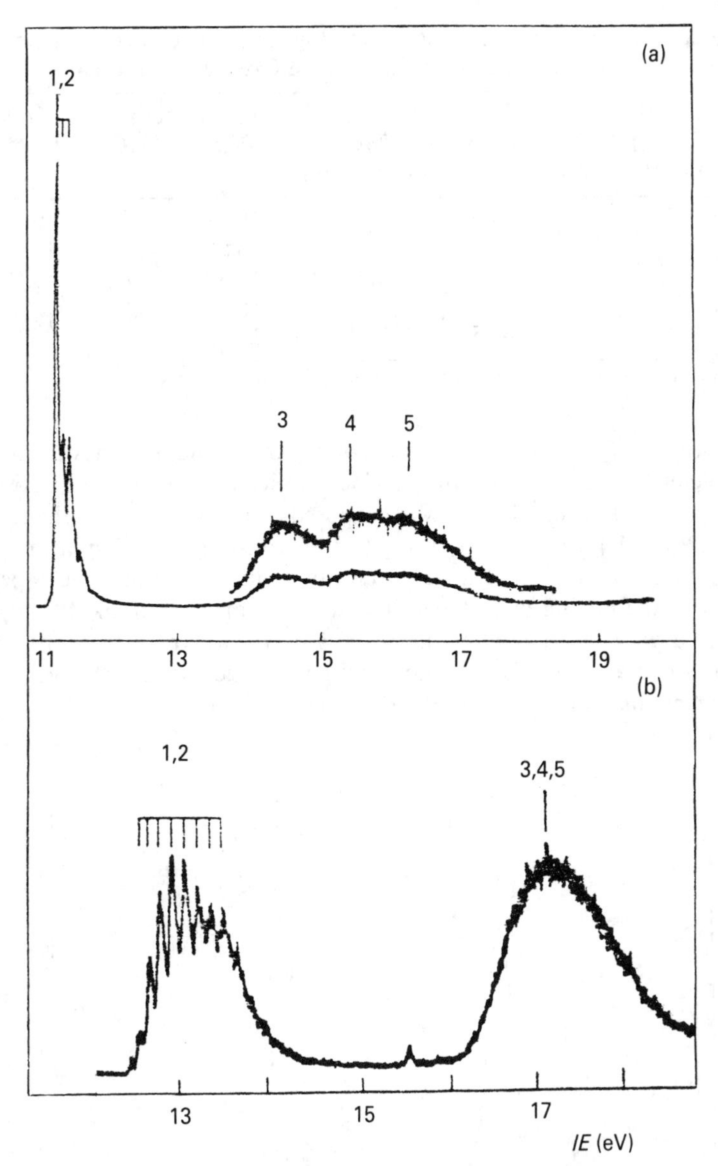


Fig. 7.2. He(I) PE spectra of fluoromethane (a) and chloromethane (b); (band numbers correspond to those of ionization potentials in Table 7.1). (From Kimura *et al.* [38])

suggestion of effective interaction between the chlorine lone pairs and (C—H) orbitals [53]. This suggestion has not been confirmed by later publications in which accurate quantum chemical calculations were reported: *ab initio* calculations in the 4-31G basis set assign the first band in the PES of chloromethane to ionization of the chlorine lone pairs [38] (Table 7.1).

At the same time, the interaction of lone pair orbitals of the chlorine and bromine

**Table 7.1.** Ionization potentials and higher occupied orbitals of halomethanes (PES-data and MO LCAO calculations)

| Fluoromethane | | Chloromethane | | Bromomethane | | Iodomethane | |
|---|---|---|---|---|---|---|---|
| $IP_i$ (eV) | MO (localization) | $IP_i$ (eV) | MO (localization) | $IP_i$ (eV) | (MO) | $IP_i$ (eV) | (MO) |
| 13.04<br>13.04 | $2e(\pi_{CH_3})$ | 11.29<br>11.29 | $3e(n_{Cl})$ | 10.53<br>10.85 | $(e)$ | 9.54<br>10.16 | $(e)$ |
| 17.06 | $5a_1(\sigma_{CF})$ | 14.42 | $7a_1(\sigma_{CCl})$ | 13.52 | $(a_1)$ | 12.5 | $(a_1)$ |
| (17.1)<br>(17.1) | $1e(n_F)$ | 15.47<br>16.25 | $2e(\pi_{CH_3})$ | 15.14<br>15.85 | $(e)$ | 13.8 | (—) |

(From Kimura *et al.* [38] and Cradock and Whiteford [332])

atoms with the (C—H) orbitals in the corresponding haloalkanes probably cannot be completely excluded. Thus, the specific character of the first ionization bands in the PE spectra of bromoalkanes was proposed to be explained by the interaction of $4p$(Br) and (C—H) orbitals. According to Ballard, spin–orbital splitting of the band corresponding to electron removal from 4p orbitals of bromine compete in these spectra with the mixing of these orbitals with the orbitals of alkyl groups [22]. An important argument for this mixing is the vibrational structure of the first band in the spectra of bromides. The second band has no such structure; it remains narrow and corresponds to the non-bonding character of the other $4p$ orbital of bromine (Fig. 7.3).

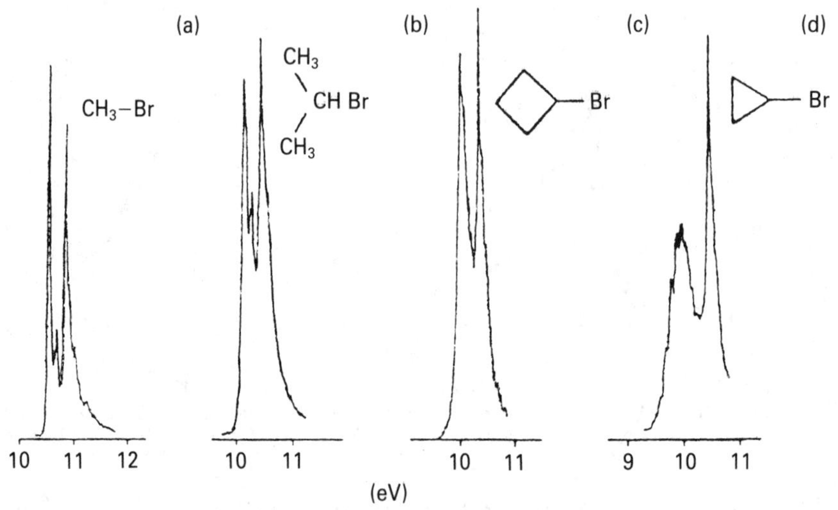


Fig. 7.3. First ionization bands in He(I) PE spectra of bromoalkanes: (a) bromomethane; (b) 2-bromopropane; (c) bromocyclobutane; (d) bromocyclopropane. (From Ballard [22])

It is difficult to estimate objectively Ballard's hypotheses without *ab initio* calculations of bromomethane. It should, however, be borne in mind that the requisite conditions for the interaction of $4p$ orbitals of bromine with the (C—H) orbitals (increasing energy difference of these orbitals and their decreasing overlapping) are less favourable than for the $3p$(Cl) and (C—H) orbitals.

The electron structure of halomethanes helps understanding of the reactivity of haloalkanes, e.g. stabilization of the positively charged reaction centre by the adjacent halogen atoms. For example, 2-fluoropropyl cation $(CH_3)_2CF$ was obtained by dissolution of 2,2-difluoropropane in superacids ($HSO_3F + SbF_5 + SO_2$ or $SbF_5 + SO_2$). It turned out to be the most stable of all other 2-halogen propyl cations $(CH_3)_2CE$ (where E = F, Cl, Br or I) [140, 145]. Besides, strong screening of the fluorine resonance was discovered in the spectrum of $^{19}$F—NMR of 2-fluoropropyl cation; this also indicates pronounced $2p$(C)-$2p$(F) interaction:

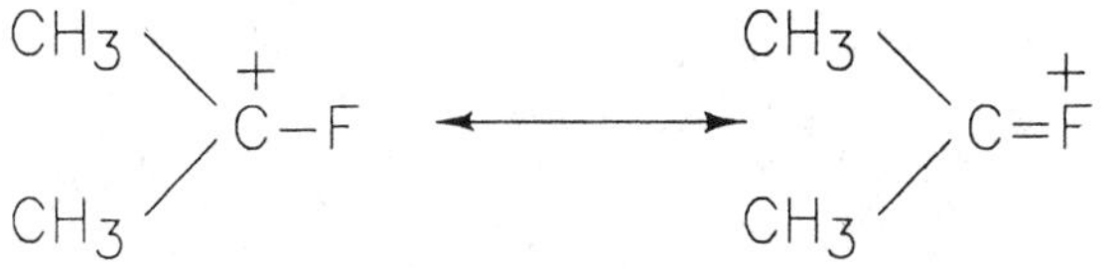

$$(CH_3)_2\overset{+}{C}-F \longleftrightarrow (CH_3)_2C=\overset{+}{F}$$

These data show unambiguously that the ability to donate lone pairs to the adjacent positively charged centre in a thermodynamically controlled system is the strongest for the fluorine atom. As the halogen atom grows in size, the stabilizing effect of its lone pairs decreases in the series F > Cl > Br > I (see also the results of *ab initio* calculations of fluoro- and chloromethyl cations [333]).

The high tendency of the fluorine atom to $\pi$ bonding with the adjacent carbon atom becomes evident not only by the interpretation of the PE spectra but also in calculations of various properties of the ground state of fluoroalkanes [334]. According to the calculations, the hydrogen atoms in the molecule of fluoromethane are more negative than in the methane molecule, while in fluoroethane, the $\beta$ carbon atom is more negative than in ethane ($Z \times 10$ values are given in electron charge units):

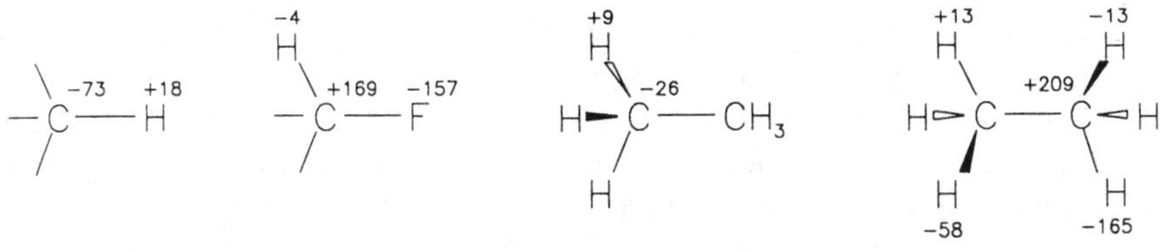

−73 +18
—C—H

−4
H
+169 −157
—C—F

+9
H
−26
H—C—CH₃
H

+13 −13
H H
+209
H—C—C—H
H H
−58 −165

This alternation of charges was explained by $\pi$ donation from the $2p$ orbitals of the fluorine atom. This effect largely explains why the energy of dissociation of the C—Cl, C—Br and C—I bonds in the corresponding halomethanes is much lower than that of the C—F bond dissociation in fluoromethane:

| | H | F | Cl | Br | I |
|---|---|---|---|---|---|
| $E_{diss}$ (kJ/mol) | 435.4 | 456.4 | 349.6 | 293 | 234.5 |

The same effects of the halogens were discovered in kinetically controlled systems. High asymmetry of the intermediate complex formed during nucleophilic substitution of fluorine in fluoromethane was seen from the calculations: the replaced fluorine atom does not immediately break its bond with the carbon atom

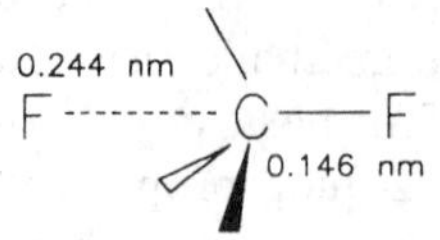

0.244 nm
F---------C—F
0.146 nm

and this is due to the ability of fluorine to stabilize the positive charge induced on the carbon at substitution [82, 335]. Rate constants of decarbonylation of acid chlorides of the general formula $(CH_3)_2C(E)COCl$ in the presence of aluminium chloride (the process is attended by formation of carbocations):

$$(CH_3)_2C(E)COCl + AlCl_3 \rightarrow [(CH_3)_2C(E)]^+[AlCl_4]^- + CO$$

also confirm the ability of the halogen atoms to stabilize the adjacent positively charged reaction centre [336]:

| | H | Cl | Br |
|---|---|---|---|
| $k_2$ (l/(mol/min)) | 0.091 | 0.680 | 0.096 |

The interaction of the $2p$ orbitals of the fluorine atom with the carbon atom orbitals that is seen in the electron structure of fluoromethane agrees with a less pronounced ability of fluorine to take part in transannular interactions (1,2, 1,3 and 1,4 types) with the positively charged centre in the ions. Ionization, for example, of 1,2-dihalogen alkanes of the general structure $(CH_3)_2CE—CF(CH_3)_2$ in $SbF_5 + SO_2$ at $-60°C$ gives bridged 1,2-halonium ions [140]:

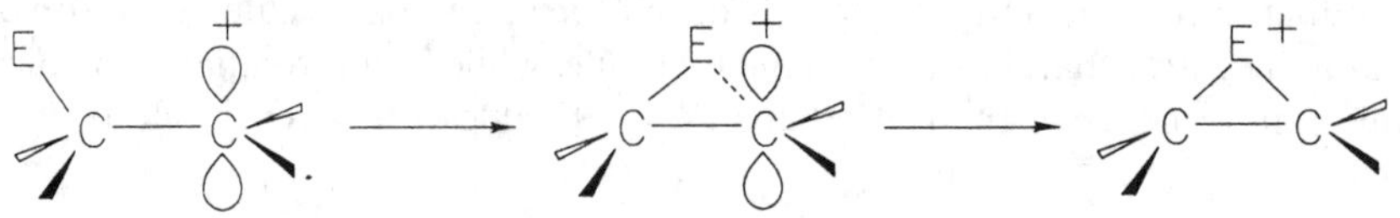

Stability of these ions increases in the series F < Cl < Br < I. The atoms of heavy halogens connected with the carbon atoms preserve unmixed their lone pair orbitals and therefore have a greater tendency to formation of transannular bonds than the fluorine atom has (because the fluorine atom has already utilized its lone pairs for $\pi$ type bonding with the adjacent carbon atom).

The different electron structure of fluoroalkanes compared with other haloalkanes also becomes evident during the study of their vacant electron levels. Fluoroalkanes have no affinity for electrons with energy under 5 eV. By contrast, all chloromethanes give one or two resonance signals in the appropriate region of the ET spectrum. These signals may be assigned to the capture of electrons by the vacant $\sigma^*(C—Cl)$ orbitals. The following values of electron affinity were thus determined for chloromethanes [337]:

| | $CH_3Cl$ | $CH_2Cl_2$ | $CHCl_3$ | $CCl_4$ |
|---|---|---|---|---|
| $EA_1$ (eV) | $-3.5$ | $-1.0$ | $-0.5$ | 1.0 |

In the interpretation of UV spectra of halomethanes, the opinion prevails that the longest-wave band corresponds to the $n \rightarrow \sigma^*$ transition while two subsequent bands correspond to the Rydberg transitions. Thus, the first band is due to excitation of the halogen lone pairs to the antibonding carbon–halogen orbital, while subsequent bands correspond to electron excitation of the lone pair to the orbitals localized predominantly over the halogen atom.

## 7.2 HALOETHYLENES

Haloethylenes have planar molecules. This is confirmed by the data of the microwave spectrum of vinyl chloride [338]:

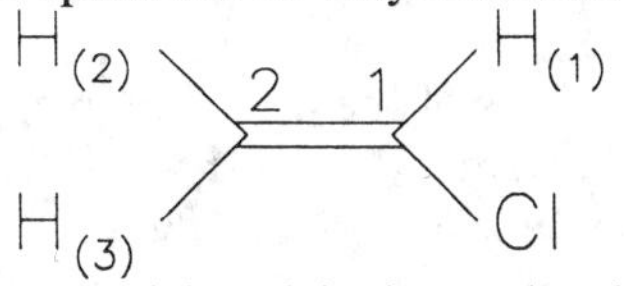

$H_{(2)}H_{(3)}C_{2}{=}C_{1}H_{(1)}Cl$

$r_{C-C} = 0.133\,(2)$ nm; $\angle C_{(2)}C_{(1)}Cl = 122°$

$r_{C-Cl} = 0.172\,(6)$ nm $\angle C_{(2)}C_{(1)}H_{(1)} = 124°$

$r_{C-H} = 0.108$ nm $\angle H_{(3)}C_{(2)}C_{(1)} = 121°$

The C—Cl bond in its molecule is shorter than in chloromethane because of the interaction of the adjacent $\pi$(C=C) and $p\pi$(Cl) orbitals. Mixing of these orbitals is proved by the analysis of the PE spectra (Table 7.2).

**Table 7.2.** Ionization potentials and higher occupied MO of haloethylenes (PES-data)

| Chloroethylene $IP_i$ (eV) | MO number and symmetry ($C_s$ group) | Bromoethylene $IP_i$ (eV) | Iodoethylene $IP_i$ (eV) |
|---|---|---|---|
| 10.2 | $2a''$ | 9.9 | 9.1 |
| 11.7 | $7a'$ | 10.9 | 10.1 |
| 13.2 | $1a''$ | 12.3 | 11.5 |
| 13.6 | $6a'$ | 13.0 | 12.25 |
| 15.4 | $5a'$ | 15.0 | 14.40 |
| 16.3 | $4a'$ | 16.0 | 15.70 |

(From von Niessen *et al.* [339])

The degree of the $\pi$-type interaction of the $p$ orbitals of the halogen atoms with the vinyl group orbitals is sufficiently high but it varies from one halogen to another. This is explained by the changes in the corresponding steric and energy parameters of $p$ orbitals of the halogens. Thus, the energy factor indicates preference of the bromine $p$ orbitals in such interactions. But as the atomic number of the halogen increases, overlapping of the halogen lone pairs with the carbon $\pi$ orbitals decreases. The controversial effects of these factors probably explain the different degree of mixing with involvement of the $p\pi$ orbitals of the halogen atoms. The effects of the interaction of lone pairs of various halogens with the carbon $\pi$ orbitals were estimated from the PE spectra of 1,2-dihaloethylenes. Below are the values of appropriate resonance integrals $H_{p\pi}$ [340]:

| | Cl | Br | I |
|---|---|---|---|
| $H_{p\pi}$ (eV) | −1.8 | −1.6 | −1.5 |

The fact of $p\pi$(Hal) and $\pi$(C=C) orbital interaction in haloethylenes is confirmed by *ab initio* calculations [38]. Mixing of these orbitals strengthens the $C_{sp^2}$-halogen bond and is probably the main cause of its very low reactivity in the nucleophilic substitution reactions. The effect of charge alternation evoked by this mixing is especially marked in molecules of unsaturated hydrocarbons with the fluorine atom at the double bond ($Z \times 10^3$ values are indicated):

H +78, H, C=C (−156), H, H        +89 H, −302 C=C +201, H +34, H, F −110

The $\pi$ bonding of fluorine is also confirmed by the values of the dipole moments of fluorohydrocarbons. Both experimental and calculated data suggest considerable compensation for the negative inductive effect of the fluorine atom in the molecules of unsaturated compounds by its positive $\pi$ donating effect.

Substitution of the hydrogen in ethylene for the halogen transforms the unoccupied electron levels in its molecule as well. In the ET spectrum of chloroethylene, for example, the value of electron affinity $EA_1$ is assigned to the $\pi^*$ MO, while the second value of $EA_2$ to the $\sigma^*$ MO [341]:

$$EA_1 = -1.28 \text{ eV}; \qquad EA_2 = -2.84 \text{ eV}.$$

## 7.3 HALOBENZENES

Compared with haloalkanes, the reactions of nucleophilic halogen substitution in halobenzenes occur under secure conditions and require high temperature, pressure, and catalyst. Inertness of halogen aromatic compounds is explained by the ability of the halogens to mix the orbitals of their lone pairs with the $\pi$ orbitals of the adjacent hydrocarbon fragment. Dipole moments of the C—Hal bonds in halobenzenes is, for example, markedly lower than in the corresponding haloalkanes. Owing to this mixing of orbitals, the halogen atoms lead to ortho and para substitution in attack by electrophilic agents.

The $\pi$-conjugative properties of fluorine and other halogens in halobenzenes can be estimated quantitatively by comparing $\sigma^+$ and $\sigma^-$ constants of the halogens:

| | F | Cl | Br | I |
|---|---|---|---|---|
| $\sigma^+$ | −0.07 | +0.11 | +0.15 | +0.14 |
| $\sigma^-$ | +0.50 | +0.46 | +0.44 | +0.39 |

The interaction of the halogen lone pair orbitals with the $\pi$ orbitals of the benzene ring according to PES is shown in Fig. 7.4. Mixing of orbitals of the bromobenzene molecule is an example [342]. Degenerate $\pi$ orbitals of the $e_{1g}$ symmetry and degenerate $\sigma$ orbitals of the $e_{2g}$ symmetry of the benzene ring respond in a different way to the introduction of the halogen atom. The $p_\pi$ orbital of the halogen can interact with $\pi_2$(symm) and $\pi_1$ orbitals of the benzene ring: each of these orbitals has the same symmetry, viz., type $b_1$. By contrast, the $\pi_3$(asymm) orbital of the benzene fragment has a different symmetry (symmetry $a_2$) and is not involved in any orbital interactions in halobenzenes.

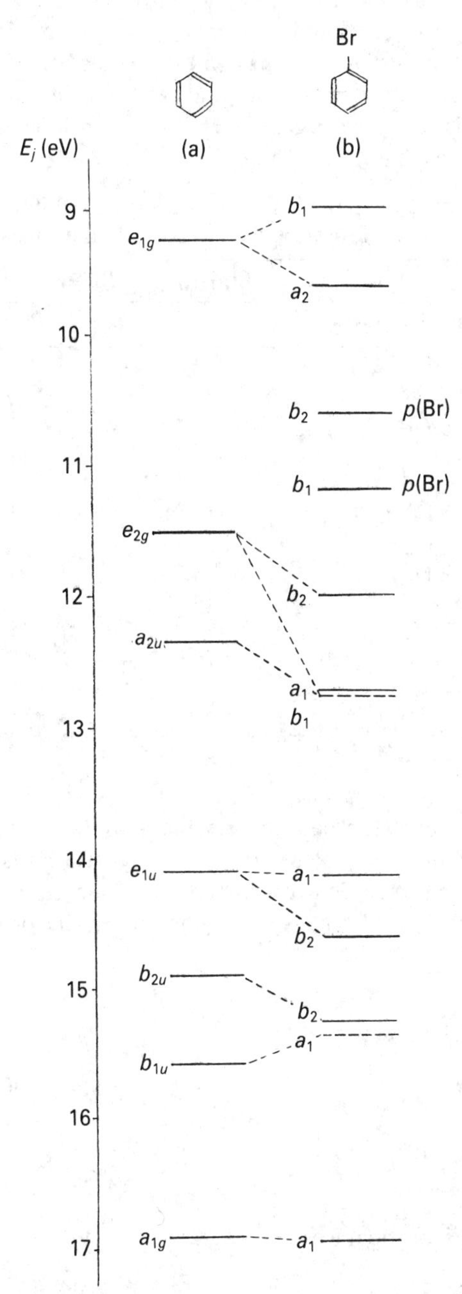


Fig. 7.4. Formation of occupied MOs of bromobenzene: (a) benzene; (b) bromobenzene. (From Cvitaš and Klasinc [342])

Of the two higher $\sigma(e_{2g})$ orbitals, only one can interact with the halogen atom by the inductive mechanism. Owing to the acceptor properties of bromine, the resultant orbitals are ionized in the region of higher potentials (about 13 eV). The other higher occupied orbitals $e_{2g}$ can (by symmetry) mix with $4p$ orbitals of the bromine atom oriented in the benzene ring plane. This orbital is transformed in bromobenzene into the $b_2$ symmetry orbital and can be seen at lower ionization potentials (about 12 eV).

Energy levels of the benzene degenerate $\pi$ and $\sigma$ orbitals are also split by mixing with lone pairs of other halogens (Table 7.3).

**Table 7.3.** Ionization potentials and occupied orbitals of halobenzenes (PES data and the 4-31G calculations)*

| Fluorobenzene | | Chlorobenzene | Bromobenzene | Iodobenzene |
|---|---|---|---|---|
| $IP_i$ (eV) | MO (localization) | $IP_i$ (eV) | | |
| 9.20 | $3b_1(\pi_2)$ | 9.10 | 9.02 | 8.79 |
| 9.81 | $1a_2(\pi_3)$ | 9.69 | 9.65 | 9.52 |
| 12.24 | $10b_2$, $B(19)$ | 11.32 | 10.63 | 9.78 |
| 12.24 | $2b_1(\pi_1)$ | 11.69 | 11.21 | 10.58 |
| 13.04 | $11a_1$, $B(18)$ | 12.26 | 11.98 | 11.63 |
| 13.89 | $9b_2$, $B(14)$, $n_F$ | (12.9) | 11.98 | 12.38 |
| 14.62 | $8b_2$, $B(16)$ | 13.15 | 12.81 | 12.60 |
| 15.17 | $10a_1$, $B(13)$ | 14.28 | 14.06 | 13.61 |
| 16.31 | $1b_1(n_F)$ | 14.68 | 14.46 | 14.33 |
| 16.31 | $9a_1$, $B(15)$ | 15.32 | 15.15 | 15.03 |
| (16.7) | $7b_2(n_F)$ | 15.9 | 15.64 | (15.5) |
| 17.82 | $8a_1(\sigma_{CF})$ | 16.89 | 16.76 | 16.66 |

*The letter B is used to designate MOs localized in the benzene ring; the numbers of these MOs in the benzene's orbita set are parenthesized.
(From Kimura *et al.* [38])

The interaction of $2p$(F) and benzene $\pi$ orbitals in fluorobenzene is most effective. This is confirmed by *ab initio* calculations in the 4-31G basis set [38]. Noteworthy are the transfer of $\pi$ electron density from the fluorine atom onto the benzene ring and the effect of charge alternation on the atoms according to the CNDO/2 calculations:

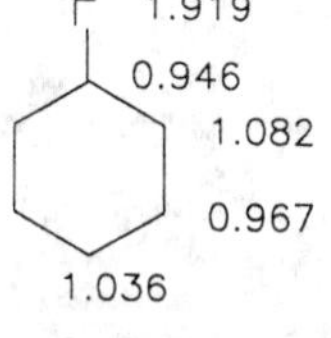

F 1.919
0.946
1.082
0.967
1.036

Total $\pi$ occupancies

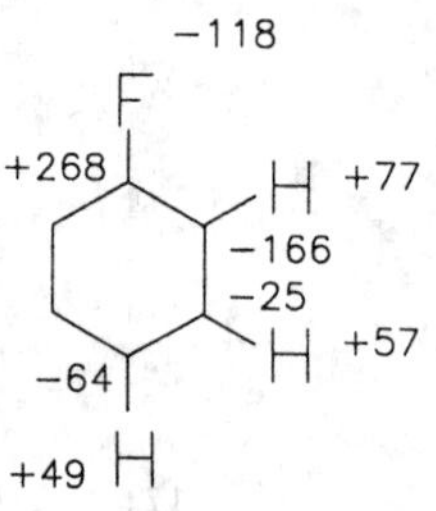

−118
F
+268
H +77
−166
−25
H +57
−64
+49 H

$Z \times 10^3$ charges on atoms

The objectiveness of these calculations is confirmed by accurate evaluation of the dipole moment of fluorobenzene, which is 1.66 D [334]. On mixing of 2*p*(F) orbitals with the benzene $\pi$ orbital, the unfavourable energy factor (great difference of the interacting orbital energies) is compensated probably by especially favourable steric conditions of orbital overlap because both fluorine and carbon belong to the second period.

The conclusion on the $\pi$ donating properties of the fluorine atom agrees with the data on the electrophilic aromatic substitution in halobenzenes, in the nitration and alkylation (Friedel–Crafts) reactions included (Table 7.4).

**Table 7.4.** Distribution of isomers (%) during nitration (nitrating mixture) and ethylation ($C_2H_5Br$, $GaBr_3$, 25°C) of halobenzenes

| Halobenzene | Nitration | | | Ethylation | | | |
|---|---|---|---|---|---|---|---|
| | ortho | meta | para | $k_{rel}$ | ortho | meta | para |
| $C_6H_5F$ | 9–13 | 0–1 | 86–91 | 0.282 | 42.9 | 13.7 | 43.4 |
| $C_6H_5Cl$ | 30–35 | 1 | 64–70 | 0.214 | 42.2 | 15.9 | 41.9 |
| $C_6H_5Br$ | 36–43 | 1 | 56–62 | 0.133 | 24.0 | 21.8 | 54.2 |
| $C_6H_5I$ | 38–45 | 1–2 | 54–60 | — | — | — | — |

(From Bethell and Gold [343])

The data on the effective transfer of $\pi$ electron density from the fluorine atom onto the benzene ring do not contradict the postulate of the greatest (compared with the other halogens) contribution of the iodine lone pair to the HOMO of the iodobenzene molecule. The contribution of the halogen $p_\pi$ orbital to the $\pi$ orbitals in various halobenzenes were evaluated by the quantum chemical calculations [62]:

| Contribution (%) of $p_\pi$ orbitals of the halogen to $\pi$ MOs | $C_6H_5F$ | $C_6H_5Cl$ | $C_6H_5Br$ | $C_6H_5I$ |
|---|---|---|---|---|
| $1b_1$ | 94.1 | 34.8 | 12.3 | 3.6 |
| $3b_1$ (HOMO) | 2.3 | 7.3 | 13.5 | 36.0 |

These data can be compared with spin densities on the halogen atoms in cation radicals of halobenzenes (unpaired electron in the cation radical is located in HOMO) calculated from ESR spectra [344]:

| | $C_6H_5F$ | $C_6H_5Cl$ | $C_6H_5Br$ | $C_6H_5I$ |
|---|---|---|---|---|
| Spin density (%) | 8 | 23 | 30 | 46 |

Other estimations of the contribution of halogen $p_\pi$ orbitals to the halobenzene $\pi$ orbitals are also reported. Thus, unique data have been obtained during the study of their Penning spectra. These spectra estimate directly the contribution of the halogen $p_\pi$ orbitals to the corresponding electron levels of halobenzenes [345]. The relative intensity of the band assigned to the HOMO ionization increases in the series $C_6H_5Cl < C_6H_5Br < C_6H_5I$ in full agreement with the increasing involvement of the halogen $p_\pi$ orbital in this MO, and its increasing steric availability for the electrophilic attack by meta-stable He* atoms.

By contrast, the relative intensity of the band assigned to ionization of the

MO $1b_1$ decreases in the series $C_6H_5Cl > C_6H_5Br > C_6H_5I$ due to decreasing contribution of the $p_\pi$ orbital of the halogen. The changes in the relative intensities of the bands assigned to ionization of halobenzene $\pi$ orbitals are illustrated in Fig. 7.5: $3b_1$ orbital intensities increase with the atomic number of the halogen in accordance with the growth of its contribution to the HOMO; $1b_1$ orbital intensities decrease with increasing atomic number of the halogen in accordance with its decreasing contribution to the $1b_1$ orbital.

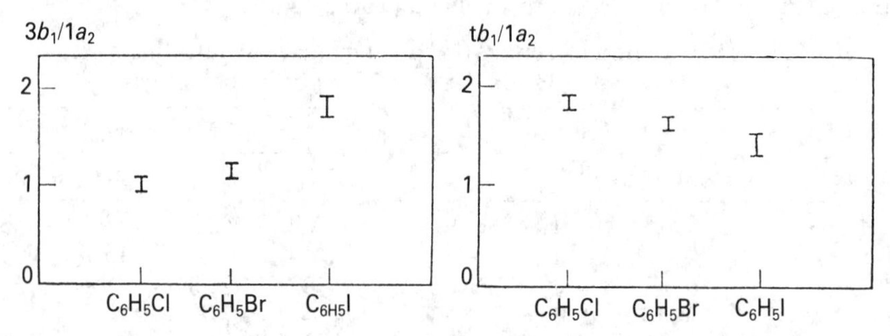


Fig. 7.5. Intensities of bands of ionization of $\pi$ orbitals in He*($2^3$S) Penning spectra of halobenzenes. (From Fujisawa *et al.* [345])

Electron transmission spectra were measured and vacant electron levels of halobenzene molecules studied. Electron affinities of all halobenzenes turned out to be higher than that of benzene (Table 7.5). Besides, the first resonance lines in the ET spectra of halobenzenes have vibrational structure, which suggests sufficiently high stability of the corresponding anion states. By contrast, the lifetime of the second anion states of halobenzenes turned out to be too small to measure their second values of electron affinity. At the same time, it can be seen from Fig. 7.6, that the third anion states of halobenzenes could be measured by the ETS method.

**Table 7.5.** Electron affinities of (eV) of halobenzenes

| Compound | $EA_1$ | $EA_2$ | $EA_3$ |
|---|---|---|---|
| Benzene | −1.15 | −1.15 (degen.) | −4.85 |
| Fluorobenzene | −0.89 | — | −4.77 |
| Chlorobenzene | −0.75 | — | −4.50 |
| Bromobenzene | −0.70 | — | −4.42 |

(From Jordan and Burrow [46])

Energy levels of frontier unoccupied MOs of halobenzenes are determined mainly by the negative inductive effect of the halogen atom. And again, despite the maximum value of the inductive effect of the fluorine atom, unoccupied levels of the fluorobenzene molecule are the closest to the benzene molecule levels: the destabilizing $\pi$ donating effect of fluorine is probably also valid for the vacant orbitals of fluorobenzene. Accumulation of the halogen atoms in the benzene nucleus increases its electron affinity to positive values. The electron affinity of each fluorine atom, for example, increases by about 0.25–0.30 eV so that $EA_1$ of hexafluorobenzene becomes 0.53 eV [349].

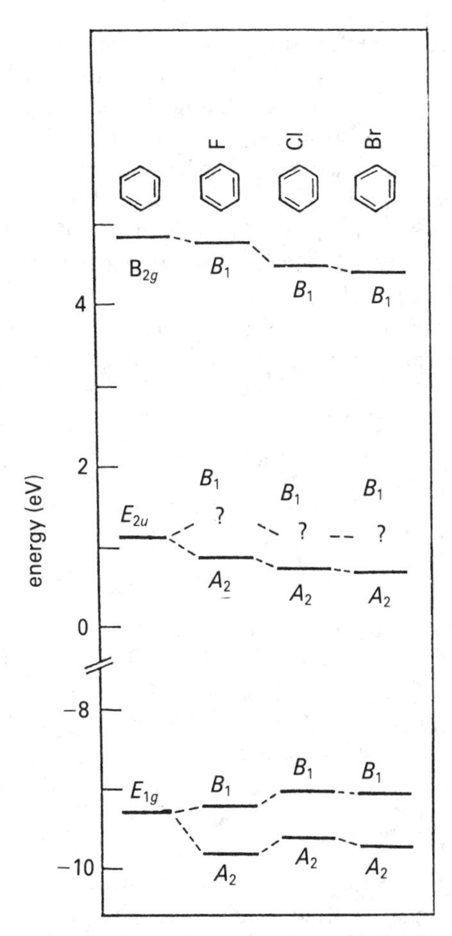


Fig. 7.6. Experimental estimations of energies of $\pi$ orbitals of halobenzenes (occupied orbitals are given at the bottom of the graph according to PES; unoccupied orbitals are given according to ETS). (By data of Jordan and Burrow [46])

Electron affinity of halobenzenes was also estimated in solvents. One-electron polarographic reduction of halobenzenes is an irreversible process. The corresponding potentials were therefore determined only at high scanning rate [45]:

| | $C_6H_5F$ | $C_6H_5Cl$ | $C_6H_5Br$ |
|---|---|---|---|
| $E^{red}_{1/2}$, V | −2.35 | −2.33 | −2.31 |

In connection with the data on the electron structure of halobenzenes, it is also necessary to mention the methods of activation of their molecules in the reactions of nucleophilic aromatic substitution. Thus, $\pi$ coordination of transition metals with halobenzenes is found to be quite effective. In the $C_6H_5F$—$Cr(CO)_3$ complex, for example, the fluorine atom is substituted for the N-piperidine group in acetonitrile with a quantitative yield at 20°C [346], while according to [347], the chlorine atom in the $C_6H_5Cl$—$Cr(CO)_3$ complex is substituted by methoxide ion at the same rate as in *p*-nitrochlorobenzene. Perfect quantum chemical calculations explaining this activation have not so far been reported. One can state, however, to a good degree of certainty, that the high reactivity of the halogen atom is explained by the changes in the energies of orbitals of the aromatic substrate. In Section 6.1.1 we have already mentioned that the highest occupied $\pi$ electron level of the benzene molecule decreases by 0.36 eV during the formation of the $\pi$ complex with $Cr^0$. It probably explains a considerable reduction of the activity of the complexes of aromatic substrates with the transition metals in the reactions of electrophilic aromatic substitution [348]. In halobenzene complexes with chromium tricarbonyl $Cr(CO)_3$, the energy of the unoccupied frontier MO of the halobenzene decreases as well. This reduction stabilizes the intermediate $\sigma$ complex and increases the rate of nucleophilic substitution of the halogen [347]:

$$\text{X-C}_6\text{H}_5\cdots M(CO)_3 \xrightarrow{RO^-} \text{(X, OR)C}_6\text{H}_5\cdots \bar{M}(CO)_3 \xrightarrow[-X^-]{} \text{RO-C}_6\text{H}_5\cdots M(CO)_3$$

***Problems***

(1) The presence of a chlorine atom makes attachment of an electron to an organic molecule much easier. Illustrate this conclusion by comparing the corresponding data for $CH_4$ and $CH_3Cl$, $CH_2{=}CH_2$ and $CH_2{=}CH\text{-}Cl$, $C_6H_6$ and $C_6H_5Cl$.

(2) The presence of a chlorine atom decreases 'Pearson hardness' of organic molecules. Explain this fact using the following data:

| | $CH_4$ | $CH_3Cl$ | $CH_2{=}CH_2$ | $CH_2{=}CH\text{-}Cl$ | $C_6H_6$ | $C_6H_5Cl$ |
|---|---|---|---|---|---|---|
| $IP_1$ (eV) | 13.0 | 11.29 | 10.51 | 10.2 | 9.24 | 9.10 |
| $EA_1$ (eV) | $< -6.0$ | $-3.5$ | $-1.78$ | $-1.28$ | $-1.15$ | $-0.75$ |

(3) Bimolecular nucleophilic substitution ($S_N^2$ reactions) can be considered as a donor–acceptor interaction. Suggest the relative reactivity of $CH_3Cl$ and $CH_3Br$ in $S_N^2$ reactions (remember that the dipole moments of these halides are 1.94 D and 1.79 D; the polarographic reduction potentials are $-2.19$ V and $-1.59$ V respectively).

(4) The presence of a halogen atom decreases 'hardness' of arenes. Arrange the following compounds in the order of their increasing hardness (polarographic reduction potentials are given in parentheses):

| | $C_6H_6$ | $C_6H_5F$ | $C_6H_5Cl$ | $C_6H_5Br$ | $C_6H_5I$ |
|---|---|---|---|---|---|
| $IP_1$ (eV) | 9.24 | 9.20 | 9.10 | 9.02 | 8.79 |
| $EA_1$ (eV) | $-1.15$ | $-0.89$ | $-0.75$ | $-0.70(-2.32$ V) | |

(5) Chlorobenzene releases an electron easier than benzene. Suggest an unpaired electron distribution in the chlorobenzene cation radical.

(6) Explain why chorobenzene is attacked by electrophiles mostly at ortho and para positions.

# 8

# Oxygen and sulphur compounds

## 8.1 ALIPHATIC ALCOHOLS, ETHERS AND SULPHIDES

The role of aliphatic alcohols in organic synthesis is probably best of all demonstrated by the prospects for using methanol: at the present time, methyl alcohol is mostly used in the manufacture of formaldehyde and acetic acid, but it is believed that all products, that are now manufactured from ethylene, will soon be produced from methanol [139]. This prognosis is based on the inexhaustibility of the natural sources of synthesis gas.

The chemical properties of alcohols depend on their amphoterity. The hydroxyl group of the alcohols can dissociate:

$$R{-}OH \rightleftharpoons R{-}O^- + H^+$$

thus giving them acid properties. Variations in the structure of the hydrocarbon fragment R change acidity within a wide range: the $pK_a$ of tert-butyl alcohol is about 19, while in the perfluorinated analogue it is 5 [350].

| R | $CH_3$ | $CH_3CH_2$ | $(CH_3)_2CH$ | $(CH_3)_3C$ |
|---|---|---|---|---|
| The value of $pK_a$ | 15.2 | 15.8 | 16.9 | 19.2 |

Acidity of alcohols in the gas phase varies in a way different from that in the liquid phase: alcohol acidity in the series $CH_3$, $CH_3CH_2$, $(CH_3)_2CH$, $(CH_3)_3C$ increases with the size of the hydrocarbon fragment.

The presence of unshared electron pairs in the oxygen atom gives basic properties to alcohols and ethers: they can be protonated and can form salts with acids. It is interesting to note that, as distinct from acidity, basicity of alcohols and ethers in the gas and liquid phase increases with the size of the alkyl group in the same way:

$$CH_3 < CH_3CH_2 < (CH_3)_2CH < (CH_3)_3C$$

The mentioned properties, the stabilizing effect of the alkyl group R on the anion centre in acid dissociation of alcohols in the gas phase included, are explained by a better ability of alkyl groups (compared with the hydrogen atom) to stabilize both positive and negative charges on the oxygen atom [140]. The reverse order of acidity of alcohols in the liquid phase is probably explained by the growth of steric hindrance to solvation of the alkoxide centre with the size of the substituent R.

Ethers are first of all characterized by their ability to form complexes. Thus, the basic properties of the oxygen atom give the ethers the ability of specific solvation of metal cations. This is why ethers (first of all diethyl ether, tetrahydrofuran, dioxane) are mostly used as solvents in reactions of organometal compounds. The ability of ethers to show specific solvation is especially emphasized by the unique properties of crown ethers [351].

### 8.1.1 Methanol and dimethyl ether

The specific electron structure of oxygen compounds is readily demonstrated by the water molecule. This molecule is another example of a different interpretation of electron distribution by the electron theory and by the MO theory.

The classical electron theory dealing with localized bonds in molecules regards the structure of the oxygen atom in a water molecule as the $sp^3$ hybrid state. This approach is based on the fact that the HOH angle is 104°, i.e. it is closer to the tetrahedral angle of 109° than to 90°, as might be expected from the involvement of unhybridized $2p$ orbitals of oxygen in bonding with the hydrogen atoms. According to this scheme, one $2s$ and three $2p$ orbitals of the oxygen atom are hybridized to give four $sp^3$ orbitals: two of them are utilized in bonding with the hydrogen atoms while the others remain in the molecule as two equivalent unshared electron pairs localized over the oxygen atom.

As the hydrogens in the water molecule are replaced by hydrocarbon fragments, the valence angle at the oxygen atom increases to about the value of the tetrahedral angle:

| | | |
|---|---|---|
| $H_2O$ | ∠HOH | 104° |
| $CH_3OH$ | ∠COH | 107–109° |
| $CH_3OCH_3$ | ∠COC | 112° |

The electron theory of organic substances thus suggests the presence of two equivalent lone pairs over the oxygen atom in both alcohols and ethers.

The MO theory leads to other conclusions on the electron structure of water, alcohols and ethers. Thus, the electron levels of the water molecule are, according to this theory, formed from the orbitals of similar symmetry with involvement of the $2s$ and $2p$ orbitals of the oxygen atom, and bonding (+) and antibonding (−) combinations of $1s$ orbitals of the hydrogen atoms. Within the framework of the point group $C_{2v}$, in which the water molecule belongs, these basis orbitals are transformed as shown in Table 8.1.

The interaction of the basis orbitals of the same symmetry gives six MOs of different energies and symmetries. As eight valence electrons are inserted into the orbitals in the order of their increasing energies, four non-equivalent occupied MOs are formed. These are the following (in the order of increasing energies): $1a_1$ orbital which is a predominantly $2s$ orbital of the oxygen slightly stabilized by the (+) combination of $1s$(H) orbitals; $1b_2$ and $2a_1$ orbitals oriented (like $1a_1$ orbital) in the

**Table 8.1.** Transformations of basis AOs of the water molecule ($C_{2v}$ point group)

| Atomic orbital | Character | | | | Symmetry |
|---|---|---|---|---|---|
| | $E$ | $C_2^z$ | $\sigma^{xz}$ | $\sigma^{yz}$ | |
| $2p_x$(O) | +1 | −1 | +1 | −1 | $b_1$ |
| $2p_z$(O) | +1 | +1 | +1 | +1 | $a_1$ |
| $2p_y$(O) | +1 | −1 | −1 | +1 | $b_2$ |
| (−)1s(H) | +1 | −1 | −1 | +1 | $b_2$ |
| (+)1s(H) | +1 | +1 | +1 | +1 | $a_1$ |
| 2s(O) | +1 | +1 | +1 | +1 | $a_1$ |

molecular plane, which are the orbitals of the O—H bonds; $1b_1$ orbitals, perpendicular to the molecular plane, formed only by the $2p_\pi$ orbitals of the oxygen atom [53].

According to the MO theory, only one unshared electron pair (in the traditional meaning of this term) is retained at the oxygen atom of the water molecule. This lone pair is localized in the $1b_1$ orbital of $\pi$ symmetry. The other orbitals of the molecule are involved (to different extent) in the H—O bonding and cannot be regarded as lone pairs. At the same time, the involvement of the $1a_1$, $1b_2$ and $2a_1$ orbitals in bonding of the oxygen and hydrogen atoms differs. According to *ab initio* calculations, their involvement in the O—H bonding is estimated as follows [22]:

| $1a_1$ | $1b_2$ | $2a_1$ | $1b_1$ |
|---|---|---|---|
| 0.2022 | 0.331 | 0.0339 | 0.0 |

It can be seen that the contribution of the $2a_1$ orbital to the H—O bonding is insignificant. The $2a_1$ orbital is therefore regarded as the second (conventional) lone pair. In order to differentiate lone pair orbitals of the oxygen atom within the framework of the MO theory, one of these obitals ($1b_1$) is designated $n$(O) and the second $2a_1$ orbital (conventional lone pair) $\bar{n}$(O).

The diagram of electron levels of the water molecule suggested by the MO theory agrees well with experiment: the PE spectrum of water has four bands assigned to ionization of four occupied MOs of the valence shell (Fig. 8.1 gives only three bands; the fourth band, corresponding to ionization of the $1a_1$ orbital, is seen at 32 eV). Only one (the first) band is narrow, in agreement with the ionization of the $1b_1$ non-bonding electron level–lone pair of the oxygen atom.

As the hydrogen atom is substituted for a hydrocarbon fragment, the conditions of formation of the lone pair levels are altered. Not only the $2p_x$ orbital of the oxygen atom but also the corresponding combination of $2p_x$ orbitals of the carbon atoms have the $b_1$ symmetry in the molecule of dimethyl ether. Strictly speaking, the HOMO of dimethyl ether with the symmetry $b_1$ cannot be regarded as a non-bonding one because it includes orbitals of several atoms. But the $2p_x(p_\pi)$ orbital of oxygen remains dominant in this MO.

Three higher MOs of dimethyl ether are shown in Fig. 8.2. According to quantum chemical calculations, the energies of these MOs largely depend on the molecular conformation [352]. This molecule is not so uniform conformationally as the water

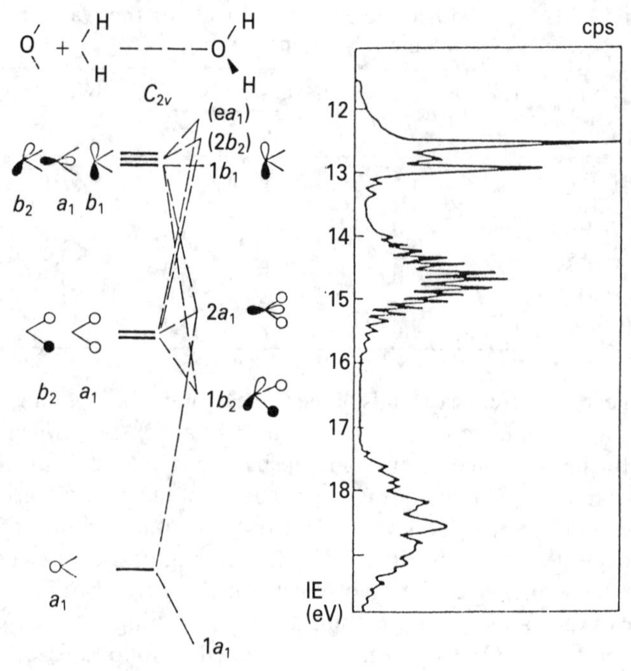


Fig. 8.1 Formation of molecular orbitals of water and its He(I) PE spectrum (From Bock and Ramsey [53])

molecule is: rotation of the methyl group about the C—O bonds results in formation of various conformers of this molecule. Two of them are shown in Fig. 8.2, viz., doubly eclipsed and singly eclipsed conformers. According to quantum chemical calculations (CNDO/2 and EHMO) and electron diffraction measurements, the doubly eclipsed conformer is the most stable. The CNDO/2 calculations predict the following order of occupied MOs for this conformer (in the order of their increasing energy): $2b_2$, $3a_1$, $1a_2$, $3b_2$, $4a_1$, and $2b_1$. The same sequence was predicted by the *ab initio* calculations in the 6-31G basis set (Table 8.2) [38].

Increasing delocalization of the frontier occupied electron level in the water–methanol–dimethyl ether series changes the form of the first ionization band which broadens regularly as the methyl groups are attached to the oxygen atom [38]. Substitution of H for $CH_3$ increases the energies of the occupied MOs. This is probably explained by the positive inductive effect of the methyl group because, according to *ab initio* quantum chemical calculations, the interaction of the $2p\pi$(O) and (C—H) orbitals results in an appreciable transfer of the $\pi$ density from the oxygen atom to the methyl groups [154, 335]. Consider the $Z \times 10^3$ values:

$H_2O$: O −406; H +203

$CH_3OH$: O −350; H(O) +211; H +14; H −61

$CH_3OMe$: O −279; C +161; H +4; H −13

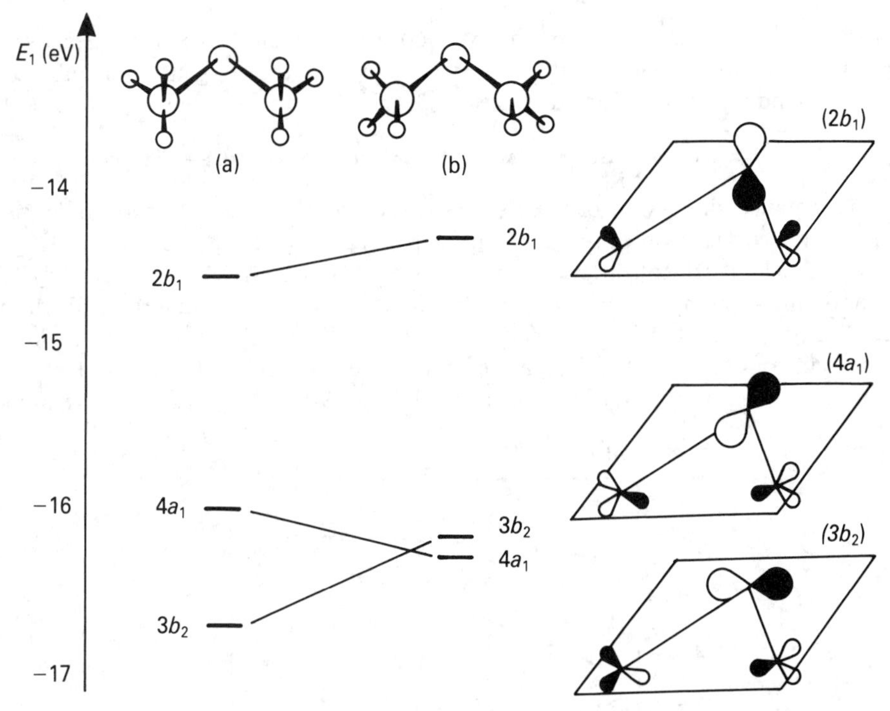


Fig. 8.2. Dependence of sequence of higher occupied MOs of dimethyl ether on conformation: (a) doubly eclipsed; (b) singly eclipsed. (From Bock *et al.* [352]).

**Table 8.2.** Ionization potentials and higher occupied orbitals of the molecules of water, methanol, and dimethyl ether (PES data and 6-31G calculations)

| Water | | Methanol | | Dimethyl ether | |
|---|---|---|---|---|---|
| $IP_i$ (eV) | MO (localization), group $C_{2v}$ | $IP_i$ (eV) | MO (localization), group $C_s$ | $IP_i$ (eV) | MO (localization), group $C_{(2v)}$ |
| 12.62 | $1b_1(n_O)$ | 10.94 | $2a''(n_O)$ | 10.04 | $2b_2(n_O)$ |
| 14.74 | $3a_1(\bar{n}_O)$ | 12.68 | $7a'(\bar{n}_O)$ | 11.91 | $6a_1(\bar{n}_O)$ |
| 18.51 | $1b_2(\sigma_{OH})$ | 15.19 | $6a'(\sigma_{CO})$ | 13.43 | $4b_1(\pi^-_{CH_3})$ |
| | | (15.7) | $1a''(\pi_{CH_3})$ | 14.20 | $1a_2(\pi^-_{CH_3})$ |
| | | 17.50 | $5a'(\sigma_{OH})$ | 16.0– | $3b_1(\sigma_{CO})$ |
| | | | | 16.5 | $5a_1(\pi^+_{CH_3})$ |
| | | | | | $1b_1(\pi^+_{CH_3})$ |

(From Kimura *et al.* [38])

It can be seen that the oxygen atom has a lower negative charge in the molecules of methyl alcohol and dimethyl ether than in the water molecule.

The conclusions of the spectral and quantum chemical analysis of the electron structure agree with other experimental findings. The hydroxyl group, for example, sensitizes the adjacent C—H bond to electrophilic agents, oxidation reactions included. When at the α-position to the positive centre, the oxygen atom stabilizes the corresponding carbocations to a greater degree than the fluorine atom does [9]:

$$CH_3—O—CH_2X + SbF_5 \rightarrow [CH_3—O—\overset{+}{C}H_2 \leftrightarrow CH_3—\overset{+}{O}{=}CH_2]SbF_5X$$

For a detailed discussion of the electron structure of α alkoxy carbocations see [24], and for comparison of the donor properties of RO and RS groups see [333].

The increasing delocalization of the occupied frontier MO of $b_1$ symmetry in methanol and dimethyl ether does not alter its π character. The diagram shown in Fig. 8.3 gives the comparison of energies of MOs with the changing COC angle in the molecule of dimethyl ether. As in the water molecule, whose $1b_1$ orbital does not respond to changes in the HOH valence angle from 90 to 180° [22], the energy of

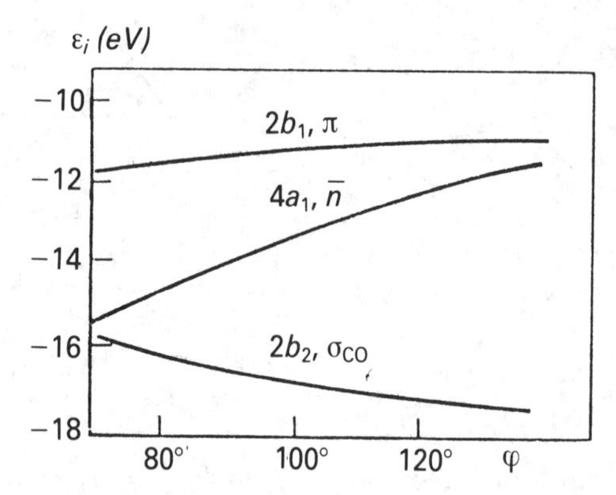


Fig. 8.3. Dependence of dimethyl ether MOs energies on COC valence angle (see Fig. 8.2). (From Traven, V.F. et al. [58])

the $2b_1$ orbital of dimethyl ether molecule increases only insignificantly as the COC angle changes from 70 to 140° (according to MNDO calculations [58]). At the same time, the $\bar{n}$(O) orbitals of the $a_1$ symmetry of both molecules are stabilized appreciably with decreasing angle, while the $b_2$ symmetry orbital, on the contrary, is destabilized because this change in the form of the molecule increases the antibonding interaction in this orbital.

It is important to note that the $\bar{n}$(O) orbital, the second (conventional) lone pair of the oxygen atom, mainly preserves its $2p$(O) character in the transition from water to methyl alcohol and dimethyl ether. This is confirmed by the ratio of intensities of the bands in the methanol spectra recorded with the He(I) and He(II) sources:

| Band no. | 1 | 2 | 3,4 |
|---|---|---|---|
| Assignment | $n$(O) | $\bar{n}$(O) | σ,π |
| Relative intensity, He(I)/He(II) | 0.64 | 0.85 | 1.1 |

The intensity of the first band, corresponding to the MO with the maximum contribution of the $2p$(O) orbital, increases mostly in the spectrum recorded with He(II); the intensities of bands 3 and 4, assigned to ionization of the σ(OH) orbital do not increase at all; and the intensity of the band corresponding to $\bar{n}$(O) orbital increases only slightly [176].

### 8.1.2 Organoelement analogues of dimethyl ether

The steric parameters of $(CH_3)_2E$ molecules change appreciably in the transition to dimethyl derivatives of heavier elements of the group VIA. Consider the C—E bond distance and CEC valence angles [62]:

| Element | O | S | Se | Te |
|---|---|---|---|---|
| C—E bond distance, nm | 0.141 | 0.180 | 0.198 | 0.211 |
| $\angle$CEC, ° | 111.7 | 98.9 | 98 | 93 |

These changes are not sufficient to exclude the preference of the doubly eclipsed conformation discovered for dimethyl ether. Nevertheless, the increase in the C—E distance is sufficient to eliminate the dependence of the electron level of dimethyl sulphide molecule on conformation (Fig. 8.4). Mollere and Bock calculated the molecule of dimethyl sulphide by the CNDO/2 method to arrive at this conclusion.

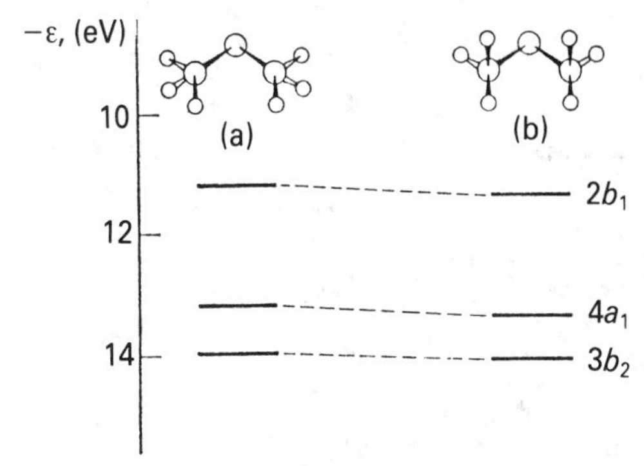


Fig. 8.4. Independence of dimethyl sulphide HOMO energies of the conformations: (a) singly eclipsed; (b) doubly eclipsed. (Data by Mollere, P. *et al.* [59])

In the transition from dimethyl ether to dimethyl sulphide, the MO sequence is on the whole preserved. The only exception is the (C—S) bond orbital of the $b_2$ symmetry. In the $(CH_3)_2S$ molecule, it becomes the third from top, above the pseudo-$\pi$ orbitals of the methyl groups [38, 59] (Table 8.3).

Changes in the HOMO energies in the series of $(CH_3)_2E$ and $H_2E$ compounds are shown in Fig. 8.5 [353]. Since HOMOs of $b_1$ symmetry in both series of compounds are largely formed by the $p_\pi$ orbitals of the element E atoms, the consecutive reduction of the first and second ionization potentials in the O,S,Se,Te series is explained by the reduction of the ionization potential of $p$ orbitals of the valence shell of free atoms E in the same series.

The regularities observed in dimethyl ether (see Fig. 8.3) and dimethyl sulphide (Fig. 2.13) can also be seen in the Walsh diagram constructed for dimethyl derivatives of selenium and tellurium. The HOMO formed by the lone pair of the atom E remains insensitive to the changes in the valence CEC angle in these compounds.

**Table 8.3.** Ionization potentials and occupied orbitals of dimethyl sulphide molecule (PES data and calculations in the 4-31G basis set)

| $IP_i$ (eV) | MO number and symmetry ($C_{2v}$ group) | MO localization | $IP_i$ (eV) | MO number and symmetry ($C_{2v}$ group) | MO localization |
|---|---|---|---|---|---|
| 8.72 | $3b_1$ | $n$(S) | 14.73 | $4b_2$ | $\pi_{CH_3}^-$ |
| 11.30 | $8a_1$ | $\bar{n}$(S) | 15.25 | $7a_1$ | $\pi_{CH_3}^+$ |
| 12.68 | $5b_2$ | $\sigma$(CS) | (15.7) | $2b_1$ | $\pi_{CH_3}^+$ |
| 14.07 | $1a_2$ | $\pi_{CH_3}^-$ | | | |

(From Kimura *et al.* [38])

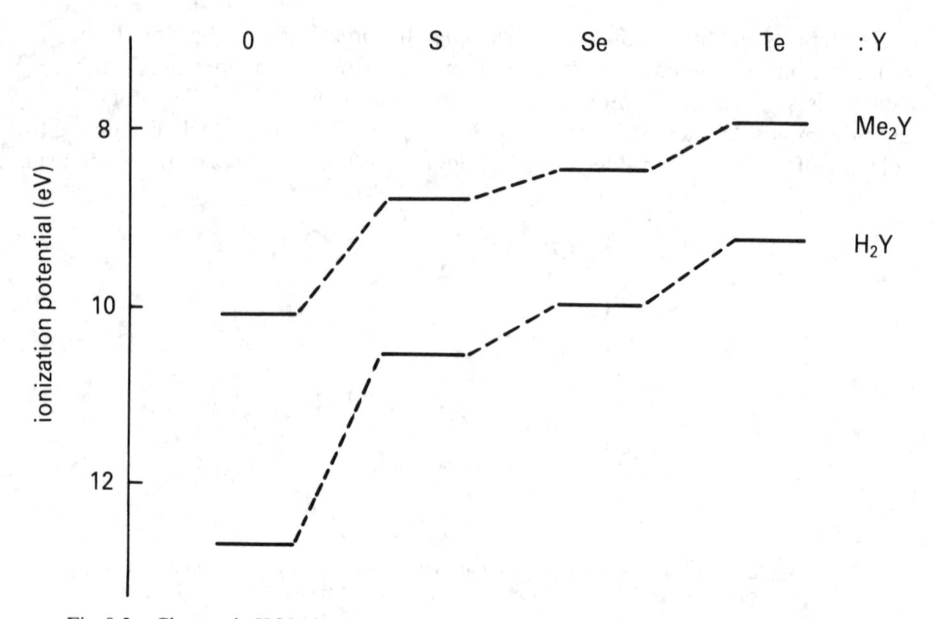


Fig. 8.5. Changes in HOMO energies in the series of chalcogen hydrides $H_2E$ and dimethyl chalcogenides $(CH_3)_2E$ according to PES. (From Vovna and Vilesov [36] and Elbel *et al.* [353])

The second lone pair of the $a_1$ symmetry is stabilized with decreasing angle, while the $\sigma$(C—E) orbital of the $b_2$ symmetry is destabilized [58].

### 8.1.3 Linear and cyclic ethers and sulphides

Enlargement of the alkyl group at the oxygen atom in the molecules of alcohols and ethers decreases regularly the ionization potential of HOMO with increasing energy of the oxygen lone pair (Table 8.4). Basicity of alcohols and ethers in the gas and liquid phases increases in the same direction.

The lone pair energy increases with the size of the alkyl group in dialkyl sulphides as well [53]: the HOMO energy increases markedly with the substitution of the methyl group at the sulphur atom for the isopropyl or tert-butyl group. It is believed

**Table 8.4.** Ionization potentials of UEP of oxygen atom in some alcohols and ethers

| R | $IP_{UEP}$ (eV) | | R | $IP_{UEP}$ (eV) | |
|---|---|---|---|---|---|
| | R—OH alcohols | $R_2O$ ethers | | R—OH alcohols | $R_2O$ ethers |
| $CH_3$ | 10.85 | 9.94 | $(CH_3)_2CH$ | 10.29 | 9.32 |
| $CH_3CH_2$ | 10.46 | 9.50 | $(CH_3)_3C$ | 10.09 | 8.94 |
| $CH_3(CH_2)_3$ | 10.32 | — | | | |

(From Cocksey *et al.* [354])

that the observed effects are (at least partly) explained within the framework of the $n$(S)–$\sigma$ conjugation.

The interaction of the oxygen and sulphur lone pairs with $\sigma$ orbitals of substituents seems to be responsible for the changes in the energies of CT bands in electron absorption spectra of TCNE complexes with the corresponding donors. Thus, ionization potentials of sulphides and ethers containing $(CH_3)_3SiCH_2$ and $(CH_3)_3Si(CH_3)_2Si$ groups are much lower than one might expect from their inductive effects. Below are ionization potentials of some ethers PhOR and sulphides $C_4H_9SR$ [355, 356]:

PhOR ethers

| R | $(CH_3)_3C$ | $CH_2Si(CH_3)_3$ | $Si(CH_3)_3$ | $Si(CH_3)_2$-$Si(CH_3)_3$ |
|---|---|---|---|---|
| $IP_1$ (eV) | 8.18 | 8.08 | 8.76 | 8.28 |

$C_4H_9SR$ sulphides

| R | $C_4H_9$ | $CH_2Si(CH_3)_3$ | $Si(CH_3)_3$ | $Si(CH_3)_2$-$Si(CH_3)_3$ |
|---|---|---|---|---|
| $IP_1$ (eV) | 8.16 | 8.07 | 8.50 | 8.36 |

The effects of orbital mixing with participation of the VIA group elements were estimated by ESR spectra of cation radicals formed from compounds with the general formula $CH_3$—E—$CH_2Si(CH_3)_3$, where E = O or S. HFI constants on (E)-methyl protons decrease markedly compared with these constants in the spectra of cation radicals formed from $(CH_3)_2E$ dimethyl derivatives. According to Sakurai *et al.*, this suggests unambiguously effective $n$(E)–$\sigma$(C—Si) interaction in the following conformation [357]:

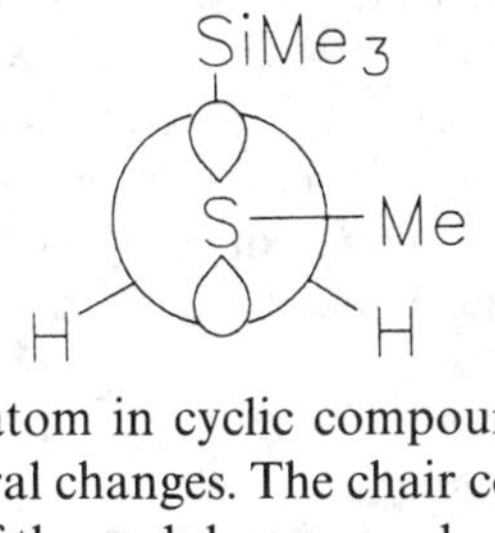

SiMe3
S
Me
H
H

Substitution of the carbon atom in cyclic compounds for the oxygen atom does not induce appropriate structural changes. The chair conformation, which is the most favourable and characteristic of the cyclohexane molecule, is preserved in hydropyran as well [358]. At the same time, this substitution can change the ratio of conjugation effects: cyclobutane has a non-planar ring while oxetane $(CH_2)_3O$, in which the number of eclipsed (C—H) bonds is lower, is planar [359].

Basicity of oxygen lone pair increases with the size of the cycle. This is confirmed by PES findings [189]:

| | $(CH_2)_2O$ | $(CH_2)_3O$ | $(CH_2)_4O$ | $(CH_2)_5O$ |
|---|---|---|---|---|
| $IP_1$ (eV) | 10.57 | 9.63 | 9.50–9.55 | 9.46 |

7-Oxabicyclo[2,2,1]heptane (XLVII) is an especially noteworthy cyclic ether. It is markedly less basic in solution than tetrahydrofuran. This was quite unexpected and was explained by hindered solvation on protonation. This is probably true because the energy of the oxygen lone pair in the gas phase turned out to be the same in both compounds:

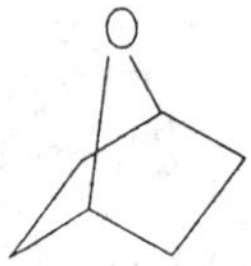

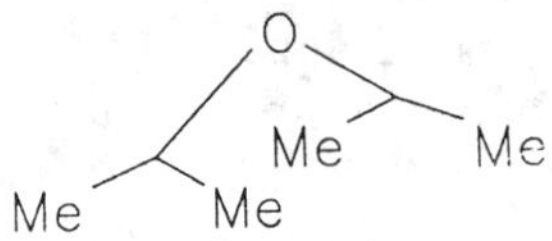

| $IP_1$ (eV) 9.50–9.55 | 9.50–9.55 | 9.35 |
|---|---|---|

At the same time, some specific properties of the electron structure of bicyclic ether XLVII were discovered. Compared with its acyclic analogue diisopropyl ether, ether XLVII has a higher ionization potential of its lone pair, while the first band of its PE spectrum is characterized by an especially narrow band, which is uncommon even for ionization of a lone pair [189]. This is explained by the fact that the alkyl groups of linear and monocyclic ether connected with the oxygen atom increase the energy of its lone pair not only by the inductive mechanism but also at the expense of lone pair orbital interactions with the C—C and C—H bond orbitals. The conditions for this mixing in the bridged ether XLVII are probably quite unfavourable: its PE spectrum has the signs of the ionization of electrons of the unmixed $2p$ orbital of oxygen whose energy increases (compared with the energy of a similar orbital, e.g. in the water molecule) only due to the inductive effect of two branched alkyl groups bonded to the oxygen atom. It appears that the unmixed orbital of the oxygen atom in the ether XLVII simulates the conditions of stabilization of 7-norbornyl cation XLVIII. This cation is highly reactive while the corresponding 7-substituted norbornanes undergo solvolysis with great difficulty [189]. This is usually explained by the change in the angular strain associated with solvolysis. At the same time, it is known that the reactivity of the tert-cyclobutyl cation XLIX is three orders higher than that of cation L, although the angle strain is, probably, even higher in the former.

The main factor of the electron structure of 7-norbornyl cation is, most probably, hindered mixing (due to its geometry) of its vacant $2p$ orbitals with $\sigma$ orbitals of cyclohexane framework as is the case with decreased orbital overlap in 7-oxanorbornane XLVII.

### 8.1.4 Crown ethers and their analogues

Cyclic oligomers of ethylene glycol $(—OCH_2CH_2—)_n$, known as crown ethers, are especially interesting saturated cyclic compounds of oxygen. Crown ethers have very

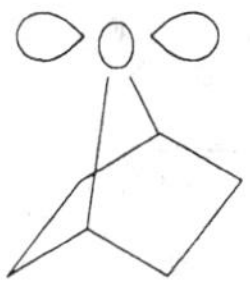

Me Me

XLVII 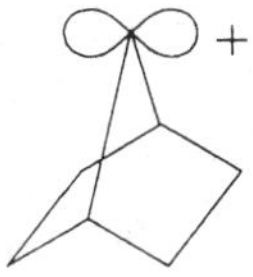 XLVIII 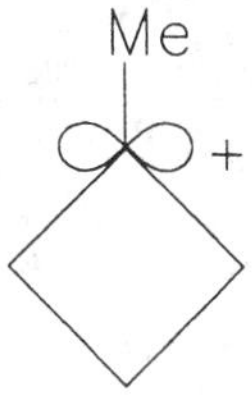 XLIX 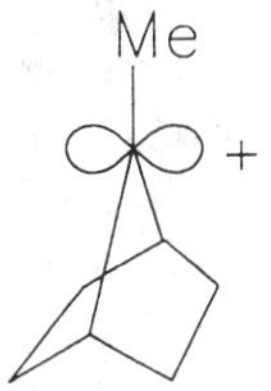 L

high selectivity in complexation with metal ions. Various – sometimes unique – problems of organic synthesis can be solved owing to this property of crown ethers.

The steric structure of molecules of crown ethers and their analogues depend to a considerable degree on the competition between the effects of interaction of lone pair orbitals of the oxygen atoms at position 1,4 and the corresponding forces of inter-electron repulsion. Thus, the simplest acyclic molecule of 1,2-dimethoxyethane, simulating the structural effects in polyethers, is present in the form of a balanced mixture of conformers *a* and *b*; the gauche conformer *a* is more stable [351]:

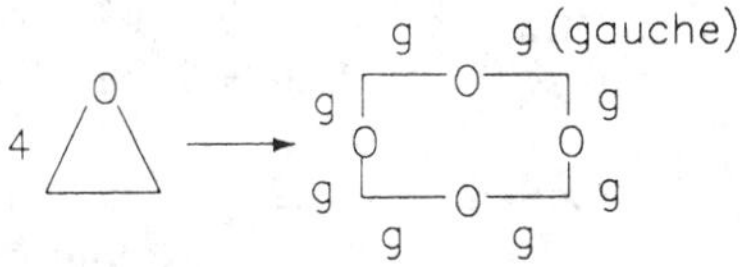

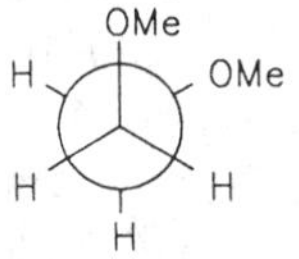

g g (gauche) OMe OMe H OMe H H 4 O g g O O g g H H H H g g H OMe a 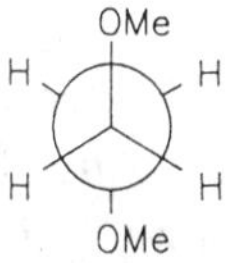 b

This conformation is most advantageous in polyoxyethylenes as well. Its advantage is manifest in synthesis: for example, the main product of ethylene oxide oligomerization in the presence of $BF_3$ is [12]crown-4 that exists completely in the gauche conformation.

Sweigart and Turner [358] were among the first to study the electron structure of cyclic ethers. They studied the PE spectra of tetrahydropyran, 1,3-dioxolane, 1,3-dioxane, 1,4-dioxane and some of their S-analogues. The first band at 9.5 eV in the spectrum of tetrahydropyran, as well as in the spectra of acyclic ethers, was assigned to ionization of oxygen's lone pair. The presence of the second oxygen atom in the six-membered ring causes splitting of the band of lone pair ionization, which becomes quite overt in the low-energy region of the spectrum (Table 8.5).

**Table 8.5.** Mixing of UEP orbitals in cyclic ethers and sulphides

| Compound | $IP_1$ (eV) | $IP_2$ (eV) | $IP_{1,2}$ (eV) |
|---|---|---|---|
| Tetrahydropyran | 9.50 | — | — |
| Pentamethylene sulphide | 8.45 | — | — |
| 1,4-Dioxane | 9.43 | 10.65 | 1.22 |
| 1,4-Dithiane | 8.58 | 9.03 | 0.45 |
| 1,3-Dioxane | 10.1 | 10.35 | 0.25 |
| 1,3-Dithiane | 8.54 | 8.95 | 0.41 |

(From Sweigart and Turner [358])

In the spectrum of 1,4-dioxane, for example, this splitting is 1.22 eV. Owing to the geometry of its molecule (which preserves the chair conformation characteristic of cyclohexanes), direct overlap of lone pair orbitals of two oxygen atoms through space is hardly probable. Mixing of the lone pair orbitals through the C—C bonds (through-bonds effect) is more advantageous for considerations of geometry:

Through-bonds(TB) effect Through-space (TS) effect

It is noteworthy that the effect of analogous interaction of lone-pair orbitals of two sulphur atoms is much lower in 1,4-dithiane; splitting of the first band is only 0.45 eV.

1,3-Dioxane also has the chair conformation. Direct overlap of $2p$ orbitals of two oxygens is also possible in its molecule. The TS effect is estimated as a splitting of 0.25 eV. The conditions for realization of this effect are more advantageous in 1,3-dithiane: the splitting value is 0.41 eV.

Orbital overlap effects, which are potentially responsible for the conformation (anomeric effects) were studied in bicyclic structures [102, 106]. The cis conformer LI*a*, for example, has one stabilizing anomeric effect and is 0.71 kJ/mol more stable than the trans conformer *b* which has no stabilizing effect [102] (for more detail see Section 3.3):

LI: a b

## 8.2 DISULPHIDES

An organic molecule containing a disulphide bridge has high conformational lability. The properties of the disulphide bond are unique in chemical respects as well: this bond is easily formed and broken:

$$2R{-}SH \underset{\text{reduction}}{\overset{\text{oxidation}}{\rightleftharpoons}} R{-}S{-}S{-}R$$

These physical and chemical properties are very important for biological systems, and possibly for the mechanism of memory.

The studies of the electron structure of disulphides usually include recording of ionization potentials corresponding to splitting of lone-pair levels of the adjacent sulphur atoms, and evaluation of their steric structure [37, 53, 73]. The planar (relative to the CSSC fragment) conformation of linear disulphides is not realized due to the repulsion of electrons in the $3p$ orbitals of the sulphur atoms. Complete withdrawal of sulphur lone pairs from the plane is evidenced, for example, by the PE spectrum of dimethyl disulphide (Fig. 8.6): splitting of the lone-pair bands is only 0.24 eV. This agrees well with the value of dihedral angle (84.7°) between the planes of the lone pairs localized over the sulphur atoms [73].

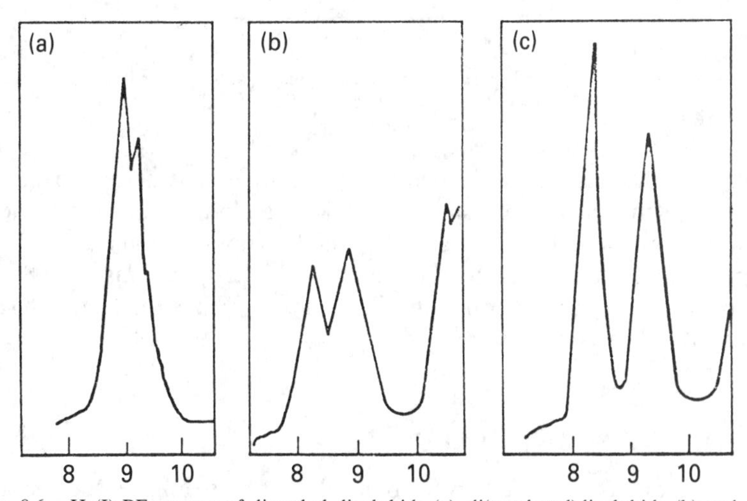


Fig. 8.6. He(I) PE spectra of dimethyl disulphide (a), di(tert-butyl)disulphide (b) and 1,2-dithiane (c). (From Wagner and Bock [73])

Small values of lone pair splitting are also characteristic of the spectra of other dialkyl disulphides, in which the dihedral angle is about 85°. Splitting of lone-pair bands in the spectrum of 1,2-dithiane, having more rigid molecular structure, is 0.95 eV which corresponds to the angle of 60° between the lone-pair axes. As in other bridged molecules, the value of the lone-pair band splitting in the PE spectra of disulphides is related linearly to the value of the cosine of the dihedral angle:

$$\Delta IP = a + b \cos \theta.$$

The structures of some R—S—S—R disulphides, have been estimated by PES[73, 360]:

| R | H | $CH_3$ | $C_2H_5$ | $C(CH_3)_3$ | $-(CH_2)_4-$ | Cl | Br | F |
|---|---|---|---|---|---|---|---|---|
| $IP_1$ (eV) | 10.01 | 8.97 | 8.70 | 8.17 | 8.36 | 10.1 | 9.6 | 10.84 |
| $IP_2$ (eV) | 10.28 | 9.27 | 8.92 | 8.82 | 9.31 | 10.3 | 9.85 | 11.25 |
| $\Delta IP_{1,2}$ (eV) | 0.27 | 0.24 | 0.22 | 0.65 | 0.95 | 0.2 | 0.25 | 0.41 |
| $\theta$ (degrees) | 90.6 | 84.5 | — | 110 | 60 | 84.8 | 83.5 | 87.9 |

Estimation of the structures of the molecules according to PE spectra correlates with the findings of independent physico-chemical measurements and quantum chemical calculations [73] (Fig. 8.7).

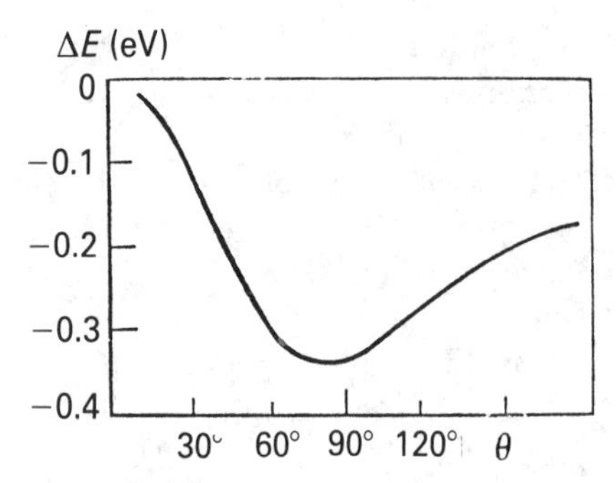


Fig. 8.7. Dependence of total energy (EHM calculations) of dimethyldisulphide on dihedral angle $\theta$. (From Wagner and Bock [73])

It is noteworthy that the lone pairs of the S–S fragment can be involved in complexation. Thus, the interaction of aromatic disulphides with strong $n$ acceptors, e.g., halides of aluminium, is associated with the rupture of the S—S bond and formation of stable complexes of thiyl radicals [361]. By contrast, the S—S bond does not break down during formation of complexes of linear and cyclic disulphides with $\pi$ acceptors [306, 362].

## 8.3 VINYL ETHERS AND SULPHIDES

The oxygen atom of vinyl ethers is bonded with the fragment of unsaturated hydrocarbons. This neighbourhood creates conditions for effective polarization of the double C—C bond. As a result, vinyl ethers are highly reactive.

The electron effects in vinyl ethers are to a considerable extent determined by their steric structure [363, 364]. *Ab initio* quantum chemical calculations indicate that methyl vinyl ether exists in two stable conformations, viz., cis and trans conformations, with the methyl group rotating in each of them:

cis–skewed
(cis–s): $\varphi=0°$; $\Theta=60°$

cis–eclipsed
$\varphi=0°$; $\Theta=0°$

trans–skewed
$\varphi=180°$; $\Theta=60°$

trans–eclipsed
$\varphi=180°$; $\Theta=0°$

The most unexpected result of the calculations was that the cis-skewed conformation, which was usually excluded due to steric hindrance, was predicted as the most stable one. To explain this feature of the steric structure of methyl vinyl ether, the aromaticity concept was used: when in the cis-skewed conformation, methyl vinyl ether becomes a pseudo-6$\pi$-electron cyclic system obeying the $(4n + 2)$ Hückel rule [363].

In the series of alkyl vinyl ethers $CH_2{=}CH{-}O{-}R$ (LII), where R = $CH_3$, $C_2H_5$ or $n$-$C_4H_9$, cis and trans conformers are also stable, while sterically hindered ethers R = $(CH_3)_2CH$ and $(CH_3)_3C$, can exist in the planar trans and less stable gauche conformation ($\theta = 90°$).

The ionization potentials of alkyl vinyl ethers $CH_2{=}CH{-}O{-}R$ are given below [364]:

| R | $CH_3$ | $C_2H_5$ | $n$-$C_4H_9$ | $(CH_3)_2CH$ | $(CH_3)_3C$ |
|---|---|---|---|---|---|
| $IP_1$ (eV) | 9.14 | 9.15 | 9.10 | 8.34 | 8.77 |
| $IP_2$ (eV) | 12.13 | 11.68 | 11.32 | 11.34 | 11.02 |
| $\Delta IP_{1,2}$ (eV) | 2.99 | 2.53 | 2.22 | 2.50 | 2.25 |

It can be seen that the first ionization potentials in the series of ethers, where R = $CH_3$, $C_2H_5$, and $n$-$C_4H_9$, remain constant while the second ionization potentials decrease in agreement with the inductive effect of the alkyl group R on the oxygen lone-pair energy. The comparison of isopropyl and $n$-butyl ethers shows that the second ionization potentials are practically the same although the inductive effect of the isopropyl group is higher than that of the $n$-butyl group. Only the ionization of planar trans and non-planar gauche conformers can probably be seen in the spectra of isopropyl and tert-butyl ethers.

In accordance with the scheme of interaction of $\pi$(C=C)— and $p_\pi$(O) orbitals (Fig. 8.8), the first ionization potentials in the spectra of alkyl vinyl ethers are assigned to the orbital, which is localized mainly over the vinyl group, while the second potentials are assigned to the orbital localized mainly over the oxygen lone pair. The difference of these potentials is very sensitive to conformational changes; this is demonstrated, for example, in Fig. 8.9 [277].

The conclusions derived in the study of PE spectra of alkyl vinyl ethers correlate with their reactivity [365]. As a rule, alkenes do not react with TCNE except a relatively slow addition to some exocyclic and strained double bonds. On the contrary, enol ethers can enter the reaction of cyclo-addition with TCNE; the enol structure produces an appreciable effect on the rate of formation of the adduct. The relative rate constants of the reactions of the ethers $CH_2{=}CH{-}O{-}R$ with TCNE are the following:

| R | Ph | $CH_2CH_2Cl$ | $C_2H_5$ | $n$-$C_4H_9$ | cyclo-$C_6H_{11}$ | $(CH_3)_3C$ |
|---|---|---|---|---|---|---|
| $k_{rel}$ | $4.7 \times 10^{-3}$ | 1 | 20.1 | 21.7 | $1.2 \times 10^2$ | $2.8 \times 10^2$ |

Higher reaction rates of cyclohexyl and tert-butyl ethers, compared with ethyl and $n$-butyl ethers, are unexplainable within the framework of correlation analysis by comparing the reaction rates with $\sigma^*$ constants. It is believed that the observed effect is due to the higher energies of occupied frontier MOs of cyclohexyl and tert-butyl ethers. This follows from the above values of $IP_1$ and from the position of CT bands

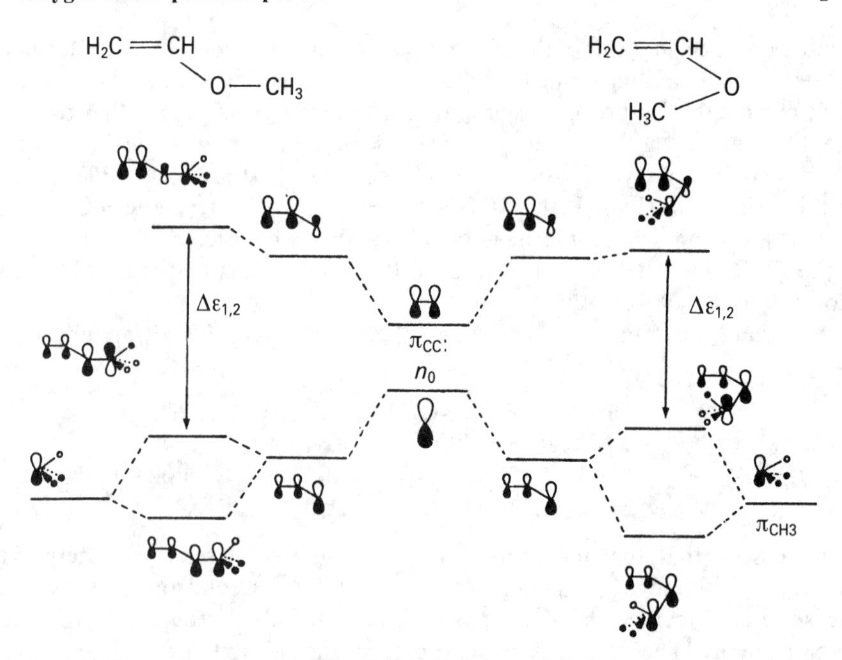


Fig. 8.8. Interaction of occupied fragment orbitals in methyl vinyl ether molecule; in the centre: $\pi$(C═C)— and $n$(O) orbital levels in the absence ($\theta = 90°$) and presence ($\Theta = 0°$ and $\theta = 180°$) of interaction. Effect of interaction with $\pi(CH_3)$ orbitals in the trans form (left); accounting for additional through-space interaction of $\pi$(CC) and $\pi(CH_3)$ orbitals in the cis form (right). (From Klessinger and Rademacher [277])

in the absorption spectra of the corresponding TCNE complexes [365].

A similar explanation probably holds also for other examples of different reactivity of vinyl ethers in cyclo-addition reactions, in Hg (II)- catalyzed peretherification, and homogeneous polymerization [365].

As in vinyl ethers, the conditions for $p_\pi$(S)- and $\pi$(C—C) orbital overlap determine the electron structure of vinyl sulphides [56]. According to electron diffraction data, methyl vinyl sulphide and its homologues in the gas phase are also found in the form of the cis- and gauche conformers (with the torsion angle of 106°). The ionization potentials of some methylthioethylenes are as follows [56]:

| | *IP* (eV) | | | | | |
|---|---|---|---|---|---|---|
| 1-Methylthioethylene | 8.45 | 11.0 | 11.5 | 12.5 | 13.8 | 14.9 |
| 1,2-bis(methylthio)ethylene | 7.85 | 9.2 | 10.8 | 11.2 | 11.8 | 12.6 |
| Tetra(methylthio)ethylene | 7.75 | 8.58 | 8.85 | 9.2 | 10.3 | |

In the discussion of the electron structure of methylthioethylenes, special attention is given to the fact that delocalization of electron density is detected not only by consecutive decrease of the first ionization potential with increasing number of methyl groups at the ethylene fragment, but also by appropriate changes in higher ionization potentials.

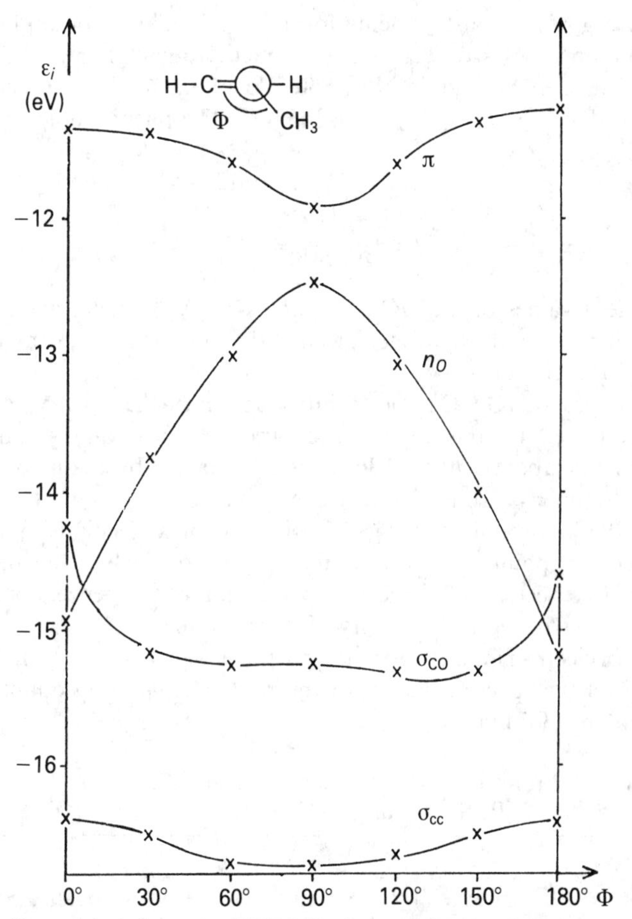


Fig. 8.9. Dependence of energies (CNDO/S) of occupied MOs of methyl vinyl ether on dihedral angle $\phi$. (From Klessinger and Rademacher [277])

## 8.4 AROMATIC HYDROXY COMPOUNDS AND ETHERS

### 8.4.1 Phenol

The benzene ring bound with the hydroxy group is a fragment found in many biologically active substances, many organic dyes and natural compounds, in phototropic compounds, inhibitors of radical reactions, etc. The electron structure of the Ph—O fragment is important for understanding of the properties of phenol, its derivatives and analogues. The MO analysis of various phenols has shown that improvement of their pharmacological properties can be associated with increasing energies of the frontier occupied electron level and occupancy of orbitals on the oxygen atom [351].

The difference between electron interactions of the benzene ring with the oxygen atom and the electron effects of the oxygen atom bonded with the aliphatic group is well known in organic chemistry. Being formally an analogue of aliphatic alcohols, phenol is a million times stronger acid than methyl alcohol ($pK_a = 9.98$ and 15.5 respectively). The C—O bond in the molecule of phenol is appreciably shorter (0.138 nm) than that of methyl alcohol (0.143 nm). The phenol molecule is planar:

$$\angle C_{(2)}C_{(1)}C_{(3)} = 121.4^\circ \quad \angle OC_{(1)}C_{(2)} = 121.3^\circ \quad \angle OC_{(1)}C_{(3)} = 117.3^\circ$$

while the different values of the $OC_{(1)}C_{(2)}$ and $OC_{(1)}C_{(3)}$ angles are believed to be due to repulsion between the hydrogen atom of the hydroxyl group and the hydrogen at $C_{(2)}$ atom [351] (cf. [244]).

Splitting of benzene HOMOs due to introduction of the hydroxy group into its molecule is estimated by the difference between the first and second ionization potentials of phenol (about 0.70 eV) [366] and is explained by mixing of the $p_\pi$ orbital of oxygen with the $\pi_2$ orbital of benzene. Like lone pairs of the halogens in halobenzenes, the $p_\pi$ orbitals of oxygen in the phenol molecule is assigned (within the framework of the point group $C_{2v}$) to the $b_1$ type, to which the $\pi_1$ and $\pi_2$ benzene orbitals belong (see Section 7.3). Effective interaction of the benzene $\pi_2$ and oxygen $p_\pi$ orbitals is also favoured by similarity of their energies.

This scheme of electron interaction in the phenol molecule was repeatedly confirmed by comparison of the spectral data with the results of *ab initio* quantum chemical calculations [38, 60, 367] (Table 8.6).

**Table 8.6.** Ionization potentials and occupied orbitals of phenol (PES data and 4-31G calculations)

| $IP_i$ (eV) | MO | | |
|---|---|---|---|
| | $-\varepsilon_i$ (eV) | No. and symmetry ($C_s$ group) | Localization* |
| 8.70 | 8.55 | $4a''$ | $\pi_2$ |
| 9.39 | 9.31 | $3a''$ | $\pi_3$ |
| 11.59 | 13.06 | $2a''$ | $\pi_1$, no |
| 12.02 | 13.32 | $21a'$ | $B(19)$ |
| 12.61 | 13.93 | $20a'$ | $B(18)$ |
| 13.44 | 15.16 | $19a'$ | $\bar{n}_O$ |
| 14.21 | 15.98 | $1a''$ | $n_O$ |
| 14.21 | 16.15 | $18a'$ | $B(16)$ |
| (14.7) | 16.68 | $17a'$ | $B(15)$ |
| 15.51 | 17.67 | $16a'$ | $B(14)$ |

*The letter *B* is used to designate MOs localized in the benzene ring; the numbers of these MOs in the benzene's orbital set are parenthesized.
(From Kumura *et al.* [38])

The analysis of $\pi$ orbital populations shows transfer of $\pi$ electron density from the oxygen atom to the benzene nucleus [24, 335]:

HO
0.975
1.068
0.976
1.039

This transfer stabilizes the anion centre formed on the oxygen atom on dissociation of phenols and explains their high acidity. The pKa of phenols para—X—$C_6H_4$—OH are as follows [8]:

| X | H | $CH_3$ | $NO_2$ | OH | I | Br | Cl | F |
|---|---|---|---|---|---|---|---|---|
| pKa | 9.98 | 10.14 | 7.15 | 9.96 | 9.31 | 9.36 | 9.38 | 9.95 |

Acidity of para-halophenols is, on the whole, higher than that of phenol because of the pronounced negative inductive effect of the halogens, but it decreases with increasing interaction of the $p_\pi$ orbital of the halogen with the benzene $\pi$ orbitals. As a result, acidities of phenol and para-fluorophenol are practically the same.

Substitution of the hydrogen atom for the hydroxy group also effectively splits vacant degenerate $\pi$ electron levels of the benzene ring. The character of splitting

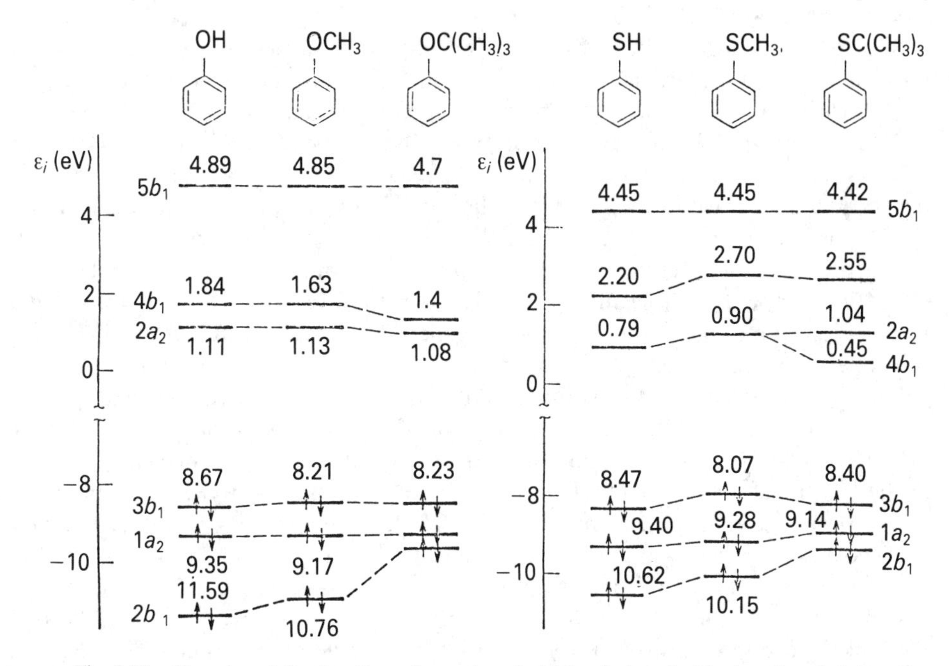


Fig. 8.10. Experimental estimation of energies of $\pi$MOs of phenol, thiophenol and their ethers (occupied MOs according to PES and vacant according to ETS findings). (Data by Dewar *et al.* [60] and Modelli *et al.* [368])

shown in Fig. 8.10 is confirmed by *ab initio* calculations in the STO-3*G* basis set. The first anion state of the phenol molecule recorded by ET spectroscopy is characterized by long life (the first resonance line in the spectrum has a well resolved vibrational structure): $EA_1 = -1.01$ eV. Higher values of electron affinity were also estimated: $EA_2 = -1.73$ eV; $EA_3 = -4.92$ eV [368].

### 8.4.2 Anisole and its homologues

The mixing effects of the $p_\pi$ orbitals of the VIA group elements with benzene $p_\pi$ orbitals have been studied in detail in the series of aryl ethers, their homologues and analogues. Unlike in halobenzenes, formation of aryl ether MOs depends substantially on the stereoelectronic factor.

The molecules of anisole and the related compounds can have a planar conformation in which the alkyl group lies in the benzene ring plane, while the $p_\pi$ orbital of a heteroatom is parallel to the benzene $\pi$ orbitals. This conformation is characterized by a dihedral angle of 0° and the most favourable conditions for $p_\pi$–$\pi$ interaction.

The molecules of anisole and its analogues can also exist in a perpendicular conformation: the alkyl group occupies the plane perpendicular to the benzene ring plane (the dihedral angle $\theta$ is then 90°). This conformation is less sterically hindered but the $p_\pi$–$\pi$ interaction is ruled out.

Planar and perpendicular conformers should differ by their occupied orbital energies, and many attempts were therefore undertaken to study them by PE spectroscopy [60, 277, 367, 369–71].

Replacement of the hydrogen in the OH group of phenol by the methyl group provides persistence of the planar conformation of the molecule and splits the energies of the two higher $\pi$ orbitals to 0.83 eV [366]. On the contrary, replacement of the hydrogen atom by the tert-butyl group decreases the energy difference $\Delta IP_{1,2}$ to 0.58 eV, probably due to deviation of the molecule from its planar conformation: the oxygen lone pair is withdrawn from the plane of effective interaction with the benzene $\pi$ orbitals. It is noteworthy that replacement of the hydrogen in phenol by trifluoromethyl and pentafluoroethyl groups rules out splitting of the benzene HOMOs: the spectra of the trifluoroanisole and pentafluorophenetol have only single bands corresponding to ionization of degenerate $\pi$ orbitals (at 10.0 and 9.97 eV respectively) in the region of low energies.

Substituents attached at position 2 and 6 of the anisole molecule increase steric hindrance to the interaction of the 2$p$ orbital of the methoxy group and benzene $\pi$ orbitals [60, 367]. Splitting of the higher $\pi$ orbitals of the benzene nucleus decreases in the spectra of all 2,6-dimethylphenyl ethers in the low energy region (Table 8.7).

The parameters of the electron structure of phenol and its ethers, as determined by electron spectroscopy and quantum chemical calculations, explain some contradictory (at first sight) data on their reactivity.

Kinetic measurements have shown the following order of the activation effect of the alkyl groups in oxidation of alkylaryl ethers by manganic acetate: $CH_3 < CH_3CH_2 < (CH_3)_2CH > (CH_3)_3C$. The same series has been found for the activation effect of the alkyl group in bromination of phenol ethers. These data can be explained by suggesting that both oxidation and bromination reactions depend mostly on the energy of the ether frontier occupied MO, as estimated by its first ionization potential [60]. According to Table 8.8, energy accessibility of HOMOs of

**Table 8.7.** Ionization potentials of some phenols and anisoles

| Compound | $IP_i$ (eV) | | | |
|---|---|---|---|---|
| Phenol | 8.73 | 9.40 | 11.59 | 11.99 |
| o-Cresol | 8.48 | 9.08 | 11.42 | 11.67 |
| 2,6-Dimethylphenol | 8.26 | 8.78 | 11.13 | 11.51 |
| Anisole | 8.39 | 9.22 | 11.06 | 11.52 |
| o-Methylanisole | 8.24 | 8.93 | 10.92 | 11.37 |
| 2,6-Dimethylanisole | 8.51 | 8.71 | 9.87 | 11.31 |

(From Kobayashi and Nagakura [367])

the compounds $CH_3—C_6H_4—O—R$ (LIII) varies in the same series as in the reactions with electrophiles: $CH_3 < CH_3CH_2 < (CH_3)_2CH > (CH_3)_3C$.

**Table 8.8.** Relative oxidation rates $k_{rel}$, oxidation potentials $E^{ox}_{1/2}$, ionization potentials $IP_1$ and UV spectral data of ethers LIII

| *R* | $k_{rel}$ | $E^{ox}_{1/2}$ (eV) | $IP_1$ (eV) | UV spectrum | |
|---|---|---|---|---|---|
| | | | | $\lambda_{max}$ (nm) | $\varepsilon$ |
| $CH_3$ | 1.00 | 1.18 | 8.16 | 280 | 2590 |
| $C_2H_5$ | 1.30 | 1.12 | 8.13 | 280 | 1750 |
| $(CH_3)_2CH$ | 1.89 | 1.09 | 8.09 | 281.5 | 1860 |
| $(CH_3)_3C$ | 0.64 | 1.17 | 8.23 | 277 | 810 |

(From Dewar *et al.* [60])

Tert-butyl phenyl ether is not planar and this hinders the interaction of the oxygen lone pair and benzene $\pi$ orbitals. This ether has the maximum value of the first ionization potential among the other ethers. As the energy of the ether HOMO decreases, its availability for electrophilic agents decreases as well, and the reactivity lowers.

The order of reactivity of para-alkoxybenzyl chlorides para-RO-$C_6H_4$-$CH_2Cl$ in the solvolysis in 90% alcohol is probably explained in a similar way [371]: $OCH_3 < OC_2H_5 < OCH(CH_3)_2 > OC(CH_3)_3$. The *p*-isopropoxyphenyl group with a higher occupied electron level decreases more effectively the energy of the transition carbocation state by mixing of the highest occupied orbital of the aryl fragment and vacant $p_\pi$ orbital of the C atom, thus accelerating the reaction rate. Long-wave absorption in the electron spectra of aryl ethers is largely due to electron transitions from the frontier occupied MO: a bathochromic shift of the long-wave band is observed in the absorption spectrum of phenyl isopropylate and a hypsochromic shift, in the spectrum of tert-butyl (phenyl) ether (see Table 8.8).

### 8.4.3 Organoelement analogues of anisole

First reports did not contain information on the conformational heterogeneity of thioanisole, but later investigators [60, 372] succeeded in recording bands assigned to ionization of various conformers in the PE spectra (Table 8.9).

**Table 8.9.** Ionization potentials of thioethers $C_6H_5SR$ (PES data)

| | $IP_i$ (eV) | | | | |
|---|---|---|---|---|---|
| | Planar conformer | | | Non-planar conformer | |
| | $\pi^-(b_1)$ | $\pi(a_2)$ | $\pi^+(b_1)$ | $n(S)$ | $\pi(b_1) + \pi(a_2)$ |
| H | 8.47 | 9.40 | 10.62 | — | — |
| $CH_3$ | 8.07 | 9.30 | 10.15 | 8.60 | * |
| $C_2H_5$ | 8.00 | * | 10.12 | 8.53 | 9.29 |
| $CH(CH_3)_2$ | * | * | * | 8.46 | 9.24 |
| $C(CH_3)_3$ | — | — | — | 8.40 | 9.14 |

*The attempt to determine $IP_i$ was a failure.
(From Dewar *et al.* [60])

The study of PE spectra of selenoanisole and the related compounds gave definite data on different conformers of at least some of them [373]. It was thus reported that the ratio of the planar to perpendicular conformers of selenoanisole with $IP_1$ of 8.0 and 8.3 eV was 1:1 (this ratio for thioanisole is believed to be 3:2).

The PE spectra of telluroanisole have no bands assigned to ionization of separate conformers [374]. The most stable conformer has the dihedral angle of 60° and the first ionization potential of 7.83 eV. The ionization potential of planar telluroanisole is estimated as 7.60 eV.

The analysis of the PE spectra of anisole analogues made it possible to estimate objectively the degree of mixing of the lone pair orbitals of various group VIA elements with benzene $\pi$ orbitals. The corresponding data, the results of calculations within the framework of the HMO method included, are given in Table 8.10.

As in the halogen series, the mixing effect of the $p_\pi$ orbitals and benzene $\pi$ orbitals is the highest with oxygen, the element of the second period. According to calculations, the methoxy group is the strongest in ortho and para orientation.

The vacant electron levels of the molecules of alkyl(aryl) chalcogenides, were also studied [368, 375, 376]. The ET spectra of compounds of the general formula Ph-ER, where E = O or S, R = H, $CH_3$, $C(CH_3)_3$, were compared [376]. It turned out that the benzene $\pi^*$ MO $4b_1$ is much destabilized in sterically non-hindered ethers (due to the $p$–$\pi^*$ interaction), while the first electron affinity value, estimating the energy of frontier vacant MO $2a_2$, is close to the corresponding value of the benzene molecule (see Fig. 8.10). The following electron affinities were found for anisole: $EA_1 = -1.13$ eV, $EA_2 = -1.63$ eV, $EA_3 = -4.85$ eV.

Vacant electron levels in thioether molecules are of great interest [376, 377]: for thioanisole, $EA_1 = -0.90$ eV, $EA_2 = -2.70$ eV, $EA_3 = -4.45$ eV. The appearance of a new vacant level with electron affinity from $-2.20$ to $-2.70$ eV (depending on the substituent at the sulphur atom) indicates additional interaction between the $\pi^*$ orbitals of the benzene ring and vacant orbitals of the substituent – the interaction that exceeds the effect of $p_\pi$–$\pi^*$ interaction. These states of anion-radicals of thiophenol and thioanisole appear to be formed with involvement of the vacant $3d$ orbitals of sulphur. If planar conformations of neutral molecules were disturbed during formation of anions, it should be necessary to account for the $\sigma$(C—S)–$\pi^*$ interaction as well.

**Table 8.10.** Ionization potentials (PES), position of long-wave absorption (EAS), and parameters of HOMO of PhEMe organoelement compounds

| Element E | $IP$(E) (eV) | $IP_i$(PhEMe) (eV) | | | | | | $H_{E\pi}$ (eV) | HOMO of PhEMe[†] | | | | | UV spectrum | |
|---|---|---|---|---|---|---|---|---|---|---|---|---|---|---|---|
| | | | | | | | | | $\varepsilon$ (eV) | $c_{ortho}$ | $c_{meta}$ | $c_{para}$ | $c_{EMe}$ | $\lambda$ (nm) | log $\varepsilon$ |
| O | 10.04 | 8.21 | 9.17 | 10.76 | 12.14 | 13.67 | 15.37 | −1.37 | −8.33 | 0.38 | −0.15 | −0.47 | −0.54 | 220 | 3.89 |
| | | | | | | | | | | | | | | 265 | 3.13 |
| | | | | | | | | | | | | | | 271 | 3.30 |
| | | | | | | | | | | | | | | 278 | 3.27 |
| S | 8.67 | 8.07 | (8.87) | 9.28 | 10.15 | | | −0.84 | −8.16 | 0.30 | −0.10 | −0.36 | −0.77 | 254.5 | 3.99 |
| Se | 8.40 | 7.99 | 8.29 | 9.24 | 9.81 | 10.77 | | −0.72 | −8.07 | 0.27 | −0.08 | −0.31 | −0.84 | 250 | 3.79 |
| | | | | | | | | | | | | | | 271 | 3.51 |
| Te | 7.89 | (7.60–7.83) | | 9.20 | (9.45) | 10.27 | | −0.69 | −7.70 | 0.24 | −0.05 | −0.26 | −0.89 | 223 | 3.99 |
| | | | | | | | | | | | | | | 269 | 3.69 |
| | | | | | | | | | | | | | | 331 | 2.84 |

*Note*: $IP(E)$ is ionization potential of electrons of $n(E)$ orbital of $Me_2E$;
*HOMO parameters of PhEMe (energy and coefficients $c_\mu$) as calculated by HMO method with PES parametrization.
(From Traven and Stepanov [62])

In order to confirm this suggestion, electron affinities of tert-butyl derivatives, in which $\sigma^*$–$\pi^*$ interaction is possible, were measured. In fact, their $4b_1$ orbital is stabilized. The difference in the energies of the anion states of $(CH_3)_3C$ derivatives on the one hand and SH- and SMe derivatives on the other hand shows that the anion states of the latter are planar. In this case, their parameters are most likely determined by the $3d$–$\pi^*$ interaction. Splitting of the degenerate benzene frontier $\pi$ orbitals, shown in Fig. 8.10 (associated with the introduction of the methoxy group into the benzene molecule), is confirmed also by the ESR spectra of the corresponding ion radicals [87]:

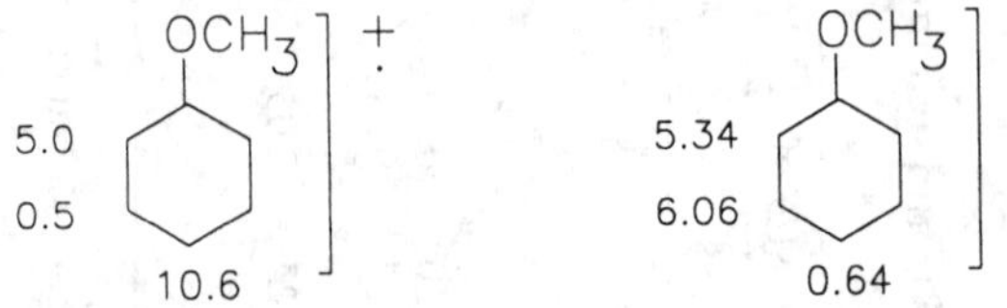

$OCH_3$ ]$^{+\cdot}$ 5.0 0.5 10.6

$OCH_3$ ]$^{-\cdot}$ 5.34 6.06 0.64

In agreement with the HFI constants of the ion radicals, the frontier occupied orbital of anisole is a transformed benzene $\pi$ orbital of $b_1$ symmetry, while the frontier unoccupied MO of anisole is only a slightly transformed benzene $\pi$ orbital of $a_2$ symmetry.

## 8.5 ALDEHYDES AND KETONES

Carbonyl compounds are very important organic substances. Formaldehyde and acetaldehyde are among the most common raw materials used in industrial organic synthesis. Many aldehydes and ketones (primarily, unsaturated ones) occur in nature; aromatic quinones are fragments of valuable organic dyes, etc.

In terms of the model of hybridized orbitals, the carbonyl group has a simple electron structure: a pair of electrons located on the $sp^2$–$sp^2$ carbon–oxygen $\sigma$ bond, a pair of electrons located on the $2p_x$–$2p_x$ carbon–oxygen $\pi$ bond and two equivalent lone pairs located in the oxygen $sp^2$ orbitals.

Owing to the higher electronegativity of oxygen, the carbonyl group is strongly polarized so that a significant negative charge is found on the oxygen atom and the positive charge on the carbon atom. It is generally believed that the nature of the bond in the carbonyl group of ketones is essentially the same as in aldehydes [350].

This interpretation of the electron structure of carbonyl compounds seems to be oversimplified since it disregards completely the experimental data obtained by new methods of electron spectroscopy (ETS in particular). Besides, it does not account for the ample information available on the different reactivity of carbonyl compounds of various types. Aldehydes, for example, are on the whole characterized by much higher reactivity of their carbonyl group, and are more susceptible to formation of addition products. (Owing to significant chemical inertness, some ketones are used as solvents in many reactions.) In contrast to aldehydes, the corresponding equilibria of ketones are usually strongly shifted toward the starting reagents:

$$R(R')C{=}O + H_2O \rightleftharpoons R(R')C(OH)_2$$

$$K_d = \frac{[R(R')C{=}O]}{[R(R')C(OH)_2]}$$

Thus, in the transition from formaldehyde to acetone, the concentration of the addition product decreases a million times [378]:

| | $CH_2O$ | $CH_3CHO$ | $ClCH_2CHO$ | $CH_3COCH_3$ | $(ClCH_2)_2CO$ |
|---|---|---|---|---|---|
| $K_d$ | $5.0 \times 10^{-4}$ | 0.7 | $2.7 \times 10^{-2}$ | $5.0 \times 10^{2}$ | 0.1 |

Steric effects seem to be not very important since, e.g. dichloroacetone forms a far stronger adduct than acetaldehyde.

Spectral quantum chemical methods have revealed substantial differences in the electron structures of aldehydes and ketones.

### 8.5.1 Formaldehyde

The molecule of formaldehyde is planar (experimental and calculated data on its geometric parameters are given in [154]):

The symmetry analysis of atomic orbitals shows why the formaldehyde molecule cannot have two lone pairs of electrons on oxygen. Its $2p_x$ orbital and the carbon $2p_x$ orbital (both of $b_1$ symmetry) form the $\pi$ orbital of the carbonyl group. The $2p_z$ orbital of the oxygen atom (which is only partly involved in the (O—H) bonds and is therefore regarded as a conventional lone pair of the oxygen atom in the water molecule) strongly interacts in the formaldehyde molecule with the carbon $2p_z$ orbital in the $\sigma$(CO) bond:

| Basis orbitals | C | | | | O | | | | H | |
|---|---|---|---|---|---|---|---|---|---|---|
| of formaldehyde | $2s$ | $2p_z$ | $2p_y$ | $2p_x$ | $2s$ | $2p_z$ | $2p_y$ | $2p_x$ | $1_s^+$ | $1_s^-$ |
| Symmetry type | $a_1$ | $a_1$ | $b_2$ | $b_1$ | $a_1$ | $a_1$ | $b_2$ | $b_1$ | $a_1$ | $b_2$ |

The analysis of MO occupancy shows that the $2p_y$ orbital of the oxygen atom interacts only insignificantly with the (C—H) bond orbitals and it is the only orbital that can, in terms of the MO theory, be regarded as a lone pair of the oxygen atom.

Assignment of the bands of the formaldehyde PE spectrum by *ab initio* calculations in the 6-31G basis set is given in Table 8.11. Below are the frontier MOs and the $\pi$(C=O) MO which is adjacent to the HOMO:

2b1, π* (C=O) — LUMO

2b2, nO — HOMO

1b1, π(C=O)

**Table 8.11.** Ionization potentials and occupied molecular orbitals of formaldehyde (PES data and calculations in 6-31G basis set)

| $IP_i$ (eV) | Molecular orbitals | | |
|---|---|---|---|
| | $-\varepsilon_i$ (eV) | No. and symmetry point (group $C_{2v}$) | Localization |
| 10.88 | 11.96 | $2b_2$ | $n_O$ |
| 14.5 | 14.53 | $1b_1$ | $\pi_{C=O}$ |
| 16.0 | 17.51 | $5a_1$ | $\sigma_{C=O}$ |
| 16.6 | 19.06 | $1b_2$ | $n_{CH_2}$ |

(From Kimura *et al.* [38])

The proposed assignment is confirmed also by the relative intensities of bands in PE spectra of formaldehyde with He(I) and He(II) sources of radiation [176]:

| No. of band (assignment) | $1(n_O)$ | $2(\pi_{C=O})$ | 3.4 $(\sigma, \pi)$ |
|---|---|---|---|
| Relative intensity, He(II)/He(I) | 1.89 | 1.69 | 0.83 |

Not only the $\pi$(C=C) orbitals (as can be seen from the comparison of unsaturated hydrocarbon spectra) but also $\pi$(C=O) and $n_O$ orbitals double their band intensities with the He(II) source.

The comparison of the electron structures of formaldehyde and ethylene shows that the highest electron level of the formaldehyde molecule is formed by the $n_O$ orbital and has $\sigma$ symmetry, while the occupied $\pi$(C=O) orbital is stabilized (compared with the $\pi$(C=C) orbital) by almost 4 eV (the electron is excited from this orbital at 14.5 eV).

Replacement of one carbon atom by a more electronegative oxygen atom in the transition from ethylene to formaldehyde preserves the $\pi$ character of the LUMO and stabilizes it by almost 1 eV (ETS): first values of electron affinities of ethylene and formaldehyde are $-1.78$ and $-0.86$ eV respectively [46]. Vacant electron levels of formaldehyde are characterized by another feature: lowering the level of its LUMO increases significantly the stability of the anion state. This is confirmed by a well-resolved vibrational structure of the resonance signal in the ET spectrum of formaldehyde which is commonly seen in cases with protracted life-span of the anion formed in the gas phase (Fig. 2.8).

The data on the orbital structure agree with the assignment of the long-wave band in the electron absorption spectrum of formaldehyde to the $n \rightarrow \pi^*$ transition. This band is close to 310 nm. The low intensity of this band ($\varepsilon \approx 5$) is another confirmation of the HOMO assignment: transitions involving lone pairs correspond to low intensity bands in electron absorption spectra because they are symmetry forbidden.

### 8.5.2 Acetaldehyde

Substitution of the hydrogen atom in formaldehyde for the methyl group appreciably destabilizes the $n$(O) and $\pi$(C=O) levels by 0.62 and 1.26 eV respectively. The energy rise is especially notable with the $\pi$(C=O) orbital, which is probably due to its $\pi$ type interaction with the $\sigma$(C—H) orbitals of the methyl group.

The steric structure of acetaldehyde is advantageous for this $\sigma$–$\pi$ interaction: the carbonyl group and one of the C—H bonds lie in one plane and are eclipsed; two other C—H bonds of the methyl group are in the gauche position relative to the carbonyl group and can be partly overlapped by its $\pi$ orbital. The barrier to rotation of the methyl group in the acetaldehyde molecule is small, 4.86 kJ/mol (experiment) and 4.56 kJ/mol (quantum chemical calculations). The enol form, which is less accessible for experimental studies, was also calculated. This form turned out to be 54 kJ/mol less stable than the keto form [24].

**Table 8.12.** Ionization potentials and occupied molecular orbitals of acetaldehyde (PES data and 6-31G calculations)

| $IP_i$ (eV) | Molecular orbitals | | |
|---|---|---|---|
| | $-\varepsilon_i$ (eV) | No. and symmetry (group $C_s$) | Localization |
| 10.26 | 11.57 | $10a'$ | $n_O$ |
| 13.24 | 13.75 | $2a''$ | $\pi_{C=O}$ |
| 14.15 | 15.17 | $9a'$ | $\pi_{CH_3}$ |
| 15.34 | 16.63 | $1a''$ | $\pi_{CH_3}$ |
| (15.6) | 16.94 | $8a'$ | $\sigma_{C-C}$ |
| 16.47 | 18.44 | $7a'$ | $\sigma_{C=O}$. |

(From Kimura *et al.* [38])

Assignment of the PE spectral bands of acetaldehyde is given in Table 8.12. The first band in the spectrum is narrow and corresponds to an oxygen lone-pair ionization energy that increases partly due to the positive inductive effect of the methyl group and partly due to the overlap of the lone-pair orbital with the (C—H) orbital located in the carbonyl fragment plane. The energy of the electrons of the $\pi$ bond of the carbonyl group of acetaldehyde is almost 3 eV lower than its lone-pair energy.

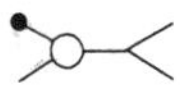

3a'' , π* (C=O)    10a' , n$_O$    2a'' , π(C=O)

LUMO    HOMO

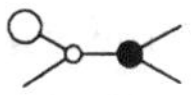

The assignment of the acetaldehyde PES bands is confirmed also by the comparison of their intensities with He(I) and He(II) sources [176]:

| Band No. (assignment) | $1n_O$ | $2\pi_{C=O}$ | $3\pi CH_3$ | $4.5(\pi,\sigma)$ | $6(\sigma_{CO})$ |
|---|---|---|---|---|---|
| Relative intensity, He(II)/He(I) | 2.0 | 1.82 | 0.91 | 0.9 | 0.71 |

### 8.5.3 Acetone and related compounds

The importance of acetone for organic synthesis derives from its ability to take part in condensation reactions attended by formation of the C—C bond. In addition, acetone has some other specific properties that explain the interest in its molecular structure. It is, for example, contained in the human body. The acetone content of blood in a healthy person does not exceed 1 mg/100 ml while in diabetic patients or in starvation (the process is attended by accelerated degradation of fats), its concentration (as well as of other ketones) increases to a level that may cause death.

Like acetaldehyde, the acetone molecule can exist in two tautomeric forms:

$CH_3$—C(=O)—$CH_3$ keto–

$CH_3$—C(OH)=$CH_2$ enol–

but the content of the enol form does not exceed 0.0001% under normal conditions. The geometric parameters of the keto form were determined by microwave spectroscopy: $r_{CC} = 0.1507$ nm; $r_{CO} = 0.1222$ nm; $r_{CH} = 0.1085$ nm. The CCC angle is 117°12′ while the angle between the symmetry axes of methyl groups is 119°54′. The methyl groups in the acetone molecule are thus slightly drawn (each to about 1.5°) toward the carbonyl group [379].

PES succeeded only in detecting the ketone form of acetone [38, 179]. The assignment of bands in the spectrum is given in Table 8.13.

**Table 8.13.** Ionization potentials and higher occupied MOs of acetone (PES data and 4-31G calculations)

| $IP_i$ (eV) | MO | | $IP_i$ (eV) | MO | |
|---|---|---|---|---|---|
| | $-\varepsilon_i$ (eV) (Number) | Localization | | $-\varepsilon_i$ (eV) (Number) | Localization |
| 9.70 | 11.20($5b_2$) | $n$(O) | (14.8) | 15.46($1a_1$) | $\pi^-_{CH_3}$ |
| 12.59 | 13.02($2b_1$) | $\pi$(C=O) | 15.60 | 17.18($1b_1$) | $\pi^+_{CH_3}$ |
| 13.41 | 14.62($4b_2$) | $\sigma^-_{CH_3}$ | (16.1) | 17.23($7a_1$) | $\sigma_{C-O}$ |
| 14.04 | 15.09($8a_1$) | $\sigma_{C-O}, \sigma^+_{CC}$ | (16.6) | 17.69($3b_2$) | $\sigma^-_{CC}, n_O$ |

(From Kimura *et al.* [38])

Attachment of the second methyl group to the carbonyl likewise increases the energies of $n$(O) and $\pi$(C=O) orbitals. Enlargement of the alkyl group attached to the carbonyl in the series of aldehydes (as in the series of ketones) destabilizes HOMOs probably due to the increasing positive inductive effect. Ionization potentials of the oxygen atom lone pair in some aldehydes and ketones are given below [354]:

| R | $CH_3$ | $C_2H_5$ | $C_3H_7$ | iso-$C_3H_7$ | $C_4H_9$ | tert-$C_4H_9$ |
|---|---|---|---|---|---|---|
| $IP_{n(O)}$(eV) | | | | | | |
| RCHO | 10.22 | 9.77 | 9.73 | 9.69 | 9.77 | 9.51 |
| RC(O)$CH_3$ | 9.71 | 9.54 | 9.40 | 9.30 | 9.36 | 9.14 |

In the comparison of the electron structures of formaldehyde, acetaldehyde and acetone, noteworthy are the results of calculations of electron density distribution in these molecules [335]. Substitution of the hydrogens at the carbonyl by the methyl groups is attended by a marked growth of polarity of the carbonyl group ($Z \times 10^3$ values are indicated):

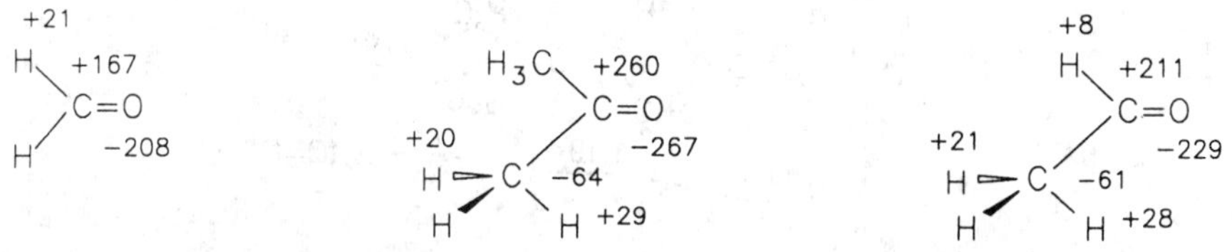

$H_2C=O$: +21 (H), +167 (C), −208 (O)

$CH_3CHO$: +260 (C), −267 (O), −64 ($CH_3$ carbon), +20 (H), +29 (H)

$CH_3COCH_3$: +8 (H), +211 (C), −229 (O), −61 ($CH_3$ carbon), +21 (H), +28 (H)

These data seem to be reasonably accurate since the dipole moments calculated on this basis reproduce adequately the trends in the experimental findings (below are the values of dipole moments $\mu$,D):

| | STO-3G | CNDO/2 | Experiment |
|---|---|---|---|
| $CH_2O$ | 1.53 | 1.98 | 2.34 |
| $CH_3CHO$ | 1.66 | 2.53 | 2.68 |
| $(CH_3)_2CO$ | 1.87 | 2.90 | 2.90 |

The higher energy of the $n$(O) orbital in the acetone (and other ketones) compared with acetaldehyde, and also the higher polarity of the acetone molecule cannot account for the above-mentioned differences of various carbonyl compounds in their reactivity to nucleophilic agents. This is especially important because the ease of nucleophile attachment to the carbonyl compounds is traditionally associated with polarization of the carbonyl group.

It appears that the reactions of aldehydes and ketones with nucleophilic agents are probably more dependent on the parameters of the LUMO rather than on the total charges on atoms. In this connection, aldehyde and ketone reactivity should be better estimated on the consideration of the energies and coefficients of their frontier vacant orbitals.

Quantum chemical calculations show that in the transition from formaldehyde to acetaldehyde and to acetone, the symmetry of their LUMOs does not change: this is the $\pi^*$(C=O) orbital in each case. Figure 8.11 shows that attachment of the methyl group to the carbonyl group destabilizes not only occupied but also vacant molecular orbitals. In the transition from formaldehyde to acetone, the energy of the LUMO increases by 0.65 eV. In the terms of the frontier orbital concept, this increase in the energy of an frontier unoccupied MO must decrease significantly its availability in reactions with nucleophiles. This parameter of the electron structure probably determines inertness of ketones compared with aldehydes in nucleophilic reactions.

While discussing the reactivity of aldehydes and ketones toward nucleophiles, consider the results of *ab initio* calculations of the model reaction between the hydride ion and the formaldehyde molecule:

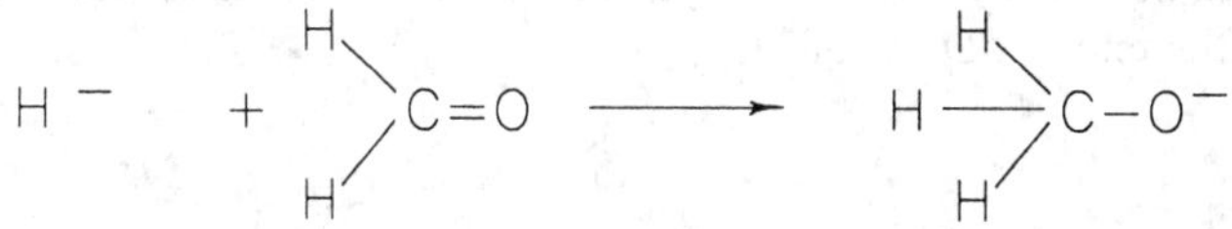

$$H^- + H_2C=O \longrightarrow H-CH_2-O^-$$

It follows from these calculations that the anion approaches the carbon atom in the bisector plane to an H···C distance of 0.3 nm, at an angle of 180° to the C=O bond.

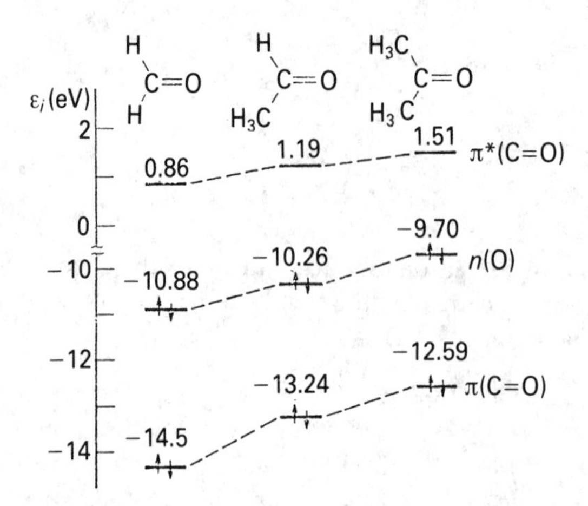


Fig. 8.11. Experimental estimation of energies of MOs of formaldehyde, acetaldehyde and acetone (occupied MOs according to PES and vacant according to ETS findings). (From Jordan and Burrow [46])

Within the range of distances from 0.3 to 0.15 nm, this angle decreases from 180° to 125° while the formaldehyde molecule remains planar. Further approach of the hydride ion to a distance of 0.112 nm causes transformation of the angles at the carbon atom to tetrahedral ones. It is believed that during the entire interaction, the mixing of the $1s(H^-)$ orbital with the vacant $\pi^*(C{=}O)$ orbital is of decisive importance. This orbital has the prevalent coefficient at the carbon atom to which the hydride ion is attached. The $\pi$ and $\pi^*$ orbitals of the carbonyl group transform into lone pairs over the oxygen atom and unoccupied orbital over the carbon atom (see the discussion in [1]), since the RR′CO fragment becomes non-planar.

The electron structure of the thio analogues of carbonyl compounds is of interest as well. First of all, replacement of the oxygen atom by the sulphur atom increases substantially the stability of the anion state. Thus, the O → S replacement in di(tert-butyl)ketone decreases the energy of the frontier vacant MO by about 1.5 eV: the first value of electron affinity of di(tert-butyl)thioketone is 0.5 eV. It is believed that this effect is explained largely by the $\pi^*$–$3d(S)$ orbital mixing [380].

The low value of the first unoccupied electron level of thiocarbonyl compounds accounts for their higher activity toward nucleophilic agents while prevalent localization on the vacant level over the sulphur atom accounts probably for their preferable attack by nucleophiles at the sulphur atom. Ketones attach the nucleophilic hydrocarbon fragment of an organometal reagent at the carbon atom, while the electrophilic fragment attaches at the oxygen atom with formation of the corresponding alcohol as the end product [381]:

$$R_2C{=}O + R^-M^+ \longrightarrow R_2\underset{}{\overset{R}{\overset{|}{C}}}{-}OM \xrightarrow{H_3O^+} R_2\overset{R}{\overset{|}{C}}{-}OH$$

By contrast, thioketones often react by the scheme of thiophilic addition. For example, in the reaction of thiobenzophenones with phenyllithium, phenylmagnesium bromide and butyllithium, the nucleophile attacks the sulphur atom rather than the carbon atom [382]:

$$Ph_2C{=}S + R^-M^+ \qquad Ph_2\overset{\overset{\displaystyle M}{|}}{C}{-}SR \quad \xrightarrow{H_3O^+} \quad Ph_2\overset{\overset{\displaystyle H}{|}}{C}{-}SR$$

Examples of nucleophile attachment to the sulphur atom are known for non-aromatic ketones as well. They interact not only with organometal reagents, but e.g. with bisulphite ion:

$$(CF_3)_2C{=}S + HSO_3^- \qquad (CF_3)\overset{\overset{\displaystyle H}{|}}{C}{-}S{-}SO_3^-$$

Acid catalysed reactions of carbonyl compounds probably depend on the energies of their frontier occupied MOs, i.e. the state of the oxygen lone pair in their molecules. In accordance with the increasing energy of lone pairs in ketones, ketones are more basic than aldehydes:

| | $CH_3CH{=}O$ | $(CH_3)_2C{=}O$ |
|---|---|---|
| pKa $(RR'C{=}\overset{+}{O}H)$ | $-8.0$ | $-7.2$ |

The data on the reactivity of cyclic thiocarbonyl compounds with methyl iodide illustrate the role played by the lone-pair orbital of the sulphur atom in thioketones [383]:

$$\text{(cyclic X, Y)}C{=}S + CH_3I \longrightarrow \text{(cyclic X, Y)}\overset{+}{C}{-}S{-}CH_3 + I^-$$

It has been established that a decrease in the ionization potential of the sulphur lone pair from 8.3 to 7.8 eV corresponds to the increase in the reaction rate by three orders.

Other data estimating the effect of structure on the energy of frontier occupied MO of ketone should also be mentioned. Thus, substitution of the methyl group by a more electropositive $Si(CH_3)_3$ or $Ge(CH_3)_3$ group increases significantly the energy of the oxygen lone pair [384]:

| | $H_2C{=}O$ | $Me(H)C{=}O$ | $Me_2C{=}O$ | $Me(Me_3Si)C{=}O$ | $Me(Me_3Ge)C{=}O$ |
|---|---|---|---|---|---|
| $IP_{n(O)}$ (eV) | 10.88 | 10.26 | 9.7 | 8.6 | 8.5 |

It can be seen from Fig. 8.12 that, compared with the first band of acetone PE spectrum, the first bands in the spectra of organoelement ketones (the bands assigned to ionization of oxygen lone pair) are much broader. CNDO/2 calculations show

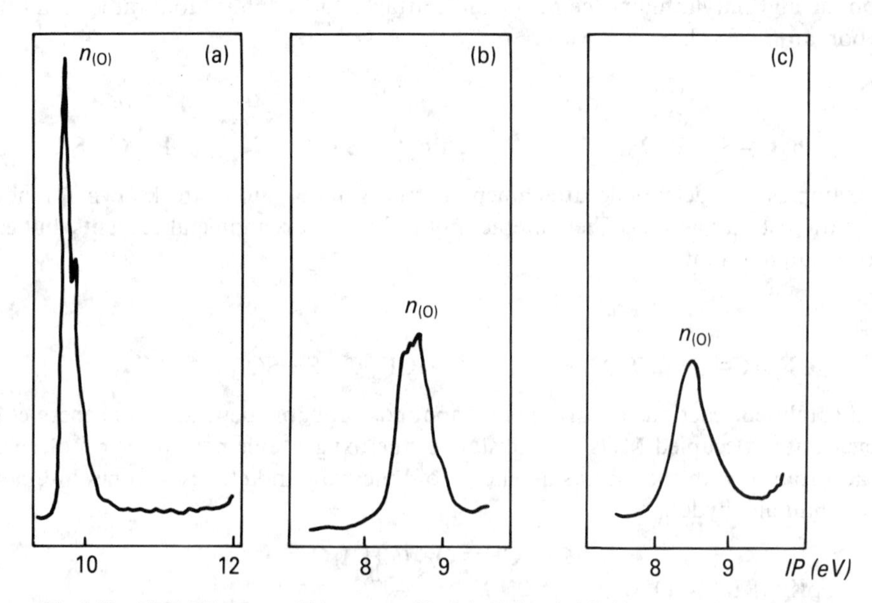


Fig. 8.12. He(I) PE spectra of (a) acetone, (b) methyl(trimethylsilyl)ketone and (c) methyl(trimethylgermyl)ketone. (From Ramsey *et al.* [348])

that the cause of the observed spectral changes is the mixing of the $n$(O) orbital with the $\sigma$(C—E) orbital. According to calculations, the lone pair of acetone is made up of the corresponding $2p$(O) orbital by 64%, while the lone pair of trimethylsilyl-(methyl)ketone is made of this orbital by only 44%. It seems quite natural because $\alpha$-organoelement ketones have such a conformation in which the $n$(O) and $\sigma$(C—E) orbitals are located in the same plane.

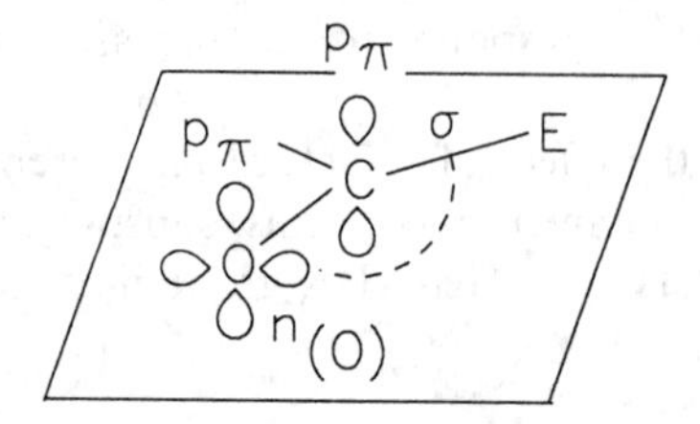


Destabilization of the lone pair in silyl- and germylketones is so high that they absorb light in the visible region of the spectrum: acetone absorbs at 280 nm ($\log \varepsilon = 4.5$) while ketones $(CH_3)_3SiC(O)CH_3$ and $[(CH_3)_3Si]_2CO$ absorb at 370 and 500 nm respectively ($\log \varepsilon$ about 3.5). The difference in the energies of electron transitions in the absorption spectra of acetone and trimethylsilyl(methyl)ketone is equal to the difference of the energies of the lone pairs of the oxygen atoms in their molecules, i.e. 1.1 eV (PES). These data confirm the assignment of long-wave absorption in the UV spectra of ketones to the $n(O) \rightarrow \pi^*$ transition.

### 8.5.4 Acrolein

Acrolein is used as a starting material or is formed as an intermediate in the synthesis of many important organic substances, e.g. methionine, glutaric aldehyde, $\delta$-valerolactone, glycerol, quinoline, benzanthrone, etc. The reactivity of acrolein is high since the $\pi$(C=O) bond of the carbonyl group is conjugated with the C—C $\pi$ bond in its molecule.

Acrolein is an iso-electronic analogue of buta-1,3-diene. It is noteworthy that acrolein and butadiene-1.3 molecules have similar geometrical parameters as well:

H, 0.1213, C=O, 119.9 °, 124.0 °, 0.1483, $H_2C$=C, H, 0.1343

H, C=$CH_2$, 0.1480, $H_2C$=C, H, 0.1330

The role of the frontier molecular orbitals of buta-1,3-diene was discussed in detail in Chapter 5, but only electrocyclic reactions of this compound were considered. Meanwhile, alkadienes enter some other reactions characteristic of conjugated systems. Buta-1,3-diene can undergo both 1,2- and 1,4-addition with electrophiles. Thus, 1,4-addition prevails in thermodynamically controlled conditions:

$$H_2C{=}CH{-}CH{=}CH_2 \xrightarrow[60^\circ]{HBr}$$

$$\underset{\text{1-bromo-butene-2}}{CH_3{-}CH{=}CH{-}CH_2B_2} > \underset{\text{3-bromo-butene-1}}{CH_3{-}CHBr{-}CH{=}CH_2}$$

Acrolein can also enter the reactions of 1,2- and 1,4-addition. But while buta-1,3-diene reacts, by 1,4-addition, mostly with electrophiles, acrolein tends to 1,4-addition reactions with nucleophiles:

$$\overset{4}{H_2C}{=}\overset{3}{CH}{-}\overset{2}{CH}{=}\overset{1}{O} \xrightarrow[\text{1,2-addition}]{R{-}MgBr} \overset{4}{H_2C}{=}\overset{3}{CH}{-}\underset{R}{\overset{2}{CH}}{-}\overset{1}{O}{-}MgBr \xrightarrow{H_2O} H_2C{=}CH{-}\underset{R}{CH}{-}OH$$

$$\overset{4}{H_2C}{=}\overset{3}{CH}{-}\overset{2}{CH}{=}\overset{1}{O} \xrightarrow[\text{1,4-addition}]{R{-}NH_2} \underset{R{-}NH}{\overset{4}{H_2C}}{-}\overset{3}{CH}{=}\overset{2}{CH}{-}\overset{1}{OH} \rightleftharpoons \underset{R{-}NH}{H_2C}{-}CH_2{-}CH{=}O$$

This difference is probably explained by a considerable stabilization of both occupied and unoccupied $\pi$ electron levels as a result of substitution of a more electronegative oxygen atom for the $CH_2$ fragment in the buta-1,3-diene molecule. This stabilization can be noted, e.g. from the results of calculations of buta-1,3-diene and acrolein molecules by the HMO method.

The higher ionization potentials of the occupied molecular $\pi$ orbitals of acrolein – compared with those of buta-1,3-diene – were confirmed by PE spectroscopy [385,

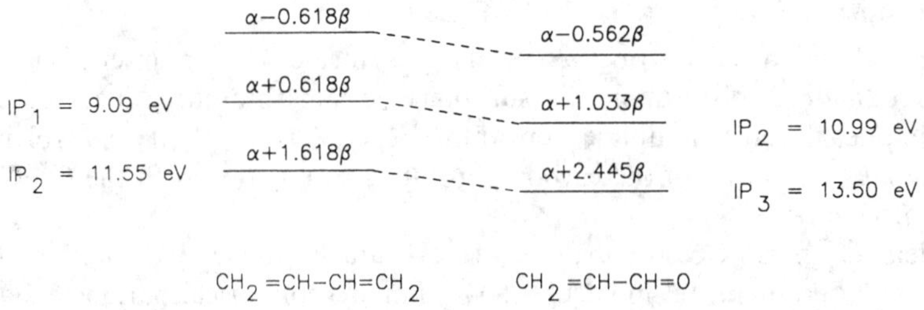

$\alpha-0.618\beta$ $\alpha-0.562\beta$

$IP_1 = 9.09$ eV $\alpha+0.618\beta$ $\alpha+1.033\beta$ $IP_2 = 10.99$ eV

$IP_2 = 11.55$ eV $\alpha+1.618\beta$ $\alpha+2.445\beta$ $IP_3 = 13.50$ eV

$CH_2=CH-CH=CH_2$ $CH_2=CH-CH=O$

386]. As in formaldehyde, the HOMO of acrolein is localized predominantly over the oxygen atom, $n_O$ orbital. The first band in the PE spectrum of acrolein corresponding to the ionization of this orbital has vibrational structure. The second potential is assigned to ionization of the higher occupied $\pi$ orbital of acrolein; this $\pi$ orbital is localized predominantly over the C=C bond. As distinct from the spectra of saturated aldehydes, the spectrum of acrolein has a third band with vibrational structure as well. The third ionization potential refers to the $\pi$ MO localized mostly over the C=O bond of acrolein. The diagram of the acrolein molecular orbitals is shown below:

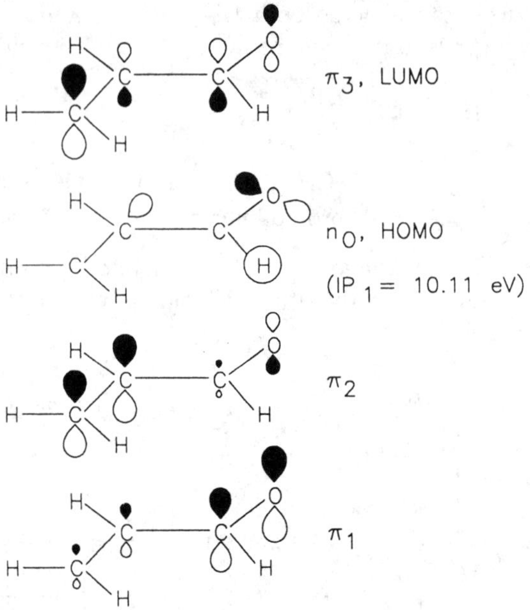

$\pi_3$, LUMO

$n_O$, HOMO

($IP_1 = 10.11$ eV)

$\pi_2$

$\pi_1$

The LUMO of acrolein is the highest of the MOs shown. Its lower energy – compared with the buta-1,3-diene LUMO – is probably the cause of the higher tendency of acrolein to reactions with nucleophiles. Its coefficients show that as an electron is inserted into the LUMO, greater spin densities should be seen at positions 2 and 4. The acrolein molecule is attacked by amines at position 4, since the coefficient of the LUMO is a maximum here.

The corresponding reaction systems can be regarded as 'soft' (see [90] and Section

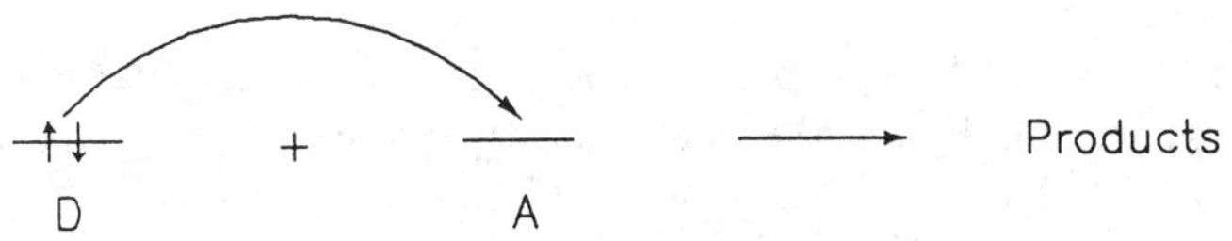

$$D \;(\uparrow\downarrow) + A \longrightarrow \text{Products}$$

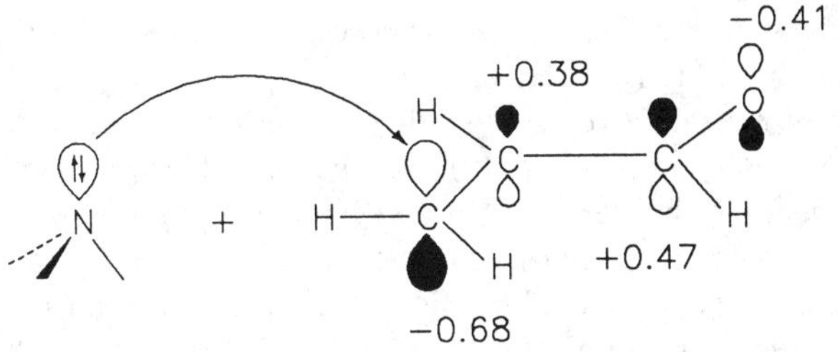

3.1); they are orbital controlled: the amine molecule acts as a donor when it attacks the acrolein molecule (acceptor) at the carbon atom with the maximum coefficient in the LUMO. Conjugated 1,4-addition occurs in a similar way in the reaction of nucleophiles with other α,β-unsaturated carbonyl compounds:

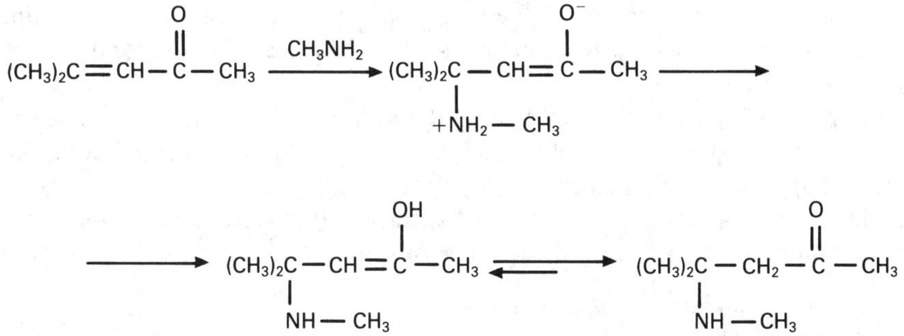

$$(CH_3)_2C{=}CH{-}\overset{O}{\overset{\|}{C}}{-}CH_3 \xrightarrow{CH_3NH_2} (CH_3)_2\underset{{}^+NH_2-CH_3}{\underset{|}{C}}{-}CH{=}\overset{O^-}{\overset{|}{C}}{-}CH_3 \longrightarrow$$

$$\longrightarrow (CH_3)_2\underset{NH-CH_3}{\underset{|}{C}}{-}CH{=}\overset{OH}{\overset{|}{C}}{-}CH_3 \rightleftharpoons (CH_3)_2\underset{NH-CH_3}{\underset{|}{C}}{-}CH_2{-}\overset{O}{\overset{\|}{C}}{-}CH_3$$

By contrast, the interactions of acrolein with Grignard reagents are charge controlled: the nucleophile is attached at the C atom of the carbonyl group having the highest positive charge, while the metal atom is attached to the negatively charged oxygen of the acrolein molecule. The reaction system should probably be classified as 'hard':

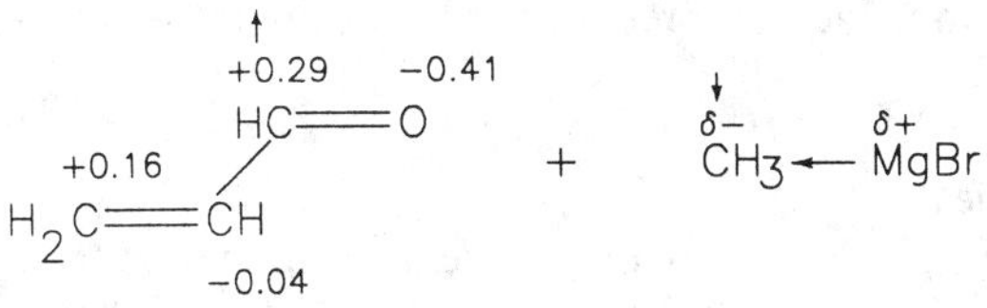

$$\underset{+0.16}{H_2C}{=}\underset{-0.04}{CH}{-}\underset{+0.29}{HC}{=}\underset{-0.41}{O} + \overset{\delta-}{CH_3}\leftarrow\overset{\delta+}{MgBr}$$

Reactions of acrolein with electrophiles, e.g. with HBr, begin with the proton attack at the oxygen atom of the substrate. It has already been said that the acrolein HOMO is localized mainly over the oxygen lone pair (this oxygen atom has a higher negative charge as well).

α,β-Unsaturated ketones have similar orbital structures and undergo similar transformations in their reactions with electrophiles and nucleophiles.

### 8.5.5 Benzaldehyde

The molecule of benzaldehyde is planar. The preference for the planar conformation is estimated experimentally as 20 kJ/mol, and by quantum chemical calculations as 27.63 kJ/mol [237]. The advantages of the planar conformation are quite evident: this conformation ensures effective interaction of the $\pi$ orbitals of the carbonyl group and the benzene nucleus, and this, in turn, is the necessary condition for delocalization of the charge induced by a highly polar carbonyl group.

The PE spectrum of benzaldehyde was studied by only a few authors [263, 366]. Turner *et al.* failed to discover band splitting in the benzaldehyde spectrum due to ionization of the degenerate benzene $\pi$ orbitals. The relatively high intensity of the first band and its shift toward the region of higher ionization energies were, however, noted. The high intensity was explained by the fact that, in addition to the ionization of the electron from the $\pi$(ph) orbital, this band includes also the ionization from the $n$(ph) orbital. The shift of the ionization of the benzene $\pi$ orbital toward the region of higher energies was assigned to the acceptor properties of the formyl group. Ionization of the $\pi$(C=O) orbital in the benzaldehyde spectrum is observed at 14.09 eV.

The acceptor properties of the formyl group affect the frontier vacant MOs of aromatic aldehydes. The energy of the first anion state of benzaldehyde is definitely lower than that of the benzene molecule: their $EA_1$ are $-0.76$ and $-1.15$ eV respectively [46].

Anion radicals of aromatic aldehydes and ketones, the ketyl anions, are of special interest since they are intermediate products of reduction of carbonyl compounds. Their ESR spectra are characterized by splitting of degenerate benzene $\pi^*$ orbitals. HFI constants $a_H$, measured from the ESR spectra of the benzaldehyde anion radical, experimental and calculated (HMO) values of spin densities are as follows [50]:

| | $a_H$ (Oe) | $\rho_{exp}$ | $\rho_{HMO}$ |
|---|---|---|---|
| ortho | 3.39–4.69 | 0.123 | 0.123 |
| meta | 1.31–0.75 | 0.031 | 0.018 |
| para | 6.47 | 0.197 | 0.188 |
| CHO | 8.51 | 0.260 | 0.235 |

The spin density distribution in the anion radical of benzaldehyde indicates that the unpaired electron is located predominantly in the transformed benzene $\pi$(symm) orbital.

### 8.5.6 Acetophenone

The acetophenone molecule is planar. The measured and calculated barriers to rotation of the acetyl group are 13.0 and 18.4 kJ/mol respectively [237].

From the PE spectrum of acetophenone, it follows that changes in the electron levels of the benzene ring induced by the introduction of the acetyl group are similar to the effects of the formyl group. These changes are illustrated in Fig. 8.13. It is supposed that the frontier occupied MO of acetophenone is the oxygen lone pair orbital. Owing to a considerable negative inductive effect of the acetyl group, the energies of benzene $\pi$ orbitals are markedly lowered. In addition, the benzene

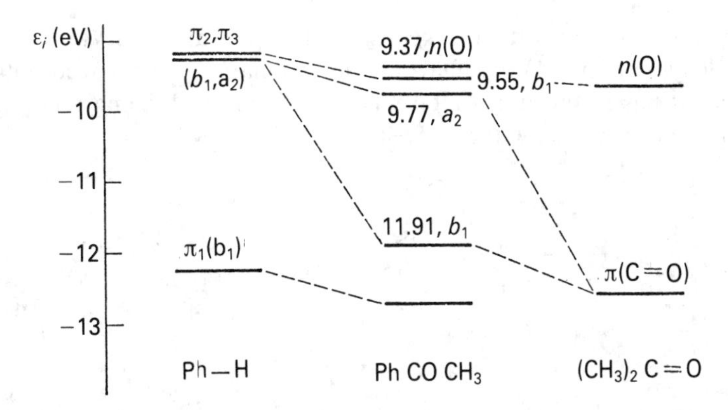


Fig. 8.13. Higher occupied MOs of acetophenone according to PES. (Data by Kobayashi and Nagakura [367])

$\pi_2$(symm) orbital has a symmetry which is the same as that of the $\pi$ orbital of the carbonyl group,and the interaction of these orbitals forms higher occupied electron levels of acetophenone which are next to the HOMO (see the Figure). Based on their CNDO/2 calculations and empirical rules, Kobayashi and Nagakura propose that the first band in the PE spectrum of this compound corresponds to ionization of three closely spaced electron levels. The proposed structures of the occupied MOs agree with the character of the changes in their energies associated with the attachment of the methyl groups to the acetophenone molecule [367].

The energy of the oxygen lone pair is decisive for the complexation reaction with involvement of the carbonyl group. The well-known examples include formation of intramolecular coordination bonds in the hydrazone forms of o-hydroxyazo dyes (for details see Section 9.3.5):

H---O
N
N=

O---SiF3
X–Ph–C
O–CH2

and transannular bonds between oxygen and silicon atoms in benzoyloxymethyl(trifluoro)silanes [387].

### 8.5.7 Phenalenone

The phenalenone fragment occurs in natural plant pigments. It has been isolated from about seventeen species of Australian plants. Several fungal pigments based on the phenalenone unit have been extracted from the genus Penicillium.

The Hückel rule, according to which only planar conjugated systems containing $(4n + 2)$ $\pi$ electrons can be aromatic, is inapplicable to compounds having atoms that are common for three cycles. The phenalenyl cation is an example of the aromatic system that does not obey the Hückel rule. While phenalene (LIVa) is non-aromatic, the phenalenyl cation (LIVb) behaves as an aromatic species. According to HMO calculations it has six occupied molecular $\pi$ orbitals, each being a bonding orbital, i.e. having the energy below the $\alpha$ level:

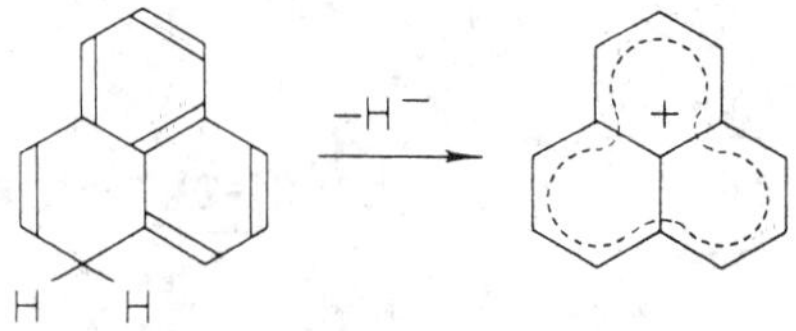

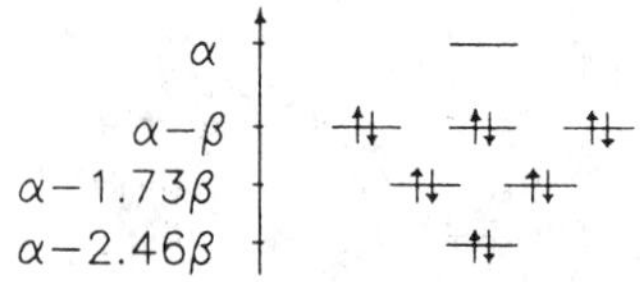

$-H^-$

LIVa LIVb

$\alpha$

$\alpha-\beta$

$\alpha-1.73\beta$

$\alpha-2.46\beta$

Phenalenone (LVa) gives both a stable cation radical and a stable anion radical. The high basicity of phenalenone, compared with the non-cyclic ketones (the basicity of phenalenone is six logarithmic orders higher than acetone, and five orders higher than benzophenone) [388] is probably explained by high stability of the cation (LVb) formed on protonation:

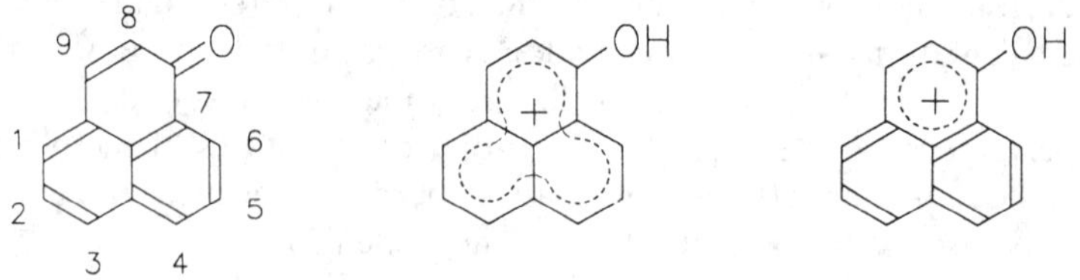

LV: a b c

The suggestion that the positive charge can be delocalized only over one ring (LVc) is based on the fact that the bridged annulenones (XLI) have high basicity as well (see Section 6.1.3). Nevertheless, the results of quantum chemical calculations show that phenalenone is characterized by a considerable delocalization of molecular $\pi$ orbitals over three benzene rings since its 14 $\pi$ electrons are found only in the bonding MOs.

The parameters of the occupied MOs of phenalenone are available also from comparison of its PE spectrum with the CNDO calculations [389]. Phenalenone is expected to reveal first ionization bands arising from the uppermost of the occupied $\pi$ molecular orbitals and the non-bonding n orbital of the carbonyl O atom.

Within the framework of the fragment approach, phenalenone MOs can be considered as combinations of molecular $\pi$ orbitals of naphthalene and the molecular orbitals ($n_O$ and $\pi$) of acrolein. The frontier MOs of phenalenone are shown below. Experimental values of ionization potentials of the corresponding occupied MOs are parenthesized.

In good agreement with the values of phenalenone HOMO coefficients, the electrophilic reactions of chlorination, benzoylation and nitration occur, first of all, at positions 3 and 8 [390]. According to the ESR spectrum, spin density in the anion radical of phenalenone has maximum (and approximately similar) values at positions

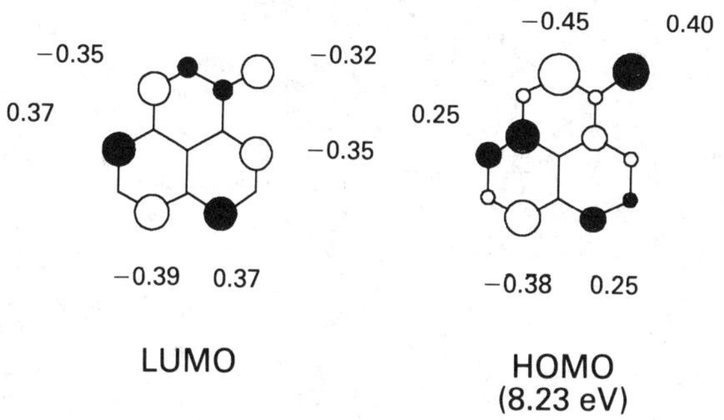

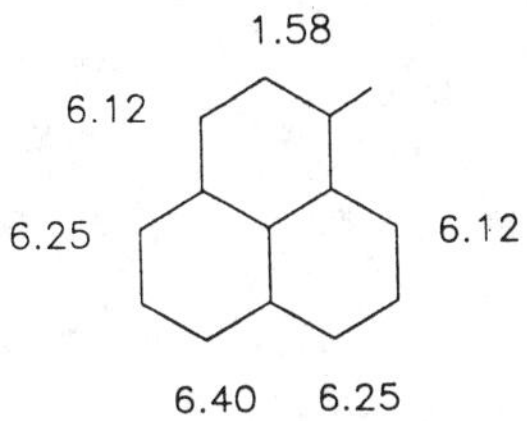

LUMO

HOMO
(8.23 eV)

HOMO-1
(8.84 eV)

1, 3, 4, 6, and 9; this finding is well explained by the parameters of its frontier unoccupied MO.
The highest spin densities at positions 3 and 4 of phenalenone anion radical agree well with its tendency to condensation reaction at these positions in the presence of bases:

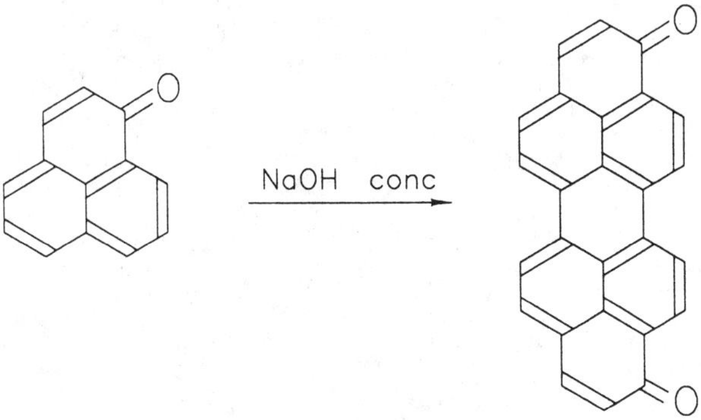

NaOH conc

### 8.5.8 Benzanthrone

Benzanthrone is used in industrial manufacture of valuable vat dyes. During recent years, benzanthrone and the products of its condensation have been intensively studied as valuable materials for molecular electronics. The hydroxy and amino derivatives of benzanthrone have shown laser properties.

The basicity of benzanthrone (LVIa) is much lower than that of phenalenone. This suggests less beneficial conditions for delocalization of the positive charge in the carbocation (LVIb) which is formed during protonation.

LVI: a $\xrightarrow{H^+}$ b

$pK_{a(BH^+)} = -3.94$

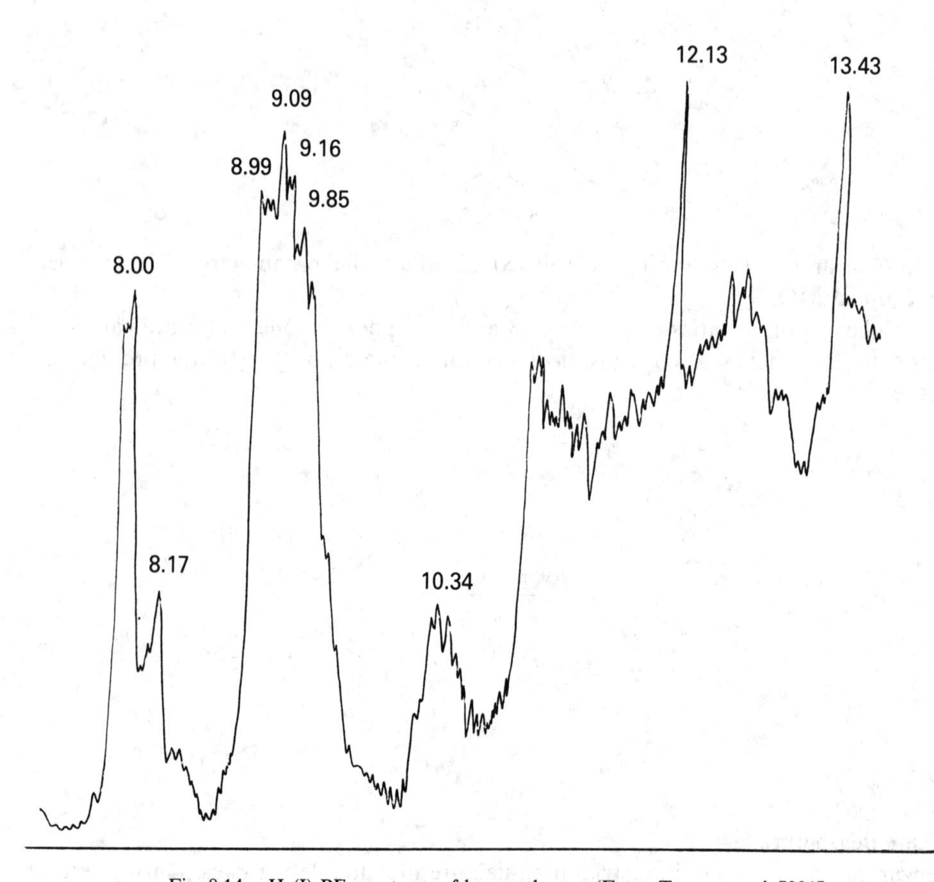


Fig. 8.14. He(I) PE spectrum of benzanthrone. (From Traven *et al.* [391]

According to PE spectroscopy [391], the highest occupied MO of benzanthrone is well separated in energy from subsequent occupied MOs. It is the $\pi$ orbital, delocalized over the entire molecule although the contribution of the $\pi$ orbital of the carbonyl group to it is quite small. This assignment is confirmed by the presence of a distinct vibrational structure of the first band at 8.00 eV, the energy difference between the vibrational levels being 0.17 eV (1371 cm$^{-1}$) (Fig. 8.14). This value correlates with the parameters of the vibrational structure of the first ionization bands measured in PE spectra of polycyclic aromatic hydrocarbons for which the $\pi$ character of HOMO is quite obvious, while the vibrational structure is assigned to the vibrations of aromatic C=C bonds.

The assignment of the second band at 8.99 eV is complicated by ionization of several orbitals of similar energy in this region. According to INDO/S calculations, the second band includes ionization of four molecular orbitals: $n_O$ orbital, $\pi$ orbitals localized over the benzene, naphthalene, and anthracene fragments respectively. The schemes of the frontier molecular orbitals of benzanthrone are shown below:

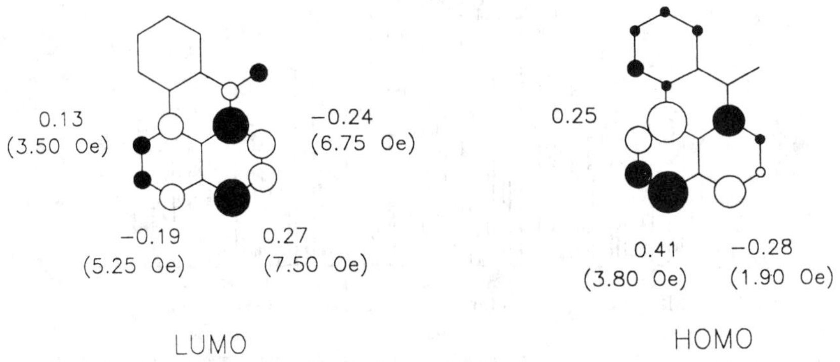

0.13 (3.50 Oe)
−0.24 (6.75 Oe)
−0.19 (5.25 Oe)
0.27 (7.50 Oe)
LUMO

0.25
0.41 (3.80 Oe)
−0.28 (1.90 Oe)
HOMO

It can be seen that both frontier MOs of benzanthrone are mostly localized over the naphthalene fragment. This probably explains the fact that the first ionization potentials of naphthalene and benzanthrone are about the same: 8.15 and 8.00 eV respectively.

The reactions of electrophilic substitution, e.g. halogenation and nitration, occur mostly at position 3 of benzanthrone where the coefficient in the HOMO is of maximum value. Spin density in the benzanthrone cation radical is also maximum at position 3. Parenthesized in the HOMO scheme are the values of $a_H$(Oe) measured from the ESR spectrum of the cation radical obtained by treatment of benzanthrone with chromium oxide $CrO_3$ in concentrated sulphuric acid.

The maximum value of the coefficient in the LUMO is at position 4. Spin density in the anion radical of benzanthrone is also maximum at this position. Parenthesized on the scheme of LUMO are values of $a_H$(Oe) measured from the ESR spectrum of the anion radical obtained by the reaction of benzanthrone with potassium isobutoxide (in the absence of oxygen). Figure 8.15 shows the comparison of the experimental and simulated ESR spectra of the benzanthrone anion radical (the INDO calculations for open-shell particles were used in the assignment of findings of ESR spectrum).

It is important to note that as in the case with a significant separation of the HOMO from subsequent occupied MOs, the LUMO of benzanthrone is also markedly separated from subsequent higher unoccupied MO: the first and second half-reduction potentials $E_{1/2}$ are $-1.20$ and $-1.90$ V respectively. Within the framework of the frontier orbital concept, it is the LUMO that determines the

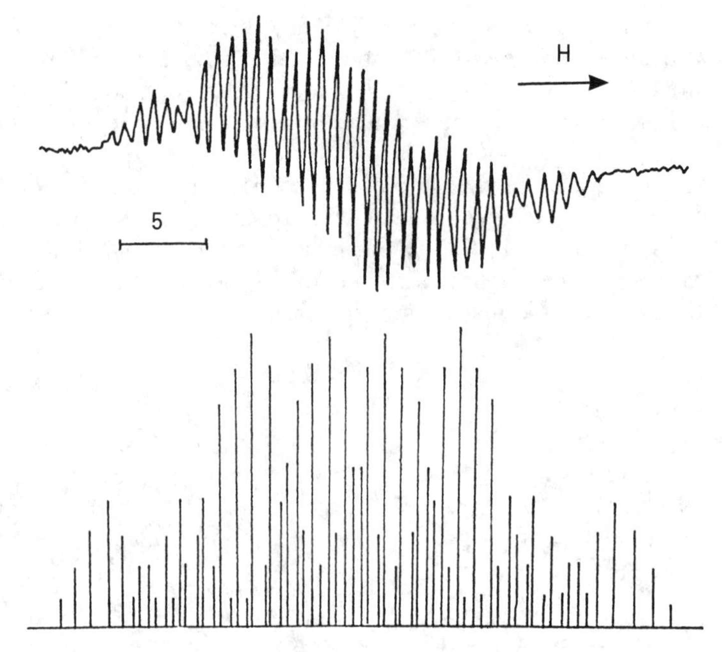


Fig. 8.15. ESR spectra of benzanthrone anion-radical: (a) experimental; (b) simulated.

reactivity of benzanthrone toward bases and nucleophiles. Thus, dimerization of two anion radicals of benzanthrone in alcoholic solution to dibenzanthronyl (an important intermediate for vat dyestuffs) occurs at position 4.

The low energy of the benzanthrone LUMO probably explains the fact that its complex with TCNE ($IP_1 = 8.00$ eV) absorbs at 500 nm while the naphthalene complex ($IP_1 = 8.15$ eV) with the same acceptor absorbs at 550 nm [392].

### 8.5.9 Diketones

Diketones are common natural substances. Some of them are used in medicine. They are traditional intermediate products in the synthesis of various carbo- and heterocyclic compounds.

It can be seen from the previous Section that the carbonyl fragment is important for the electron structure of organic molecules. The presence of two carbonyl groups in the molecule accounts for some special properties of diketones. Thus, if the carbonyl groups are adjacent as in α-diketones, the lone pairs of two oxygen atoms can overlap effectively since they are located close to one another in space:

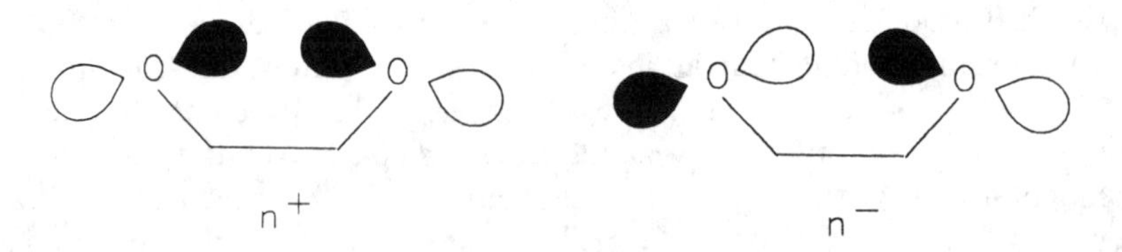

The appropriate interaction leads to appreciable changes in the energies of frontier occupied MOs. The degree of splitting of $n$(O) levels in the PE spectra of α-diketones LVIIa-d is about 2 eV [393]:

| α-diketones LVII: | (a) | (b) | (c) | (d) |
|---|---|---|---|---|
| $IP_{n^+}$ (eV) | 9.55 | 9.61 | 10.45 | 8.85 |
| $IP_{n^-}$ (eV) | 11.46 | 11.71 | 13.04 | 10.90 |
| $\Delta IP$ (eV) | 1.91 | 2.10 | 2.59 | 2.05 |
| $\lambda_n^+ \rightarrow \pi^*$ | 417 | 487 | 645 | 533 |
| Colour | yellow | orange | blue | red |

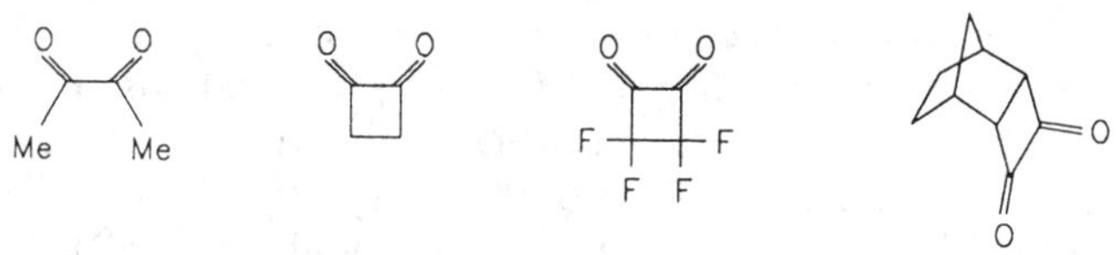

LVII: a b c d

The causes of the marked shift of the long-wave bands to the visible region of the spectrum of α-diketones are not so unambiguous because the reported data do not include the energies of the vacant electron levels. The low excitation energies of electrons in α-diketones can probably be due to a corresponding decrease in the energies of their frontier vacant MOs.

The pronounced CH acidity of α-Hs relative to hydrocarbons is probably most characteristic of β-diketones [193]:

| | Ethane | Acetone | Acetylacetone |
|---|---|---|---|
| $pK_a$ | 42 | 24.6 | 8.9 |

This property is due to the ability of the carbonyl group to decrease π electron density over the α-carbon atom and thus to stabilize the carbanion that is formed during dissociation. The conditions for stabilization of such carbanions have been studied in detail using various quantum chemical calculations [24, 140, 145].

Acetylacetone and some of its derivatives have equilibrium concentrations of both keto and enol forms which are sufficiently high that their PE spectra could be studied in the gas phase [33].

### 8.5.10 1,4-Benzoquinone, 1,4-naphthoquinone, 9,10-anthraquinone

The electron structure of quinones is of special interest. These compounds are important mostly due to their ability to participate in reversible oxidation–reduction reactions, in both chemical and biological systems. Quinones are common in plants, e.g. in mushrooms. Polycyclic quinones form a large class of organic dyes [115, 394].

The molecules of all the three quinones are planar in all phases. The geometric parameters of quinones are determined by conjugation of their C=C and C=O bonds. Given below for the purpose of comparison, are bond distances (nm) and endocyclic angles in the carbonyl group, measured for 1,4-benzoquinone by electron diffraction (the gas phase) [365] and by x-ray (the solid phase) [395]:

| | $r_{C=O}$ | $r_{C=C}$ | $r_{C-C}$ | $\angle C_{(2)}C_{(1)}C_{(6)}$ |
|---|---|---|---|---|
| Gas phase | $0.1225 \pm 0.0002$ | $0.1344 \pm 0.0003$ | $0.1481 \pm 0.0002$ | 118.1 |
| Solid phase | $0.1222 \pm 0.0008$ | $0.1322 \pm 0.0008$ | $0.1477 \pm 0.0006$ | |

The PE spectra of quinones are difficult to interpret, especially in the low-energy region where ionization of both $n$(O) and $\pi$ orbitals should be observed. The first bands in the benzoquinone PE spectrum were first assigned by Turner: $n^-$(O), $n^+$(O), $\pi_1$ and $\pi_2$. Later reports on the sequence of occupied MOs in a benzoquinone molecule were contradictory [221, 223, 396]. The cause of controversy was studied in detail by Schweig *et al.* They proposed that due to the low energy of its LUMO, ionization of 1,4-benzoquinone, as well as of carbon monoxide, is described inadequately by Koopmans' theorem. CNDO/S calculations, including those of configurational interactions, as well as calculations on the basis of the Green functions, were successfully used for the analysis of the PE spectra of quinones [397].

The PE spectra of 1,4-benzoquinone, 1,4-naphthoquinone, and 9,10-anthraquinone are shown in Fig. 8.16. The sequences of their higher OMOs are also shown. Only the first bands of the 1,4-benzoquinone spectrum in the region of low ionization energies are well resolved, and their assignment is trustworthy (this assignment agrees with Turner's interpretation on the basis of purely qualitative analysis). In the spectra of 1,4-naphthoquinone and 9,10-anthraquinone, the first bands strongly overlap. A progressive decrease in the ionization energies of $n$(O) and $\pi$ orbitals in the series benzoquinone > naphthoquinone > anthraquinone can, however, be noted.

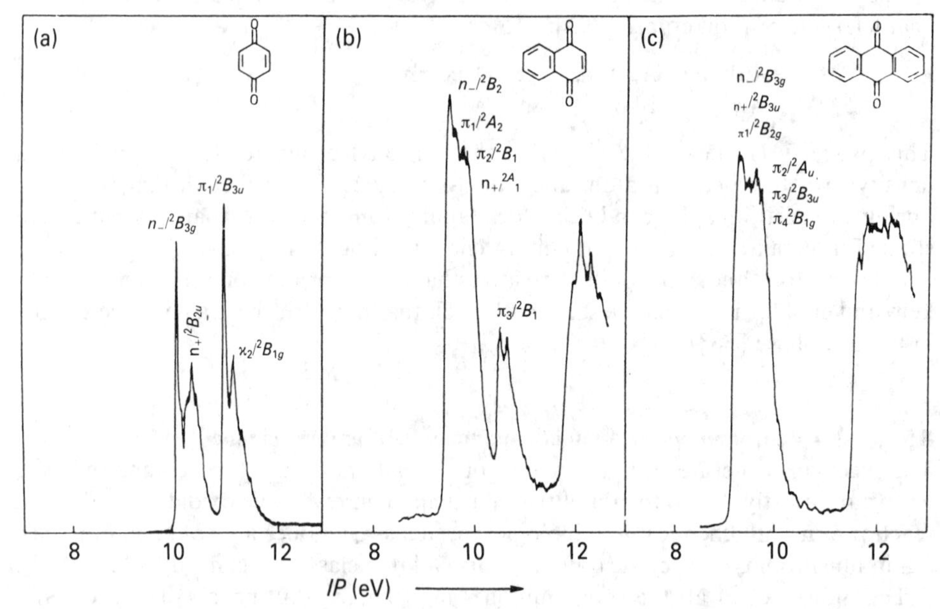


Fig. 8.16. He(I) PE spectra of (a) 1,4-benzoquinone, (b) 1,4-naphthoquinone and (c) 9,10-anthraquinone. (From Lauer *et al.* [396])

The effect of substituents on the electron structure of 1,4-benzoquinone was studied. Methyl groups appreciably destabilize all four higher occupied MOs, while fluorine atoms, on the other hand, stabilize them. The transition to tetramethyl- and tetrafluoro-1,4-benzoquinones alters the sequence of some orbitals, but one of the combinations of oxygen lone pairs remains the HOMO in all molecules [398].

The energies of vacant electron levels are important for the chemistry of all carbonyl compounds, quinones in particular. High electron affinities, which are characteristic of quinones, decrease the informative value of the ETS method for quinones. Among the quantitative characteristics of frontier vacant MOs of quinones, noteworthy are the values of their reduction potentials measured polarographically [67, 310]:

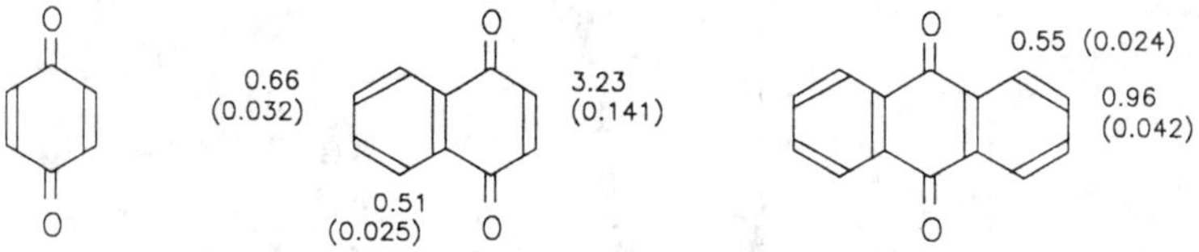

0.66 (0.032) 3.23 (0.141) 0.51 (0.025)

0.55 (0.024) 0.96 (0.042)

| | 1,4-benzoquinone | 1,4-naphthoquinone | 9,10-anthraquinone |
|---|---|---|---|
| $E^{red}_{1/2}$, V | +0.711 | +0.493 | +0.155 |

The ESR spectra of anion radicals of 1,4-benzoquinone, 1,4-naphthoquinone and 9,10-anthraquinone were studied [50]. The measured values of their HFI constants are given in the structural formulae of quinones at the appropriate positions; spin densities calculated by the HMO method are included in parentheses.

## 8.6 CARBOXYLIC ACIDS

### 8.6.1 Acetic acid

The tendency to dimerization is the common property of carboxylic acids. The heat of dimerization of formic acid, for example, in the gas phase is 58.62 kJ/mol [399]. Acetic acid also exists as a dimer in the crystal, liquid, and gas phases (under moderate pressure):

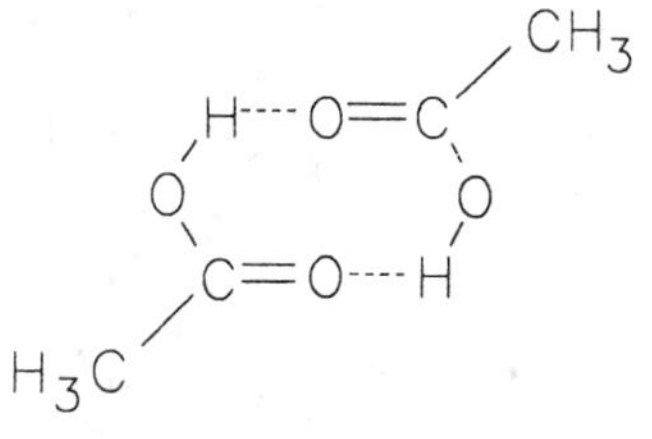

$CH_3$

H····O=C

O O

C=O····H

$H_3C$

The geometric parameters of acetic acid in the monomer and dimer forms, measured by electron diffraction in the gas phase, are as follows [400]:

| Bond distance (nm) and valence angles (% degrees): | $r_{C-C}$ | $r_{C-H}$ | $r_{C=O}$ | $r_{C-O}$ | $\angle C-C=O$ | $\angle C-C-O$ |
|---|---|---|---|---|---|---|
| monomer | 0.1520 | 0.1102 | 0.1214 | 0.1364 | 126.6 | 110.6 |
| dimer | 0.1506 | 0.1102 | 0.1231 | 0.1334 | 123.6 | 113.0 |

It can be seen that dimerization does not involve the C—H bonds but the (C=O) and (C—OH) bond distances become appreciably closer.

The PE spectra of both the monomer and dimer forms of acetic acid were recorded (Fig. 8.17) [38, 131].

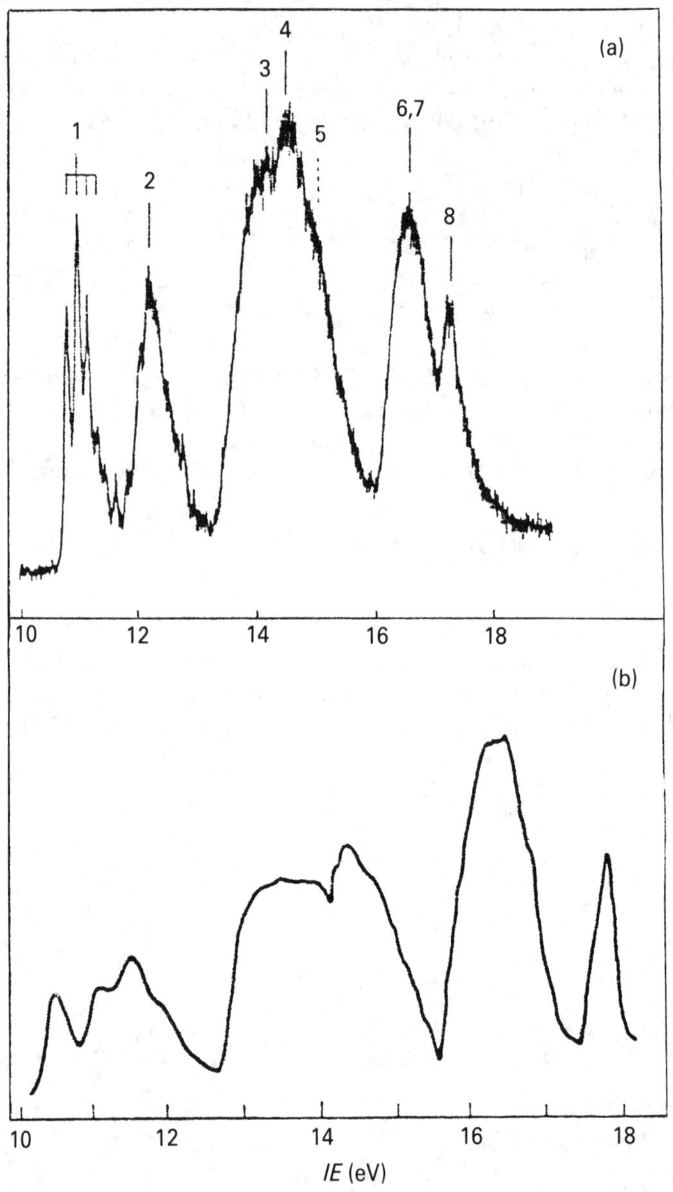


Fig. 8.17. He(I) PE spectra of acetic acid [38]: (a) monomer; (b) dimer (band numbers correspond to ionization potential numbers in Table 8.14). (From Kimuza *et al.* [38])

The $n_O$ (C=O) orbital is the HOMO of acetic acid in the monomer form, as it is in other carbonyl compounds. The orbital of the lone pair of the oxygen atom of the hydroxyl group is the third from top. Unlike the HOMO, this orbital has $\pi$ symmetry.

The dimer has the same electron structure, but its electron levels are located closer to one another than in the monomer. While studying acetic acid in the gas and solid phase by x-ray electron spectroscopy, Baker and Betteridge concluded that the oxygen atoms of the carbonyl group in the gas phase are non-equivalent but become equivalent in the solid phase, giving only one (but broad) 1*s*(O) line in the spectrum [401] (see Section 3.5).

*Ab initio* calculations show the non-equivalence of the oxygen atoms in the monomer form of the acid (Table 8.14) in agreement with observation in the gas phase.

**Table 8.14.** Ionization potentials and higher occupied molecular orbitals of acetic acid (PES data and 4-31G calculations)

| Nos. | $IP_i$ (eV) | Molecular orbitals | | |
|---|---|---|---|---|
| | | $-\varepsilon_i$ (eV) | No. and symmetry (group $C_s$) | Localization |
| 1 | 10.87 | 12.41 | $13a'$ | $n_O$(C=O) |
| 2 | 12.05 | 13.11 | $3a''$ | $\pi_{C=O}, n_O$ |
| 3 | 14.05 | 15.34 | $2a''$ | $\pi_{CH_3}, n_O$ |
| 4 | 14.40 | 15.47 | $12a'$ | $\pi_{CH_3}$ |
| 5 | (14.9) | 15.69 | $11a'$ | $\bar{n}_O, \sigma_{C=O}$ |
| 6 | 16.42 | 17.83 | $1a''$ | $\pi_{C=O}, n_O, \pi_{CH_3}$ |
| 7 | 16.42 | 18.56 | $10a'$ | $\sigma_{C-C}$ |
| 8 | 17.08 | 19.09 | $9a'$ | $\sigma_{O-H}$ |

(From Kimura *et al.* [38])

The energies of the MOs of dimeric acetic acid are higher than in the monomer, but this increase of the occupied electron levels upon dimerization does not hold for all carboxylic acids. According to [131], dimerization, for example, of trifluoroacetic acid, stabilizes all occupied electron levels. The energy level diagrams of the monomeric and dimeric forms of acetic and trifluoroacetic acids are compared in Fig. 8.18.

In the discussion of the mechanisms of acid-catalysed reactions of carboxylic acids and their derivatives, special attention is given to the site of proton attachment to the acid molecule. Both the oxygen atom of the carbonyl group and the hydroxyl group oxygen can attach a proton. According to PES and quantum chemical calculations [24], the lone pair of the oxygen atom of the carbonyl group is a preferable site for proton attachment.

### 8.6.2 Benzoic acid

The molecule of benzoic acid is planar. In the solid phase, the acid exists in the form of a dimer with the carbonyl groups separated by a distance of 0.264 nm (hydrogen bond distance). The C—C bond distances in the benzene nucleus change insignificantly upon dimerization (by not more than 0.003 nm), while two C—O bonds in the

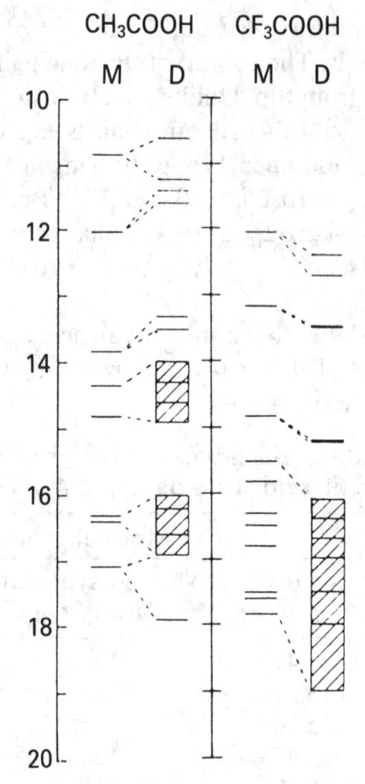


Fig. 8.18. Experimental estimation of energies of occupied MOs of monomeric (M) and dimeric (D) forms of acetic and trifluoroacetic acids. (From Carnovale *et al.* [131])

carboxyl group become almost equivalent: 0.124 and 0.129 nm. The similarity of the C—O bonds of the carboxyl groups in the dimer of benzoic acid is greater than in the dimer of acetic acid. This might be due to effective mixing of the orbitals of the carboxyl group and the $\pi$ orbitals of the benzene nucleus: the rotational barrier of the carboxyl group in the benzoic acid molecule is estimated as 21.12 kJ/mol by calculations in the STO-3G basis set [237]. It is noteworthy that in benzoic acid derivatives, e.g. in its amide, the degree of $\pi$ bonding, and hence the rotational barrier are much lower. The molecular plane becomes distorted: the dihedral angle between the benzene ring and the amide group, for example, is 30° (26° according to x-ray structural analysis).

The $\pi$ electrons are noticeably drawn from the benzene nucleus in the benzoic acid molecule. $\pi$-Electron densities in the molecules of benzoic acid, its amide, benzaldehyde, and acetophenone calculated by quantum chemical methods are given below [237]:

$CHO$ 1.020 0.981 1.002 0.982

$CH_3CO$ 1.020 0.981 1.003 0.984

$COOH$ 1.040 0.969 1.006 0.976

$H_2NCO$ 1.016 0.985 1.000 0.988

It can be seen that all carbonyl-containing groups should orient the electrophilic attack at the meta position in the corresponding benzene derivatives. The distribution of the isomer formed, e.g. by nitration of carbonyl-containing benzene derivatives is given in Table 8.15. (For a detailed discussion of isomer distribution in electrophilic reactions of benzene derivatives see Chapter 10).

**Table 8.15.** The distribution of the isomers in nitration of carbonyl-containing benzene derivatives

| Compound | Isomer distribution (%) | | |
|---|---|---|---|
| | ortho | meta | para |
| Acetophenone | 26 | 72 | 0–2 |
| Benzaldehyde | — | 84–91 | — |
| Benzoic acid | 15–20 | 75–85 | 1 |
| Ethyl benzoate | 24–28 | 66–73 | 1–6 |

(From Jones [346])

The distribution of electron density in the frontier occupied MO of benzoic acid is illustrated by the HFI constants measured from the ESR spectra of its cation radical: $a_H$(ortho) = 9.05 Oe; $a_H$(meta) = 11.85 Oe, $a_H$(para) = 0.75 Oe [402]. These values can be used to determine the symmetry of the benzoic acid HOMO as $a_2$ (conventionally within the framework of the point group $C_{2v}$). Thus, total $\pi$ electron occupancies (charge characteristics) agree in the determination of prevalent meta orientation of electrophilic substitution in the benzoic acid molecule.

***Problems***

(1) Explain why hydroxy- and alkoxy groups increase the ability of organic molecules to react with electrophiles:
$CH_4 + HCl \rightarrow$ no reaction;
$CH_3—O—CH_3 + HCl \rightleftharpoons CH_3—\overset{H}{\overset{|+}{O}}—CH_3 + Cl^-$;
$CH_2{=}CH_2 + TCNE \rightarrow$ no reaction;
$CH_2{=}CH—O—C_2H_5 + TCNE \rightarrow$ cycloaddition;
$C_6H_6 + Br_2 \rightarrow$ no substitution;
$C_6H_5OH + Br_2$ 30° (no catalyst) 2,4-$Br_2C_6H_3OH$.
Which of the frontier orbitals of each oxygen compound is responsible for the reaction?

(2) Why are alcohols more easily oxidized than alkanes? For example, a primary alcohol releases two electrons and is transformed to an aldehyde:

$$R—CH_2OH \rightarrow R—CHO + 2e^- + 2H^{\oplus}$$

Compare $CH_4$ and $CH_3OH$ to illustrate your answer.

(3) Explain why benzene, phenol, and anisole are very different in their ability to release an electron ($IP_1$s are 9.24 eV, 8.70 eV and 8.21 eV), but are very similar in their ability to accept an electron ($EA_1$s are −1.15 eV, −1.11 eV and

$-1.13$ eV).

(4) Phenol reacts with electrophiles mostly in the ortho and para positions. Explain this fact in terms of the frontier orbitals concept.

(5) Suggest an unpaired electron distribution in the phenol anion radical.

(6) Anisole forms a cation radical when reacted with electrophiles. Suggest an unpaired electron distribution in the anisole cation radical.

(7) Compare the ability of ethylene, formaldehyde, and acrolein to accept an electron (the values of $E_{1/2}^{red}$ are $-1.87$, $-1.48$, and $-0.83$). Consider the nature of the frontier unoccupied orbitals and explain this fact.

(8) Benzaldehyde accepts an electron more easily than benzene; explain this fact in terms of the frontier orbitals concept using appropriate experimental data.

(9) Suggest an unpaired electron distribution in the benzaldehyde anion radical.

(10) The ability to accept an electron increases from biphenyl through benzaldehyde to benzophenone (the $E_{1/2}^{red}$ values are $-2.70$ eV, $-1.48$ V, and $-1.35$ V). Explain this trend.

(11) Predict what product is formed in the reaction of acrolein with ethylamine. Assume the 'soft' interaction.

(12) Predict what product is formed in the reaction of acrolein with ethylmagnesium bromide. Assume the 'hard' interaction.

(13) 1,4-Naphthoquinone undergoes dimerization in the presence of a base. Suggest which position is more reactive.

(14) Suggest how the reactivity to nucleophiles changes from 9,10-anthraquinone through 1,4-naphthoquinone to 1,4-benzoquinone.

# 9

# Nitrogen compounds

## 9.1 AMINES

Amines are organic compounds containing non-bonding electron pair on the heteroatom. The properties of a lone electron pair localized over the $sp^3$ hybridized nitrogen atom determine largely the reactivity and biological potency of amines. The analysis of the electron structure of organic amines should begin with the discussion of molecular electron levels of ammonia and its analogues in the VA group.

### 9.1.1 Ammonia and its analogues

The molecule of gaseous ammonia has a pyramidal structure: $r_{H—N} = 0.1015$ nm; the HNH angle = 106.6° [22].

Within the framework of the point group $C_{3v}$

| AO | $2s$ | $2p_z$ | $2p_x$ | $2p_y$ | $(+)1s$ | $(-)1s$ | $(-)1s$ |
|---|---|---|---|---|---|---|---|
| Symmetry | $a_1$ | $a_1$ | $e$ | $e$ | $a_1$ | $e$ | $e$ |

$2p_z$ and $2s$ orbitals of the nitrogen atom belong to the same type of symmetry $a_1$. The (+) combination of $1s$ orbitals of the three hydrogens have this symmetry as well. These orbitals form the HOMO of symmetry $a_1$. The $2p_x$ and $2p_y$ orbitals of the nitrogen and (−) combinations of $1s$(H) orbitals give orbitals of symmetry $e$ in the ammonia molecule.

Table 9.1 presents the results of calculations of ammonia by the *ab initio* method. According to these calculations, the $1a_1$ orbital is the $1s$ orbital of nitrogen which is not involved in the chemical bonding. The frontier occupied MO $3a_1$ is by almost 90% made up of the $2p_z$ orbital of the nitrogen atom and is also only partly involved in the N—H and H ... H interactions. The contribution of the $1e$ orbitals to the N—H bonding is the greatest.

The Walsh diagram of ammonia should be mentioned as well (Fig. 9.1). As distinct from water and dimethyl ether whose HOH and COC valence angles change without altering the HOMO energies, the change in the HNH valence angle in the ammonia molecule is important for the energy of the nitrogen lone pair. In the transition from

**Table 9.1.** Molecular orbitals of ammonia (*ab initio* calculations)

| MO | ε (eV) | MO coefficients | | | | | | | |
|---|---|---|---|---|---|---|---|---|---|
| | | 1s ($H_1$) | 1s ($H_2$) | 1s ($H_3$) | 1s (N) | 2s (N) | $2p_x$ (N) | $2p_y$ (N) | $2p_z$ (N) |
| $1a_1$ | −422.3 | 0.00 | 0.00 | 0.00 | 1.00 | 0.00 | 0.00 | 0.00 | 0.00 |
| $2a_1$ | −31.16 | 0.19 | 0.19 | 0.19 | −0.30 | 0.87 | 0.00 | 0.00 | 0.18 |
| $3a_1$ | −11.22 | 0.07 | 0.07 | 0.07 | 0.08 | −0.30 | 0.00 | 0.00 | 0.94 |
| $1e$ | −16.97 | 0.00 | 0.43 | −0.43 | 0.00 | 0.00 | 0.00 | 0.80 | 0.00 |
| $2e$ | 9.34 | 0.49 | −0.20 | −0.20 | 0.00 | 0.00 | 0.80 | 0.00 | 0.00 |

(From Ballard [22])

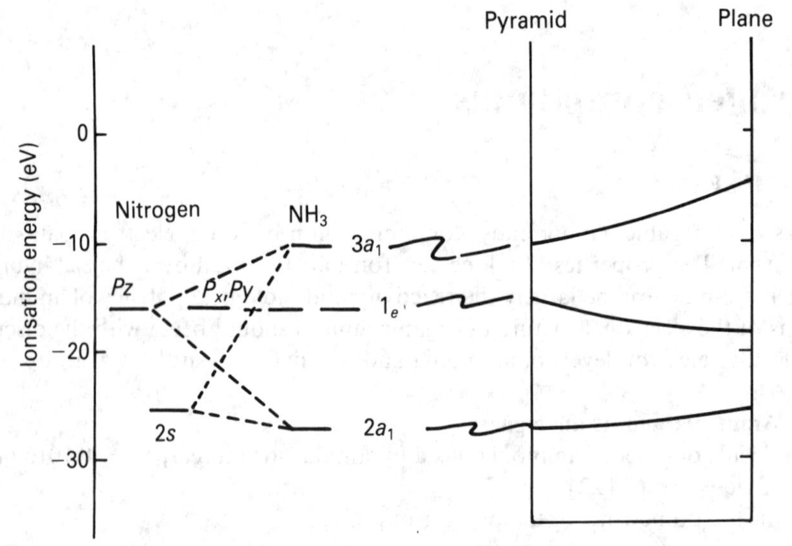


Fig. 9.1. Occupied MOs of ammonia and dependence of their energies on molecular conformation (Walsh diagram). (From Ballard [22])

the pyramidal to planar conformation, the $3a_1$ orbital destabilizes appreciably with decreasing contribution of the nitrogen 2*s* orbital; the $2a_1$ orbital destabilizes to a slightly smaller extent. On the contrary, the electron levels of symmetry *e* in the planar conformation are stabilized.

In the transition to heavier elements, the pyramidal structure of the $E(V)H_3$ hydride molecule is conserved but the pyramid becomes more pronounced: in a molecule of, for example, phosphine, $r_{P-H} = 0.14206$ nm, while the HPH angle is 93.8°. As the pyramidal character of the phosphine molecule increases, the contribution of the $3p_z$ orbital of the phosphorus atom to HOMO decreases. Below follow the basis AOs and the coefficients estimating their contribution to phosphine HOMOs [21]:

| AO | 3s | $3p_z$ | $3p_x$ | $3p_y$ | 1s(H) | 1s(H) | 1s(H) |
|---|---|---|---|---|---|---|---|
| $c_\mu$ | −0.4122 | −0.8373 | 0.0 | 0.0 | 0.1410 | 0.1410 | 0.1410 |

The increasing involvement of the valence *s* orbital of the element in the HOMO of the corresponding hydride is probably the cause of unusually small changes in the first ionization potential in the series of compounds $NH_3$, $PH_3$, $AsH_3$, $SbH_3$ (Fig. 9.2).

The steric structure of small molecules, derivatives of the VA group elements, has been studied in detail by *ab initio* quantum chemical calculations. It has been established that the tendency to the pyramidal structure seen in the $sp^3$ hybridized phosphorus atom is so high that (by contrast to ammonia) the planar conformation is disadvantageous for phosphine even in the cation-radical state [403].

### 9.1.2 Trimethylamine and its organoelement analogues

The marked tendency to the pyramidal configuration is seen in the series of alkyl derivatives of the group V elements as well. Even substitution of all the three hydrogens in ammonia for methyl groups increases the valence angle at the nitrogen atom only from 106.6° to 108°. The methyl groups, however, decrease appreciably the inversion barrier: $NH_3$–24.28 kJ/mol, $(CH_3)_2NH$–18.42 kJ/mol.

The PE spectra of trimethylamine and its organoelement analogues were analysed [353]. The comparison of the PE spectrum of trimethylamine, for example, with the results obtained by INDO, CNDO and EHMO calculations give reliable information on the electron structure of its molecule.

The HOMO of trimethylamine is formed not only by $2p_z$ and $2s$ orbitals of the nitrogen but also by (+) combination of $2p_z$ orbitals of carbons of the three methyl groups. The first band in the PE spectrum of trimethylamine, corresponding to the ionization of the nitrogen lone pair, is therefore quite broad. It differs substantially in form and intensity from the ionization bands of lone pairs of the group VI and VII elements in the spectra of ethers and haloalkanes.

The first ionization potentials of trimethyl derivatives of the VA group elements are practically the same (See Fig. 9.2). As in the case with the corresponding hydrides,

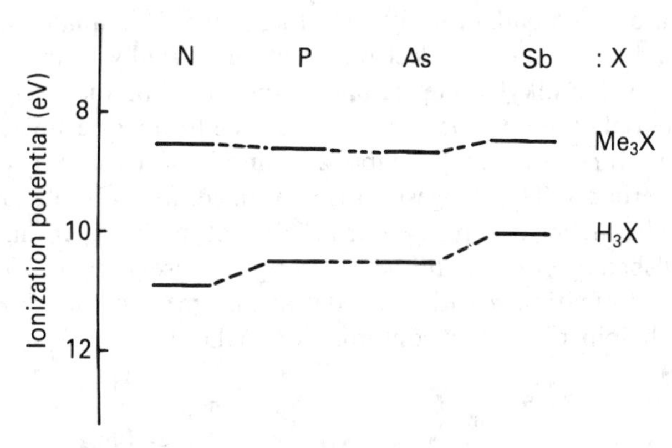


Fig. 9.2. Energies of HOMOs of $H_3E$ hybrid [36] and trimethyl derivatives of VA group elements $(CH_3)_3E$ according to PES. (From Elbel *et al.* [353])

this tendency characterizes specifically the $E(V)(CH_3)_3$ compounds to differentiate them from their analogues $E(VI)(CH_3)_2$. This can be explained by the different mechanism of formation of the HOMOs. It has already been said that the HOMO in the series of dimethyl chalcogenides is localized largely in the $p_\pi$ orbital of the chalcogen atom. Its composition does not practically change with the changes in the CEC valence angle while the ionization potential of its electrons definitely decreases with increasing atomic number of the element. In compounds of the group VA elements, the frontier occupied MOs include an appreciable contribution of the *ns* orbital of the heteroatom. This contribution increases in the series N < P < As. The corresponding changes are associated with the geometry of the molecules. The valence angles in the trimethylphosphine and trimethylarsine molecules are lower (99 and 96° respectively [353]) than in the trimethylamine molecule. The large contribution of the *ns* orbitals of the group VA elements is also confirmed by *ab initio* quantum chemical calculations.

### 9.1.3 Trialkylamines and Azocycloalkanes

The electron structure of the amine molecule changes insignificantly with enlargement of the hydrocarbon fragment at the nitrogen atom. The energy level increases consecutively, the difference between the first and second ionization potentials (as in the trimethylamine molecule) being 2–3 eV. The values of the first ionization potentials of the amines $NR_3$ are given below:

| R | H | $CH_3$ | $C_2H_5$ | $n$-$C_3H_7$ | $n$-$C_4H_9$ |
|---|---|---|---|---|---|
| $IP_1$ (eV) | 10.92 | 8.53 | 8.08 | 7.92 | 7.90 |

Likewise, the energy of HOMO increases with the enlargement of the cycle in the azacycloalkane molecule. The first ionization potentials of the amines $NH(CH_2)_n$ are as follows [405]:

| $n$ | 2 | 3 | 4 | 5 | 6 |
|---|---|---|---|---|---|
| $IP_1$ (eV) | 9.85 | 9.04 | 8.77 | 8.64 | 8.41 |

As in the case of haloalkanes, dialkyl ethers and their analogues, it would be incorrect in the discussion of the electron structure of trialkylamines to state that the inductive effect of the alkyl group is only responsible for the decrease in the first ionization potential of amine. Destabilization of the frontier occupied MO of amine, of the cyclic structure included, probably also implies $\pi$ (pseudo) overlapping of the corresponding orbitals. This suggestion is confirmed, for example, by the results of the studies on the higher occupied electron levels of methylpiperidines [406].

It was established by electron diffraction (the gas phase) that 60–70% of piperidine exists in the conformation *a* with the axial arrangement of the nitrogen lone pair while in N-methylpiperidine this conformation makes 99%:

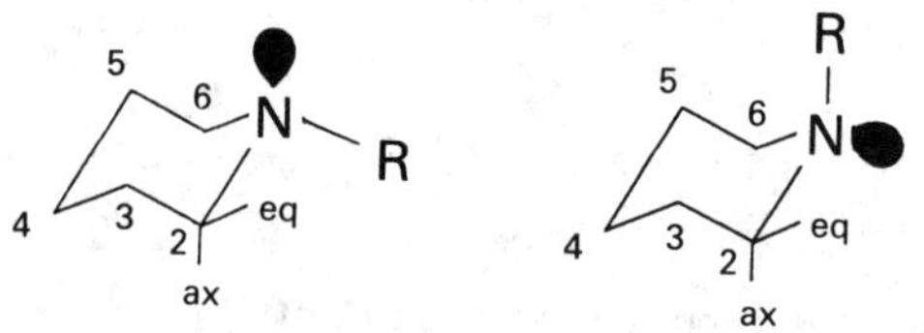

5 6 N R 4 3 2 eq ax

a

R 5 6 N 4 3 2 eq ax

b

Mutual orientation of the nitrogen lone pair and the C—$CH_3$ bond produces a noticeable effect on the energy of the frontier occupied MO in the series of 2(6)-methylpiperidines (Table 9.2).

**Table 9.2.** Ionization potentials of nitrogen lone pair in some methylpiperidines

| $CH_3$ group position in piperidine | $IP_{n(N)}$ (eV) | |
|---|---|---|
| | R = H | R = $CH_3$ |
| — | 8.70 | 8.37 |
| 2-$CH_3$ (equ) | 8.63 | 8.23 |
| 3-$CH_3$ (equ) | 8.63 | 8.35 |
| 4-$CH_3$ (equ) | 8.61 | 8.33 |
| cis-2,6-$(CH_3)_2$ (equ, equ) | 8.53 | 8.22 |
| 2,2,6,6-$(CH_3)_4$-(equ, ax, equ, ax) | 8.04 | 7.68 |

(From Rozenboom and Houk [406])

It follows from the table that the methyl group at position 2(6) of conformer *a* in the axial orientation, decreases the first ionization potential by about 0.26 eV, while in the equatorial position it decreases the potential only by 0.07 eV. This is explained by overlapping of the nitrogen lone pair and $\sigma(C—CH_3)$ orbitals. This overlapping is possible only in the axial position of the methyl group. Accurate *ab initio* calculations in the STO-3G basis set have confirmed the scheme of $n(N)—\sigma(C—CH_3)$ mixing. According to the same calculations, the preference of the conformation *a* for 2(6)-methylpiperidines is estimated by at 4.8 kJ/mol (at R = H) and 22–45 kJ/mol (R = $CH_3$).

The $sp^3$ hybridized nitrogen atom can assume the planar conformation with enlargement of the azacycloalkane cycle. Manxine (LVIII) is an example of such an amine. The steric and electron structures of its derivatives are quite specific. Manxyl chloride (LVIII, R = Cl) has high solvolytic activity. It is believed that in the transition to the planar (relative to the nitrogen atom) manxyl cation, this energy gain is 25–34 kJ/mol, which exceeds the inversion barrier [404, 407]:

LVIII, $IP_1$=7.05eV (R=H) LIXa, $IP_1$=8.06eV LIXb LXa, $IP_1$=6.85eV

The effect of the planar configuration of the nitrogen atom in the manxine molecule consists in a marked increase in the energy of its frontier occupied MO compared with quinuclidine (LIXa): the first ionization potentials are 7.05 and 8.06 eV respectively. In the planar form (the one that the manxine molecule has) the $2p_z$ and $2s$ orbitals of the nitrogen atom cannot have the same symmetry: HOMO becomes an unhybridized $2p_z$ orbital of the nitrogen atom. As a result, the stabilizing contribution of the $2s$ orbital is absent.

The same explanation holds probably for the diamine LXa. In this bicyclic diamine, the NCCC fragments are also planar and the first ionization potential of its molecule is even lower.

Further enlargement of the bicyclic amine creates conditions for the transition of the amine fragment to the end-pyramidal configuration. This configuration occurs, for example, in amine LIXb. In the gas phase, its proton affinity is 83.74 kJ/mol lower than in tributylamine. Basicity is even lower in solution: the value of $pK_a$ of the conjugated acid of amine LIXb is +0.6. It is believed that in both the gas and liquid phase, this amine is only protonated 'on the outside'.The energy of inversion of this amine is 71.18 kJ/mol, which is actually equal to the decrease in its proton affinity [407].

### 9.1.4 Transannular effects in bicyclic amines and diamines

The electron structure of bicyclic amines and diamines is interesting in many respects.

First, many alkaloids contain fragments of bicylic amines such as quinuclidine, tropane, pyrrolizidine, quinolizidine, etc. Most of such alkaloids are characterized by high biological activity and some of them are known as strong medicines (atropine, cocaine,quinine, morphine, codeine, etc.). The mechanisms of the physiological effects of alkaloids are unclear. It is not clear either, why some plants contain these compounds in comparatively high concentrations.

Second, bicyclic amines and diamines are especially convenient models for the study of the ability of various heteroatoms to participate in transannular orbital interactions.

The specific electron structure of manxine (compared with its more rigid structural analogue quinuclidine) has already been mentioned. Incorporation of the second 'head' nitrogen atom into the molecule, and transition to larger bicyclic structures account for the appearance of additional effects. Thus, the orbitals of nitrogen lone pairs in the molecule of 1,4-diazabicyclo[2,2,2]octane and in the molecules of its higher homologues, can interact through space and through bonds. These interactions form two new levels in bicyclic diamines. The levels correspond to bonding and antibonding combinations of nitrogen lone pairs respectively:

$$n^- = 1\sqrt{2}(\chi_n - \chi_n)$$

$$n^+ = 1\sqrt{2}(\chi_n + \chi_n)$$

Prevalent through-space interaction (TS-effect) leads to a 'normal' sequence of combinations of lone pairs: $n^-$, $n^+$. If two lone pair orbitals interact through bonds (TB-effect), the resultant sequence of their combinations is inverted: $n^+$, $n^-$.

The proportion of the TS- and TB-effects depends on the structure of the corresponding diamines and varies in the series of compounds LX (a)–(f):

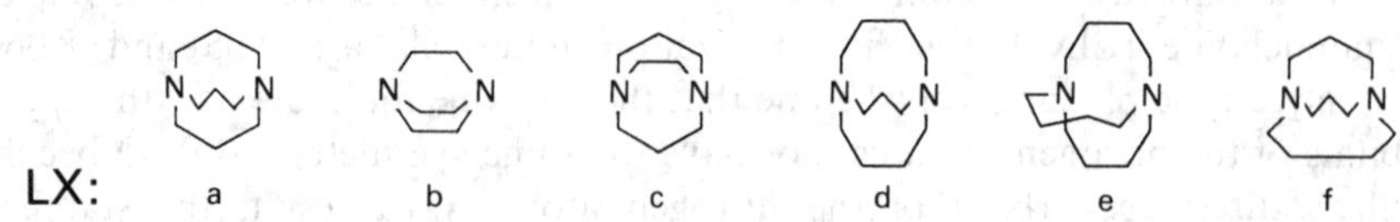


Consider, first of all, the data on the steric structure of the mentioned diamines. Thus, the geometry of 1,4-diazabicyclooctane LXb (like that of quinuclidine) has quasi $D_{3h}$ symmetry with the amplitude of torsional vibrations about the $C_3$ axis of about 7°. Steric structures of other diamines also favours interaction of the lone pairs of two nitrogen atoms [1, 198, 404].

Diamine LXb was studied in detail by electron microscopy methods. The first studies on its PE spectra were reported in 1969, while a detailed analysis of the sequence of the higher occupied electron levels of its molecule was published a year later [408]. Heilbronner *et al.* used the EHM calculations and detailed analysis of the vibrational structure of the spectral bands of diamine LXb to arrive at a conclusion that its two higher levels belong to the bonding (+) and antibonding (−) combinations of *n*(N) orbitals. Strong destabilization of the (+) level was explained by its antibonding interaction with σ(C—C) orbital (TB effect). The energy diagram of molecular orbitals of 1,4-diazabicyclo[2,2,2]octane is shown in Fig. 9.3.

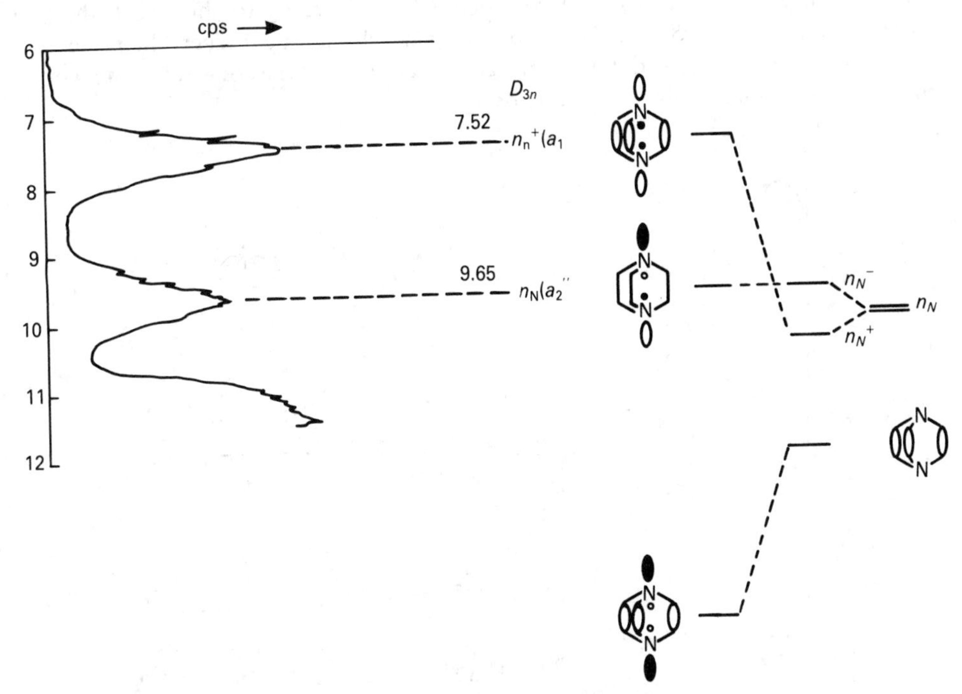


Fig. 9.3. He(I) PE spectrum and scheme of orbital interaction in 1,4-diazabicyclooctane. (From Bock and Ramsey [53])

The PE spectra of other diamines LX also have two bands of nitrogen lone pairs in the low-energy region. The first and second ionization potentials of these compounds are lower than in compound LXb:

| Diamine LX | (a) | (b) | (d) | (e) | (f) |
|---|---|---|---|---|---|
| $IP_1$ (eV) | 6.85 | 7.52 | 6.64 | 6.75 | 6.75 |
| $IP_2$ (eV) | 7.90 | 9.65 | 7.77 | 7.87 | 7.80 |

Bicyclic diamines turned out, on the whole, to be the most easily ionizable saturated organic molecules. They readily form cation radicals by their oxidation, but the stability of these cation radicals is quite different. The stability of quinuclidine cation radical was first estimated to be very high, but the stability of the diamine cation radical under the same conditions turned out to be much higher: half-life of cation radicals of diamines LXb,a, and e (measured in acetonitrile at 25°C) was 1 second, 1 day, and 1 year respectively. The conditions for the formation of frontier occupied MOs are probably even more favourable in these diamines: two electrons in their cation radicals are located in the bonding and one electron in the antibonding combination of $n$(N) orbitals. This filling of MOs becomes possible if the said compounds have a 'normal' sequence of the higher occupied MOs in agreement with the through-space interaction of the orbitals.

The values of the two first ionization potentials of some diamines are given in Fig. 9.4. The TB effect prevails in the compound Xb: according to this effect, the (+) combination of $n$(N) orbitals of two nitrogens should be higher than the (−) combination for $(CH_2)_2$ and $(CH_2)_4$ bridges. By contrast, if the conformation favours, the (−) level is higher for the $(CH_2)_3$ bridge.

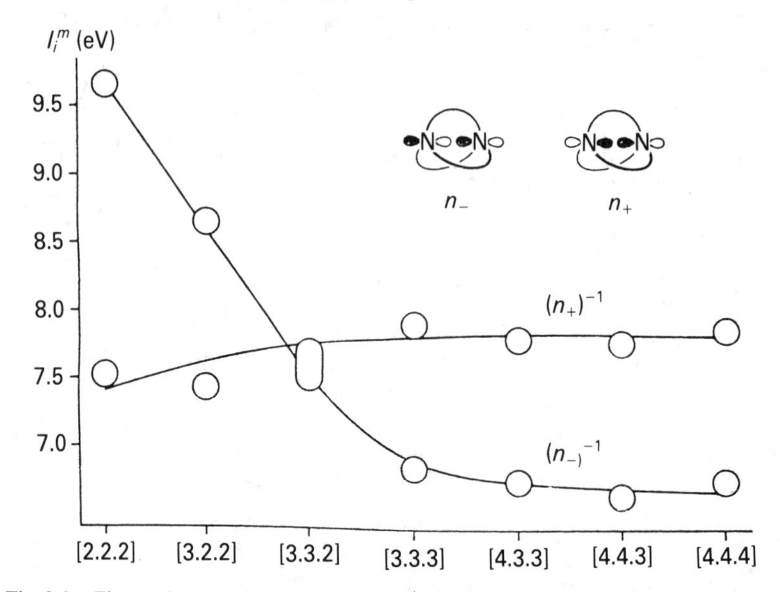


Fig. 9.4. First and second ionization potentials in He(I) PE spectra of diazabicycloalkanes LVI and estimation of TP and TS interactions in their molecules. (From Alder *et al.* [404])

As the length of the hydrocarbon bridge connecting two nitrogen atoms increases, the probability of a favourable conformation of the studied bicyclic diamines decreases rapidly. The geometry of large bicycles is such that the conformation that is necessary for the interaction of adjacent $\sigma$ orbitals is not realized: beginning with diamine LXa, lone pairs of two nitrogens interact through space only.

The electron structure of 1,5-diazabicyclo[3,3,2] decane LXc is especially interesting since the (+) and (−) combinations of its $n$(N) orbitals are close to degeneracy. The cation radical of this compound therefore has a very short life span in solution and the attempt to record its ESR spectrum was a failure. The cation radical of diamines LXa, b, d, and f are, on the other hand, stable: their ESR and electron absorption spectra were successfully recorded [289].

Nitrogen's lone-pair orbital in bicyclic compounds can overlap not only with the similar orbital of the second nitrogen but also with orbitals of other atoms of suitable symmetry [409]. It turned out that there are two ionization bands in the low-energy region of the PE spectrum of 1-azabicyclo[4,4,4]tetradec-5-ene LXII as in the spectra of bicylic diamines (see Fig. 9.5). Comparison of its ionization potentials with those

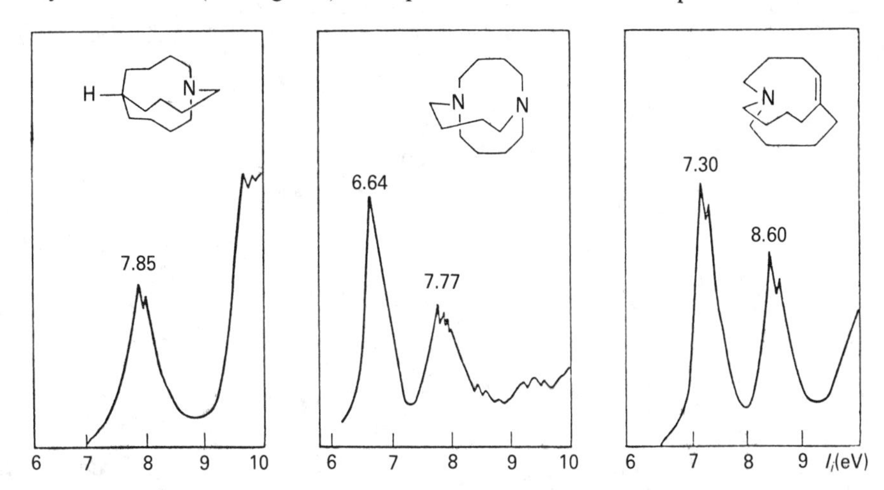


Fig. 9.5. He(I) PE spectra of some bicyclic amines (to estimation of transannular effects in their molecules). (From Alder *et al.* [404])

of related compounds indicates that the higher occupied electron levels in amine LXII are formed by the interaction of nitrogen lone-pair orbitals (its energy level is estimated by the value of the ionization potential of a lone pair of 1-azabicyclo[4,4,4]tetradecane LXI) with the $\pi$(C=C) orbital (its energy is estimated by the first ionization potential of bicyclo[4,4,4]tetradec-1-ene LXIII):

The method of molecular mechanics was used to determine preferred conformations of bicyclic compounds LXI and LXII. Thus, the conformation of alkene LXII with 'external' orientation of nitrogen's lone pair was by 41 kJ/mol less stable than the conformation with 'internal' orientation, which is favourable for overlapping of $n$(N) and $\pi$(C=C) orbitals. The corresponding estimation within the framework of *ab initio* calculations (STO-3G basis set) showed that the interaction of these orbitals occurs almost completely through space and can be estimated as about 0.64 eV.

It is noteworthy that both the electron levels and the chemical behaviour of 1-azabicyclo[4,4,4]tetradec-5-ene are determined by the transannular interaction of nitrogen's lone pair and $\pi$ orbital. This compound behaves actually like an enamine. It is oxidized more readily than its saturated analogues, while on protonation it rapidly forms the salt of saturated propellane LXIV (the product of C protonation). Even ethanol is a strong acid for amine LXII: when dissolved in this solvent, it is completely converted into LXIV in 9 minutes at 25°C [409].:

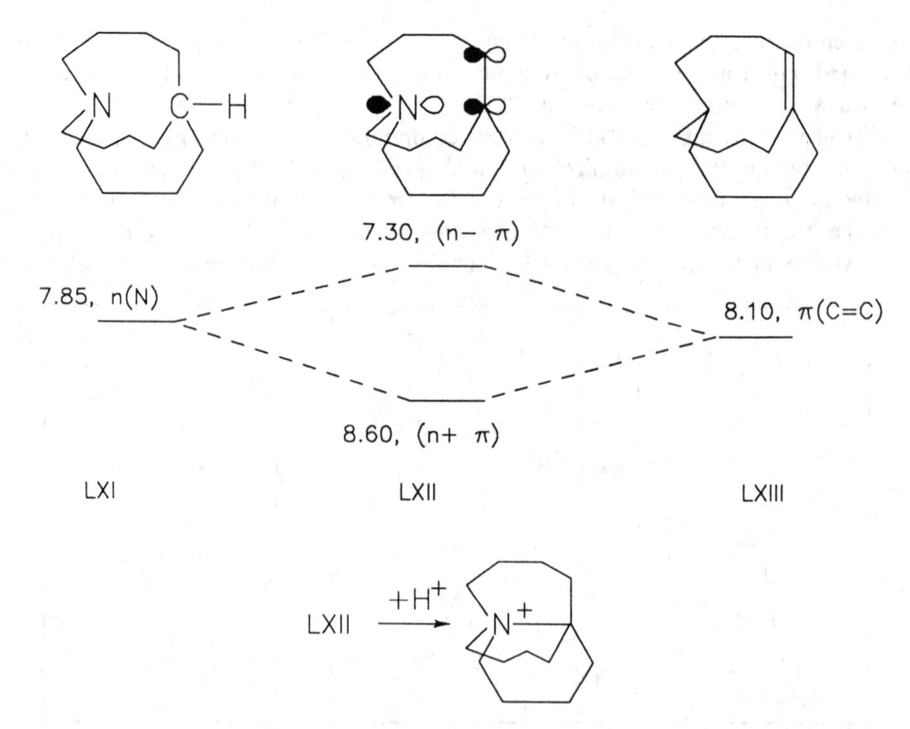


Some alkaloids show similar properties. The half-life of N-methyltetrahydroberberine LXV in ethanol is 2 minutes [1]. The presence of the transannular bond between the nitrogen atom and the carbon of the carbonyl group in the molecule of alkaloid LXVIa was confirmed by the IR spectrum. The frequency of valence vibrations of the carbonyl group at 1675 $cm^{-1}$ corresponds to the carbonyl group of an amide rather than arylketones (the band of the arylketone C=O group vibrations in the IR spectra is located at 1693 $cm^{-1}$). On protonation of alkaloid LXVIa, the band of valency vibrations of the carbonyl group disappears: the proton attaches to the oxygen atom (structure LXVIb).

X-ray structural analysis of alkaloids showed that nitrogen's lone pair in these molecules is usually directed inside the cycle while the nitrogen and carbon atoms are close to one another. Distances of the transannular N ... C bonds in some alkaloids are as follows:

| | | | |
|---|---|---|---|
| Cryptopine | 0.258 nm | Retazamine | 0.164 nm |
| Protopine | 0.255 nm | N-Brosylmitomycin | 0.149 nm |
| Clivorine | 0.199 nm | | |

These alkaloids turned out to be particularly interesting; crystallographic studies of their molecules determined the spatial pathways for the addition of the nucleophile to the carbonyl group. Distances between the nucleophile (amino group) and electrophile (carbonyl group) vary within a wide range in these compounds: from a weak transannular interaction (0.30 nm) to covalent bonding (0.15 nm) [1]. Nearing of the nitrogen and carbon atoms in alkaloid molecules is associated with elongation

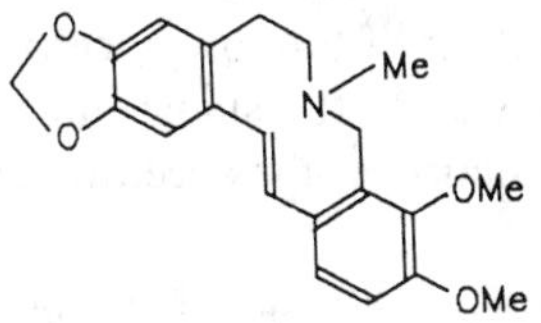

LXV

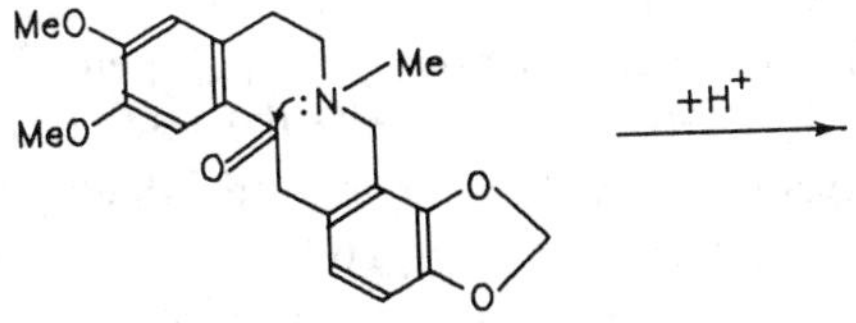

LXVI a

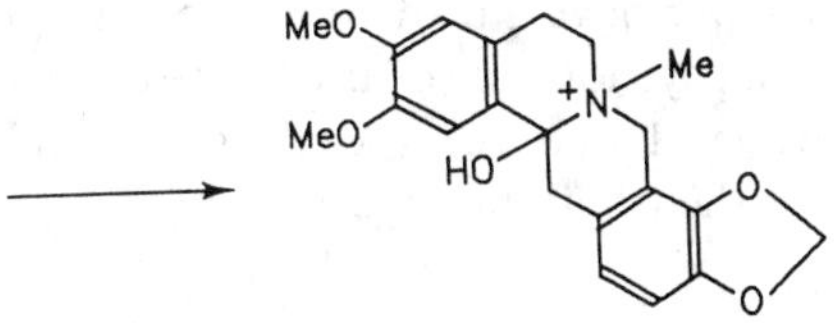

$+H^+$

LXVI b

of the C=O bond and withdrawal of the carbonyl group from the plane. But perhaps the most surprising fact is that the nucleophile approaches the carbonyl group not in the CCC=O plane, and not perpendicular to it, but at an angle of 107° to the C=O bond axis, as if the nucleophile might be 'aware' of the forthcoming conversion of the trigonal configuration of the carbonyl carbon to its tetragonal configuration. Being determined, on the whole, by x-ray studies, the stereochemistry of attachment of the amino group to the carbonyl fragment turned out to be similar to that predicted by *ab initio* calculations of the model reaction between the molecule of formaldehyde and the hydride ion in the gas phase (see Section 8.5.3).

The transannular interactions between amino- and carbonyl groups in amino-ketones were studied by PE spectrometry as well. Pronounced stabilization of the *n*(N) orbital and destabilization of the *n*(O) level was established in aminoketone LXVII (as compared to amine LXVIII and ketone LXIX) [410].

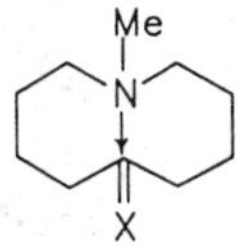

LXVII: X=O
LXVIII: X=NH

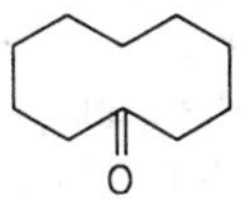

LXIX

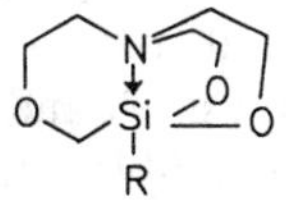

LXX

The silatranes LXX are an interesting illustration of nitrogen's ability to transannular effects. Voronkov *et al.* studied thoroughly the steric and electronic structure of these compounds and their analogues to show that the lone pair of the nitrogen atom interacts in these molecules with the vacant electron level of the silicon fragment [411, 412]. The length of the Si ... N transannular bond in silatranes is about 0.200 nm, which is far shorter than the sum of van der Waals' radii of nitrogen and silicon (0.365 nm). On the whole, the structure of silatranes is a good illustration of the 'work' of the vacant electron levels of the tetrahedral silicon atom. The acceptor properties of the carbonyl group depend on the localization of the low vacant level over the carbon atom, while the low vacant electron levels of the silane molecules are localized over the silicon atom which performs the function of the acceptor centre (see Section 6.2.2 and Fig. 6.6).

The mentioned transannular effects in bicyclic amines and the related compounds are interesting not only in the aspect of structure of organic compounds. The study of psychotropic drugs by PES and quantum chemical calculations shows that the interactions involving amino groups both inside the corresponding molecules and with specific receptors of the central nervous system can be decisive for their pharmacological properties. We have already discussed (Section 3.2) the dependence of pharmacological effects of psychotropic drugs on the energies of frontier molecular orbitals [33, 90]. It is believed that the specific steric and electron structure of silatranes also accounts for their high biological activity.

### 9.1.5 Enamines

In enamines, the amino group is joined with the fragment of an unsaturated hydrocarbon. Almost all enamines exist in the conformation that is not planar because the $sp^3$ hybridized nitrogen atom remains pyramidal in these compounds. Like an unsaturated vinylamine, sterically unhindered enamines mostly take the conformation A, in which the bisector plane of the $NR_2$ fragment is perpendicular to the C=C—N plane [413]:

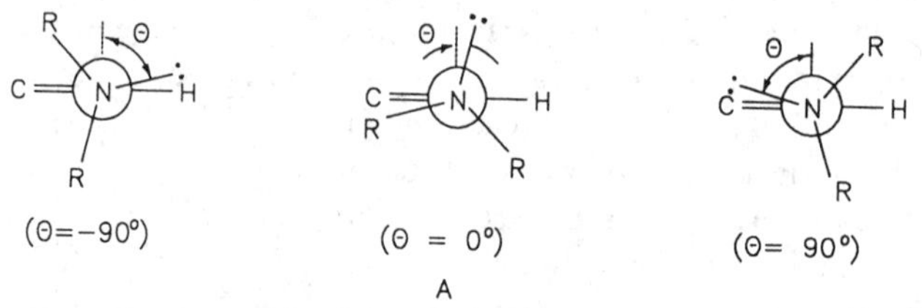

($\Theta = -90°$) ($\Theta = 0°$) ($\Theta = 90°$)

A

Strictly speaking, the nitrogen's lone pair axis somewhat deviates from the appropriate plane in almost all enamines. The deviation of angle $\theta$ varies with substituents at the nitrogen angle, but it remains negative in agreement with the conformational shift toward the cis isomer ($\theta = -90°$). The absolute values of the angle $\theta$ are usually small and one can assume that the $\pi$(C=C) and $n$(N) orbitals in enamines have favourable orientation for mixing. Some steric parameters of enamines are given in Table 9.3 (PRDDO).

The tabulated data agree with the parameters of enamine fragments determined by x-ray studies of the corresponding compounds.

Ionization of $n$(N) and $\pi$ orbitals in PE spectra of enamines is well separated from the region of ionization of $\sigma$ orbitals. This makes it possible to estimate reliably the efficiency of the interaction of the mentioned orbitals in enamine molecules. Under normal conditions, vinylamine is unstable and its PE spectrum was therefore recorded only recently [414]. But the substituted enamines have been studied in detail. The spectra of N-propenyl and N-isobutenylpyrrolidines are shown in Fig. 9.6, while the first and second ionization potentials of many other enamines are given in Table 9.3.

**Table 9.3.** Steric parameters (angle $\theta$ and sum of angles at N atom, in degrees) and first and second ionization potentials (eV) of some enamines

| | $CH_2{=}CH{-}NR_2$ | | $H_3C{-}CH{=}CH{-}NR_2$ | | | $(H_3C)_2C{=}CH{-}NR_2$ | | |
|---|---|---|---|---|---|---|---|---|
| $NR_2$ | $\theta$ | $\sum \angle$ | $I_1$ | $I_2$ | $\Delta I_{1,2}$ | $I_1$ | $I_2$ | $\Delta I_{1,2}$ |
| $NH_2$ | −8.1 | 337.2 | | | | | | |
| $N(CH_3)_2$ | −14.3 | 346.2 | 7.57 | 10.28 | 2.71 | 8.15 | 9.5 | 1.35 |
| $N(CH_2)_2$ | −8.5 | 300.5 | 8.26 | 10.58 | 2.32 | 8.2 | 10.5 | 2.3 |
| $N(CH_2)_3$ | −9.5 | 341.3 | 7.62 | 10.17 | 2.55 | 7.56 | 9.97 | 2.41 |
| $N(CH_2)_4$ | −8.6 | 344.9 | 7.24 | 10.04 | 2.80 | 7.75 | 9.6 | 1.85 |
| $N(CH_2)_5$ | −13.7 | 349.9 | 7.46 | 9.95 | 2.49 | 8.0 | 9.3 | 1.3 |

(From Müller *et al.* [413])

If we assume the difference $\Delta IP_{1,2}$ as a measure of the interaction of *n*(N) and $\pi$ orbitals, it is possible to note especially high donor properties of the N atom of the pyrrolidine ring, which well correlate with many experimental findings.

In the N-propenylaziridine molecule, the $\pi$(C=C) and *n*(N) orbitals are the closest as regards their energies; the most stable conformation of this compound is the conformation A (the angle $\theta$ is about 0°), which is the most favourable for orbital interaction. At the same time, the N atom of aziridines is highly pyramidal and this probably accounts for the least splitting of these first two electron levels.

The frontier occupied MOs of aziridine derivatives are high in their energies. This probably accounts for the observed chemical behaviour of aziridines: while N-vinyl derivatives of secondary amines readily polymerize (both as individual substances and in solution) at room temperature, N-vinylaziridine remains stable even at elevated temperatures. The N atom of the aziridine cycle has relatively low $\pi$ donor properties even when bound with strong acceptors. This is demonstrated by x-ray analysis of N-acyl and N-nitroaziridines, by their chemical and spectral properties [415].

The changes in the geometry and electronic structure of enamines are appreciable on the introduction of two substituents at the $\beta$ position. It can be seen from Table 9.3 that the presence of two methyl groups at the $\beta$ carbon atom of the ethylene fragment leads to a notable decrease in the splitting of two higher electron levels in the corresponding molecules. In the case of a piperidine enamine, $\Delta IP_{1,2}$ decreases, for example, from 2.49 to 1.3 eV. This is explained by steric hindrance to the interaction between *n*(N) and $\pi$(C=C) orbitals in isobutenylamines. The PRDDO calculations show that the structure of these enamines is transformed toward the cis conformer. According to the calculations, the angle $\theta$ in N-(isobutenyl)piperidine is −56.5°, while according to x-ray structural analysis of 1,4-bis(3-methyl-2-buten-2-yl)piperazine, the same angle is −65°.

In the series of aziridine enamines, the transition from the butenyl to the isobutenyl derivative is attended by a considerable decrease in the energy barrier between the conformers: ionization of two conformers can be seen in the PE spectrum [415].

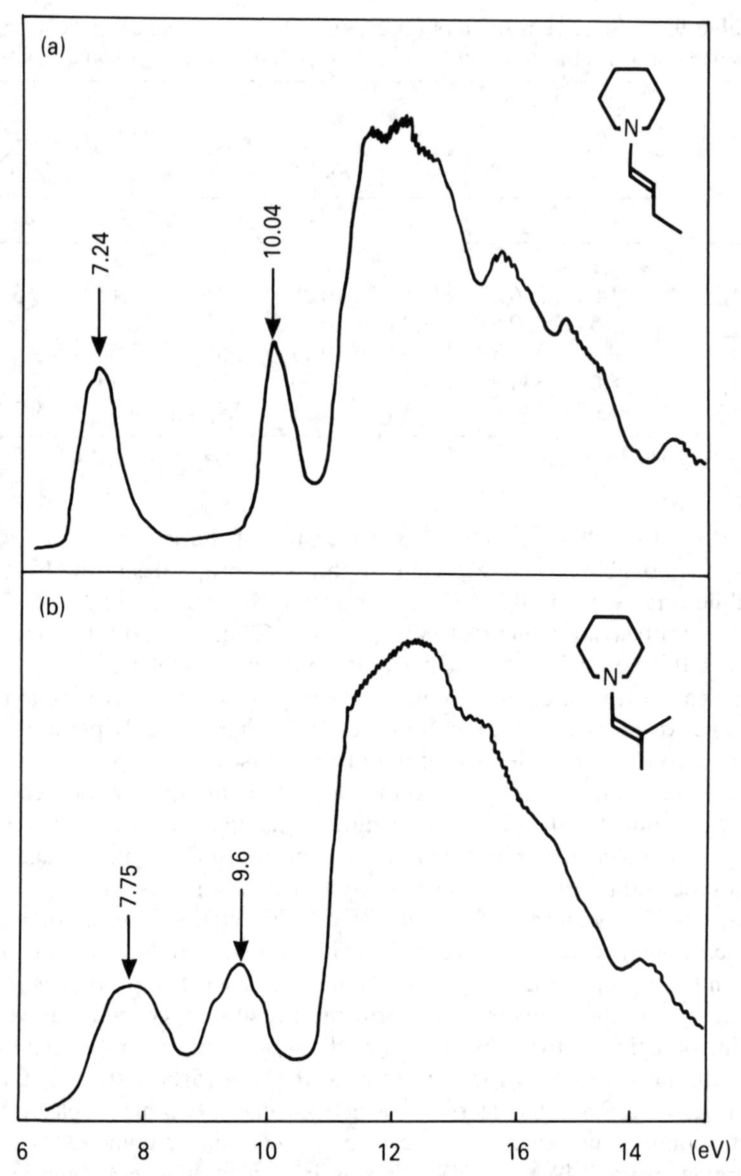


Fig. 9.6. He(I) PE spectra of (a) N-propenyl- and (b) N-(2-methylpropenyl) pyrrolidines (lesser splitting of the first and second bands in spectrum b indicates steric hindrances to the interaction of $n$(N) and $\pi$ orbitals). (From Müller *et al.* [413])

Interesting properties were discovered in azetidine enamines. Substitution of the butenyl group at the N atom by the isobutenyl group decreases splitting of two higher electron levels only insignificantly: from 2.55 to 2.41 eV. According to calculations, the angle $\theta$ in the N-isobutenylazetidine molecule is $-24°$. At this angle, the molecular conformation probably does not rule out the mixing of the $n$(N) and $\pi$(C=C) orbitals.

### 9.1.6 Aniline

Attachment of the benzene ring to the amino group sharply decreases its basic properties: aniline is a million times weaker base than methylamine [8]:

| | $CH_3NH_2$ | $PhNH_2$ | $PhCH_2NH_2$ |
|---|---|---|---|
| pKa($R—\overset{+}{N}H_3$) | 10.62 | 4.58 | 9.34 |

The value of pKa of the conjugated acid of benzylamine suggests that the low basicity of aniline cannot be explained by the negative inductive effect of the phenyl group. Delocalization of the nitrogen lone pair by the mechanism of $n$(N) and $\pi$(Ph) orbital

**Table 9.4.** Ionization potentials and occupied orbitals of aniline (PES data and 4-31G-calculations)

| $IP_i$ (eV) | Molecular orbitals | | | MO, No. and symmetry [417] |
|---|---|---|---|---|
| | $-\varepsilon_i$ (eV) | No. and symmetry (group $C_{2v}$) | Localization* | |
| 8.00 | 7.95 | $3b_1$ | $\pi_2$ | $3b_1$ |
| 9.21 | 9.10 | $1a_2$ | $\pi_3$ | $1a_2$ |
| 10.80 | 11.91 | $2b_1$ | $n_N$ | $2b_1$ |
| 11.39 | 13.23 | $10b_2$ | $B(19)$ | $13a_1$ |
| 11.39 | 13.54 | $11a_1$ | $B(18)$ | $1b_1$ |
| 12.39 | 14.35 | $1b_1$ | $\pi_1, n_N$ | $6b_2$ |
| 14.04 | 15.82 | $9b_2$ | $B(16)$ | $11a_1$ |
| 14.04 | 16.15 | $10a_1$ | $B(15)$ | $5b_1$ |
| 14.04 | 16.21 | $8b_2$ | $B(14)$ | $10a_1$ |
| 15.52 | 17.53 | $9a_1$ | $B(13)$ | |

*$B$ stands for MOs localized over the benzene nucleus. Parenthesized is the No. of MO in benzene orbital set.
(From Kimura *et al.* [38])

interaction is decisive for the electron structure of aniline (see Table 9.4). When estimating this effect it should be remembered that the aniline molecule is not planar. According to microwave spectroscopy, the angle $\theta$ between the axis of the C—N bond and the bisector of the HNH angle is $37 \pm 2°$ [416]. Among other structural parameters of the aniline molecule are: much shorter C—N bond (to 0.1402 nm) compared with the length of this bond in the methylamine molecule (0.1474 nm), the N—H bond distance (0.1001 nm), HNH valence angles (113°), and the endocyclic CC(N)C angle (119°).

Although there is a slight discrepancy in the results of *ab initio* calculations of aniline as regards the sequence of MOs (Table 9.4), the findings of the PE spectra and quantum chemical calculations on the whole indicate effective interaction of the $n$(N) and $\pi$(Ph) orbitals in the aniline molecule. This interaction has been estimated quantitatively. Thus, calculations by the MINDO/3 method gave the following estimation of the nitrogen's $2p\pi$ orbital contribution to the four higher occupied

MOs of aniline (%): 40, 0, 48, and 17. The effect of fluorine on the electron structure of aniline was estimated as well. This estimation is based on the fact that aniline and pentafluoroaniline have the same stable conformation [418, 419].

The results of *ab initio* calculations are in full agreement with the $\pi$ donation N $\rightarrow$ Ph and with attraction of the electron to the N atom through the chain of $\sigma$ bonds [237]. Effective transfer of $\pi$ electron density from the amino group to the benzene nucleus (the results of calculation of the planar molecule are shown on the left, and of the molecule in which the HNH angle is 112°, on the right) is attended by an appreciable decrease in the inversion barrier observed in compounds with the pyramidal nitrogen: in aniline this barrier is only 6.70 against the 18.42 kJ/mol in dimethylamine:

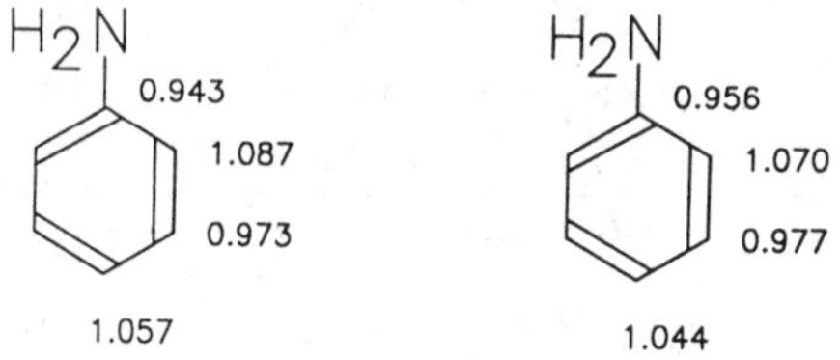

$H_2N$ 0.943 1.087 0.973 1.057

$H_2N$ 0.956 1.070 0.977 1.044

As a result of the $\pi$ electron transfer to the benzene nucleus aniline is highly reactive toward electrophiles. Its activity in bromination is so high that projects were proposed for recovery of bromine from sea water in which it is contained in the concentration of 0.015%. The measured electron affinities of aniline are: $EA_1 = -1.13$, $EA_2 = -1.85$, and $EA_3 = -5.07$ eV [368]. As can be seen from Fig. 9.7, both vacant and occupied benzene degenerate $\pi$ orbitals are split in a similar way by the introduction of the amino group into the benzene molecule: the energy of the orbitals of $b_1$ symmetry increases appreciably; the energy of symmetry $a_2$ orbitals with the node at the site of substituent attachment, practically does not change.

For estimation of the aniline molecular orbitals, it is important to compare ionization potentials measured by PES in the gas phase and the electron transfer energies measured from the electron absorption spectrum of the cation-radical of aniline generated in the solid phase [289].

### 9.1.7 N,N-dimethylaniline and its organoelement analogues

The conformation of N,N-dimethylaniline ensures even more effective mixing of the $n$(N) and benzene $\pi$ orbitals. It was shown by electron diffraction that the molecule of N,N-dimethylaniline in the gas phase is almost planar: the $C(H_3)NC(H_3)$ angle is 116° against the 113° found for the HNH angle in the aniline molecule [416].

Effective removal of degeneracy of benzene $\pi$ orbitals in the N,N-dimethylaniline molecule is confirmed by a much higher flexibility of the first and third bands compared with the position of the second band, which is due to ionization of unperturbed benzene's $\pi$(asymm) orbital [277, 366, 367]. The difference in the first and second ionization potentials is further evidence: it is 1.07 for aniline, 1.30 for N-methylaniline, and 1.52 eV for N,N-dimethylaniline.

The $h$–$\pi$ orbital interaction is much weaker in analogous derivatives of other elements of group VA. The interaction effects are extremely small between the $n$(P)

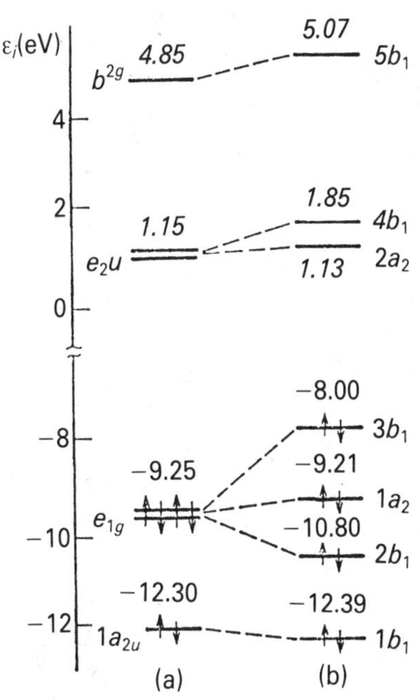


Fig. 9.7. Experimental estimation of energies of occupied (PES) and vacant (ETS) πMOs of (a) benzene and (b) aniline. (Data by Kimura *et al.* [38] and Jordan *et al.* [376])

and π orbitals in, e.g., P,P-dimethyl(phenyl)phosphine and other dialkyl(phenyl) phosphines. This result does not look unexpected if we consider the nature of the phosphorus' lone pair (see Section 9.1.2) and the fact that, according to [420], dialkyl(aryl)phosphines are present in the conformation where the phosphorus' lone pair, the bisector of the $C(H_3)PC(H_3)$ angle and the benzene ring are located in the same plane. The rules of mixing of the *n*(P) and π orbitals in arylphosphines have been discussed in [421–5], and the donor properties of $NH_2$ and $PH_2$ groups in the corresponding carbocations, as determined by *ab initio* quantum chemical calculations, are evaluated in [333]. According to PE spectra and CNDO/2 calculations, the *n*(As) and π(CC) orbitals do not mix at all in dimethyl (vinyl) arsine and dimethyl(phenyl)arsine. The *n* orbital of the arsenic atom and the benzene ring in the latter compound lie in the bisector plane [426]. The data estimating the interaction of fragment orbitals in dimethylaniline and its organoelement analogue molecules are given in Table 9.5.

The absence of a π donor effect of the phosphorus and arsenic atoms on the occupied π electron levels in P,P-dimethyl(phenyl)phosphine and dimethyl(phenyl)arsine agrees with their stabilizing influence on the vacant electron levels. Thus, the acceptor properties toward vacant benzene π orbitals are obvious in the dimethylphosphine group attached to the benzene ring:

According to the HFI constants $a_H$, found for various annular protons from ESR spectra, the unpaired electron in the anion radical of P,P-dimethyl(phenyl)phosphine occupies the transformed benzene π(symm) orbital with $b_1$ symmetry (as in anion

**Table 9.5.** Ionization potentials (PES data), position of long-wave absorption bands (EAS data), and parameters of HOMOs of compounds of group VA elements $PhEMe_2$

| Element | $IP(E)$ (eV) | | $IP(PhEMe_2)$ (eV) | | $H_E$ (eV) | Parameters of $PhEMe_2$ HOMO* | | | | | $\lambda$ (nm) | ($\log \varepsilon$) |
|---|---|---|---|---|---|---|---|---|---|---|---|---|
| | | | | | | $\varepsilon$ (eV) | $c_{ortho}$ | $c_{meta}$ | $c_{para}$ | $c_{EMe2}$ | | |
| N | 8.50 | 7.60 | 9.20 | 10.00 | −1.21 | −7.82 | −0.36 | 0.08 | 0.39 | 0.72 | 251 | (4.15) |
| | | | | | | | | | | | 299 | (3.31) |
| P | 8.60 | 8.45($n$) | 9.20($a_2 + b_1$) | | −0.34 | −8.47 | −0.18 | 0.08 | 0.23 | 0.91 | 251 | (3.54) |
| As | 8.65 | 8.67($n$) | 9.31($a_2 + b_1$) | 10.70–11.0 | 0.0 | — | — | — | — | — | 241.5 | (3.83) |
| Sb | 8.48 | | | | | — | — | — | — | — | 250 | (2.57) |

*Note*: $IP(E)$ – ionization potential of the lone pair $n(E)$ in $Me_3E$ compounds (PES); $IP(PhEMe_2)$ – ionization potentials of N,N-dimethylaniline analogues (PES).

*Parameters of $PhEMe_2$ HOMOs (energies $\varepsilon$ and coefficients $c_\mu$) are calculated by the HMO method with PES parametrization.

(From Traven and Stepanov [62])

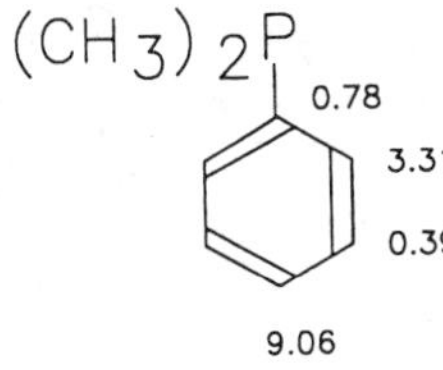

$(CH_3)_2P$

0.78

3.31

0.39

9.06

radicals of benzene derivatives containing substituents with marked acceptor properties) [40].

### 9.1.8 Other arylamines

The substitution of the methyl groups at the nitrogen atom in dimethylaniline by other alkyl groups produces only an insignificant effect on the energy of molecular $\pi$ orbitals. This is confirmed, for example, by estimations of the donor properties of pyrrolidinyl-, piperidinyl-, morpholinyl and dimethylamino groups in the corresponding substituted benzenes. Below are substituents R in Ph—R compounds and frequencies of CT bands in electron absorption spectra of Ph—R complexes with 1,3,5-trinitrobenzene [427]:

| R | $N(CH_3)_2$ | $N(CH_2)_4$ | $N(CH_2)_5$ | 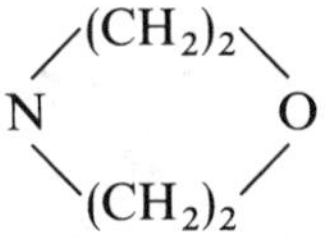 $N{<}^{(CH_2)_2}_{(CH_2)_2}{>}O$ |
|---|---|---|---|---|
| $\lambda_{CT}$ ($cm^{-1}$) | 20 700 | 19 500 | 21 500 | 23 600 |

It can be seen that the N-pyrrolidinyl group interacts with the benzene $\pi_2$ orbital more effectively than the other mentioned groups. The level of the frontier occupied MO of N-(pyrrolidinyl)benzene is therefore the highest. This group produces the same effect in enamine molecules (see Section 9.1.5) as well as in compounds with a much more developed $\pi$ electron structure. This is confirmed, for example, by the comparison of the position of long-wave absorption bands in the spectra of *n*-dialkylaminoazobenzenes [428].

The amino group attached to phenyl or another aryl fragment can also be involved in the intramolecular transannular interactions with an acceptor atom or a group of atoms [1]. But these interactions are lower than in the above aliphatic amines. Thus, the pyramidal character of the nitrogen atom in LXXI is far weaker than in trialkylamines: the sum of the angles at the N atom in the molecule of this compound is 356 against 330–40° in aliphatic amines. The $n$(N)—$\pi$(Ph) interaction is considerable, which is confirmed by the fact that the $r_{N-C(Ph)}$ distance in this compound is only 0.139 against 0.147 nm in aliphatic amines. Since the basicity of the aniline nitrogen is lower, the transannular bond formed is weaker: the N ... CO distance in LXXI is 0.276 nm:

The transannular effects with involvement of amino groups were studied in many other compounds. Transannular orbital interactions were discovered, for example, in perisubstituted naphthalene LXXII. 1,8-bis(dimethylamino)naphthalene LXXII proved to have stronger basicity (pKa of the conjugated acid is 12.34) than aliphatic

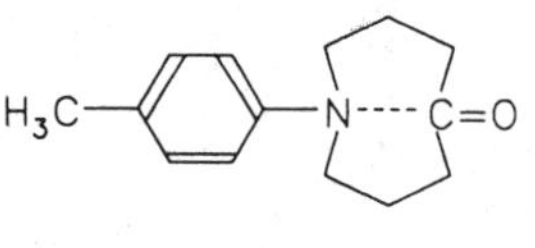

LXXI

amines [393]. The PE spectra of this compound and its 1,5-bis(dimethylamino)-analogue LXXIII suggest that the TS interaction of *n*(N) orbitals of two nitrogens of the adjacent amino groups in the diamine LXXII molecule should be taken into account. This interaction increases the energy of the (−) combination of *n*(N) orbitals and thus increases the basicity of LXXII. The energies of the occupied MOs of 1,8- and 1,5-bis(dimethylamino)naphthalenes are compared in Fig. 9.8. The following properties of the molecular orbitals of the two isomers should be noted: (a) in the 1,8-isomer, the $n^-$ and $n^+$ levels are split to a greater extent than in the 1,5-isomer; (b) although direct TS overlapping of the *n*(N) orbitals has the greatest contribution to the said splitting, the TB interaction with involvement of deeper $\sigma$ orbital of suitable symmetry should also be taken into consideration in these diamines.

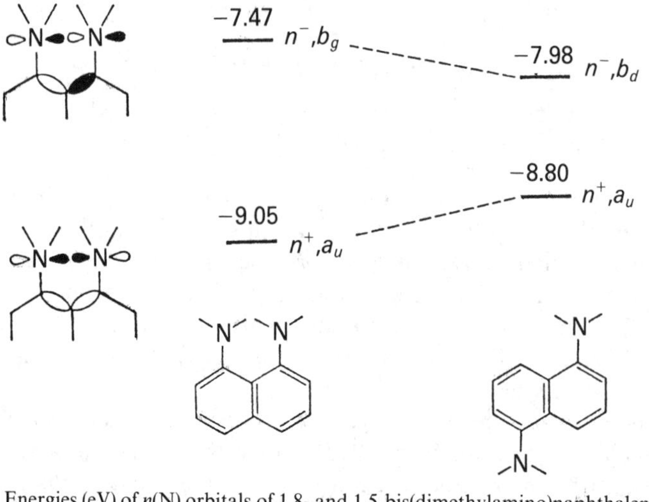

−7.47 $n^-, b_g$
−7.98 $n^-, b_d$
−8.80 $n^+, a_u$
−9.05 $n^+, a_u$

Fig. 9.8. Energies (eV) of *n*(N) orbitals of 1,8- and 1,5-bis(dimethylamino)naphthalene (PES) in the Koopmans' theorem approximation. (From Martin and Mayer [393])

The analysis of the electronic effects in molecules, where appropriate substituents are brought close to one another, should be considered in detail because they are superimposed inevitably on the corresponding steric interactions of the substituents. Dunitz *et al.* studied the structure of 1-amino-8(COR)naphthalenes LXXIV and gave a good example of the estimation of the role of orbital interactions in compounds with sterically close substituents:

It has been established, e.g., that the molecules of the acid LXXIV (R=OH) exist in two crystal forms. The first one (a) corresponds to the zwitter-ion state. Its structure is typical of molecules with steric repulsion of two substituents at positions 1 and 8.

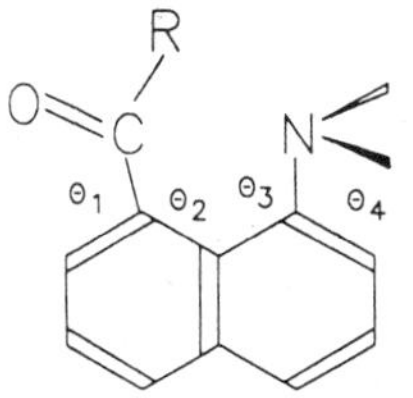

LXXIV: 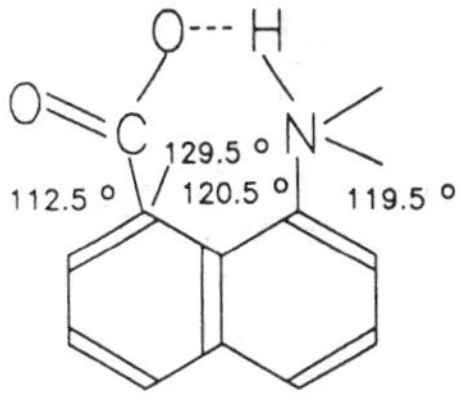 R= OH, a 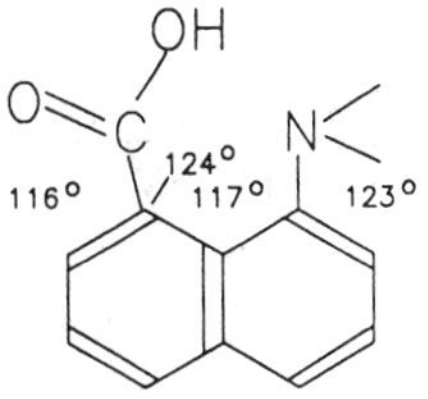 R= OH, b

The other form (b) corresponds to un-ionized amino acids; it is characterized by a marked transannular interaction of amino and carbonyl groups. Similar interactions associated with overlapping of vacant orbitals of the carbonyl fragment and nitrogen lone pair were discovered in other compounds with the general formula of LXXIV. The values of the angle $\theta$ for various R (in degrees) are given below [1]:

| | $\theta_1$ | $\theta_2$ | $\theta_3$ | $\theta_4$ |
|---|---|---|---|---|
| OH | 116 | 124 | 117 | 123 |
| $OCH_3$ | 118 | 122 | 117 | 124 |
| $CH_3$ | 117 | 123 | 116 | 123 |

## 9.2 HYDRAZO COMPOUNDS

### 9.2.1 Aliphatic hydrazo compounds

Nitrogen lone-pair orbitals in hydrazines are brought to the shortest possible distance since they are localized over directly bonded atoms. The conditions of formation of frontier occupied MOs in hydrazine molecules are in many respects similar to those described earlier for diphenyl disulphides. The geometric parameters of hydrazine are as follows [429]: $r_{NN} = 0.1449 \pm 0.0004$ nm; $r_{NH} = 0.1022 \pm 0.0006$ nm; the NNH angle $= 112 \pm 1.5°$.

As in disulphides with restricted rotation about the S—S bond, the dialkylamino groups can rotate about the N—N bond in hydrazines $R_2N—NR_2$:

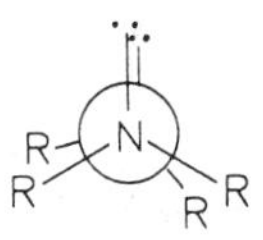

a (sin) 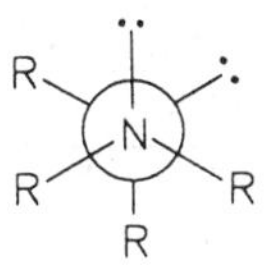 b (gauche) 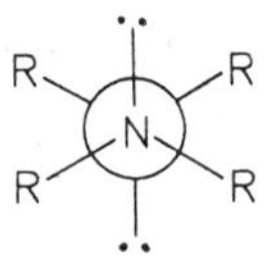 c (anti)

It has been shown by experiment and quantum chemical calculations that the total energy of hydrazine and methyl hydrazine in conformation (b) is minimal, while the energies of the conformations (a) and (c) are 50 and 15 kJ/mol higher [277].

The preferable conformation for hydrazine and methyl hydrazine molecules is

**Table 9.6.** Ionization potentials and higher occupied molecular orbitals of hydrazine (PES data and calculations in the 6-31G basis set)

| $IP_i$ (eV) | Molecular orbitals | | |
|---|---|---|---|
| | $-\varepsilon_i$ (eV) | No. and symmetry (group $C_2$) | Localization |
| 9.91 | 10.82 | 5*a* | $n^+$(N) |
| 10.64 | 11.15 | 4*b* | $n^-$(N) |
| 15.61 | 16.62 | 4*a* | $\sigma_{NN}$ |
| 16.66 | 17.81 | 3*b* | $\pi^-_{NH_2}$ (pseudo) |
| (16.7) | 18.07 | 3*a* | $\pi^+_{NH_2}$ (pseudo) |

(From Kimura *et al.* [38])

confirmed by PES data, given in Table 9.6 [227, 430]. Accurate assignment of the two first potentials of hydrazine to ionization of MOs formed as linear combinations of $n$(N) orbitals:

$$n^+(N) = 1/\sqrt{2}(n_1 + n_2), \quad n^-(N) = 1/\sqrt{2}(n_1 - n_2)$$

is the basis for the study of their steric structure using PES data. The difference of the first and second ionization potentials in the hydrazine spectrum determines splitting of the $n^+$(N) and $n^-$(N) hydrazine levels and, in turn, depends unambiguously on the angle $\theta$. An example of such dependence is given in [277]:

$$\Delta IP = 2.20 \cos\theta - 0.25 \tag{9.1}$$

In agreement with this equation, at $\theta = 90°$, overlap of adjacent nitrogen lone pairs is ruled out, while the value of $\Delta IP$ turned out to be minimal. By contrast, at $\theta = 0$ and 180°, the overlap of the adjacent lone pairs is maximum and splitting of these MOs is the highest.

When estimating the effect of substituent R on the electron structure of N,N,N′,N′-substituted hydrazines, it is important to note that the potential energy curve of molecules relative to angle $\theta$ depends only insignificantly on the substituent. It is only the gauche form that is always present as a stable conformer with various R (H, Alk, $CF_3$, $SiH_3$, etc.) [430]. Table 9.7 contains the values of ionization potentials of lone pairs in the hydrazines $R_2N{-}NR'_2$.

**Table 9.7.** Ionization potentials of some tetra-alkylhydrazines $R_2N{-}NR'_2$

| $R_2$ | $R'_2$ | $IP_1$ (eV) | $IP_2$ (eV) | $\Delta IP_{1,2}$ (eV) |
|---|---|---|---|---|
| $H_2$ | $H_2$ | 9.90 | 10.75 | 0.85 |
| H, Me | H, Me | 9.00 | 9.73 | 0.73 |
| $Me_2$ | $Me_2$ | 8.27 | 8.83 | 0.55 |
| $Et_2$ | $Me_2$ | 8.10 | 8.63 | 0.53 |
| Me, $CMe_3$ | Me, $CMe_3$ | 7.67 | 8.17 | 0.50 |

(From Nelsen and Buschek [430])

The mutual position of lone pairs of two nitrogens in cyclic and bicyclic hydrazines is fixed, and the angle $\theta$ varies from 30 to 164° [431]. Splitting of the energy levels of (+) and (−) combinations of the nitrogen lone pairs increases significantly to values from +1.57 to −2.32 eV. Maximum values of lone pair splitting are seen in 1,6-diazabicyclo[4,4,0]decane LXXV and in the central hydrazine fragment of 1,3,4,6,8,9-hexaazabicyclo[4,4,0]decane LXXVI: $\Delta IP_{1,2}$ are −2.35 and −2.40 eV respectively. This agrees with the value of the angle $\theta$ in these compounds (180°):

LXXV LXXVI

Similar dependences between the splitting of the frontier occupied MOs in hydrazines and their steric structure, were calculated for some other hydrazines. Thus, $\Delta\varepsilon = \varepsilon_{n^-} - \varepsilon_{n^+}$ values were calculated by the MINDO/2 method and the following equation was derived for the known conformations of some hydrazines [432]:

$$\Delta\varepsilon = 2.17 \cos \theta - 0.35 \qquad (9.2)$$

Hydrazine and tetramethyl hydrazine were calculated by the INDO method. It was found that the reversion of orbitals occurs at 90° as shown in Fig. 9.9, where the dependence of $\Delta\varepsilon$ on $\theta$ is estimated.

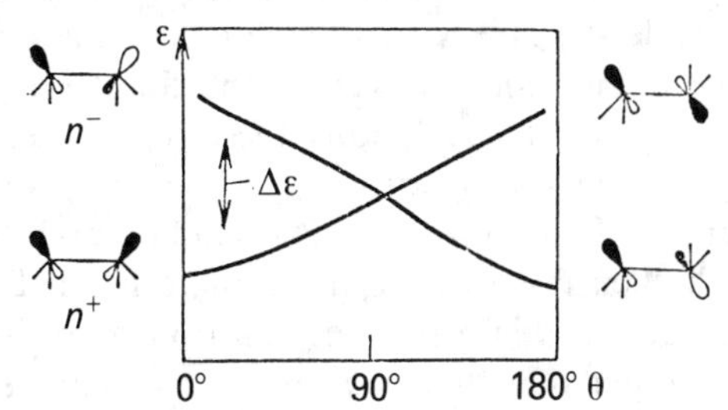


Fig. 9.9. Dependence of interaction effects of $n$(N) orbitals on dihedral angle $\theta$ in a hydrazine molecule. (From Klessinger and Rademacher [277])

The reliability of the PES data on the dependence between the character of lone-pair splitting in hydrazines and their steric structure is positively confirmed by the analysis of the steric structure of 1,2,4,5-tetramethyltetrahydro-1,2,4,5-tetrazine LXXVII [433]. It follows from the PMR spectrum that this compound has conformation (a) with axial and two equatorial methyl groups, i.e. one of the (a)–(c) conformations:

LXXVII: a b c

Selection between these conformers was made on the basis of the PE spectrum: four ionization bands of about the same intensity with $\Delta IP$ splitting of $-2.47$ eV ($\theta = 180°$) and $+0.55$ eV ($\theta = 70°$) were assigned to diequatorial and diaxial N,N-dimethylhydrazine fragments of tetrazine in the conformation (b).

### 9.2.2 Hydrazobenzenes

Hydrazobenzene and its analogues have high tendency to oxidation with formation of coloured azo compounds and are easily rearranged in the presence of acids.

Rotation about the N—N bond in hydrazobenzenes might be expected to have the same torsion potential as in aliphatic hydrazines. But the situation becomes more complicated in hydrazobenzenes due to the presence of the *n*(N) and *π*(Ph) orbital interaction and possible spatial interaction between two benzene rings. The corresponding effects for hydrazobenzene LXXVIII were considered within the framework of two conformations, (**a**) and (**b**) [434]:

LXXV: a b

Objective estimation of their stabilities (in addition to rotation about the N—N and C—N bonds) requires accounting for spatial interaction between the halves of the molecules. Parameters of the aniline molecule were used for calculation of phenylamino groups, while the N—N bond distance was assumed to be 0.145 nm. Optimum values of the torsion angle $\theta$ were found for the (a) form; these are 20–150° and 280–310°. According to calculations, the form (b) is stable at $\theta = 180°$ (the anti form) and 90 or 270° (the gauche form).

Maximum total energies of form (a) at $\theta = 200°$ and of form (b) at 0° are explained by the interaction of the benzene rings in these conformers. The interaction of the H(N) protons and the adjacent benzene rings contributes to the total energies. The results [434] can be compared with the *ab initio* calculations of hydrazobenzene [237].

A qualitative diagram of the hydrazobenzene molecular orbitals is shown in Fig. 9.10. The scheme is based on splitting of all levels of MOs of two aniline fragments in the hydrazobenzene molecule. The degree of splitting is proportional to the overlap of the two halves of the molecule: the orbitals with an appreciable contribution of the atomic *n*(N) orbitals are split to a greater degree than the orbitals incorporating more remote fragments, i.e. the $\pi$ and $\sigma$ orbitals. The first and third bands in the spectra of hydrazo compounds are split. The energy levels of unmixed $\pi$ MOs (benzene $\pi_2$ orbitals in particular) remain unchanged irrespective of the conformation.

The corresponding effects are especially overt in the spectra of cyclic hydrazines in which the most favourable steric conditions for the *n*(N)–*n*(N) mixing are fixed (the spectrum of 1,2-diphenylpyrazolidine is shown in Fig. 9.11 by way of illustration).

The steric structure of some hydrazobenzenes was estimated by the degree of band splitting in PE spectra (Table 9.8). Use was made of Eq. (9.3), which is similar to that

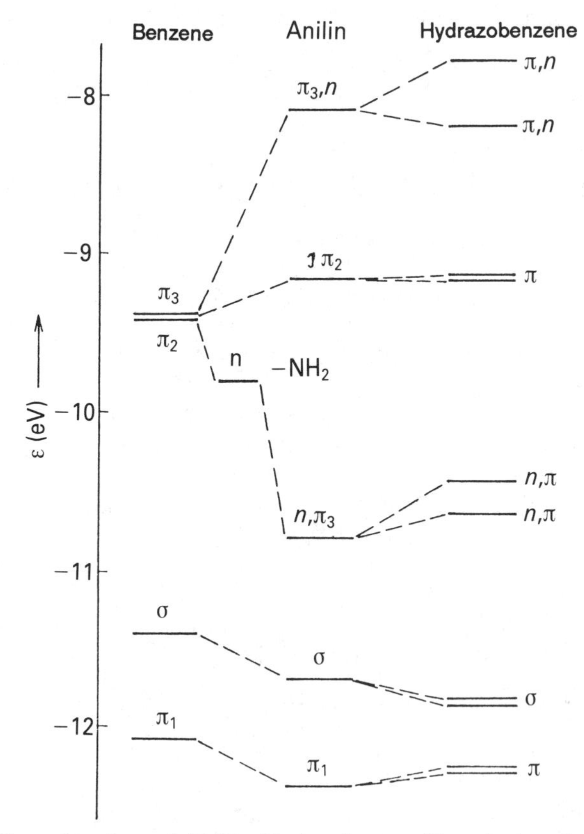


Fig. 9.10. Formation of occupied MOs of hydrazobenzene. (From Rademacher *et al.* [434])

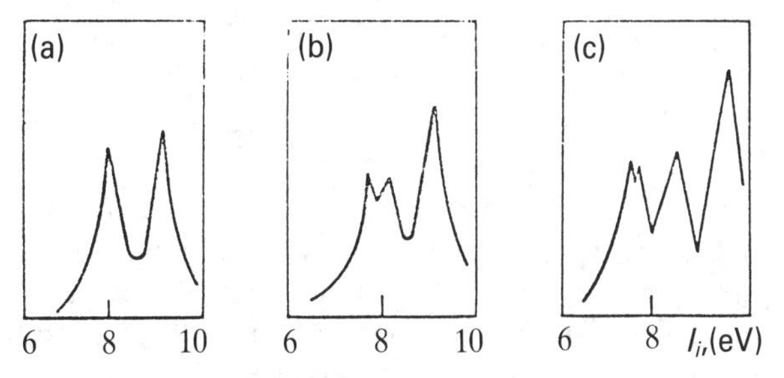


Fig. 9.11. Estimation of $n$(N) orbitals interaction in hydroazobenzenes (PES): splitting of the first band increases in transition from aniline ((a) the first band is not split) to hydrazobenzene ((b) the first band is split insignificantly), and further to 1,2-diphenylpyrazolidine ((c), splitting of 0.6 eV). (From Rademacher *et al.* [434])

derived earlier for aliphatic hydrazines (9.2):

$$\Delta IP = a \cos \theta + b \tag{9.3}$$

One more property of hydrazobenzenes was discussed from the orbital structure

**Table 9.8.** Ionization potentials (eV) and steric structure of some hydrazobenzenes

| Hydrazo compound | $IP_1$ | $IP_2$ | $IP_3$ | $IP_4$ | $IP_5$ | $\triangle IP_{1,2}$ | $\triangle IP_{4,5}$ | Angle $\theta°$ (a) | Angle $\theta°$ (b) |
|---|---|---|---|---|---|---|---|---|---|
| Hydrazobenzene | 7.78 | 8.19 | 9.19 | 10.45 sh. | 10.66 | 0.41 | 0.21 | 66 | 66 |
| N,N′-dimethyl-hydrazobenzene | 7.30 | 7.80 | 8.95 | 10.00 | 10.25 sh. | 0.50 | 0.25 | 60 | 62 |
| N,N′-diisopropyl-hydrazobenzene | 7.20 | 7.53 | 8.87 | 9.77 | 9.95 sh. | 0.33 | 0.18 | 72 | 68 |
| 1,2-diphenylpy-razolidine | 7.50 | 8.10 | 9.10 | 9.82 | 10.40 | −0.60 | −0.58 | 136 | 130 |
| 2,3,-diphenyl-trans-decahydro-phthalazine | 7.01 | 7.51 | 8.81 | 9.60 | 9.75 | 0.50 | 0.15 | 60 | 71 |

Note: (a) Angle values were calculated from Eq. (9.3): $A = 0.90$ eV; $B = 0.05$ eV.
(b) Angle values were calculated from Eq. (9.3): $A = 0.75$ eV; $B = -0.10$ eV.
(From Rademacher *et al.* [434])

standpoints, viz. the applicability of the Woodward–Hoffmann rule to benzidine rearrangement [231]. The formation of benzidine:

$$\text{Ph—NH—NH—Ph} \rightarrow H_2N\text{—}C_6H_4\text{—}C_6H_4\text{—}NH_2$$

from hydrazobenzenes could be regarded as a [5,5]-sigmatropic shift: the signs of HOMO of two ion radicals formed by rupture of the N—N bond shows that this synchronous rearrangement to benzidine might be symmetry-allowed:

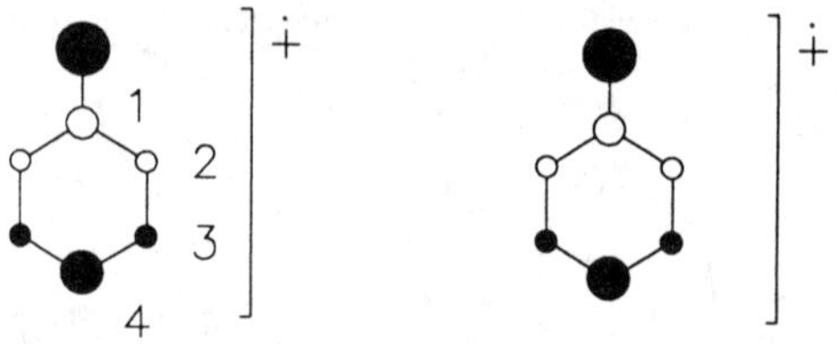

But as in Claisen's rearrangement, the signs of these HOMOs permit 2,2′-bonding as well. At the same time, examples of formation of 2,2′-diaminodiphenyl in benzidine rearrangement are extremely rare. It is quite possible that the analysis of MO coefficients at the corresponding fragment atoms can give additional objective information to explain why the [5,5] shift completely inhibits the [3,3] shift in this unusual reaction: at position 4, the frontier MOs of aniline have the coefficient 0.495, while at position 2 the value is only 0.378.

## 9.3 AZO COMPOUNDS

As distinct from amines, azo compounds contain $sp^2$ hybridized nitrogen atoms in their molecules. These compounds have specific properties: (a) aliphatic azo compounds are highly reactive – they split off the nitrogen molecule and generate active particles, radicals, biradicals, or carbenes; (b) aromatic azo compounds absorb light in the visible region of the spectrum and are the representatives of the most common class of organic dyes.

### 9.3.1 Azomethane and other azoalkanes

Since the nitrogen atoms in azo compounds are attached at the double bond, the CN=NC fragment is planar. Prevalence of cis or trans forms depends on the nature of the substituents at the nitrogen atoms. Azomethane exists in the form of the trans isomer:

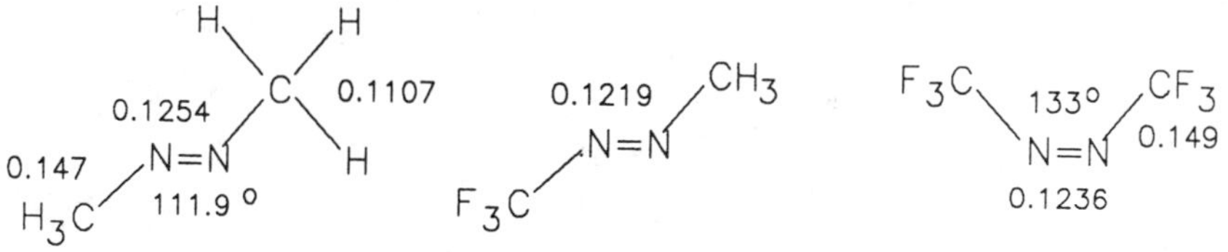

$H_3C$–N=N–$CH_3$ (H, H, H on C): 0.147; 0.1254; 0.1107; 111.9°

$F_3C$–N=N–$CH_3$: 0.1219

$F_3C$–N=N–$CF_3$: 133°; 0.149; 0.1236

1,1,1-Trifluoroazomethane also has the trans configuration while hexafluoroazomethane has the cis form. Diazene $N_2H_2$ exists in the trans configuration and difluorodiazene, in the cis form.

The properties of the substituent determine the CNN angle in azo compounds: in trans azomethane, this angle is under 120° (characteristic of the $sp^2$ hybrid state); it is actually 111.9° (electron diffraction in the gas phase). But the same angle in hexafluoroazomethane increases to 133°. In diazene, the HNN angle is only 106.8° [435].

The PE spectra of azomethanes were recorded first by Heilbronner *et al.* [436, 437]. The assignment of the occupied MOs of azomethane was later confirmed by the study on the perfluoro effect and by *ab initio* calculations [179].

The data on the electron structure of other azoalkanes, both linear and cyclic, were also obtained [236]. These data were compared with $\sigma$ constants, with the values of the dihedral angle between nitrogen lone pairs, with the energies of $n \rightarrow \pi^*$ transition, and with the ability of azo compounds to split off the nitrogen molecule on heating [438].

Figure 9.12 illustrates the effect of consecutive $H \rightarrow CH_3$ substitution in diazene LXXIX on the electron structure of the azo compound. Incorporation of the methyl groups does not change the sequence of MOs but destabilizes the higher electron levels (see Table 9.9). Destabilization of $\pi$ and $n^+$ levels is more pronounced than destabilization of the $n^-$ level: in the transition from diazene to azomethane, the difference of the energies of $n^+$ orbitals is 2.7 eV, while the difference of energies of $n^-$ orbitals is only 1.3 eV. It is supposed that the $\pi$ and $n^+$ MOs with low energies interact with the $\sigma$ orbitals more intensively and are therefore more sensitive to the introduction of the methyl groups than the frontier $n^-$ orbital. The degree of splitting of $n^-$ and $n^+$ orbitals therefore decreases in the series of diazenes LXXIX: $a = 5.0$; $b = 3.8$; $c = 3.6$ eV.

The sequence of higher occupied MOs in aliphatic azo compounds is the same as in azomethane: HOMO is the (−) combination of lone pairs of two nitrogen atoms, next follow the $\pi$(N=N) orbital and the (+) combination of nitrogen lone pairs. The

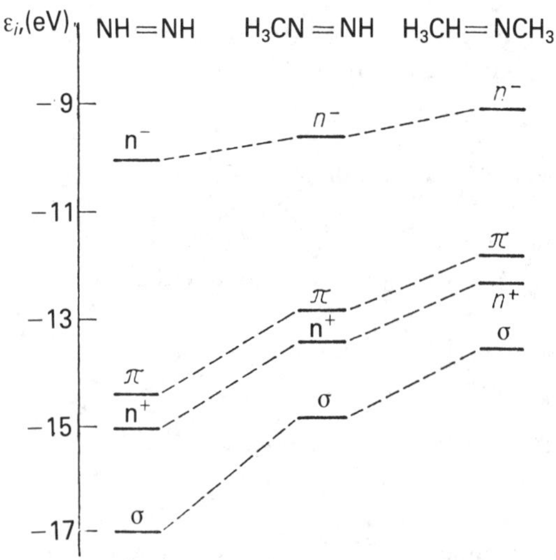


Fig. 9.12. Energy of diazene, methyldiazene and azomethane HOMOs (PES) in Koopmans' theorem approximation. (From Haselbach and Heilbronner [436] and Haselbach and Schmelzer [437])

**Table 9.9.** Ionization potentials and occupied MO of diazenes (PES data and MO LCAO calculations)

| H—N=N—H (a) | | $CH_3$—N=N—H (b) | | $CH_3$—N=N—$CH_3$ (c) | |
|---|---|---|---|---|---|
| $IP_i$ (eV) | MO | $IP_i$ (eV) | MO | $IP_i$ (eV) | MO |
| 10.02 | $4a_g, n^-$ | 9.57 | $a', n^-$ | 8.7 | $n^-$ |
| 14.39 | $1a_u, \pi$ | 12.9 | $a'', \pi$ | 8.98 | $\pi$ |
| | | | | 11.38 | |
| | | | | 11.53; 11.68 | |
| 15.03 | $3b_u, n^+$ | 13.4 | $a', n^+$ | 12.3 | $n^+$ |
| 16.9 | $3a_g$ | 14.7 | $a', \sigma$(CN) | 13.5 | |
| | | 15.6 | $a'', \pi$(CN) | 14.6 | |
| | | 16.7 | $a', \sigma$(NN) | 15.8 | |

(From Chong *et al.* [435] and Haselbach and Heilbronner [436])

frontier occupied MO in all diazenes is located much higher than subsequent orbitals: the difference between the first and second ionization potentials is not less than 2-3 eV. The second bands are broad because of overlapping ionizations of $\pi$(N=N) and $n^+$ orbitals even in the spectra of the simplest azo compounds. The overlap of the appropriate bonds is even greater in the spectra of higher azoalkanes. This should be remembered when estimating the second and third ionization potentials of the corresponding compounds (Table 9.10). As in methyldiazenes, the highest occupied electron levels of azoalkanes are destabilized consecutively with enlargement of the alkyl group. Owing to the absence of trustworthy calculations, the information on deeper occupied orbitals is insufficient. It can be stated that ionization of $\sigma_{NN}$, $\sigma_{CN}$, and $\sigma_{CH}$ orbitals is seen in the region of 13–17 eV.

**Table 9.10.** Ionization potentials of some trans-azoalkanes R—N=N—R (PES) and their assignment

| R | $IP_1, n^-$ | $IP_2, \pi(N{=}N)$ | $IP_3, n^+$ |
|---|---|---|---|
| $CH_3$ | 8.98 | 11.84 | 12.3 |
| $C_2H_5$ | 8.77 | 11.43 | 11.79 |
| n-$C_3H_7$ | 8.61 | (11.36) | (11.78) |
| $CH(CH_3)_2$ | 8.47 | 11.1 | 11.5 |
| $C(CH_3)_3$ | 8.20 | 10.83 | 11.28 |
| $C(CH_3)_2CH_2C(CH_3)_3$ | 8.00 | 9.6 | (10.96) |

(From Houk *et al.* [438])

The properties of the organoelement analogues of azoalkanes, e.g., silicon- and germanium-containing diazenes, should be noted [439]. According to x-ray structural analysis, the attachment of the trimethylsilyl group to the azo group shortens considerably the N=N bond:

| | $N_2$ | $Me_3CN{=}NCMe_3$ | $Me_3SiN{=}NSiMe_3$ |
|---|---|---|---|
| $r_{N=N}$ (nm) | 0.111 | 0.123–0.127 | 0.117 |

This fact indicates a considerable electron-donating effect of the trimethylsilyl groups. Increasing electron density over the azo group in silicon-containing diazenes is demonstrated also by the CNDO calculations:

$Me_3C$(0.14)–N(−0.12)=N–$CMe_3$ $\qquad$ $Me_3Si$(0.53)–N(−0.24)=N(−0.09)–$CMe_3$(0.14) $\qquad$ $Me_3Si$(0.53)–N(−0.20)=N–$SiMe_3$

From the analysis of PE spectra it follows that subsequent replacement of the $CMe_3$ group by the $SiMe_3$ group destabilizes the HOMO of azo compounds by an average of 0.5–0.6 eV. It can be seen from Table 9.11 that the first ionization potential of bis(trimethylsilyl)diimine is 1.1 eV lower than the first ionization potential of its isoelectronic analogue bis(tert-butyl)diimine.

The increase in the energies of frontier occupied MOs of silicon-containing diazenes is attended by a considerable decrease in the energies of their frontier vacant MOs. A substantial long-wave shift of the $n^- \rightarrow \pi^*$ transition in the electron absorption spectra is an indirect confirmation of this suggestion: from Table 9.11 it follows that the decrease in the $h\nu$ energy is more pronounced than the increase in the HOMO levels as recorded by PES.

The ability of vacant $\pi^*$ orbitals located between various elements to accept electrons is discussed in [380]: electron affinities according to ETS, and ionization potentials according to PES of the ethylene and some tert-butyl derivatives, whose structures are shown in Fig. 9.13, are compared. The frontier electron levels of ethylene (the key molecule) are shown on the left. The replacement of two hydrogens at the 1,1 position of ethylene by tert-butyl groups destabilizes considerably (by about 1.5 eV) the HOMO and slightly destabilizes (by 0.22 eV) the LUMO.

**Table 9.11.** Ionization potentials (eV), position ($\lambda$, nm) and energies ($h\nu$, eV) of long-wave absorption bands of azo compounds R—N=N—R′

| R | R′ | $IP_1$ | $IP_2$ | $IP_3$ | $IP_4$ | $\lambda(\varepsilon)$ | $h\nu$ |
|---|---|---|---|---|---|---|---|
| $CMe_3$ | $CMe_3$ | 8.2 | 10.9 | 11.5 | 12.5 | 367.6(12) | 3.37 |
| $CMe_3$ | $SiMe_3$ | 7.6 | 10.4 | 11.0 | 12.6 | 500.0(9) | 2.48 |
| $SiMe_e$ | $SiMe_3$ | 7.1 | 10.6 | 12.45 | 13.5 | 784.3(5) | 1.58 |

(From Bock [439])

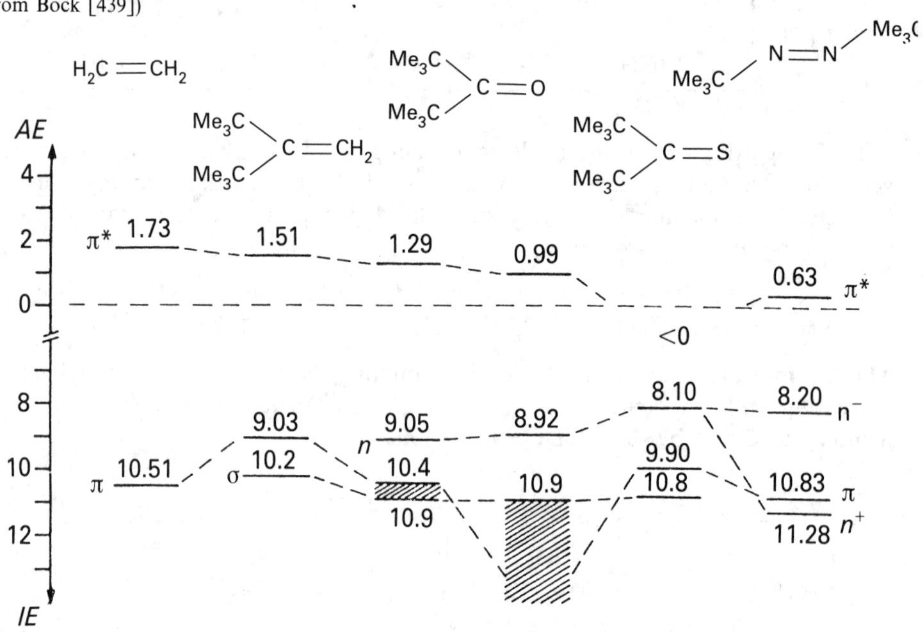


Fig. 9.13. Energy diagram of occupied (PES) and vacant (ETS) electron levels in some unsaturated compounds. (From Modelli *et al.* [380])

Substitution of the $CH_2$ fragment in ethylene by a more electronegative NH group stabilizes both frontier MOs. The occupied frontier orbital is stabilized by about 1.4 eV and the vacant FO by 0.22 eV. Replacement of the second $CH_2$ fragment by a nitrogen atom stabilizes the lowest vacant level to a greater extent: the first value of electron affinity of di(tert-butyl)diazene is −0.63 eV. Stabilization of a frontier vacant electron level is significant in the transition from benzene to azines as well. It is also attended by replacement of the $sp^2$ carbon atom by the $sp^2$ nitrogen atom (see Chapter 11).

### 9.3.2 Diazirines

Diazirines are the simplest azo compounds in which the azo group is incorporated into a ring. These compounds are highly reactive; they readily decompose with evolution of $N_2$ and formation of carbenes. The electron structure of diazirine is determined by the replacement of the 'ethylene' carbons by the $sp^2$ nitrogen atoms [440] (see Table 9.12). The cyclopropene molecule has two occupied orbitals of $\pi$

symmetry: HOMO, the $2b_2$ orbital localized over the region of the double C—C bond, and the $1b_2$ orbital formed mostly by the (C—H) orbitals of the methylene group.

**Table 9.12.** Ionization potentials and higher occupied MOs of cyclopropene and diazirine (PES data and MO LCAO calculations)

| Cyclopropene | | | Diazirine | | |
|---|---|---|---|---|---|
| $IP_i$ (eV) | MO | | $IP_i$ (eV) | MO | |
| | No. and symmetry (group $C_{2v}$) | Localization* | | No. and symmetry (group $C_{2v}$) | Localization* |
| 9.86 | $2b_2$ | $0.526\pi$(C=C) | 10.75 | $3b_1$ | $-0.228n^-$(N) |
| 10.89 | $3b_1$ | $-0.559\sigma$(C—C) | 13.25 | $2b_2$ | $0.352\pi$(N=N) |
| 12.7 | $6a_1$ | $0.348\sigma$(C—C) | 14.15 | $6a_1$ | $-0.001n^+$(N) $0.057\sigma$(C—N) |
| 15.09 | $1b_2$ | $0.288\pi(C_{(3)}$—H) | 16.5 | $5a_1$ | $0.186\sigma$(N—N) |
| 16.68 | $5a_1$ | $0.268\sigma(C_{(2)}$—H) | 17.5 | $1b_2$ | $0.210\pi$(C—H) |

*The value of the fragment orbital coefficient in the MO is given.
(From Robin *et al.* [440])

Diazirine has the same number of $\pi$ electrons as the cyclopropene but owing to the higher electronegativity of nitrogen both its occupied $\pi$ electron levels are much stabilized: the highest level from $-9.86$ to $-13.25$ eV, and the lowest from $-15.09$ to $-17.5$ eV. Furthermore, compared with the azomethane molecule, the $\pi$(N=N) orbital and combinations of the nitrogen lone pairs in the diazirine molecule have far lower energies.

The energy levels of the diazirine, difluorodiazirine, cyclopropene, and cyclopropane HOMOs are given in Fig. 9.14. The frontier occupied MO of diazirine (and of other azo compounds) is the antibonding combination of the nitrogen lone pairs. Ionization of the $n^+$ level occurs at 14.15 eV. Energy splitting of the $n^-$ and $n^+$ levels in this molecule is thus 3.4 eV (3.3 eV in trans-azomethane and 3.55 eV in dimethyldiazirine). Difluorodiazirine has the same splitting: owing to the high electronegativity of the fluorine atoms, the electron levels of difluorodiazirine are much stabilized although the sequence of MOs does not change.

### 9.3.3 Cyclic azoalkanes

The electron structure of the compounds, in which the azo group is incorporated into the middle ring, has been studied in detail [103, 393]:

As in linear diazenes and diazirines, the highest occupied electron levels in compounds LXXX are formed by the (−) combinations of lone pairs of adjacent nitrogen atoms. It is important to note that the chemical behaviour of some diazenes (diazenes containing the cyclopròpane ring in particular) is determined not only by the state of their HOMO but also by the parameters of their deeper electron levels. The rate of thermal evolution of nitrogen, for example, in the series of azo compounds LXXX (e)–(g), increases with $n$ (prismane (e) shows an unusual stability in thermolysis)

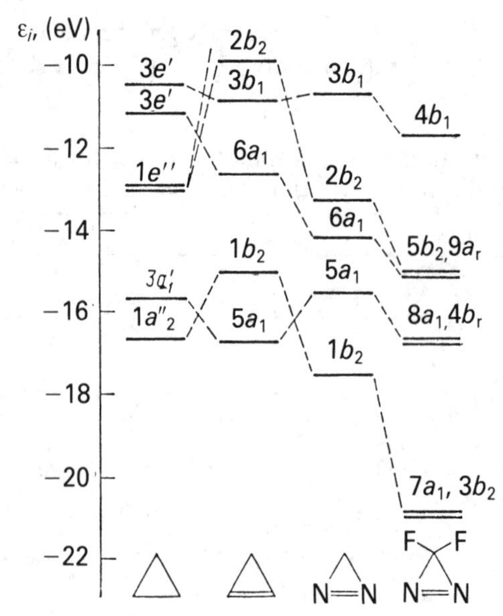


Fig. 9.14. Energy of diazirine, difluorodiazirine, cyclopropene and cyclopropane HOMOs (PES). (Data by Robin *et al.* [440])

LXXX: a b c d

LXXX:
e, n=0
b, n=1
f, n=2
g, n=3

$(CH_2)_n$ → $[(CH_2)_n]$ → $(CH_2)_n$ + $N_2$

It can be seen from the scheme that instability of the diazenes mentioned is accompanied by the breakdown of the three-membered exocycle. Since the main structural difference between these compounds concerns the angle $\theta$ formed by the three-membered cycle and diazanorbornene fragment, it was suggested that the value of this angle mainly determines their thermal lability. The growing stability of azo compounds with decreasing angle $\theta$ becomes clear if we consider the character of orbitals of the bonds broken and formed in the transition state. Since a cyclopropane bond breaks in the transition state, the increasing energy of the $\omega_S$ orbital ($\omega_S$ and $\omega_A$ are used to designate Walsh orbitals of cyclopropane) must destabilize the ring:

the higher its level, the lower energy is required to break the cyclopropane bond. On the contrary, the N=N bond strengthens in the transition state and stabilization of the $\pi$(N=N) and $n^+$ levels must therefore promote thermolysis.

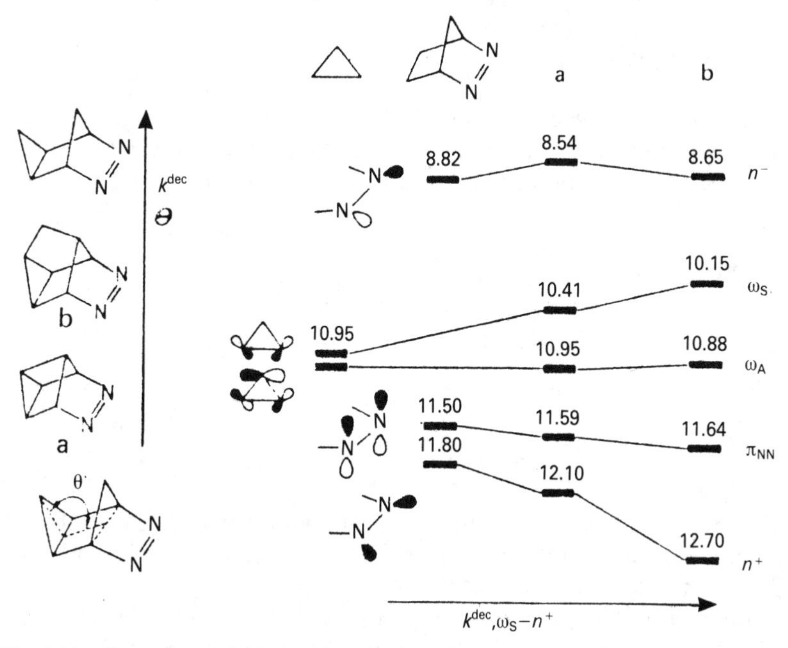


Fig. 9.15. Dependence of relative kinetic (thermal) stability $k$ of azo compounds LXXVII on the angle $\theta$ and energy difference of $\omega_s$ and $n^+$ orbitals. (From Martin and Mayer [393])

The shapes of orbitals of separate fragments in compounds LXXXb, a–g show that the increase in angle $\theta$ ensures a better overlap of the orbitals involved in thermolysis (Fig. 9.15). It should be noted that within the symmetry group $C_s$, effective overlap of the three-membered ring and the diazanorbornene fragment can be provided only by orbitals of the same symmetry: $\omega_S$, $n^+$ and $\pi$(N=N). The better the overlap of the three-membered ring with the C—N=N—C fragment in the corresponding azo compound (e.g. in LXXXb), the greater is the difference in the $\omega_S$ and $n^+$ levels (the $\omega_S$ level is destabilized while $\pi$(N=N) and $n^+$ levels are stabilized consecutively with increasing degree of overlap), the smaller is the energy of activation of thermolysis, and the higher its rate.

By contrast, the thermal stability of compound LXXXe is manifested even in photolysis. In the $S_1$ state, in order to split the nitrogen molecule, another 25.1 kJ/mol is necessary. If, at low temperature, this energy is not available for the molecule, it is isomerized to diazacyclooctatetraene.

### 9.3.4 Azobenzenes

Azobenzene absorbs in the visible region of the electron absorption spectrum (bright orange crystals) but it is not used as a dye. Meanwhile its electron structure is very important for understanding of colour variations in this large group of organic dyes.

Like dialkyldiazenes, aromatic azo compounds can exist in cis and trans forms. The structural parameters of both azobenzene isomers are the same:

| | $r_{N=N}$ (nm) | NNC angle (degree°) |
|---|---|---|
| trans | $0.123 \pm 0.005$ | $121.5 \pm 3$ |
| cis | $0.123 \pm 0.005$ | $121 \pm 3$ |

The trans isomers are usually more stable. Thus, trans azobenzene can partly be converted to the cis isomer by photolysis, and the energy of reversion to the trans form is not great: the reaction is 50% completed at room temperature within a few hours [4]:

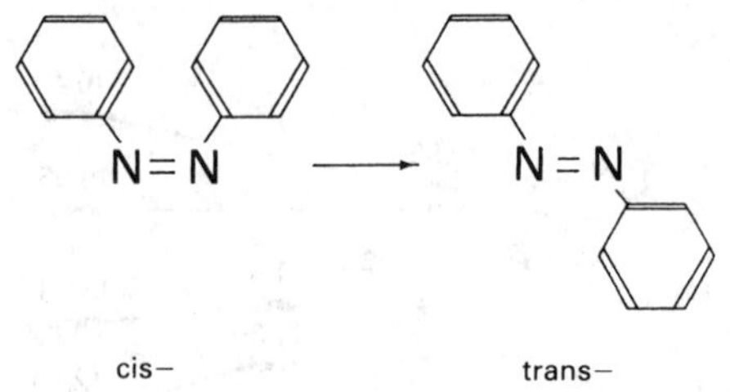

N=N ⟶ N=N

cis- trans-

The rapid isomerization of cis azobenzene to the trans form interferes with a detailed electron diffraction study of the cis isomer. For this reason, the trans isomer is better studied by PE spectroscopy [51, 441].

As in azomethane, the energy of the (−) combination of azobenzene nitrogen lone pairs is above the $n^+$ level. But in contrast to azoalkanes, the ionization band of the $n^-$ orbital is strongly overlapping with the bands assigned to the benzene $\pi$ orbitals: the difference of the corresponding ionization potentials does not exceed 0.3 eV. The band at 9.30 eV in the spectrum of trans azobenzene is assigned to degenerate $2a_u(\pi_2)$ and $1b_g(\pi_3)$ orbitals, the band at 9.77 eV, to the $2b_g(\pi_4)$ orbital. Since azobenzene is isoelectronic with stilbene, it should be noted that, compared to stilbene, all occupied $\pi$ levels of the azobenzene molecule are markedly stabilized (Fig. 9.16).

The sequence of the higher occupied MOs of azobenzene specified in [441] is doubted by some authors. Thus, according to INDO/S calculations of azobenzene and some *p*-methoxyazobenzenes, reversal of the higher occupied MOs is expected:

| | $\pi_1$ | $n^-$(N) | $\pi_2$–$\pi_4$ |
|---|---|---|---|
| Azobenzene | 8.46 | 8.77 | 9.30 |
| *p*-Methoxyazobenzene | 7.95 | 8.38 | 9.16 |

This assignment is also confirmed by PES, EAS, polarographic reduction, and calculations of some *p*′-substituted *p*-methoxyazobenzenes [51]. Among other evidence are intensities of separate bands in the PE spectra; the third band, corresponding to overlap of ionization of three $\pi$ orbitals, is especially intense.

The substituent effects on the occupied electron levels of azobenzene were studied in detail. The first and second ionization potentials of azobenzenes Ph—N=N—$C_6H_4$—X—*p*, and also the differences between the ionization potentials

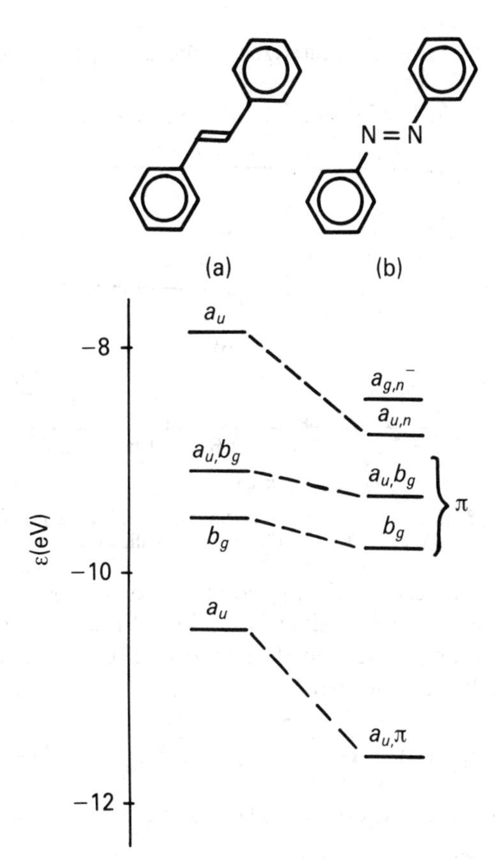


Fig. 9.16. HOMOs of (a) stilbene and (b) azobenzene. (From Kobayashi *et al.* [441])

of $n^-$ orbitals and the polarographic reduction potentials $E^{\text{red}}_{1/2}$ are the following [51]:

| X | $NH_2$ | $OCH_3$ | OH | $CH_3$ | H | Cl | COOH | $NO_2$ |
|---|---|---|---|---|---|---|---|---|
| $IP_1$ (eV) | 7.67 | 8.00 | 8.2 | (8.3) | 8.5 | 8.55 | 8.75 | 9.05 |
| $IP_2$ (eV) | 8.25 | 8.3 | 8.5 | (8.45) | 8.8 | 8.8 | 8.95 | 9.6 |
| $(IP_{n(-)} - E^{\text{red}}_{1/2}$ (eV) | 9.85 | 9.83 | — | 9.87 | 9.87 | 9.85 | 9.99 | 10.02 |

It is believed that the first and second ionization potentials of azobenzenes containing electron-donating substituents X ($CH_3$, OH, $OCH_3$, $NH_2$) are assigned to the $\pi$ and $n^-$ orbitals respectively: the $\pi$ orbital is the HOMO in these azobenzenes. In agreement with this assignment is the fact that the $\pi$ orbital (HOMO) is more sensitive to the change of substituents than the $n^-$ orbital. The $n^-$ orbital, however, remains the HOMO in azobenzenes containing electron withdrawing substituents X (Cl, COOH, $NO_2$).

It should be noted that the differences between ionization potentials $IP_{n^-}$ and polarographic reduction potentials $E^{\text{red}}_{1/2}$ are practically the same for all azobenzenes. This result correlates with the low sensitivity of the energy of the $n^- \rightarrow \pi^*$ transition to substituent X in the electron absorption spectra of azo compounds. It should also be noted that these differences characterize the energy gaps between the frontier MOs of azobenzenes.

Polarographic reduction of *p*′-substituted *p*-methoxyazobenzenes is noteworthy

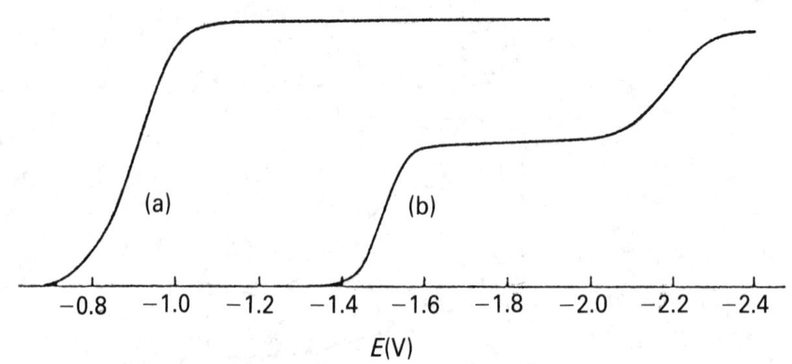


Fig. 9.17. Polarographic reduction of p-methoxyazobenzene in (a) methanol and (b) acetonitrile. (From Millefiori and Millefiori [51])

[51]. Azo compounds in acetonitrile are reduced in two steps (Fig. 9.17, curve b): the first wave corresponds to the reversible formation of the anion radical, the other to the irreversible addition of the second electron owing to the rapid protonation of the dianion (HS is the solvent molecule):

$$\mathrm{Ar{-}N{=}N{-}Ar} \xrightleftharpoons{+e^-} [\mathrm{Ar{-}N{=}N{-}Ar}]^{\overline{\cdot}} \xrightleftharpoons{+e^-} [\mathrm{Ar{-}N{=}N{-}Ar}]^{2-} \xrightarrow{+2\mathrm{HS}}$$

$$\longrightarrow \mathrm{ArNH{-}NHAr} + 2\mathrm{S}^-$$

The irreversible two-electron reaction (curve a) occurs in a proton-containing solvent (methanol):

$$\mathrm{Ar{-}N{=}N{-}Ar} + 2\mathrm{e}^- + 2\mathrm{H}^+ \rightarrow \mathrm{ArNH{-}NHAr}$$

The first reversible waves of polarographic reduction thus correspond to addition of an unpaired electron over the anion radical in the frontier vacant MO of the corresponding azobenzene. Using Koopmans' theorem, we can assume that $EA_1 = -\varepsilon_{\mathrm{LUMO}}$, and compare first half-wave potentials of polarographic reduction with the results of quantum chemical calculation of LUMO energies:

$$E_{1/2}^{\mathrm{red}} = -\varepsilon_{\mathrm{LUMO}} + \Delta G_{\mathrm{solv}}$$

The following equation was derived by the INDO/S calculations:

$$E_{1/2}^{\mathrm{red}} = (0.82 \pm 0.05)\varepsilon_{\mathrm{LUMO}} - (2.83 \pm 0.09) \qquad (9.4)$$

The comparison of the values of polarographic reduction potentials of azobenzenes with the corresponding data for aromatic hydrocarbons [18] shows that LUMOs of azobenzenes have far lower energies. Thus, the dependence for polarographic reduction of hydrocarbons in 2-methoxyethanol has the following parameters (see also Section 6.3.5):

$$E_{1/2}^{\text{red}} = (2.41 \pm 0.09)\varepsilon_{\text{LUMO}} - (0.43 \pm 0.06) \tag{9.5}$$

A detailed discussion of the electron structure of aromatic azo compounds is given in [31, 393].

### 9.3.5 Electron absorption spectra of azo compounds

The low energy transition at 345 nm, assigned to the $n^- \to \pi^*$ excitation, is well defined in the electron absorption spectrum of diazene $N_2H_2$. This transition has the ${}^1B_g \leftarrow {}^1A_g$ symmetry and very low intensity ($\varepsilon = 6$).

Haselbach and Schmelzer [437] studied the electron absorption spectrum of trans azomethane using the modified INDO/2 and CNDO/2 methods. The $n^- \to \pi^*$ transition was assigned to the longest wavelength in this spectrum as well. On the whole, the sequence of transitions found:

$$\Delta E_{n^- \to \pi^*} < \Delta E^+_{n \to \pi^*} < \Delta E_{\pi \to \pi^*}$$

was different from the transitions that might be predicted on the basis of the sequence of the occupied MOs:

$$IP_{n^-} < IP_{\pi} < IP_{n^-}$$

The results of the analysis confirmed the previous conclusion that the $\pi \to \pi^*$ transition in the azo fragment had the energy of 12 eV.

Other azo alkanes also have long-wave absorption of the $n \to \pi^*$ type in the region of 320–80 nm. According to the electron symmetry rule, this transition is forbidden in trans compounds but it is allowed in cis diazenes (Fig. 9.18). The extinction coefficients of the corresponding band increases by an order of magnitude from the trans azoisopropane to the cis isomer [442].

Substitution of the methyl groups at the diazene fragment by greater alkyl groups shifts the $n \to \pi^*$ bands in the spectra of azo compounds toward longer wavelengths. The R values in trans-diazenes R—N=N—R, the positions and energies of their long-wave absorption bands are the following [438]:

| | Me | Et | n-Pr | iso-Pr | $Me_3C$ | $Me_3Si$ |
|---|---|---|---|---|---|---|
| $\lambda$, max, nm | 353 | 356 | 359 | 359 | 368 | 784 |
| $\Delta E$, eV | 3.51 | 3.48 | 3.45 | 3.45 | 3.37 | 1.58 |

The linear dependence between the energies of $n^- \to \pi^*$ transitions of trans-dialkyldiazenes and their first ionization potentials has been found:

$$E_{n^- \to \pi^*} = 0.18 IP_{n(-)} + (1.90 \pm 0.02) \tag{9.6}$$

It follows from Eq. (9.6) that the energy of the $\pi^*$(N=N) orbital (the frontier vacant MO of diazene) increases with the size of the alkyl group at the nitrogen atom: only about 20% of the energy gain in the $n^-$ level is manifested by the bathochromic shift in the long-wave absorption band.

There are exceptions from the dependence defined in Eq. (9.6): for example, the increase in the energy of the $n^- \to \pi^*$ transition in the UV spectrum of LXXXI by 0.43 eV (compared with azo-tert-butane) almost completely coincides with the gain

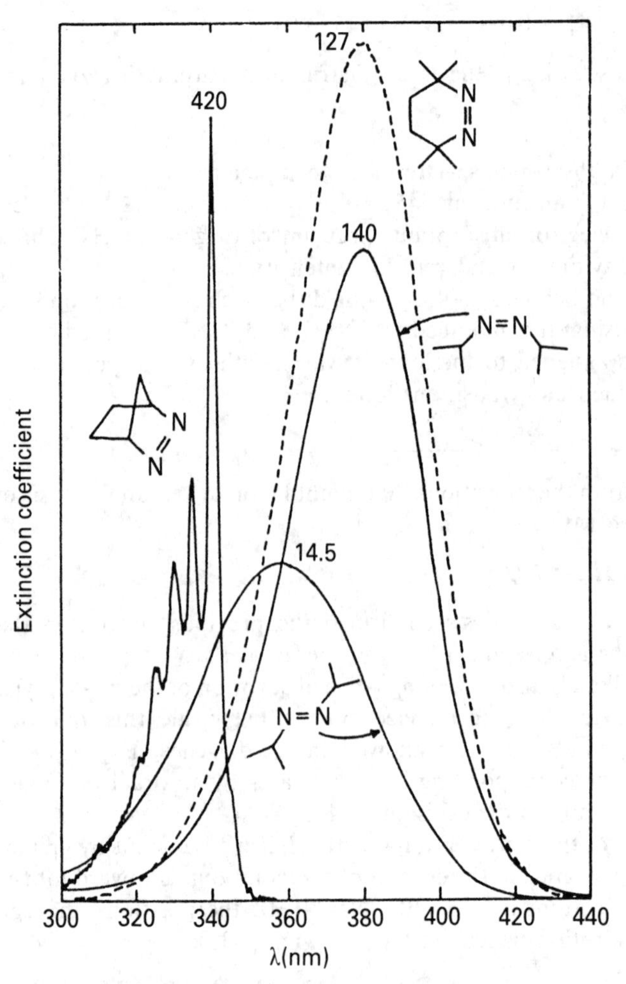


Fig. 9.18. Electron absorption spectra of some trans and cis azo compounds. (From Engel and Steel [442])

(0.38 eV) in the ionization potential of its $n^-$ level [438]:

$$\begin{array}{c} CH_2 \\ \| \\ C \\ Me_2C \quad\quad CMe_2 \\ N{=}N \end{array} \qquad \begin{array}{l} IP_1 = 8.58 \text{ eV} \\ h\nu = 3.80 \text{ eV} \end{array}$$

LXXXI

The vacant electron levels in organoelement diazenes differ substantially. It has already been mentioned in Section 9.3.1 [393] that, on substitution of tert-butyl groups at the nitrogen atom for trimethylsilyl groups, the energy of the $n^- \rightarrow \pi^*$ transition decreases by 2 eV, while the ionization potential, corresponding to the $n^-$

orbital, decreases by only 1.1 eV. This substitution does not increase (in the presence of an apparent bathochromic shift in the long-wave absorption) the energy of the frontier vacant MO in the corresponding diazenes. On the contrary, it drastically reduces this energy [117].

Electron transitions are more varied in aromatic azo compounds. Absorption bands corresponding to transitions from $\pi$ orbitals of aromatic fragments are seen

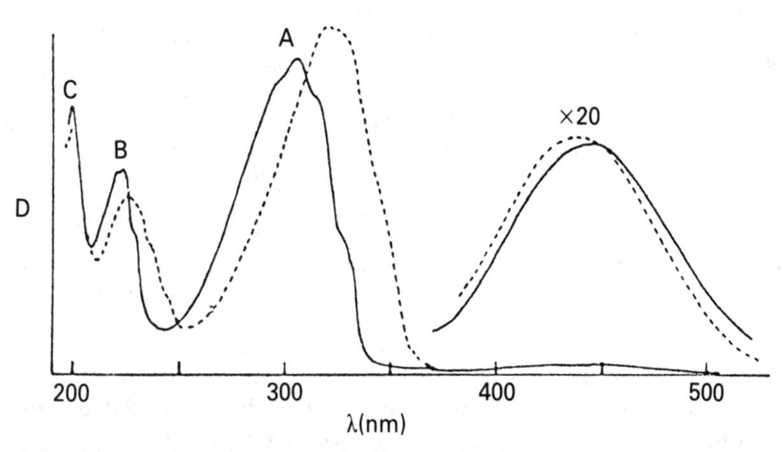


Fig. 9.19. Electron absorption spectra of azobenzene (solid line) and para-methoxyazobenzene (broken line). (From Millefiori and Millefiori [51])

in the long-wave region of their spectra [51]. Figure 9.19 shows the absorption spectra of azobenzene and *p*-methoxyazobenzene recorded in the gas phase. The spectra of both compounds have three bands of high intensity: A (at 4 eV), B (at 5.5 eV) and C (at 6.2–6.4 eV). Besides, there is a low-intensity band at 444 nm (about 2.8 eV) in the long-wave region. This band is assigned undoubtedly to an almost pure $n \rightarrow \pi^*$ transition of $B_g$ symmetry (polarized in the plane of the molecule and formally forbidden). Partly permitted, this transition is explained by its interaction with the adjacent state $B_u$ to which the intense transition (also polarized in the molecule plane) is assigned.

Band A in the azobenzene spectrum is polarized in the direction of the long axis. Both experimental findings and calculations suggest, however, that this band is due to more than one transition.

The transition predicted by calculations in the region of 6 eV is the $n^+ \rightarrow \pi^*$ transition. Bands corresponding to this transition were not yet resolved in the experimental spectrum. The results of INDO/S calculations (including CI procedure) of the absorption spectra of substituted azobenzenes, and also the experimental data, are given in Table 9.13.

The analysis of the electron absorption spectra of substituted azobenzenes show that $n^-$ and $n^+$ orbitals localized over the nitrogen atoms should not produce an appreciable effect on the colour of azo compounds: the appropriate bands occur in the long-wave region of the spectrum but their intensities are very low.

But there is a comparatively large group of azo dyes in which the state of lone

**Table 9.13.** Experimental (gas phase) and calculated (INDO/S) energies (eV) of electron transitions in absorption spectra of some para-methoxyazobenzenes X—$C_6H_4$—N=N—$C_6H_4$—OMe-*p*

| X | $B_g(n \rightarrow \pi^*)$ transition | | $B_u(\pi \rightarrow \pi^*)$ transition | | $B_u(\pi \rightarrow \pi^*)$ transition | | $B_u(\pi \rightarrow \pi^*)$ transition | |
|---|---|---|---|---|---|---|---|---|
| | exp. | calc. | exp. | calc. | exp. | calc. | exp. | calc. |
| H | 2.83 | 1.41 | 3.90 | 3.95 | 5.51 | 5.51 | 6.26 | 6.60–6.80 |
| *p′*-OMe | 2.88 | 1.45 | 3.73 | 3.85 | 5.37 | 5.46 | 6.36 | 6.40–6.80 |
| *m′*-OMe | 2.80 | 1.41 | 3.81 | 3.91 | 5.37 | 5.43 | 6.20 | 6.60–6.80 |
| *p′*-Me | 2.84 | 1.43 | 3.80 | 3.88 | 5.44 | 5.47 | 6.39 | 6.60–6.80 |
| *m′*-Me | 2.83 | 1.41 | 3.81 | 3.93 | 5.41 | 5.45 | 6.20 | 6.60–6.70 |
| *p′*-Cl | 2.83 | | 3.79 | | 5.39 | | 6.02 | |
| *m′*-Cl | 2.84 | | 3.77 | | 5.41 | | 5.99 | |
| *p′*-$NO_2$ | 2.75 | 1.42 | 3.77 | 3.79 | 5.39 | 5.30 | — | 6.80 |
| *m′*-$NO_2$ | 2.84 | 1.46 | 3.76 | 3.76 | 5.32 | 5.19 | 6.39 | 6.70–6.80 |

(From Millefiori and Millefiori [51])

pairs over the nitrogen atoms can be decisive for their colour, steric and electronic structure. These are the dyes containing the OH, $NH_2$ and related groups at the ortho position to the azo fragment. These groups are capable of forming hydrogen bonds with the azo nitrogen atom.

The azo dyes of the 1-phenylazo-2-naphthol series LXXXII are an example. They can exist in both the purely azo from and the azo form with hydrogen bonds (forms (a) and (b) respectively):

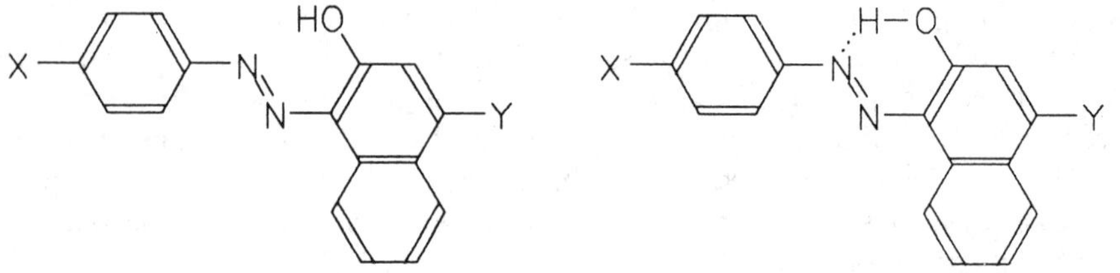

LXXXII: a b

and also in the second tautomeric form, as a quinone hydrazone and quinone hydrazone with hydrogen bonds (forms (c) and (d) respectively):

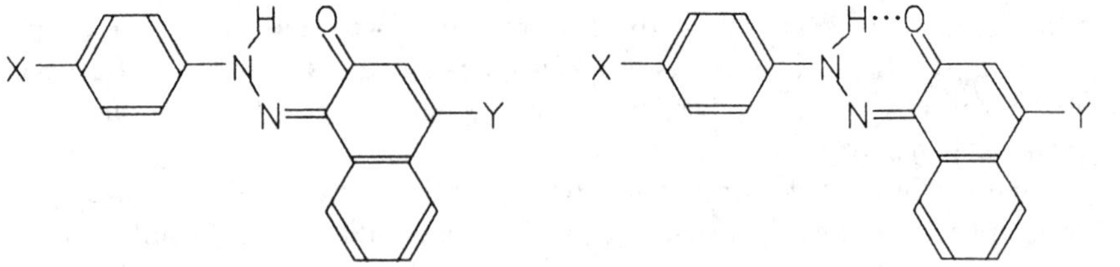

LXXXII: c d

The state of lone pairs over the nitrogen atoms must show its effect on the equilibria of the (a–d) forms and hence on the colour of a particular azo dye: formation of a six-membered ring with hydrogen bonds increases the planarity of the molecule and creates favourable overlap conditions for the formation of delocalized $\pi$ levels.

1-Phenylazo-2-naphthols were studied in detail in solution and in the solid state. It was confirmed that in both states these dyes can exist in two tautomeric forms. Both forms were calculated by the PPP method [31, 32]. The results of these

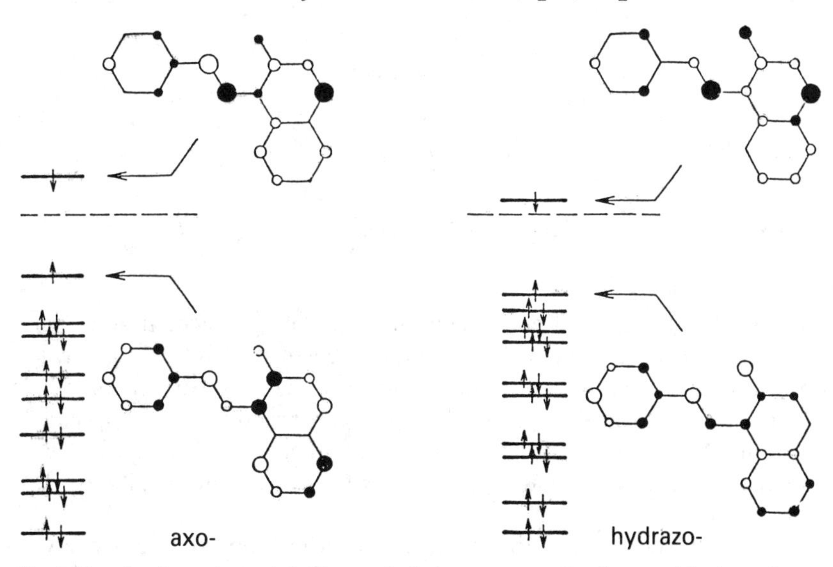


Fig. 9.20. Configurations of the first excited electron states $S_1$ of azo and hydrazo forms of 1-phenylazo-2-naphthol. (Data by Kelemen *et al.* [32])

calculations are partly shown in Fig. 9.20. It was established that in both forms, the first intensive electron transition $S_0 \rightarrow S_1$ refers to the $\pi \rightarrow \pi^*$ type. The shapes of the HOMO and LUMO show that the substituents at the para positions in the azo group must produce a strong effect on the absorption spectra of both forms since HOMO and LUMO have high coefficients at these positions in both benzene and naphthalene fragments.

The diagram in Fig. 9.20 illustrates changes in the electron density on $S_0 \rightarrow S_1$ electron excitement: the areas of the circles in these fragments are proportional to the changes in the electron density; the shaded circles show the growth of the electron density, and the open circles show decreasing electron densities over the appropriate atoms of the molecule during electron transition:

It can be seen that in the azo form, excitation increases the electron density over the $\beta$ nitrogen atom. Acceptor substituents in the aromatic group should therefore cause a shift in the absorption of the azo form toward the long-wave region of the spectrum. It is noteworthy that changes in the electron density at position 3 of the naphthalene nucleus associated with the transition of the quinone hydrazone to the first excited state are also apparently great. This result can probably explain the unexpected (at first sight) experimental fact: substituents at position 3 in 1-phenylazo-2-naphthols, which are not directly conjugated with the azo $\pi$ system, produce a

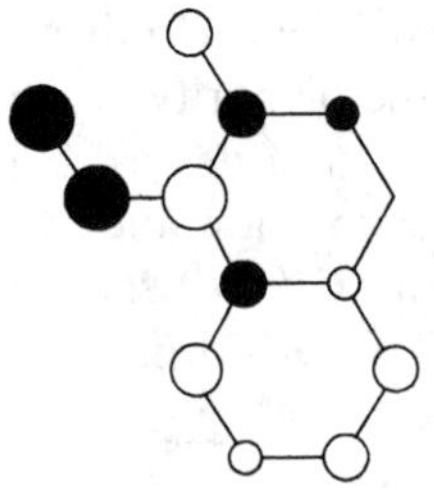

azo— 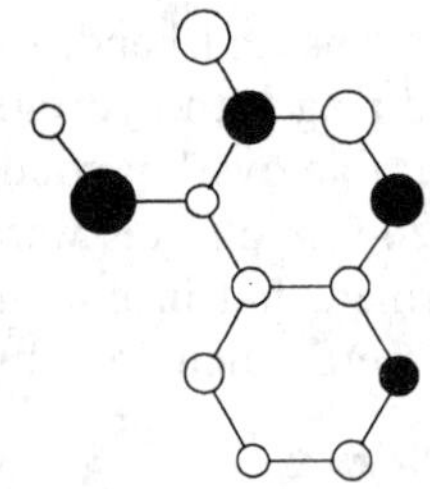　　　　hydrazo—

substantial effect on the colour of the compound [443].

## 9.4 NITROSO AND NITRO COMPOUNDS

### 9.4.1 Nitrosomethane

Nitroso compounds do not occur widely. But some of their structural properties are quite unusual and are of great interest for organic chemists. Under normal conditions, nitroso compounds usually exist in the dimer form and have low energy of electron excitation in their absorption spectra. This property accounts for the specific blue or green colour of nitroso compounds and an extremely low dissociation energy: on exposure to light, C-nitrosoalkanes easily break the C—N bond to give alkyl radicals and NO.

The stability of nitrosoalkanes to dimerization explains the controversy of the conclusions on their orbital structure that were derived in the first studies of their PE spectra [435, 444]. Subsequent measurements of their PE spectra within a wide range of temperatures made it possible to record only the parameters of monomeric forms of nitroso compounds [444]. Figure 9.21 gives an example of variation in the PE spectrum of cis nitrosomethane attending dissociation of its dimer to the monomer.

The spectrum of the nitrosomethane monomer has three ionization regions: a separate band at 9.67 eV and two intensive broad bands with the maxima at 13.9 and 16 eV. The first band is definitely assigned to a strong antibonding $n^-$ orbitals of $\pi$ type of the N—O bond. Ionizations of $2a''$ and $6a'$ orbitals overlap in the second band. Among these orbitals, the $\pi$ $2a''$ orbital is the bonding one and it is ionized at a slightly lower potential. According to this interpretation, splitting between $n^-$ and $n^+$ orbitals is about 4 eV. This value corresponds to PES data for trans azomethane in which the $n^-$ and $n^+$ levels are separated by about 3.3 eV.

The steric structure of the monomer form of nitrosomethane was studied. According to microwave spectral findings, its molecule has the following parameters:

| | $\lambda$ (nm) | | $\theta$ (degree°) |
|---|---|---|---|
| N=0 | 0.1211 | CNO | 113.2 |
| C—N | 0.1480 | H′CN | 111.0 |
| C—H | 0.1094 | H″CN | 107.2 |
| | | H″CN′ | 109.2 |

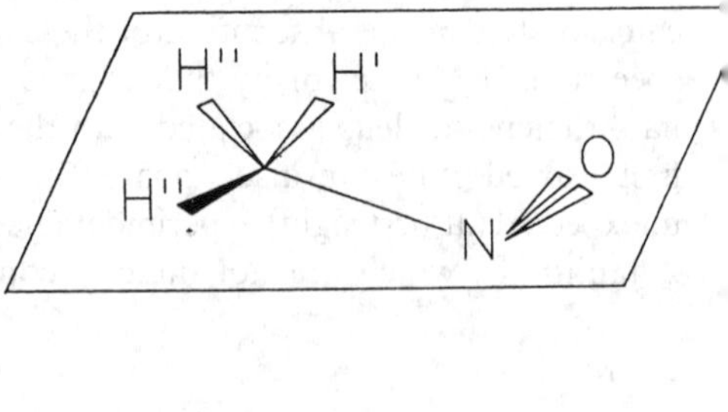

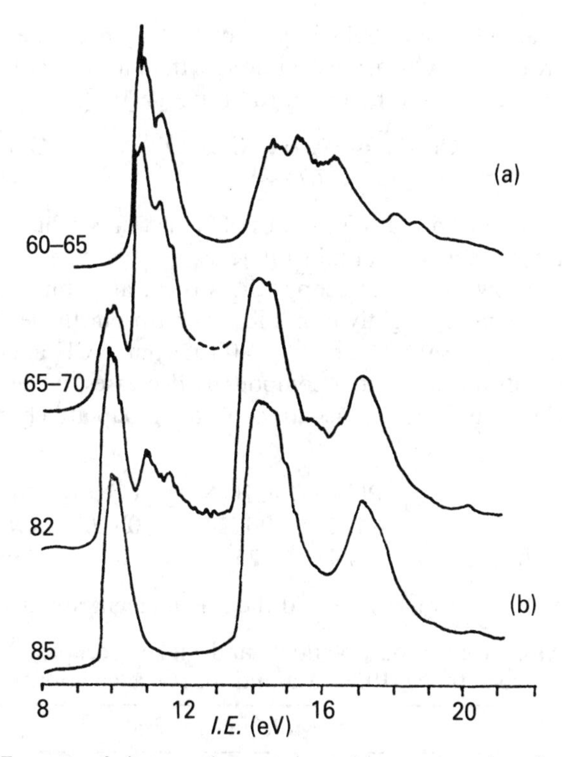


Fig. 9.21. PE spectra of nitrosomethane at elevated temperatures (are shown in degrees on the left): (a) dimer; (b) monomer. (From Bergmann *et al.* [444])

Among other noteworthy properties of the nitrosomethane molecule is the N=O bond distance which is the same as in HNO (0.1212 nm), while the C—N bond distance is similar to the same bond in methylamine (0.1472 nm).

*Ab initio* quantum chemical calculations also show weak interaction of the N—O and methyl group orbitals. These facts are probably associated with a high tendency of aliphatic nitroso compounds to break down at the C—N bond.

Absorption of aliphatic C-nitroso compounds which have relatively high ionization potentials in the visible region of the electron spectrum definitely shows that the main cause of low energies of electron transitions is the presence of low vacant orbitals in the molecule of nitroso compounds. Low energies of frontier vacant orbitals in nitroso compounds explain, to a certain measure, the difficulties in the assignment of ionization bands in PE spectra due to the defects of Koopmans' theorem [393].

### 9.4.2 Nitromethane

Aliphatic nitro compounds are not widely distributed. Only some nitro derivatives of aliphatic hydrocarbons (e.g. nitromethane, nitroethane, 2-nitropropane) are produced in large amounts and used as solvents.

The structure of these compounds has some unique properties. High C—H acidity of nitroalkanes is among them. Substitution of hydrogen by the methyl group in nitromethane increases the CH-acidity, although the increase is lower than in the case with hydrogen substitution by the nitro group [4, 193]:

| | $CH_3NO_2$ | $CH_3CH_2NO_2$ | $(CH_3)_2CHNO_2$ | $CH_2(NO_2)_2$ | $CH(NO_3)_3$ |
|---|---|---|---|---|---|
| pKa | 10.2 | 8.5 | 7.7 | 4.0 | 0 |

2-Nitropropane is an acid that dissociates 50% in water, while trinitromethane's acidity is about the same as that of mineral acids.

According to microwave spectroscopy, the structural parameters of the nitromethane molecule differ only slightly from those of nitrosomethane: $r_{NO} = 0.1224$ nm; ONO angle = 125.3°; $r_{CN} = 0.1489$ nm; $r_{CH} = 0.1088$ nm; NCH angle = 107.2°.

The N—O bond distance in nitro compounds decreases, while the ONO angle increases with the increasing electronegativity of the group attached to the nitrogen atom:

| | $PhNO_2$ | $CH_3NO_2$ | $ClNO_2$ | $FNO_2$ |
|---|---|---|---|---|
| $r_{N-O}$ (nm) | 0.1227 | 0.1224 | 0.1202 | 0.1180 |
| ONO angle (degree°) | 124.4 | 125.3 | 130.1 | 136.0 |

Ionization potentials of nitromethane and their assignment are given in Table 9.14.

**Table 9.14.** Ionization potentials and higher occupied MOs of nitromethane (PES data and 6-31G-calculations)

| $IP_i$ (eV) | Molecular orbital | | |
|---|---|---|---|
| | $-\varepsilon_i$ (eV) | No. and symmetry (group $C_{2v}$) | Localization |
| 11.28 | 12.15 | $3a_2$ | $\pi_2(NO_2)$ |
| 11.69 | 13.41 | $13a_1$ | $n_O^+$, $\sigma_{C-N}$ |
| 11.69 | 13.61 | $12a_1$ | $n_O^-$ |
| 14.72 | 16.63 | $2a_2$ | $\pi(CH_3)$ |
| 15.83 | 17.11 | $11a_1$ | $\pi(CH_3)$ |
| 17.42 | 19.89 | $10a_1$ | $\sigma_{N-O}^-$ |
| 17.42 | 20.46 | $1a_2$ | $\pi(NO_2)$, $\pi(CH_3)$ |

(From Kimura *et al.* [38])

The frontier occupied MO of nitromethane is a non-bonding $\pi$ orbital $3a_2$ formed by the orbitals of oxygen atoms with a node over the nitrogen atom. Two orbitals of $a_1$ symmetry formed by $2p(\sigma)$ orbitals of the oxygen atoms are almost degenerate (according to calculations, the difference in their energies is only 0.2 eV). As in a nitromethane molecule, the interaction of the orbitals of the methyl group and NO bonds in the nitromethane molecule is weak. The main effect of hydrogen substitution by the nitro group is a considerable shift in ionization of the C—H orbitals toward higher potentials: their ionization is manifested by a highly intense band at 17.5 eV [38]. This band is absent in this region of the PE spectrum of nitrobenzene.

On the transition from nitromethane to other nitroalkanes (nitroethane, nitropropane, nitrobutane), all ionization bands shift regularly toward lower potentials [369].

The most important feature of the electron structure of nitroalkanes is low energy of their frontier vacant MOs localized over the nitro group: according to [445], the electron affinity of nitromethane $EA_1 = 0.44 \pm 0.1$ eV. This property agrees with the high stability of carbanions formed in acid dissociation of nitroalkanes.

### 9.4.3 Nitrosobenzene

Information on PE spectra of nitrosobenzene is scarce [263, 366], but it has been definitely shown that its electron structure differs from the structures of its isoelectronic analogues styrene and benzaldehyde. The most remarkable property of nitrosobenzene is that its frontier occupied MO is completely localized over the substituent and lies higher than the highest occupied benzene $\pi$ orbitals. There are two ionization bands in the region of low potentials of the PE spectrum of nitrosobenzene: at 8.51 and 9.49–9.90 eV [263]. The first band is assigned to the non-bonding MO localized over the N and O atoms. The second band is about twice as intense and is assigned to the ionization of $\pi$ orbitals localized over the benzene nucleus: the vibrational frequency of this band (846 cm$^{-1}$) corresponds to one of the skeletal vibrations of the ring. Ionization of the second benzene $\pi$ orbital is seen at 9.90 eV as the shoulder on the second band next to the vibrational structure. The $\pi$ orbital of the nitroso group is ionized at higher energies, together with the $\sigma$ orbitals of the ring. According to CNDO/2 calculations, this $\pi$ orbital does not interact with the higher occupied benzene $\pi$ orbitals (Fig. 9.22).

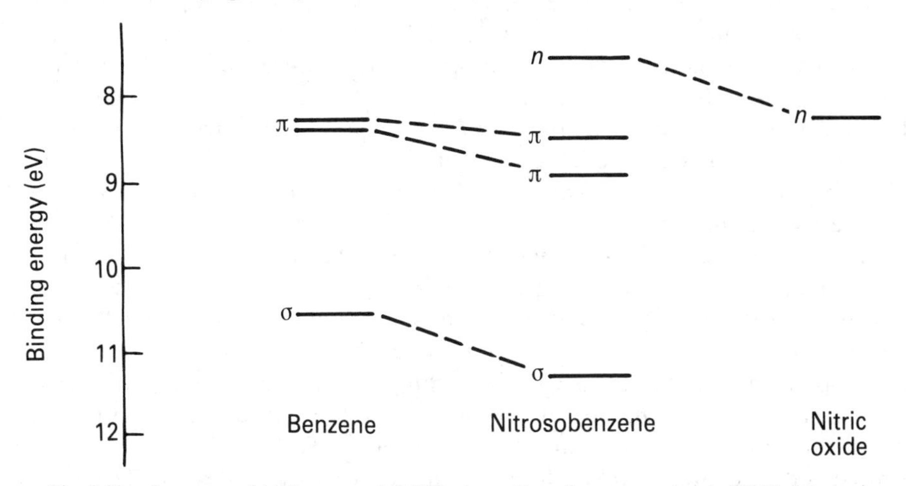


Fig. 9.22. Sequence of higher occupied MOs in a nitrosobenzene molecule (PES). (From Rabalais and Colton [263])

*Ab initio* calculations show that the nitroso group is an effective acceptor of $\pi$ electrons from the benzene ring in the nitrosobenzene molecule. Electron density can be delocalized considerably only in a planar conformer a with a parallel arrangement of $\pi$ orbitals of the benzene ring and the substituent [335]. If the $\pi$-type interaction is excluded (conformer b), the nitroso group becomes a donor. Charges are similar in both planar and orthogonal conformers:

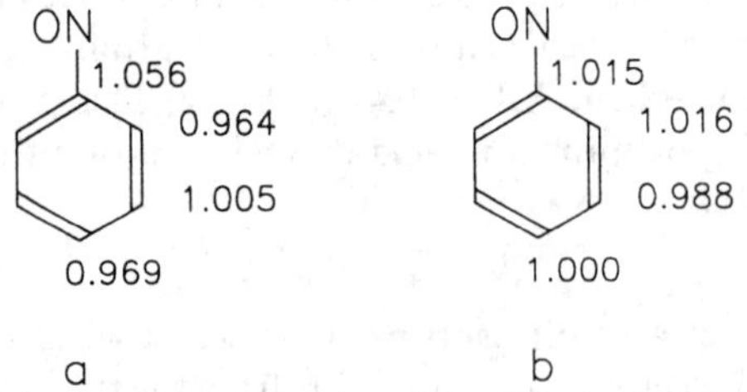

Reliability of these calculations is confirmed (to a certain degree) by the measured value (16.33 kJ/mol) of the rotational barrier of the nitroso group in the nitrosobenzene molecule, and shortening of the C—N bond (0.144 nm) compared with the length of this bond in the nitrosomethane molecule (0.149 nm) [237].

### 9.4.4 Nitrobenzene

The nitrobenzene molecule is planar. As a rule, the nitro group is found in the plane of the benzene ring in substituted benzenes as well. Deviation from this plane occurs, however, after introduction of a substituent at the ortho position to the nitro group. According to the electron diffraction data, the planes of the nitro groups in the o-dinitrobenzene molecule are parallel and form an angle of 33° to the benzene ring [446].

Conclusions on the structure of occupied electron levels in the nitrobenzene molecule are controversial. According to EHMO calculations [367] the first and second bands in the nitrobenzene PE spectrum are assigned to the ionization of benzene's $\pi$ orbitals $a_2$ and $b_1$ respectively. It is believed that the $b_1$ orbital is formed by the benzene $\pi$ (symm) orbital $a_{1g}$ and the LUMO of the nitro group. The bands at 11 eV (the third at 11.01 eV and the fourth at 11.23 eV) are assigned to ionization of nitro group orbitals; they correspond to the bands at 11.29 and 11.71 eV in the PE spectrum of nitromethane.

The results of the analysis of the nitrobenzene electron structure are given in Table 9.15. Palmer *et al.* showed that all ionization bands of the benzene nucleus in the nitrobenzene PE spectrum (compared with the benzene spectrum) are shifted toward higher potentials by about 0.7–0.9 eV (in Section 9.4.2, it is shown that a similar shift in the spectrum of nitromethane is 3.5 eV, compared with the methane PE spectrum) [446]. The highest benzene's degenerate $\pi$ level in nitrobenzene also remains degenerate but it is ionized at 9.92 eV instead of 9.24 eV. Likewise, the higher benzene's degenerate $\sigma$ level in the nitrobenzene spectrum is seen at 12.6 eV instead of 11.7 eV of the benzene spectrum.

Calculated electron densities in the planar a and orthogonal b conformations of nitrobenzene are shown below

Although $\pi$ electron density is drawn markedly from the benzene nucleus onto the nitro group, the barrier to rotation of the nitro group of nitrobenzene is small: 12.14 kJ/mol as determined experimentally, and 25–38 kJ/mol according to calculations. This is in agreement with equivalence of C—N bond distances in nitromethane and nitrobenzene molecules (0.1489 and 0.1486 nm respectively). At the same time,

**Table 9.15.** Ionization potentials and higher occupied MOs of nitrobenzene (PES data and *ab initio* calculations)

| $IP_i$ (eV) | Molecular orbitals | | | |
|---|---|---|---|---|
| | $-\varepsilon_i$ (eV) | | No. and symmetry | |
| | minimal basis | DZ basis | within framework of group $C_{2v}$ | within framework of group $D_{6h}$ |
| 9.92 | 10.98; 11.30 | 9.99; 10.27 | $a_2, b_1$ | ($e_{1g}$-symm, $e_{1g}$-asymm) |
| 11.23 | 11.81; 13.28 13.40 | 11.65; 12.76 12.98 | $a_2, a_1, b_2$ | ($\pi$-$NO_2$, $\sigma$-$NO_2$, $\sigma$-$NO_2$) |
| 12.75 | 14.77; 15.71 | 14.13; 14.58 | $b_2, b_1$ | ($e_{2g}$-asym, $a_{2u}$) |
| 13.3 | 15.54 | 14.89 | $a_1$ | $e_{2g}$-symm |
| 14.9 | 17.49 | 16.81 | $b_2$ | $e_{1u}$-asymm |
| 15.4 | 18.03; 18.20 | 17.61; 17.62 | $a_1, b_2$ | ($b_{1u}, b_{2u}$) |
| 16.6 | 19.41; 19.75 21.03 | 19.23; 19.24 19.80 | $b_2, a_1, b_1$ | $\sigma$-$NO_2$, $a_{1g}$, $\pi$-$NO_2$ |
| 18.0 | 21.37 | 21.14 | $a_1$ | $\sigma$-$NO_2$ |

(From Palmer *et al.* [446])

$O_2N$ 1.090 0.958 1.003 0.957

a

$O_2N$ 1.081 0.981 0.994 0.970

b

compared with the orthogonal conformation, the planar conformation provides a more effective acceptance of electron density by the nitro group, especially from the ortho and para positions of the benzene nucleus [237].

Inactivation of the nitrobenzene molecule for the attack by electrophiles due to the shift of the electron density onto the nitro group is so high that nitrobenzene is about $10^6$ times less active than benzene [367].

It should be noted that although ionization of oxygen lone pairs in nitro compounds occurs at comparatively high energies (about 12 eV in the nitromethane spectrum), these atoms preserve their basic properties. Thus, the nitro group is responsible for the intra- and intermolecular hydrogen bonding in para- and ortho-nitrophenols [6]: Having low vacant electron levels, the nitro group of the benzene nucleus increases the affinity of the corresponding derivatives for nucleophilic agents. This is illustrated by the data of polarographic reduction: nitro aromatic compounds are reduced at unusually low potentials, while the anion radicals formed are highly stable [50]. Spin density distribution in the anion radical of nitrobenzene (according to ESR spectrum) shows that the unpaired electron is found in the transformed benzene $\pi$ (symm) orbital of $b_1$ symmetry. In agreement with the high acceptor power of the nitro

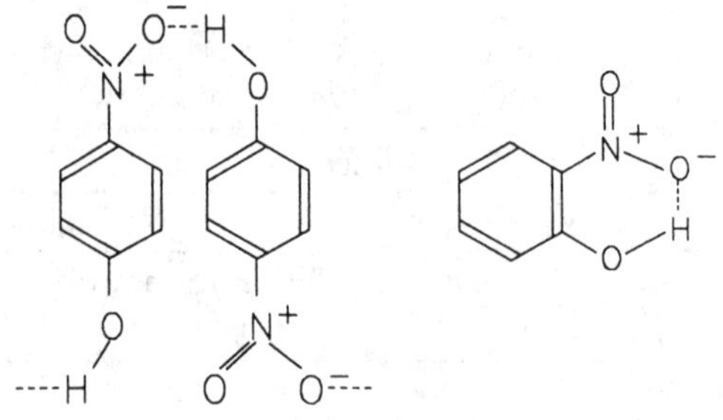

B.p. > 200°C at $9.3 \times 10^3$ Pa B.p. = 100°C

group, spin density is the highest over the nitrogen atom:

| | ortho | meta | para | N atom |
|---|---|---|---|---|
| $a_H$ (Oe) | 3.39 | 1.09 | 3.97 | 10.32 |
| $\rho_{exp}$ | 0.142 | 0.045 | 0.17 | — |
| $\rho_{HMO}$ | 0.106 | 0.005 | 0.124 | 0.226 |

## 9.5 NITRILES

### 9.5.1 Acetonitrile

The nitrile group gives specific physical and chemical properties to an organic molecule. Below are bond distances (nm) in molecules of some nitriles [198]:

$$\underset{0.1102}{\mathrm{H}\text{—}}\mathrm{CH_2}\overset{0.1466}{\text{—}}\underset{0.1159}{\mathrm{C{\equiv}N}} \qquad \mathrm{CH_2{=}CH}\underset{0.1438}{\text{—}}\overset{0.1167}{\mathrm{C{\equiv}N}} \qquad \mathrm{N{\equiv}C}\underset{0.1319}{\text{—}}\overset{0.1162}{\mathrm{C{\equiv}N}}$$

The C≡N group is isoelectronic to the ethynyl group: it also has two degenerate $\pi$ orbitals localized in two mutually perpendicular planes. The nitrogen atom bound by a triple bond with the carbon in nitriles has lower basicity than the $sp^2$ hybridized nitrogen atom [193]:

| | $CH_3$—H | $CH_2$=CH—H | CH≡C—H | $\overset{+}{N}H_3$—H | RC≡$\overset{+}{N}$—H |
|---|---|---|---|---|---|
| pKa | 48 | 44 | 25 | 9 | −10 |

In the PE spectrum of acetonitrile, ionization of degenerate $\pi$ orbitals occurs at 12.21 eV. The band at 13.14 eV is assigned to ionization of an MO consisting largely of the $2p$(N) orbital oriented along the molecule axis. The orbitals of the methyl group of acetonitrile are ionized at appreciably higher potentials (about 16 eV) compared with the occupied methane orbitals (Table 9.16).

The study of the electron structure of malonodinitrile showed the ability of the $\pi$(CN) orbitals to interact. As a result, the sequence of malonodinitrile MOs is determined by the combined TS and TB effects [393].

The frontier vacant MO of acetonitrile is localized in the region of the CN group. The first electron affinity of its molecule is about zero which suggests its high electron withdrawing power [445].

**Table 9.16.** Ionization potentials and higher occupied MOs of acetonitrile (PES data and 6-31G calculations)

| $IP_i$ (eV) | $-\varepsilon_i$ (eV) | Molecular orbitals | |
|---|---|---|---|
| | | No. and symmetry (group $D_{3h}$) | Localization |
| 12.21 | 12.56 | $2e$ | $\pi(C{\equiv}N)$ |
| 13.14 | 14.95 | $7a_1$ | $\sigma(CN)$ |
| (16.0) | 17.05 | $1e$ | $\pi(CH_2)$-pseudo |
| 16.4 | | | |
| 17.4 | 18.82 | $6a_1$ | $\sigma(C{-}C)$ |

(From Kimura *et al.* [38])

### 9.5.2 Cyanoalkenes

It has already been said (Section 5.3) that attachment of the CN group at the double bond decreases appreciably the energy of the frontier vacant MO of the corresponding ethylene (cyanoethylenes and their $EA_1$, in eV, are given): acrylonitrile, 0.02; fumaronitrile, 0.78; tetracyanoethylene, 2.88 [203].

The acceptor properties of the nitrile group show their effect on the energies of occupied MOs as well. Ionization potentials (eV) measured from the PE spectra of cyanoethylenes are given below [202]:

| | 1-cyano- | trans-1,2-dicyano | tetracyano- |
|---|---|---|---|
| $\pi(CC)[+\pi(CN)]$ | 10.84 | 11.15 | 11.79 |
| $n(N)$ | 12.98 | 13.44 | 14.1 |
| $\pi(CN)$ | 12.28 | 12.78 | 13.45 |
| | | 13.10 | 13.81 |
| | | 13.67 | 14.26 |
| $\pi(CN)[+\pi(CC)]$ | 13.51 | 14.41 | 16.12 |

### 9.5.3 Benzonitrile

As in the case with other benzene derivatives containing acceptor substituents, thorough microwave spectroscopic measurements of the molecular geometry of benzonitrile discovered the shortness of C—C bonds in the ring adjacent to the substituent and an increased $C_{(6)}C_{(1)}C_{(2)}$ angle. The exocyclic C—C bond also turned out to be short:

| | $r$ (nm) | Valence angle | (degrees) |
|---|---|---|---|
| $C_{(1)}{-}C_{(2)}$ | 0.1388 | $C_{(6)}C_{(1)}C_{(2)}$ | 121.8 |
| $C_{(2)}{-}C_{(3)}$ | 0.1396 | $C_{(1)}C_{(2)}C_{(3)}$ | 119 |
| $C_{(1)}{-}C_{(7)}$ | 0.1451 | $C_{(2)}C_{(3)}C_{(4)}$ | 120 |
| $C_{(7)}{-}N$ | 0.1158 | | |

The first and second ionization bands in the PE spectra of benzonitrile are assigned to the $\pi$ orbital $b_1$ and $a_2$ respectively. They are localized mostly over the benzene ring [263, 273, 367]. Since the $b_1$ orbital formed by the interaction of the benzene and (C≡N) $\pi$ orbitals is higher than the benzene $\pi$ level $a_2$, it can be concluded that the benzene HOMO interacts with the higher occupied orbital of the nitrile group stronger than with its LUMO. This conclusion makes it possible to join benzonitriles with phenols and anilines in the discussion of their occupied frontier MOs. The higher first ionization potential of benzonitrile, compared with benzene is explained by the strong electron-acceptor inductive effect of the nitrile group [367].

The degenerate $\pi$ orbitals of the nitrile group are also split and shifted toward greater values of energy: their ionization potentials are 11.93 and 12.18 eV respectively (against 12.21 eV for acetonitrile). On the whole, splitting of the $\pi$ levels of the benzene nucleus, as well as of the CN group, is not great. One can suggest weak $\pi$-type conjugation and delocalization of the electrons in a benzonitrile molecule. This can probably be due to a substantial difference between the energies of the $\pi$(Ph) and $\pi$(CN) orbitals. Table 9.17 shows the assignment of its ionization potentials.

**Table 9.17.** Ionization potentials and higher occupied MOs of benzonitrile (PES data and *ab initio* calculation)

| $IP_i$ (eV) | Molecular orbitals | | | |
|---|---|---|---|---|
| | $-\varepsilon_i$ (eV) | | No. and symmetry | |
| | minimal basis | DZ basis | within framework of group $C_{2v}$ | within framework of group $D_{6h}$ |
| 9.71 | 10.35 | 9.5 | $b_1$ | $e_{1g}$(symm) |
| 10.12 | 10.58 | 9.7 | $a_2$ | $e_{1g}$(asymm) |
| 11.83 | 13.22 | 12.52 | $b_2$ | $e_{2g}$(asymm) |
| 12.07 | 13.77 | 12.94 | $b_1$ | $a_{2u}$ |
| 12.61 | 14.38 | 14.04 | | |
| | 14.57 | 14.11 | $b_2, a_1$ | $e_{2g}$(symm), n(N) |
| 13.01 | 15.85 | 14.93 | $b_1$ | $a_{2u}$ |
| 13.44 | 15.05 | 15.01 | $a_1$ | $e_{2g}$(symm) |
| 14.76 | 17.08 | 16.56 | $b_2$ | $e_{1u}$(asymm) |
| 15.33 | 17.51 | 17.12 | $a_1$ | $e_{1u}$(symm) |
| 15.70 | 18.29 | 17.71 | $b_2$ | $b_2$ |

(From Palmer *et al.* [273])

*Ab initio* calculations also reliably reproduce other experimental properties of benzonitrile: considerable acceptance of $\pi$ electron density from the ring, especially from the ortho and para positions, and also the dipole moment [237]:

$\mu_{calc} = 3.65D$
$\mu_{exp} = 4.18D$

$\mu_{calc} = 3.65$ D
$\mu_{exp} = 4.18$ D

N≡C
1.056
0.976
0.999
0.972

The results of measurements of HFI constants in ESR spectra of benzonitrile cation and anion radicals lead to the following conclusions on the character of the frontier occupied MO of benzonitrile [87]:

| | $a_H$ (ortho) | $a_H$ (meta) | $a_H$ (para) |
|---|---|---|---|
| $[C_6H_5CN]^{\dot{+}}$ | 9.15 | 11.45 | 0 |
| $[C_6H_5CN]^{\dot{-}}$ | 3.63 | 0.30 | 8.42 |

It can be seen that the unpaired electron in the benzonitrile cation-radical is in the orbital of $a_2$ symmetry, while in the anion radical, it is found in the $b_1$ orbital.

The character of splitting of degenerate benzene $\pi$ orbitals, determined from the ESR spectra of ion radicals of benzonitrile, confirms the marked acceptor properties of the nitrile group. They are also evidenced by electron affinity values of benzonitrile measured by the ETS method:

$$>0, \quad -0.54, \quad -2.49, \quad -3.20, \quad -4.90 \text{ eV } [46].$$

***Problems***

(1) $S_N^2$ reaction can be considered as donor–acceptor interaction in initial steps. Compare the nucleophilic properties of $CH_3NH_2$ and $CH_3OH$; which of the two compounds is more reactive to primary alkyl bromides? Explain.

(2) Benzene and aniline are very different in their ability to release an electron ($IP_1$ are 9.24 and 8.00 eV) and are very similar in their ability to accept an electron ($EA_1$ are $-1.15$ and $-1.13$ eV respectively. Explain.

(3) Aniline is easily oxidized by electrophiles with formation of a cation radical. Suggest an unpaired electron distribution in the aniline cation radical.

(4) Reduction of nitrobenzene starts with the addition of an electron. Suggest an unpaired electron distribution in the nitrobenzene anion radical.

(5) Benzonitrile forms a cation radical by electrochemical oxidation. Suggest an unpaired electron distribution in this cation radical.

# 10

# Orbital-controlled properties of substituted aromatic hydrocarbons

The data on the electron structure of substituted benzenes given in Chapters 7–9 show that the structure of $\pi$ orbitals of the aromatic substrate changes substantially on the introduction of various substituents into the benzene molecule. The corresponding orbital effects were explained in the terms of the Dewar MO perturbation theory [137]. One of the main concepts of this theory is described by the equation:

$$\delta E_i = c_{i\mu}^2 \delta W_\mu + c_{i\mu}^2 \beta^2 \Delta E^{-1} \tag{10.1}$$

where $c_{i\mu}$ is the coefficient of the $\mu$th atom in the $i$th MO; $\delta W_\mu$ is the change of the potential of the $\mu$th atom caused by attachment of the substituent to it; $\beta$ is the resonance integral estimating the interaction of the substituent orbitals with the $i$th MO; $\Delta E$ is the energy difference of the interacting orbitals.

According to this equation, the energy of the $i$th MO changes at the expense of the inductive effect of the substituent (the first term), and at the expense of the mixing of the orbitals of the substituent and the benzene nucleus (the second term).

It is noteworthy that both effects depend on electron density at the site of substituent attachment. Thus, since the benzene $\pi$ orbitals differ by their coefficients (see the shapes of these MOs in Section 2.2.1) only the $\pi_2$ and $\pi_5$ orbitals (of the frontier degenerate occupied and vacant MOs) having the maximum coefficients at the site of substituent attachment appreciably change their energies on the incorporation of the substituent into the benzene nucleus.

Table 10.1 summarizes the data of the PE and ET spectra of substituted benzenes $C_6H_5R$. Given are substituents R, values of ionization potentials of two higher occupied $\pi$ orbitals, and electron affinities corresponding to acceptance of the electron into two lower vacant $\pi^*$ orbitals.

The following scheme can be admitted in the first approximation: donor substituents increase the energies of the orbitals of $b_1$ symmetry, while acceptor substituents decrease their energies. Both substituents produce only a weak effect on the energies of the orbitals of $a_2$ symmetry since these orbitals have zero coefficients at the site of substituent attachment.

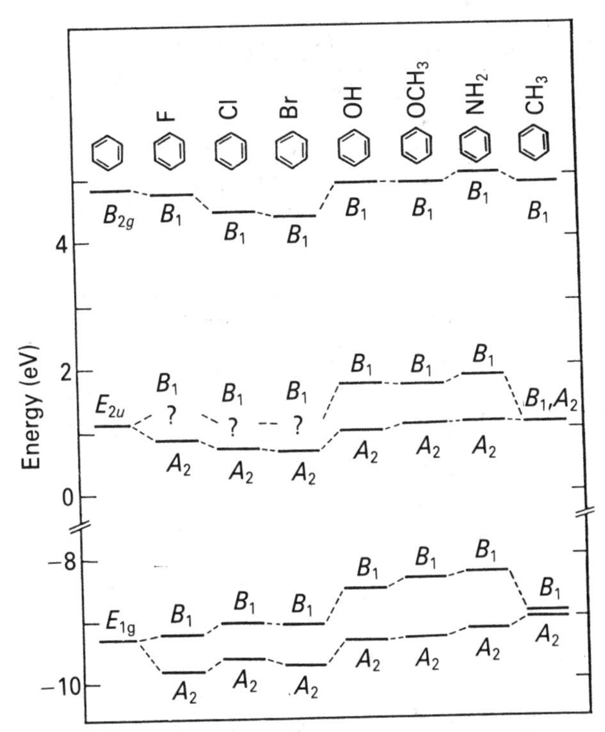


Fig. 10.1. Transformation of occupied (PES) and vacant (ETS) π orbitals of benzene caused by electron donating substituents. (From Jordan and Burrow [46])

Splitting of the degenerate benzene π electron levels on the introduction of various π donor (+*M*) substituents into the benzene molecule is also illustrated in Fig. 10.1 [46]. The splitting produced by the introduction of electron withdrawing substituents into the benzene molecule is not so pronounced. It can, however, be noted from the data given in Table 10.1 that these substituents stabilize frontier both occupied and vacant MOs.

It can also be seen from Table 10.1 that replacement of substituents in the benzene ring by other substituents can be attended by inversion of the sequence of the frontier MOs and the orbital nearest to them.

On the whole, it is necessary to note the very high sensitivity of benzene π orbitals to the introduction of a substituent. Degenerate orbitals (both occupied $e_{1g}$ and vacant $e_{2u}$) are split even by the replacement of one hydrogen isotope (H) for another (D) in the benzene molecule. This is confirmed by comparison of HFI constants of the anion radicals of benzene and monodeuterobenzene:
It is quite amusing that adding only one neutron to one of the hydrogens in the benzene molecule can produce such distinct changes in the ESR spectrum of the anion radical [87].

**Table 10.1.** Ionization potentials, electron affinities and symmetries of frontier-occupied and vacant orbitals of substituted benzenes $C_6H_5R$ (PES- and ETS data)

<table>
<tr><th>Substituent R</th><th>$IP_1(b_1)$ (eV)</th><th>$IP_2(a_2)$ (eV)</th><th>$EA_1(b_1)$ (eV)</th><th>$EA_2(a_2)$ (eV)</th></tr>
<tr><td>H</td><td colspan="2">9.24</td><td colspan="2">−1.15</td></tr>
<tr><td>$CH_3$</td><td>8.83</td><td>9.36</td><td colspan="2">−1.11</td></tr>
<tr><td>$CH_2{=}CH{-}$</td><td>8.49</td><td>9.27</td><td>−0.25</td><td>−1.05</td></tr>
<tr><td>$CH{\equiv}C{-}$</td><td>8.82</td><td>9.51</td><td colspan="2">—</td></tr>
<tr><td>Ph</td><td>8.34</td><td>9.04</td><td>−0.37</td><td>—</td></tr>
<tr><td>F</td><td>9.20</td><td>9.81</td><td colspan="2">−0.89</td></tr>
<tr><td>Cl</td><td>9.10</td><td>9.69</td><td colspan="2">−0.75</td></tr>
<tr><td>Br</td><td>9.02</td><td>9.65</td><td colspan="2">−0.70</td></tr>
<tr><td>I</td><td>8.78</td><td>9.75</td><td>−1.73</td><td>−1.01</td></tr>
<tr><td>$OCH_3$</td><td>8.21</td><td>9.17</td><td>−1.63</td><td>−1.13</td></tr>
<tr><td>$SCH_3$</td><td>8.07</td><td>9.28</td><td>−0.90</td><td>—</td></tr>
<tr><td>$NH_2$</td><td>8.00</td><td>9.21</td><td>−1.85</td><td>−1.13</td></tr>
<tr><td>$N(CH_3)_2$</td><td>7.60</td><td>9.20</td><td colspan="2">—</td></tr>
<tr><td>CHO</td><td>9.59</td><td>9.81</td><td>−0.76</td><td>−2.21</td></tr>
<tr><td>$COCH_3$</td><td>9.55</td><td>9.77</td><td colspan="2">—</td></tr>
<tr><td>CN</td><td>9.71</td><td>10.12</td><td>−0.54</td><td>−2.49</td></tr>
<tr><td>$NO_2$</td><td colspan="2">9.96</td><td colspan="2">—</td></tr>
</table>

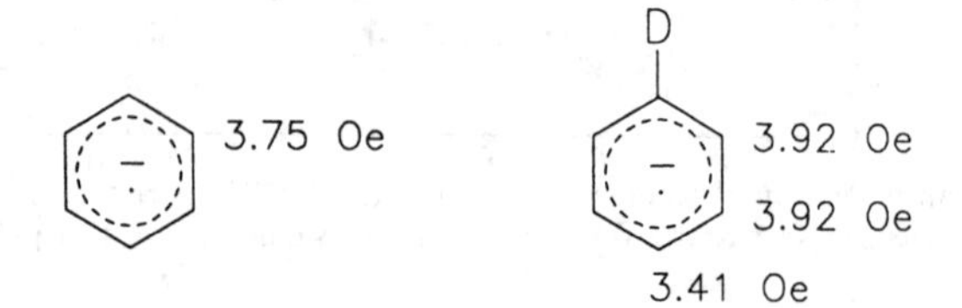


Molecular $\pi$ orbitals of polycyclic aromatic hydrocarbons are also sensitive to substituents. As distinct from benzene, these hydrocarbons usually have no degenerate HOMOs. The higher occupied $\pi$ electron levels in their molecules have, however, different coefficients at non-equivalent sites of substituent attachment. As a result, the introduction of the same substituent at different positions of a polycyclic hydrocarbon produces different changes in its electron levels. Only those MOs undergo considerable changes, which have high values of the eigencoefficients at the site of substituent attachment.

Table 10.2 contains the values of the first and second ionization potentials of 1- and 2-substituted naphthalenes. In full agreement with HOMO and HOMO-1 coefficients of naphthalene (see Section 6.3.5), the second ionization potentials do not practically change in the series of its derivatives, for example, on the introduction of an electron donating substituents at position 1, but they decrease appreciably on the introduction of the same substituent at position 2.

Polycyclic aromatic hydrocarbons with high frontier occupied MOs are far less sensitive to the substituent than benzene. The values of the first ionization potentials

**Table 10.2.** The first and second ionization potentials (eV) of substituted naphthalenes (PES data)

| Substituent | 1-substituted naphthalenes | | 2-substituted naphthalenes | |
|---|---|---|---|---|
| | $IP_1$ | $IP_2$ | $IP_1$ | $IP_2$ |
| H | 8.12 | 8.86 | 8.12 | 8.86 |
| $CH_3$ | 7.85 | 8.65 | 7.93 | 8.63 |
| $OCH_3$ | 7.70 | 8.62 | 7.82 | 8.35 |
| $NH_2$ | 7.48 | 8.62 | 7.55 | 8.27 |
| F | 8.15 | 9.07 | 8.23 | 8.89 |
| CN | 8.59 | 9.37 | 8.56 | 9.25 |
| $NO_2$ | 8.59 | 9.36 | 8.63 | 9.35 |

(From Koptyug and Salakhutdinov [447])

for some 9-substituted anthracenes are given below [199]:

| | H | $CH_3$ | $OCH_3$ | F | Cl | $COCH_3$ | COOH | CHO | CN | $NO_2$ |
|---|---|---|---|---|---|---|---|---|---|---|
| $IP_1$ (eV) | 7.43 | 7.25 | 7.21 | 7.46 | 7.48 | 7.52 | 7.56 | 7.69 | 7.84 | 7.88 |

Compare them with $IP_1$ of substituted benzenes:

$$IP_1(9-\mathrm{R}-\mathrm{C_{14}H_9}) = 0.41 \quad IP_1(\mathrm{C_6H_5R}) + 3.72\ \mathrm{eV}$$

Objective estimation of the structure of frontier occupied MOs of substituted aromatic hydrocarbons were obtained by analysis of their complexes with $\pi$ acceptors. Thus, CT bands in the electron absorption spectra of these complexes correspond to the electron transition from the HOMO of the substituted hydrocarbon to the LUMO of the acceptor. The analysis of the CT bands is simplified for those $\pi$ donors, whose HOMOs are well separated in energy from the lower occupied MOs. The electron absorption spectra of appropriate complexes have well-resolved CT bands. Below are the values of $\lambda_{CT}$ (nm) of CT bands in the spectra of TCNE complexes with some substituted benzenes (parenthesized are ionization potentials as determined by PES, eV [72]):

| | $PhCH_2SiMe_3$ | $PhC(SiMe_3)_3$ | $PhSiMe_2SiMe_3$ | $PhSi(Me)(SiMe_3)_2$ |
|---|---|---|---|---|
| $\lambda_{CT(1)}$ | 497.5(8.35) | 530(8.20) | 493(8.39) | 535(8.15) |
| $\lambda_{CT(2)}$ | 398(9.15) | 405(9.08) | 405(9.11) | 408(9.20) |

Despite the considerable size of substituents in the molecule of these donors, the positions of the CT bands in the spectra of their complexes obey the linear dependence:

$$h\nu_{CT} = a\, IP^{D} - b$$

suitable for a great number of substituted benzenes (see Eqs. (3.2)–(3.5) in Section 3.1). Steric hindrance produced by a bulky substituent in a planar molecule of hydrocarbon does not produce a significant effect on the CT band position, but it certainly decreases the stability of the complex.

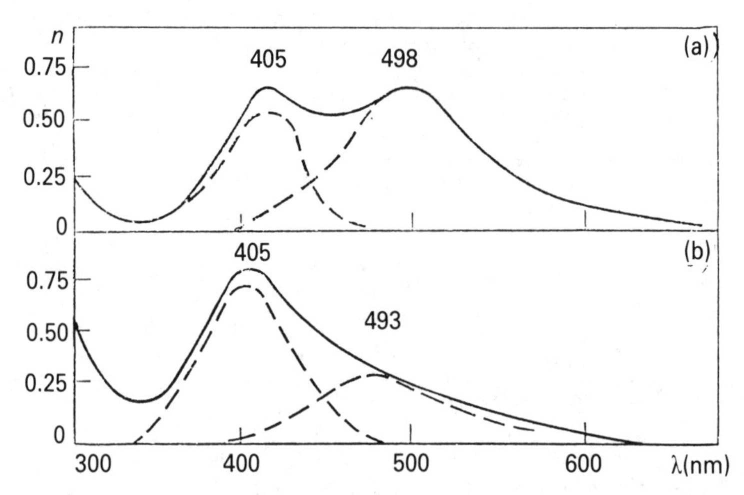


Fig. 10.2. Electron absorption spectra of TCNE complexes of (a) benzyltrimethylsilane and (b) phenylpentamethyldisilane.

The relative intensity of CT bands in the spectra of substituted-benzene complexes can vary within wide limits. Figure 10.2 shows the electron absorption spectra of the TCNE complexes of benzyl(trimethyl)silane and phenyl pentamethyl-disilane. The ratios of CT intensities $D_{CT(1)}/D_{CT(2)}$ in their spectra were 1.32 and 0.45 respectively. These ratios most probably depend on the composition of occupied MOs of the donor from which the electron is moved to the acceptor molecule. The HOMO of benzyl(trimethyl)silane is mostly formed by the $\pi$ orbital of benzene, while in the phenyl pentamethyl-disilane the HOMO is mostly formed by the $\sigma$(Si–Si) orbital. It should be added that $\pi$ donors usually give TCNE complexes whose colour is more intense than the colour of complexes formed by $\sigma$ donors [72, 75].

Fukuzumi and Kochi [88] gave unambiguous evidence of electron transfer not only from the HOMO but also from lower molecular orbitals of donors as they interact with acceptors.

Substituted polycyclic aromatic hydrocarbons can also, in a similar way, donate electrons to $\pi$ acceptors not only from the highest $\pi$ orbitals but also from the lower MOs. Consider the substituent R at position 9 of the anthracene derivatives $C_{14}H_9R$ and the energies of the first and second CT bands in the electron absorption spectra of their TCNE complexes [199]:

| | H | pH | Cl | Br | $COCH_3$ | $COOCH_3$ |
|---|---|---|---|---|---|---|
| $h\nu_{CT(1)}$ | 1.73 | 1.59 | 1.74 | 1.74 | 1.84 | 1.84 |
| $h\nu_{CT(2)}$ | 2.79 | 2.80 | 2.80 | 2.86 | 2.82 | 2.89 |

An important estimation of the role played by the frontier $\pi$ level is given by the analysis of the relative reactivity of substituted hydrocarbons to electrophiles and nucleophiles [87, 88, 448]. Kochi *et al.* compared the parameters of the interaction of aromatic hydrocarbons with various acceptors and electrophilic agents. They established a reliable correlation between the rates of electrophilic substitution reactions and the energies of long-wave CT absorption bands, observed on mixing of an aromatic hydrocarbon with appropriate electrophiles. It has been found that the decreasing optical density $D_{CT}$ of the corresponding CT bands provides information on the rate of electrophile E consumption in electrophilic benzene halogenation and mercuration [87, 88]:

$$-\frac{d[D_{CT}]}{d\tau} = -\frac{d[E]}{d\tau} = k[C_6H_6][E].$$

Relative rates $\log k/k_0$ of reactions of many substituted benzenes (alkyl, methoxy groups, and the halogen atoms as substituents) depend linearly on the energy of CT bands $h\nu_{CT}$ with benzene as the standard compound. Since the slope of each straight

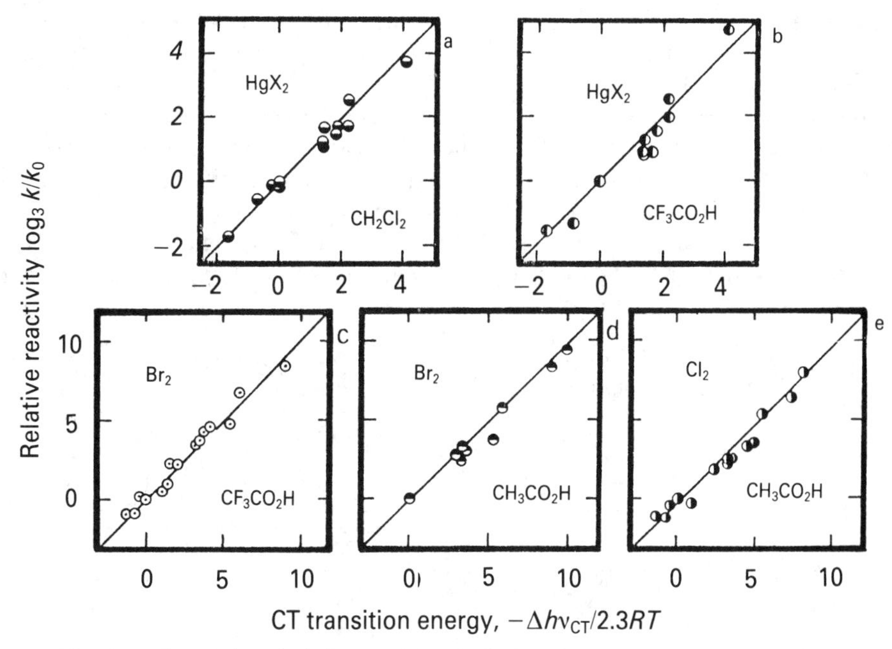


Fig. 10.3. Correlation of relative reactivity of substituted benzenes with CT band energies: (a) mercuration with $Hg(CF_3COO)_2$ in $CH_2Cl_2$; (b) same in $CF_3COOH$; (c) bromination with $Br_2$ in $CF_3COOH$; (d) same in $CH_3COOH$; (e) chlorination with $Cl_2$ in $CH_3COOH$. (From Fukuzumi and Kochi [87])

line in Fig. 10.3 is equal to 1, all points fit the same equation:

$$\log k/k_0 = \Delta h\nu_{CT}/2.3RT \quad (10.2)$$

One step in the general scheme of electrophilic substitution is thus modelled by charge transfer in terms of the Mulliken weak CT complexes. This step consists in electron transfer from aromatic hydrocarbon (donor) to electrophile E (acceptor):

$$Ar + E \rightarrow [Ar^+E^-]solv.$$

It is quite evident that according to this mechanism, the frontier orbitals of the reagents must be decisive for the result of the entire process of substitution.

While estimating the relative reaction rates of substituted hydrocarbons, it should be remembered that even insignificant (at first sight) changes in the energies of frontier orbitals can cause substantial differences in activation barriers of the corresponding reactions:

$$1 \text{ eV} = 23.2 \text{ kcal/mol} = 96.7 \text{ kJ/mol}.$$

Introduction of the nitro group into the benzene nucleus, for example, increases the first ionization potential of the molecule only by 0.72 eV. This stabilization of the frontier occupied $\pi$ electron level decreases, however, the rate of electrophilic nitration by a factor of $10^7$.

A corresponding decrease in the energy of the frontier vacant $\pi$ electron level by the introduction of the nitro group, for example, in a chlorobenzene molecule, accelerates the rate of nucleophilic replacement of the chlorine atom at the para position by about $10^5$ times.

Even simple HMO calculations reliably confirm the effectiveness of the frontier orbitals concept in the assessment of the composition of the isomers formed by electrophilic aromatic substitution in benzene derivatives: the positions with maximum electron densities in the HOMO of aromatic substrate turn out to be preferable for an electrophilic attack (see Section 3.1). The results of calculations of $C_6H_5R$ compounds by the HMO method (PES parametrization) and experimental data of their nitration are given in Table 10.3.

**Table 10.3.** Electron densities in HOMOs of substituted benzenes $C_6H_5R$ and isomer distribution in their electrophilic nitration

| R | Electron density in HOMO ($c^2_{\text{HOMO},\mu}$) [62] | | | Isomer distribution (%)* | |
|---|---|---|---|---|---|
| | ortho | meta | para | meta | para |
| F | 0.097 | 0.072 | 0.320 | 0 | 91 |
| Cl | 0.098 | 0.066 | 0.303 | 1 | 70 |
| $OCH_3$ | 0.144 | 0.023 | 0.221 | 2 | 67 |
| $NHCOCH_3$* | 0.130 | 0.006 | 0.152 | 0 | 80 |
| $C(CH_3)_3$ | 0.116 | 0.048 | 0.281 | 7 | 87 |

*$c^2_{\text{HOMO},\mu}$ values are given for R = $(N(CH_3)_2$
(From Fukuzumi and Kochi [87])

Table 10.3 includes the data for benzene derivatives with electron donating substituents only. The role of the frontier-occupied orbitals in electrophilic substitution reactions with benzene derivatives containing strong electron withdrawing substituents is not so apparent. It is believed that charge control can be decisive in some of these reactions. This is promoted by the increasing energy gap between the HOMO of the aromatic substrate (cf. $IP_1$ in Table 10.1) and the LUMO of the electrophilic reagent: the reactions, in which the frontier orbitals of the reagents differ substantially in their energies, turn out to be charge-controlled (see Section 3.10). Difficulties in differentiation between orbital- and charge-control of electrophilic substitution reactions of these aromatic substrates can also be due to low differences in the energies of their frontier and subsequent occupied MOs (see Table 10.1).

Many correlations between the ionization potentials of substituted arenes and the $\sigma^+$ constants of substituents are explained in the terms of orbital concepts. The $\sigma^+$ constants 'control' some electrophilic substitution reactions:

$$\log k/k_0 = \rho\sigma^+$$

But as has been shown above, the rates of these reactions depend on the energy levels of HOMOs and hence on the ionization potential differences of the corresponding substrates as well [87]:

$$\log k/k_0 = \Delta h\nu_{CT}/2.3RT \approx \Delta IP_X/2.3RT$$

The dependence of the $\Delta IP_X$ values of monosubstituted benzenes $C_6H_5X$ on MO coefficient at the site of substituent X attachment and on constants $\sigma_X^+$, is also shown [449]:

$$\Delta IP_X = f(c^2_{HOMO}, \mu, \sigma^+_{p-X}, \sigma^+_{m-X})$$

Koptyug *et al.* [447, 450] developed a similar approach to estimate the ionization potentials of disubstituted benzenes, naphthalenes anthracenes and pyridines.

The orbital energies are decisive for estimation of many other properties of aromatic hydrocarbons. Note that the values of proton affinity $PA_H$ depend linearly (with the slope equal to 1) on the ionization potentials (see Section 11.2.2):

$$PA_H = -IP(\text{eV}) + C \tag{10.3}$$

It has been established that there is a correlation between the potentials of anode oxidation $E^{ox}_{1/2}$ of hydrocarbons, determined by cyclic voltammetry in an organic solvent (acetonitrile or trifluoroacetic acid), and ionization potentials determined in the gas phase. One should emphasize that the slopes of the corresponding dependences are the same and are equal to unity. The differences in the energy of solvation of neutral molecules and cation radicals of various aromatic hydrocarbons should therefore be assumed as being equal and independent of the nature of the substituent in the hydrocarbon, or of the solvent [87].

The dependences of oxidation of benzolonium ions on HOMO parameters were established [451]. As in unsubstituted aromatic hydrocarbons (see Section 6.3.5), the analysis of the structure and symmetry of occupied orbitals in substituted benzenes makes it possible to estimate the transition energies in electron absorption spectra of the corresponding benzolonium ions [452].

Once the structure of occupied MOs is known, one can explain and predict the results of the reactions of aromatic hydrocarbons with electrophiles. Likewise, the knowledge of the parameters of the vacant MOs permits the analysis of the reactions of these hydrocarbons with nucleophiles.

As follows from Fig. 10.1, the introduction of an electron donating substituent (methyl group) must increase the energy of the benzene orbital of $b_1$ symmetry. In fact, as has been shown in Section 6.2.1, the lower unoccupied orbital in the toluene molecule is the orbital of $a_2$ symmetry.

By contrast, as predicted by MO theory, the introduction of an electron withdrawing substituent into the benzene molecule must decrease the energy of the vacant MO of $b_1$ symmetry, which thus turns out to be a frontier vacant MO. It can be seen from ESR spectra of anion radicals of phenylmethylsilane, benzaldehyde, and nitrobenzene, that an unpaired electron is located in their orbitals of $b_1$ symmetry (see Sections. 6.2.2, 8.5.5, and 9.4.4).

The results of Birch reduction of 1,4-disubstituted benzenes is an interesting example of useful application of the frontier orbitals concept [117]. 1,4-bis(trimethylsilyl)benzene and *p*-xylene are reduced with sodium metal in liquid ammonia with subsequent treatment in ammonium chloride solution in different ways to give 1,4-bis(trimethylsilyl)cyclohexa-2,5-diene and 1,4-dimethylcyclohexa-1,4-diene respectively.

These different results were explained by the analysis of the ESR spectra of anion-radicals formed in these reactions. A transformed benzene $\pi_5$ orbital (the so-called symmetric $\pi$ orbital $b_1$) turned out to be the LUMO of 1,4-bis(trimethylsilyl)benzene. Positions 1 and 4 in the corresponding anion radical have maximum spin densities and are protonated in subsequent treatment of the anion radical with aqueous $NH_4Cl$:

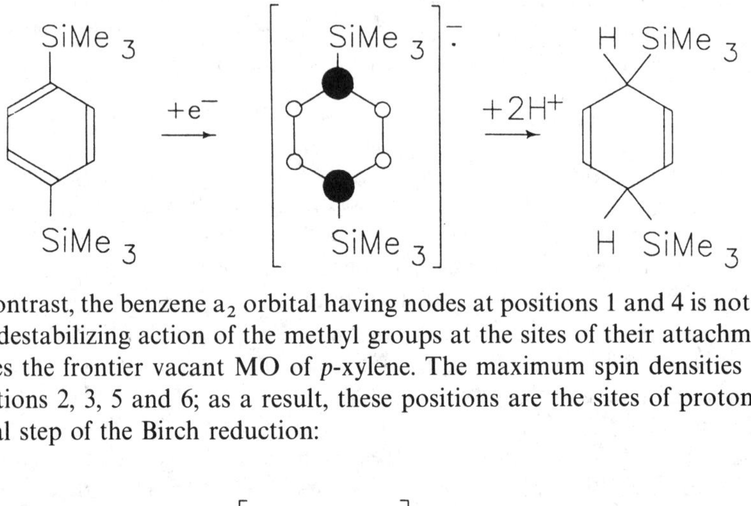

By contrast, the benzene $a_2$ orbital having nodes at positions 1 and 4 is not affected by the destabilizing action of the methyl groups at the sites of their attachment, and becomes the frontier vacant MO of *p*-xylene. The maximum spin densities are seen at positions 2, 3, 5 and 6; as a result, these positions are the sites of protonation in the final step of the Birch reduction:

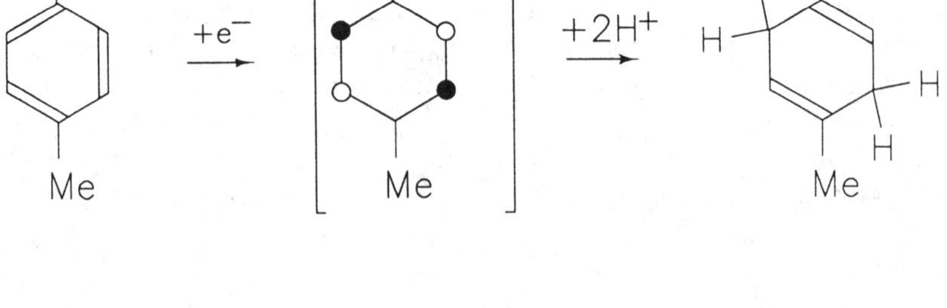

***Problems***

(1) Toluene forms a cation radical when reacted with electrophiles. Suggest an unpaired electron distribution in the cation radical.

(2) Benzoic acid forms a cation radical on polarographic oxidation. Suggest an unpaired electron distribution in this cation radical.

(3) Toluene forms an anion radical on electrochemical reduction. Suggest an unpaired electron distribution in this anion radical.

(4) Substituted benzenes can be reduced by alkali metals in a mixture of liquid ammonia and alcohol. In many cases, a regioselective reaction takes place. Explain the following results:
   (a) anisole forms 1-methoxycyclohexa-1,4-diene with a yield of 84%;
   (b) *p*-xylene forms 1,4-dimethylcyclohexa-1,4-diene in good yield;
   (c) benzoic acid forms 2,5-cyclohexa-2,5-diene-1-carboxylic acid with a yield of 89–95%;
   (d) naphthalene, anthracene and phenanthrene form 1,4-dihydronaphthalene, 9,10-dihydroanthracene, and 9,10-dihydrophenanthrene respectively.

(5) Suggest the direction of electrophilic addition, e.g. of HBr to styrene.

(6) Allyl $\beta$-naphthyl ether undergoes the Claisen rearrangement to give exclusively 1-allyl-2-naphthol. Explain why preference is given to this reaction over the alternative reaction with formation of 3-allyl-2-naphthol.

# 11

# Heterocyclic compounds

Heterocyclic compounds are a large class of organic compounds whose molecules, in addition to carbons, contain also atoms of other elements (N, O, S) in their rings. According to this definition, heterocyclic compounds also include cyclic anhydrides, amides, lactones, lactams, acetals, ethers. These saturated heterocyclic compounds have, in general, the same parameters of electron structure as their linear analogues. The molecular orbitals of saturated heterocyclic compounds have already been discussed in Chapters 8 and 9. In this chapter we shall only discuss the structure of heterocyclic compounds with delocalized MOs and a definite degree of aromaticity. These compounds are very important in biological systems. Chlorophyll, haemoglobin, fragments of vitamins, enzymes, proteins can be mentioned as examples. The specific sequence of attachment of heterocyclic molecules in long chains of nucleic acids determines, for example, the hereditary mechanisms. Fragments of heterocyclic compounds are present in many dyes, medicines, phototropic compounds, etc.

## 11.1 FIVE-MEMBERED HETEROCYCLES

### 11.1.1 Pyrrole

The pyrrole nucleus is a fragment in many natural compounds, such as chlorophyll, bilirubin, nicotine, cocaine, etc. Many of them are biologically active substances.

The molecule of pyrrole is planar. The angles and bond distances of this compound were determined by microwave spectroscopy:

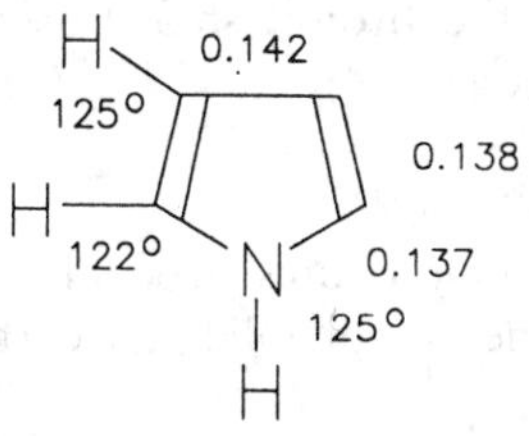

The aromaticity of pyrrole is quite high. For this reason, the lone pair of the nitrogen atom is strongly delocalized so that the basicity of pyrrole is low (pKa of the conjugated acid is $-4.4$). Hydrogenation is hindered but substitution reactions occur readily. It is admitted within the terms of the valence bond method that the electron structure of the pyrrole molecule is determined by the resonance of five resonance structures *a* through *e*, the form *a* being the main contributor:

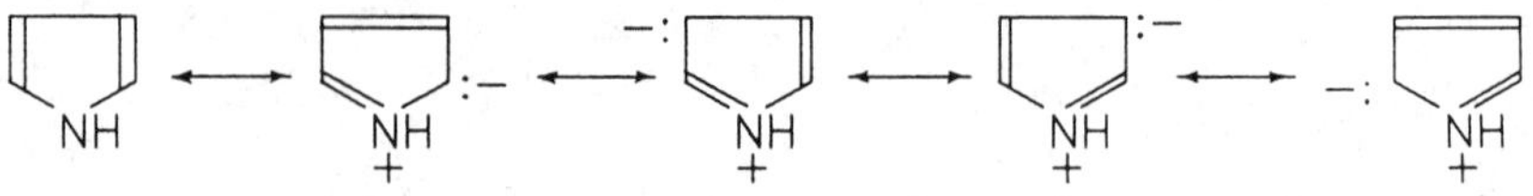

In terms of the MO theory, the electron structure of pyrrole has the fragment $\pi$ orbitals of butadiene and nitrogen atoms in the basis set (Fig. 11.1) [453]. The shapes of the occupied $\pi$ MOs and of the frontier vacant MO of pyrrole are shown below:

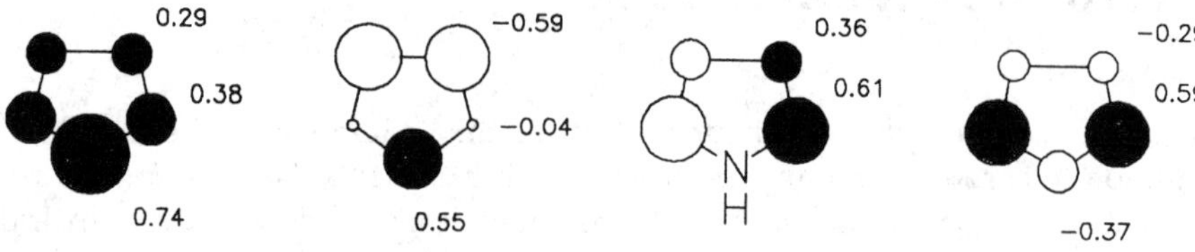

$\pi_1, b_1$ $\pi_2, b_1$ $\pi_3, a_2$ $\pi_4, b_1$

By the rules of symmetry, the $a_2$ orbital is formed without involvement of the nitrogen atom, while $b_1$ orbitals, on the other hand, contain a considerable contribution of the $p\pi$ orbital of the heteroatom.

The first two bands of the PE spectrum are assigned to ionization of $a_2(\pi_3)$ and $b_1(\pi_2)$ orbitals. Assignment of ionization of the $\pi_1$ orbital is subject to argument, and for this reason, the region of the spectrum at 12–13 eV was studied in more detail. The spectra of pyrrole, recorded with Ne(I) and Ne* sources, were, for example, compared [454]. Since there is an abrupt increase in the intensity on Penning ionization, the third band in the PE spectrum is assigned to ionization of the $\pi_1$ orbital: 8.23 eV ($\pi_3$), 9.22 eV ($\pi_2$) $\approx$ 11.2 eV ($\pi_1$).

The electron affinity of the pyrrole molecule is rather low compared with that of butadiene: $EA_1 = -2.38$ eV; $EA_2 = -3.44$ eV [46]. The position of the longest wavelength band in the UV spectrum of pyrrole does not depend on solvent: 207.5 nm ($\log \varepsilon = 3.88$) in cyclohexane, and 208 nm ($\log \varepsilon = 3.86$) in ethyl alcohol. Since the electron system of pyrrole is $\pi$ excessive, the alkyl groups produce a slight hypsochromic shift in the longest wavelength band while the electron withdrawing substituents cause a strong bathochromic shift. Formyl- and cyanopyrroles absorb, for example, at 260–300 nm ($\log \varepsilon > 4$).

### 11.1.2 Furan

The fragment of a furan molecule is present in many natural compounds, such as egonol, furaneole, sesquiterpenes, cardenolides, ascorbic acid, etc.:

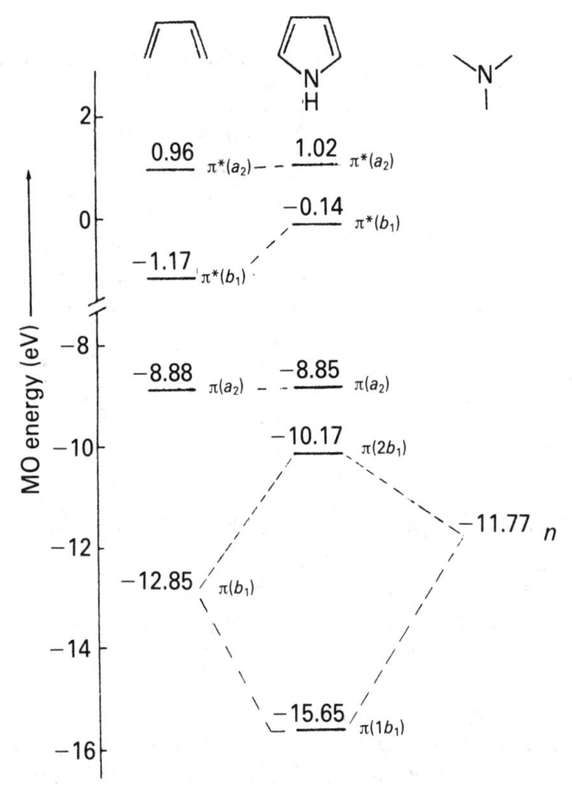


Fig. 11.1. Formation of $\pi$MOs of pyrrole in the basis of fragment orbital according to CNDO/S calculations. (From Schäfer *et al.* [453])

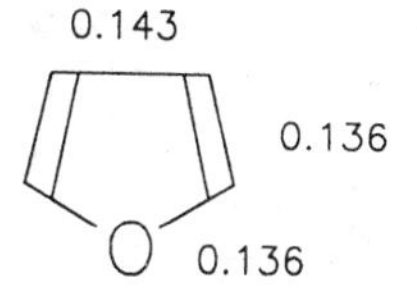


As distinct from pyrrole, the aromaticity of furan is lower: the energy of delocalization of its $\pi$ electrons is 6.7 kJ/mol (instead of 35.2 kJ/mol for pyrrole). This is also confirmed by the geometry of the furan molecule.

Its C—C bond distances are closer to those of butadiene than of the pyrrole molecule. The CCO angles are 110°, while the CCC and COC angles are 106° (bond distances and angles are determined by microwave spectroscopy).

As in pyrrole, the first two electron levels in furan are formed by the $\pi$ orbitals: $a_2(\pi_3)$ orbital is the frontier occupied MO, which is followed by the $b_1(\pi_2)$ MO. Assignment of the band corresponding to the ionization of the $\pi_1$ orbital of furan is also difficult since it is strongly overlapping in the region of the spectrum at 15–16 eV [455]. Table 11.1 gives the assignment of the bands in the furan spectrum according to calculations in the 4-31G basis set.

**Table 11.1.** Ionization potentials and higher occupied Mos of furan (PES data and calculations in the 4-31G basis set)

| $IP_i$ (eV) | Molecular orbitals | | |
|---|---|---|---|
| | $-\varepsilon_i$ (eV) | No. and symmetry (group $c_{2v}$) | Localization |
| 8.87 | 8.78 | $1a_2$ | $\pi^-$(C=C) |
| 10.38 | 10.81 | $2b_1$ | $\pi^+$(C=C) |
| 12.94 | 14.53 | $9a_1$ | $\sigma$(C—O) |
| 13.84 | 15.38 | $8a_1$ | $\sigma$(C—C), $\sigma$(C—H) |
| 14.47 | 15.76 | $6b_2$ | $\sigma^-$(C=C), $\sigma^-$(C—H) |
| 15.20 | 16.43 | $5b_2$ | $\sigma^-$(C—O), $\sigma^-$(C—H) |
| (15.2) | 17.17 | $1b_1$ | $n$(O) |

(From Kimura *et al.* [38])

The electron affinity of furan was measured by the ETS method: $EA_1 = -1.76$ eV, $EA_2 = -3.14$ eV [46]. Attachment of the oxygen atom to the butadiene fragment increases the energies of the lower vacant $\pi$ levels of the molecule ($EA_1$ of butadiene is $-0.62$ eV), but the increase is lower than in the pyrrole molecule ($EA_1 = -2.38$ eV).

### 11.1.3 Thiophene, selenophene, tellurophene

Thiophene is the nearest heteroanalogue of benzene. There are some thiophene derivatives among naturally occurring compounds and medicines.

Variations in the C—C bond distances in the thiophene molecule are less significant than in furan and more significant than in pyrrole molecules:

$\angle$CSC = 92°
$\angle$CCC = 112°

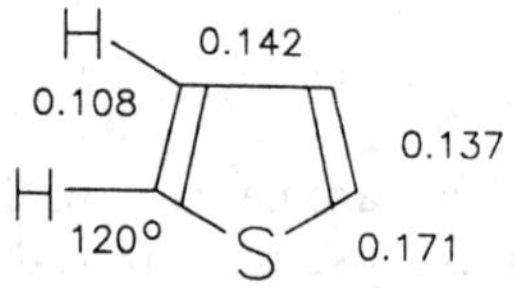

As in the other five-membered heterocycles, the main difficulty in the interpretation of thiophene PE spectrum is assignment to the ionization band of the lowest occupied $\pi_1$ orbital. The use of various sources of ionization (excited neon atoms included [454]) and comparison of intensities of the recorded bands made it possible to differentiate $\pi$ and $\sigma$ electron levels in thiophene. It can be seen from Fig. 11.2 that, of the three possible assignments – the second band, its shoulder or the third band – this is the shoulder that should be assigned to $\pi_1$ MO in the Ne(I) spectrum of thiophene, because the intensity of this band increases markedly with the Ne* source of excitation.

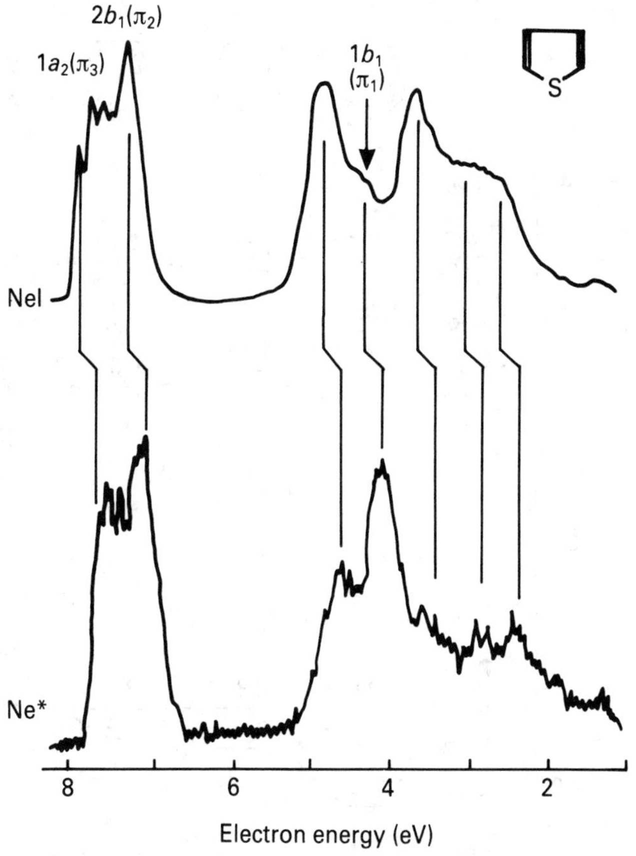


Fig. 11.2 Ne(I) PE and Ne* Penning spectra of thiophene (see explanation to Fig. 4.4). (From Munakata *et al.* [454])

It was thus found that the $\pi_1$ orbital in thiophene is separated from the $\pi_2$ and $\pi_3$ orbitals by only one $\sigma$ level, while in furan these orbitals are separated by at least four $\sigma$ levels. The comparison of the orbital structure of thiophene, selenophene, and tellurophene by the PE spectra and electron absorption spectra of their TCNE complexes showed the inversion of the higher occupied $\pi$ orbitals in tellurophene (Fig. 11.3). Since the $\pi$ orbital $1a_2$ has a node at position 1, the change of heteroatom in the series O, S, Se, Te produces a small effect on the position of the band assigned to ionization of this orbital. On the contrary, the $\pi$ orbital $2b_1$ contains the contribution of the heteroatom. Therefore, its energy increases markedly with the atomic number of the heteroatom (Table 11.2). Vacant electron levels of thiophene, selenophene, and tellurophene were studied by ETS spectroscopy. The obtained data are also given in Table 11.2 and Fig. 11.3 [455].

It can be seen from Fig. 11.3 that, as compared with the ET spectrum of furan, the spectra of selenophene and tellurophene have a new resonance signal assigned to stabilization of the corresponding anion state. The results of calculations by the

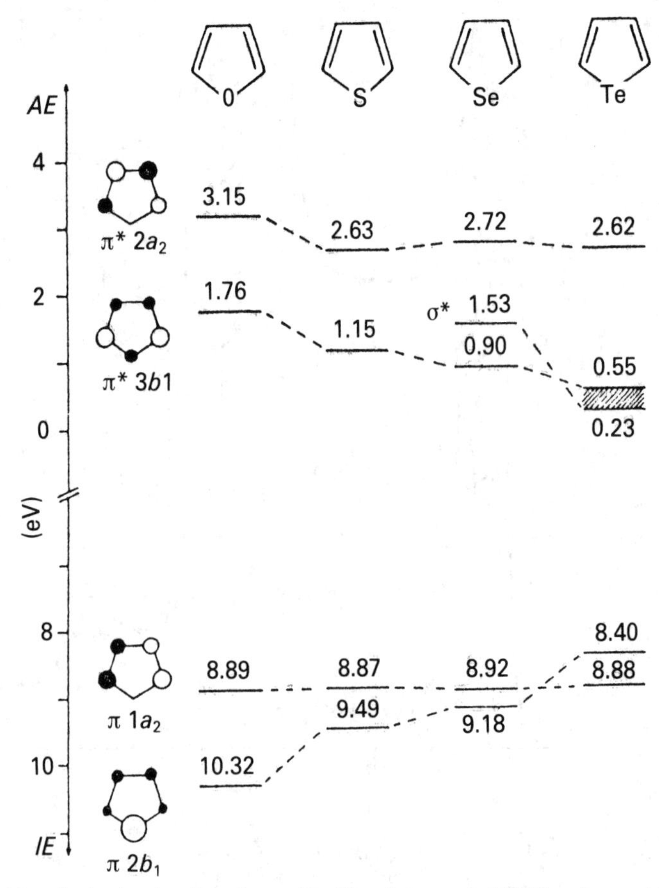


Fig. 11.3. Experimental estimation of energies of frontier occupied (PES) and vacant (ETS) MOs of chalcogenophenes. (From Schäfer *et al.* [453])

$X_\alpha$ method show that this state is formed as a result of electron capture by the $\sigma^*$ MO having a considerable contribution from the vacant *d* orbitals of the heteroatom. It seems that contributions of *d* orbitals in thiophene and selenophene are about equal.

## 11.2 SIX-MEMBERED HETEROCYCLES

### 11.2.1 Pyridine

The fragment of pyridine occurs in many drugs, dyes (e.g. anthrapyridone, diazapyranthrone) and in alkaloids (e.g. cocaine, piperine, nicotine).

The geometric parameters of pyridine are similar to those of benzene:

$$\angle C_{(3)}C_{(4)}C_{(5)} = 118^\circ$$
$$\angle C_{(2)}NC_{(6)} = 117^\circ$$

**Table 11.2.** Ionization potentials, electron affinities, higher occupied and lower unoccupied MOs of thiophene and selenophene (PES, ETC data and $X_2$ calculations)

| No. and symmetry of MO (group $C_{2v}$) | Energy ($EA_j$ for vacant and $IP_i$ for occupied MO) | | Charge distribution over atoms in % (parenthesized are AO contributions in %) | | |
|---|---|---|---|---|---|
| | exp. | calcul. | heteroatom | $C_1$ | $C_2$ |
| | | Thiophene | | | |
| $2a_2(\pi^*)$ | 2.63 | 2.81 | 3.0(d) | 3.0 | 13.0 |
| $3b_1(\pi^*)$ | 1.15 | 1.57 | 12.2(76% *p* 24% *d*) | 27.0 | 11.4 |
| $b_2(\sigma^*)$ | | 1.63 | 11.8(40% *p* 60% *d*) | 6.5 | 1.0 |
| $1a_2(\pi)$ | 8.87 | 8.40 | 2.1(d) | 45.9 | 14.6 |
| $2b_1(\pi)$ | 9.49 | 9.11 | 34.5(p) | 1.2 | 29.7 |
| | | Selenophene | | | |
| $2a_2(\pi^*)$ | 2.72 | 2.76 | 1.0(d) | 1.5 | 15.8 |
| $3b_1(\pi^*)$ | 0.90 | 1.53 | 10.8(81% *p* 19% *d*) | 26.4 | 12.0 |
| $b_2(\sigma^*)$ | 1.5 | 1.72 | 15.6(48% *p* 52% *d*) | 8.1 | 1.0 |
| $1a_2(\pi)$ | 8.92 | 8.22 | 1.4(d) | 46.2 | 15.2 |
| $2b_1(\pi)$ | 9.18 | 8.62 | 39.7(p) | 1.1 | 24.9 |

(From Modelli *et al.* [455])

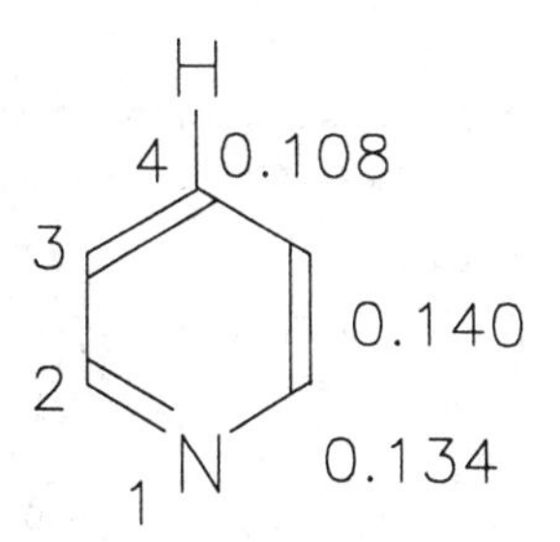


In accordance with its heteroaromatic character, pyridine has a comparatively low basicity (pKa of the conjugated acid is 5.20) and ability to electrophilic substitution reactions.

Replacement of one CH fragment in a benzene molecule by a more electronegative nitrogen atom removes degeneracy of the $b_1$ and $a_2$ orbitals. The energy of the $b_1$ orbital, having non-zero coefficient at the position of the heteroatom, thus decreases. Ionization of this orbital in the PE spectrum occurs at 10.51 eV (Fig. 11.4) [454]. Assignment of the overlapped band at 9.3–9.5 eV is more complicated. According to *ab initio* calculations, n(N) and $\pi_3$ orbitals are ionized in this region (Table 11.3).

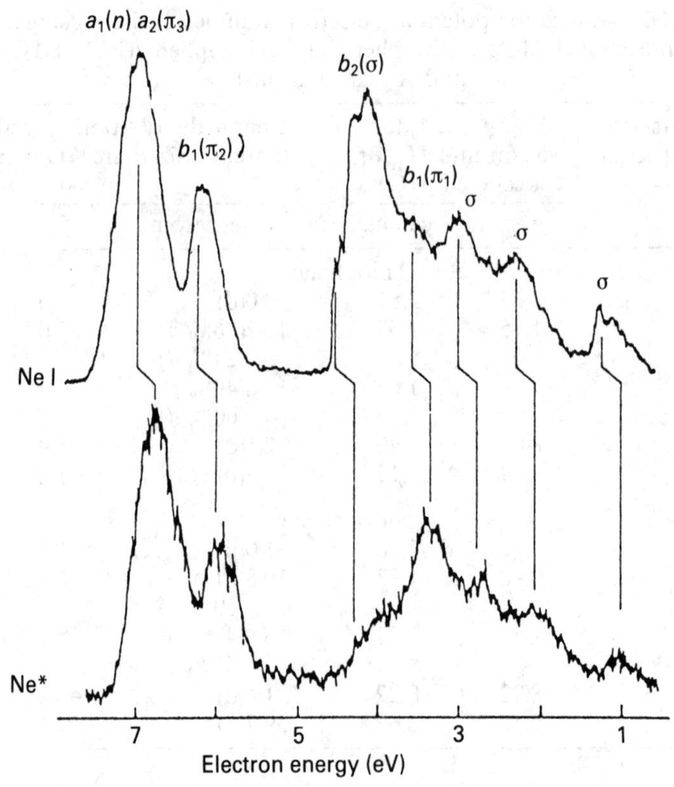


Fig. 11.4. Ne(I) PE and Ne* Penning spectra of pyridine (see explanation to Fig. 4.4). (From Munakata *et al.* [454])

**Table 11.3.** Ionization potentials and higher occupied MOs of elementa-benzenes (PES data and calculations in STC-3G basis set)

| E = N | | | E = P | | E = As | | E = Sb | |
|---|---|---|---|---|---|---|---|---|
| $IP_i$ (eV) | $-\varepsilon_i$ (eV) | MO | $IP_i$ (eV) | MO | $IP_i$ (eV) | MO | $IP_i$ (eV) | MO |
| 9.7 | 6.16 | $11a_1(\sigma, \pi)$ | 9.2 | $3b_1(\pi)$ | 8.8 | $5b_1(\pi)$ | 8.3 | $7b_1(\pi)$ |
| 9.8 | 7.72 | $1a_2(\pi)$ | 9.8 | $1a_2(\pi)$ | 9.6 | $2a_2(\pi)$ | 9.4 | $3a_2(\pi)$ |
| 10.5 | 8.03 | $2b_1(\pi)$ | 10.0 | $13a_1(\sigma, n)$ | 9.9 | $17a_1(\sigma, n)$ | 9.6 | $21a_1(\sigma, n)$ |
| 12.5 | 10.63 | $7b_2(\sigma)$ | 11.5 | $8b_2(\sigma)$ | 11.0 | $10b_2(\sigma)$ | 10.4 | $12b_2(\sigma)$ |
| 12.6 | 11.52 | $1b_1(\pi)$ | 12.1 | $2b_1(\pi)$ | 11.8 | $4b_1(\pi)$ | 11.7 | $6b_1(\pi)$ |

(From Batich *et al.* [457])

Angular distribution of photoelectrons in the PE spectrum of pyridine and ionization spectra with Ne(I) and Ne* sources were measured and compared [454]. The first, the second and the fourth ionization bands in the Penning spectrum exceed appreciably the other bands by their intensity. The first band in both spectra (ionization of $n_N$ and $\pi_3$ orbitals) is twice as intense compared with the second band.

Excitation with the Ne* atoms thus equally increases the intensity of the bands assigned to $\pi$ and $n$(N) orbitals. The shapes of the $\pi$ orbitals of pyridine are shown below:

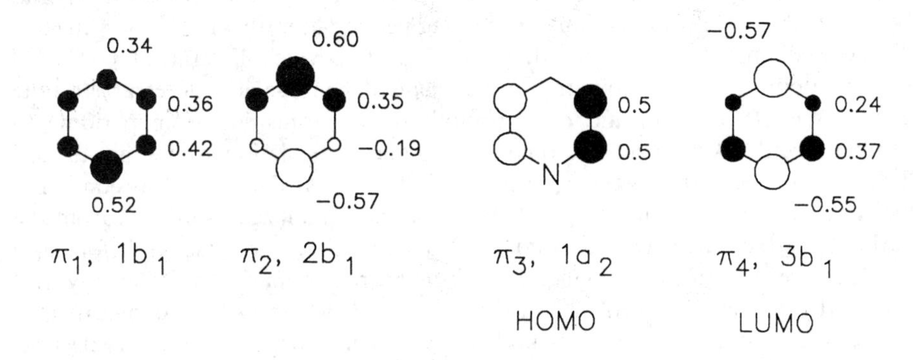


### 11.2.2 Other heterocyclic analogues of benzene

Very low interaction of the $n$(P) orbital with carbon $\pi$ orbital in phosphorus analogues of anilines is due to the $s$ character of phosphorus lone pair. An alternative explanation of this fact by geometric parameters (greater C—P bond length compared with the C—N bond distance, different size of $p$ orbitals of carbon and phosphorus) is hardly reasonable because the compounds with di-coordinated atoms of the VA group elements, up to bismuth, have been studied sufficiently well.

The analogues of benzene $C_5H_5E$ (E = N, P, As, Sb, Bi) are of special interest in this respect. These heterocyclic compounds have planar molecules. The bond distances in them are characteristic of aromatic compounds. According to [456], the P—C bond distance in 2,6-dimethyl-4-phenylphosphabenzene is 0.174 nm, while the P=C double bond distance in $Ph_3P{=}CH_2$ is 0.165 nm and the P—C ordinary bond, 0.187 nm. The C—N, C—As and C—Sb bond lengths in the corresponding elementabenzenes are 0.137, 0.185 and 0.205 nm respectively [62].

The ionization potentials of elementabenzenes are given in Table 11.3. The occupied MOs in these compounds have been considered in terms of the MO perturbation theory [117, 457]. The changes in the energy levels of benzene $\pi$ orbitals are proportional to the squares of the coefficients of these orbitals at the site of heteroatom attachment. Benzene $\pi$ orbitals and values of squares of the coefficients are:

$$1b_1, 0.17; \quad a_2, 0.0; \quad 2b_1, 0.33$$

The ionization potential of the $2b_1$ orbital must accordingly change to a greater extent than the other orbitals. The ionization potential of the $1b_1$ orbital changes less significantly, while the $a_2$ orbital does not practically change its potential on substitution of the CH fragment by the heteroatom. The dependences of the ionization potentials of $\pi$ MOs of elementabenzenes on the ionization potentials $IP_E$ of the $p$ orbitals of the elements are found to be as follows:

$$IP_{(2_{b1})} = 5.24 + 0.36IP_E$$
$$IP_{(1_{b1})} = 10.25 + 0.16IP_E$$

Slopes of these dependences ($IP_E$ values are 14.54 for N, 11.0 for P, 9.81 for As, and 8.64 eV for Sb) confirm predictions of the MO perturbation theory.

While discussing the correlation diagram (Fig. 11.5), which illustrates the effect of the heteroatom on the electron structure of elementabenzene, it is necessary to note the inversion of the $a_2(\pi)$ and $b_1(\pi)$ orbitals on the transition from pyridine to phosphabenzene. Experiment and quantum chemical calculations show that the $sp^2$ hybridized nitrogen atom is the acceptor of electrons. For this reason, as compared with benzene, the $2b_1$ orbital in pyridine is markedly stabilized as distinct from the $a_2$ orbital in which the heteroatom does not participate. The reverse holds for more electropositive elements: the higher $b_1$ orbital increases markedly its energy and becomes the frontier occupied MO in the compounds where E is P, As or Bi.

Some other properties of elementabenzenes, which depend on their electron structure, should be noted. Thus, the proton affinity $PA_H$ measured in the gas phase as the energy of heterolytic dissociation of conjugated acids $BH^+ \rightleftharpoons B + H^+$ $PA_H(B) = \Delta H^0$ (kJ/mol) turned out to be (in kJ/mol) 764.9 for benzene, 918.6 for pyridine, 819.8 for phosphabenzene, and 792.6 for arsabenzene [458].

A detailed analysis of the effects associated with protonation of elementabenzenes and their acyclic analogues showed, however, that the linear dependences of the type:

$$PA_H(B) = f(IP_{HOMO})$$

which are similar to Eq.(10.3), are probably not universal. It turned out, for example, that phospha- and arsabenzenes, having lower ionization potentials than pyridine, are characterized by lower proton affinities. The unusual behaviour of these compounds of the group VA heavy elements was explained by the specific hybridization of the lone pair localized over the heteroatom. In the pyridine molecule (the endocyclic CNC angle of about 117°), the higher $p$ character of the lone pair over the nitrogen atom is not doubted and it accounts for the high proton affinity. The decreased CEC angle in the molecule of phospha- and arsabenzenes (CPC and CAsC angles are 101° and 99° respectively) increases the $s$ character of the lone pair over the heteroatom and thus decreases the proton affinity of the molecule.

The low proton affinities of phospha- and arsabenzenes are due to their rigid cyclic structure. It has already been noted that acyclic phosphines and arsines also have high $s$ character of their lone pairs over the heteroatoms. But the lability of their molecules makes it possible to change their geometry and increase the $p$ character of the lone pairs on protonation: the basicity of trimethylphosphine $P(CH_3)_3$ in the gas phase is even higher than that of trimethylamine $N(CH_3)_3$ ($PA_H$ are 941.9 and 936.0 kJ/mol respectively [458]).

It is also necessary to note the following feature of the electron structure of elementabenzenes: the orbitals of lone electron pairs of the heteroatoms in phospha- and arsabenzenes are shifted toward the deeper regions of the electron shell to become the third from top, as a result of which the benzene $\pi$ orbitals become HOMOs (see

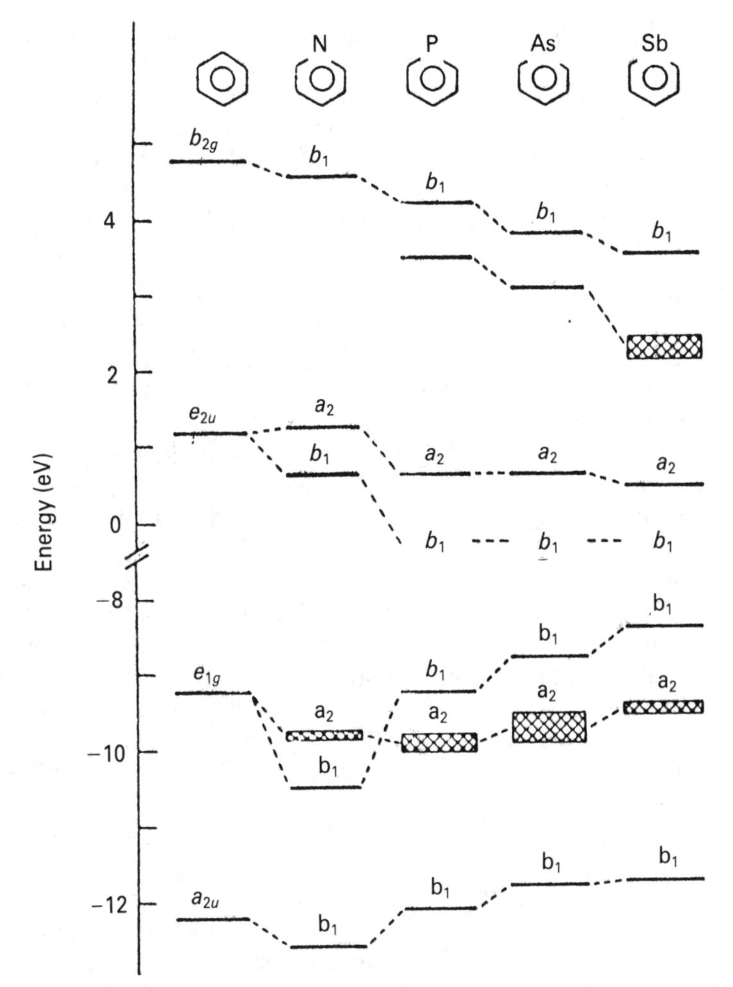


Fig. 11.5. Experimental estimation of $\pi$MO energies of element (VA)benzenes (occupied MOs are given in the lower part of the figure according to PES; vacant MOs are given according to ETS data). (From Hodges *et al.* [458] and Burrow *et al.* [459])

Table 11.3). The proton is therefore attached to the arsabenzene molecule not at the arsenic but at the carbon atom.

These features of the electron structure of elementabenzenes show themselves in solution as well. While pyridine has distinct basic properties, phospha- and arsabenzenes are not protonated by trifluoroacetic acid; nor do they form quaternary salts even with strong alkylating agents [458].

Electron affinity of elementabenzenes was studied thoroughly by ET spectroscopy as well [459]. As in the case of an occupied MO, the orbitals of $b_1$ symmetry undergo a greater decrease in their energies (see Fig. 11.5). From the pyridine spectrum it follows that the $E_2$ degenerate state is split to give anion radicals in two states: $B_1$

and $A_2$. The first resonance recorded in the spectra of phospha-, arsa-, and stilbabenzene is assigned to the anion radical in the $A_2$ state because the corresponding values of electron affinity are practically equal: $-0.64$, $-0.62$, and $-0.60$ eV. The anion radical in the $B_1$ state for these compounds cannot be detected by ET spectroscopy probably because their electron affinities may be positive. The data of the ESR spectra of the anion radical are in agreement with this suggestion: the lowest anion states of pyridine and phosphabenzene have the $B_1$ symmetry [460].

### 11.2.3 Diazines and sym-triazines

Azines occur in nature; some of them are used in medicine. Most common are pyrimidine derivatives: a uracil fragment is a part of RNA, thymine is a DNA component, while a cytosine fragment is found in both DNA and RNA. Pyridazines have not been found in natural compounds.

Since the —CH= and —N= fragments are isoelectronic, PE spectra of diazine have no new (compared with pyridine) bands, but the energies of ionization change appreciably. Figure 11.6 illustrates these tendencies using the formation of pyrazine MOs as an exemplary scheme [22].

The ionization bands in PE spectra of azines are assigned on the basis of the Green functions [461]. The obtained data are given in Table 11.4. The defect of Koopmans' theorem is noted owing to low energies of vacant electron levels in azine molecules.

The anion states of azines have been studied in both the gas and liquid phases [45]. Vibrational states of the first resonance were only recorded by ET spectroscopy because adiabatic values of $EA_1$ of azines have the positive sign. The parameters of the vibrational structure show that the vibrations of the C—C bonds are excited in the first anion state.

Electron affinities $EA_1$ of azines in the gas phase (ET spectroscopy) and in dimethyl formamide (polarography) are correlated linearly (Fig. 11.7). This makes it possible to estimate the $EA_1$ values that could not be measured by ETS in a direct experiment. Table 11.5 gives (in parentheses) the corresponding estimations. In general, accumulation of nitrogen atoms in the molecule stabilizes its vacant frontier orbital: the greater the number of nitrogens in a ring, the easier an electron attaches to the molecule.

## 11.3 POLYCYCLIC HETEROAROMATIC COMPOUNDS

Many heteroaromatic compounds were calculated by quantum chemical methods. The results of these calculations were compared with the biological activity, pharmacological effects, and other properties of the relevant compounds [91]. Objective data on the occupied and vacant electron levels are available for complicated polycyclic heteroaromatic molecules as well.

As in pyrrole and furan, in which the orbitals of five-membered monocycles are formed by the mixing of fragment orbitals of butadiene and the heteroatom, the molecular orbitals of tricyclic heteroaromatic compounds, such as carbazole and dibenzofuran, are formed by mixing of fragment orbitals of biphenyl and the

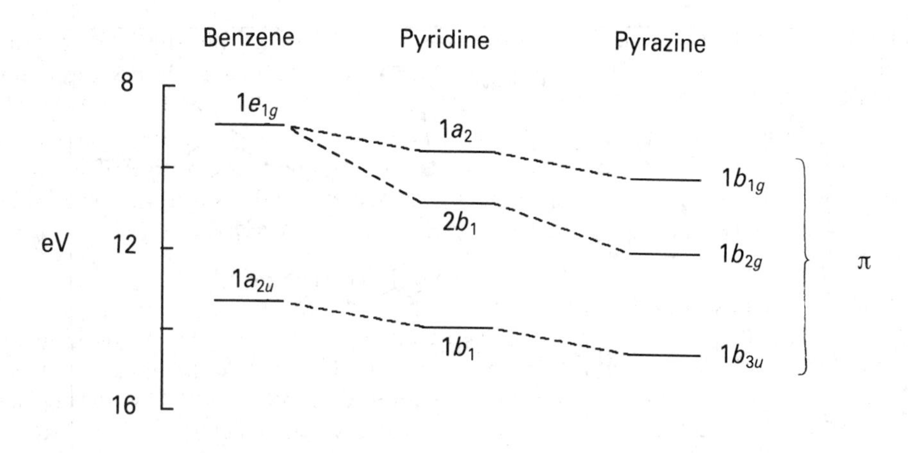


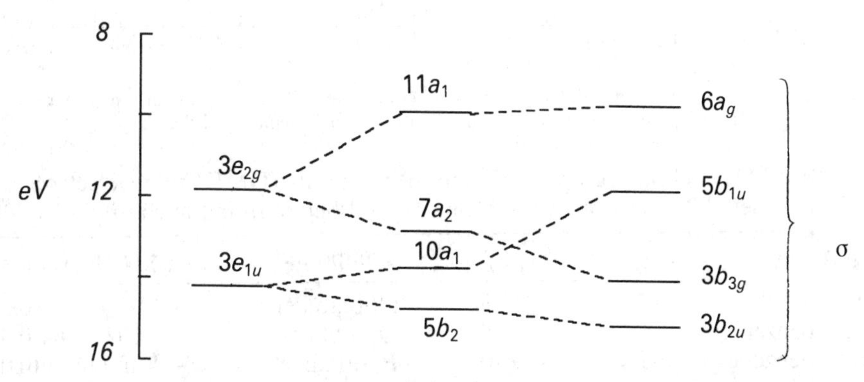


Fig. 11.6. Structure of occupied orbitals in azine molecules. (From Ballard [22])

**Table 11.4.** Ionization potentials and higher MOs of azines (PES data and calculations on the basis of the Green functions)

| Pyrazine | | Pyrimidine | | Pyridazine | | sym-Triazine | |
|---|---|---|---|---|---|---|---|
| $IP_i$ (eV) | MO | $IP_i$ (eV) | MO | $IP_i$ (eV) | MO | $IP_i$ (eV) | MO |
| 9.4 | $4a_g(n)$ | 9.7 | $5b_2(n)$ | 9.3 | $5b_2(n)$ | 10.4 | $4e'$ |
| 10.2 | $1b_{2g}(\pi)$ | 10.5 | $2b_1(\pi)$ | 10.5 | $1a_2(\pi)$ | 12.0 | $1e''(\pi)$ |
| 11.4 | $3b_{2u}(\pi)$ | 11.2 | $7a_1(n)$ | 11.3 | $2b_1(\pi)$ | 13.3 | $3a_1$ |
| 11.7 | $1b_{3g}(\pi)$ | 11.5 | $1a_2(\pi)$ | 11.3 | $7a_1(n)$ | 14.7 | $3e'$ |
| 13.3 | $2b_{1g}$ | 13.9 | $6a_1$ | 13.8 | $6a_1$ | 15.6 | $1a_2''(\pi)$ |
| 14.0 | $1b_{1u}(\pi)$ | 14.5 | $4b_2$ | 14.2 | $1b_1$ | 18.2 | $1a_2'$ |

(From von Niessen *et al.* [461])

heteroatom. In dibenzothiophene, for example, the energy is changed only in those fragment $\pi$ orbitals of biphenyl which have the same symmetry as the $\pi$ orbital of the sulphur atom, i.e. $b_1$ symmetry. Molecular $\pi$ orbitals of biphenyl, which have $a_2$

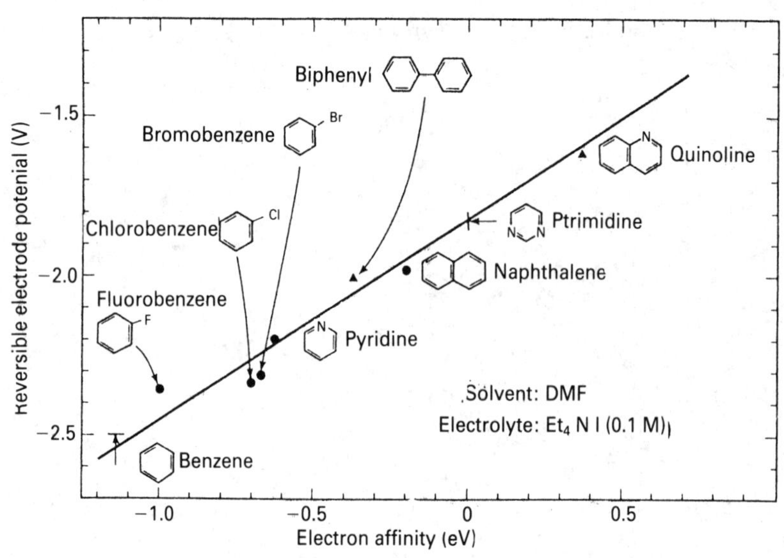


Fig. 11.7. Dependence of polarographic reduction potentials of aromatic and heteroaromatic compounds on their electron affinities. (From Nenner and Schulz [45])

**Table 11.5.** Polarographic reduction potentials (DMFA) and electron affinities (ETC, gas phase) of aromatic and heteroaromatic compounds

| Compound | $E^{red}_{1/2}$ (eV) | $EA_1$ (eV) | Compound | $E^{red}_{1/2}$ (eV) | $EA_1$ (eV) |
|---|---|---|---|---|---|
| Benzene | −2.5 | −1.15 | Naphthalene | −1.98 | −0.20 |
| Fluorobenzene | −2.35 | −1.0 | Quinoline | −1.602 | (+0.36) |
| Chlorobenzene | −2.33 | −0.70 | Pyrimidine | −1.82 | (0.0) |
| Bromobenzene | −2.31 | −0.67 | Pyridazine | −1.657 | (+0.25); |
| Pyridine | −2.20 | −0.62 | | | −0.73 |
| | | −4.58 | Pyrazine | −1.509 | (+0.4) |
| Biphenyl | −2.03 | −0.37 | sym-Triazine | −1.527 | (+0.45) |

(From Nenner and Schulz [45])

symmetry, cannot interact with a heteroatom lone pair and do not change their energies [462].

The analysis of the orbital structure of dibenzothiophene explains the regularities that govern the changes in the energy of $\pi$ electron levels of its analogues as well (compounds LXXXIII):

E

LXXXIII

where E = O, S, Se or Te. Since the heteroatom is involved only in some higher $\pi$ orbitals (Fig. 11.8), subsequent O → S, Se, Te substitution produces a different effect

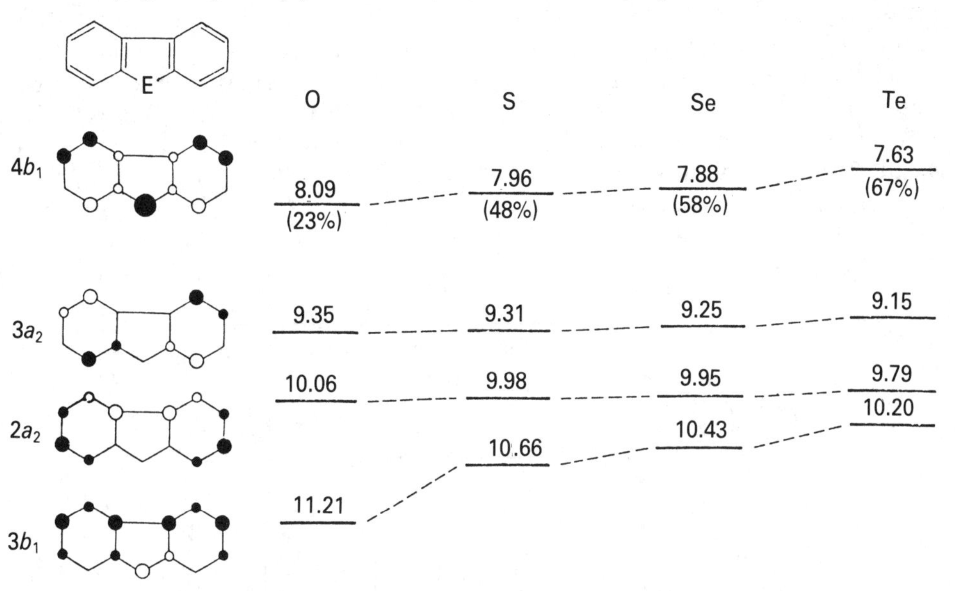


Fig. 11.8. Dependence of higher occupied $\pi$ orbitals of dibenzofuran and its analogues on the nature of the heteroatom (according to PES and simple calculations): parenthesized are contributions of $p_\pi$ orbitals of the heteroatom to HOMO. (Data by Rodin *et al.* [462])

on the energies of $\pi$ electron levels. The energies of the orbitals of $b_1$ symmetry, containing AO of the heteroatom E, increase appreciably: for example, the energy of the $3b_1$ orbital increases by 1 eV in the O → Te substitution. On the other hand, the energies of the $3a_2$ orbitals are changed insignificantly (by not more than 0.2 eV) since the orbitals of $a_2$ symmetry have no contribution of the heteroatom [462].

Likewise, the orbitals of 6,7-thiaperylene, whose molecule has five six-membered rings and the thiophene nucleus, can be regarded as combinations of perylene's $\pi$ orbitals of $b_1$ symmetry and the sulphur's $p\pi$ orbital of $b_1$ symmetry:

The energies of the 6,7-thiaperylene's orbitals of $a_2$ symmetry change only insignificantly, while the electron structure of this heterocyclic molecule is, on the whole, the same as that of its carbocyclic analogue, the corresponding benzoperylene [463]. Similar regularities were observed earlier in the study of the electron structure of thiophene, benzothiophene, dibenzothiophene and their analogues [278, 362].

Figure 11.9 illustrates the data for phenathrene and its thia analogues. The permanence of the $\pi$ electron structure of the organic molecule during substitution of the —CH=CH— fragment by the sulphur atom is, at least partly, explained by almost complete equivalence of the first ionization potentials of ethylene and hydrogen sulphide, whose molecules simulate the said fragments: 10.51 and 10.47 eV respectively. At the same time, the geometric conditions of overlap (and hence the energy effects) for the $2p$ orbitals of carbon atoms and the sulphur's $3p$ orbitals might be supposed to be similar.

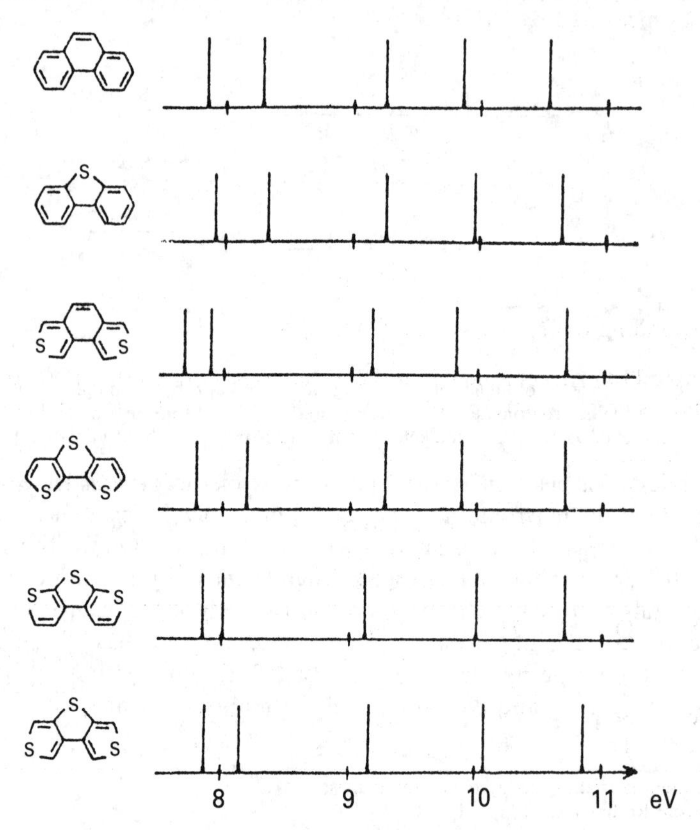


Fig. 11.9. Effect of replacement of the —CH=CH— fragment by the sulphur atom in the series of phenanthrene and its sulphur-containing analogues in their PE spectra (position and relative intensities of PE spectral bands are given). (From Gleiter and Larsen [362])

In Section 2.4, we have already given examples of reliable reproduction of values of $\pi$ electron level energies of organic molecules in calculations with PES parametrization. A multitude of such examples are available for planar conjugated cyclic molecules [58, 278, 288]. They provide convincing evidence that the study of the electron structure of conjugated heterocyclic molecules (very complicated molecules included) can be based on the estimation of the orbital interaction in simple models – substituted ethylenes and substituted benzenes. Even calculations within the framework of the simple method make it possible to predict reliably the values of the orbital energies. In the case of systematic deviations from the series of related compounds, the same calculations suggest the presence of new (disintegrated) effects in the molecule studied. Such deviations can be seen, for example, in molecules that can assume non-planar conformations. The possibility of estimation of the steric structure of organo-element analogues of dihydroanthracene LXXXIV has been studied in detail. The ionization potentials, estimated by the simple method with PES parametrization on the assumption of the planar conformation of these molecules, were noted to be regularly low for these compounds (compared with the ionization potentials as measured by PES).

E'

E

LXXXIV

The corresponding data are given in Table 11.6.

**Table 11.6.** Experimental and calculated first ionization potentials and steric structure of organo-element analogues of 9,10-dihydroanthracene

| $E$ | $E'$ | $IP_1$ (eV) | | $\triangle IP_1$ (eV) | $\theta$ (degrees)* |
|---|---|---|---|---|---|
| | | PES | calculations for planar conformer | | |
| O | O | 7.71 | 7.24 | 0.47 | 140 (176) |
| S | O | 7.75 | 7.37 | 0.38 | 136 (138) |
| S | S | 7.93 | 7.48 | 0.45 | 116 (128) |
| Se | Se | 7.89 | 7.50 | 0.39 | 120 (127) |
| Te | Te | 7.60 | 7.19 | 0.41 | 114 (124) |

*Parenthesized are X-ray structural analysis data.
(From Rodin *et al.* [61])

X-ray structural analysis and other studies show that 9,10-dihydroanthracene and its analogues usually exist in the 'butterfly' conformation:
It can be seen from the formula that folding the molecule by the E–E axis changes the $p\pi$(E) and $p\pi$(C) orbital overlap by $\cos\gamma$ times: these orbitals remain orthogonal relative to the C—E bond, but rotate through an angle $\gamma$ relative to each other. The value of the angle $\gamma$ determines unambiguously the value of the dihedral angle $\theta$, the angle between the planes of the benzene rings. The structure of the molecule suggests the following equation:

$$\theta = 180^\circ - 2\arctan(0.5\tan\gamma)$$

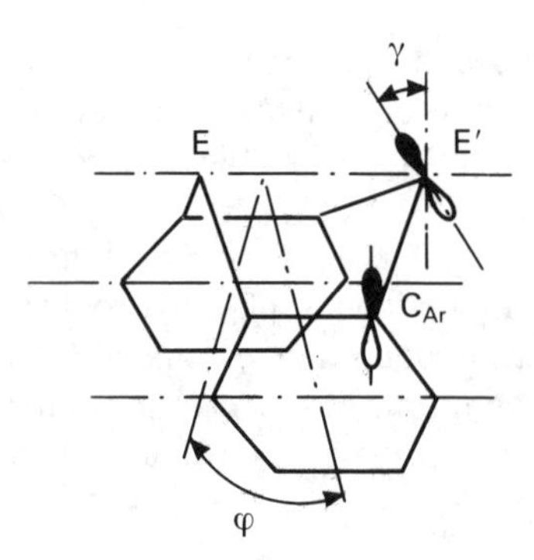


The data given in Table 11.6 show that the analysis of PE spectrum and quantum chemical calculations by the simple method adequately reproduce the orbits of complicated hetero-conjugated polycyclic molecules and provide reliable estimation of their steric structure.

## 11.4 ORBITAL-CONTROLLED PROPERTIES OF HETEROAROMATIC MOLECULES

As in carbocyclic conjugated hydrocarbons, changes in the electron energies of heteroaromatic molecules due to attachment of a substituent are determined by the symmetry of the corresponding occupied MOs of the heterocycle. In Section 11.2.1 configurations of the higher occupied $\pi$MOs of pyridine were discussed. According to the coefficients, substituents at position 2 should change the energy of the $a_2$ orbital, while the energies of $b_1$ orbitals should remain almost unchanged. By contrast, substituents at position 4 can mainly change the energy of the $2b_1$ orbital with no effect on the $a_2$ orbital. These relationships are illustrated by data for amino- and alkoxypyridines given in Table 11.7 [464]. Figure 11.10 shows the diagram for methyl- and trimethylsilyl substituted pyridines [465].

PE spectra and quantum chemical calculations give objective information on the steric structure of substituted heteroaromatic molecules and their orbitals. In pyrrole and pyridine, the $\pi$ orbital of $a_2$ symmetry (having no contribution of the nitrogen's AO) is the HOMO. (For configurations of $\pi$ OMO and coefficient values see Section 11.1.1.) Interaction of orbitals of two pyrrole fragments in N,N'-dipyrrolyls cannot therefore be estimated objectively by measuring the energy of the $a_2$ level, which is determined by the first ionization potential of pyrrole. The analysis of changes in the energy of the pyrrole orbital next to HOMO (the orbital of $b_1$ symmetry which contains a considerable contribution of the nitrogen atom) is more informative. It can be seen from Fig. 11.11 that the interactions of the $b_1$ electron levels of two pyrrole fragments in different dipyrrolyls are highly variable: splitting of the corresponding bands in PE spectra changes from 0.27 to 1.22 eV [466]. The splitting magnitude depends directly on the dihedral angle:

$$\triangle E = \Delta E_0 \cos \theta$$

where $\Delta E_0$ was determined from the spectrum of compound $a$ on the assumption

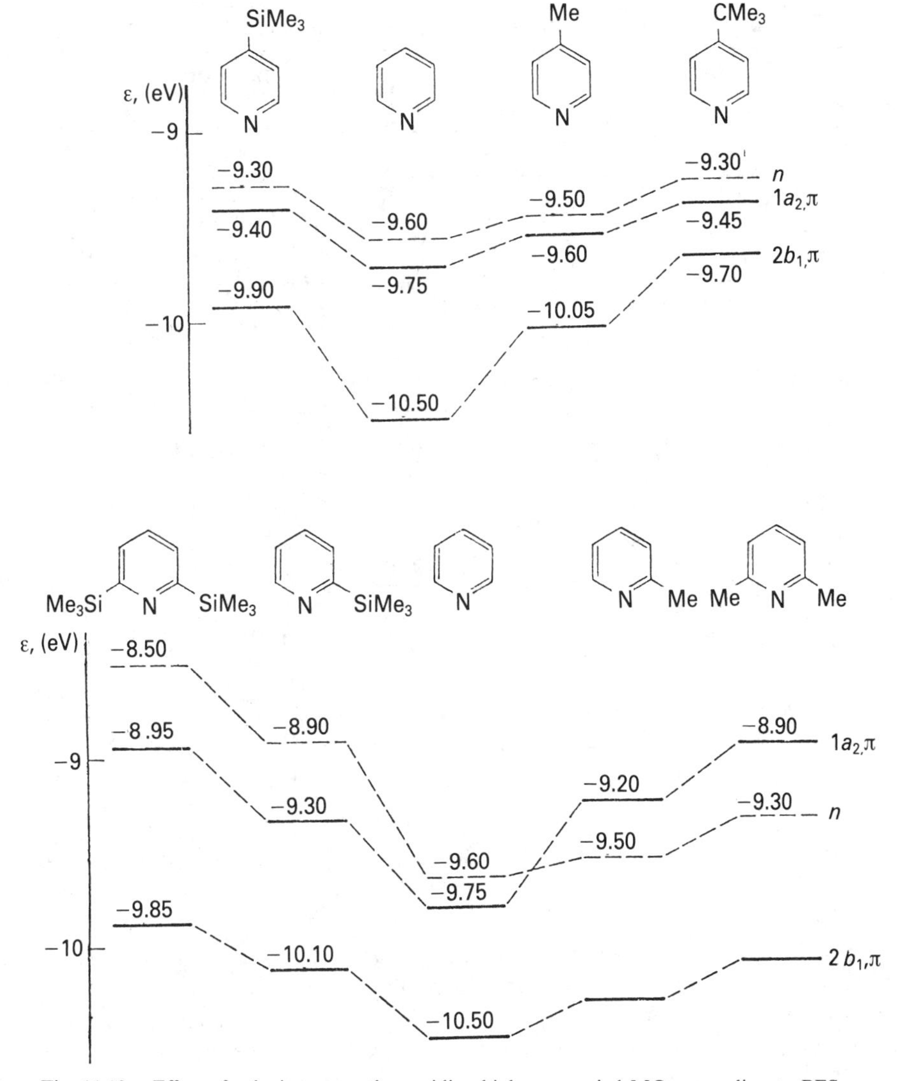


Fig. 11.10. Effect of substituent on the pyridine higher occupied MOs according to PES data. (Data by Heilbronner *et al.* [465])

that the most favourable orbital overlap conditions of two pyrrolyl fragments are provided in its almost planar molecule.

The value of the angle $\theta$ found in the N,N-dipyrrolyl molecule is 77°. The actual value of this angle is probably even higher because the $\Delta E$ value could be determined partly by the $\sigma$–$\pi$ interaction as well. According to the MINDO/2 calculations, the perpendicular conformation ($\theta = 90°$) of N,N-dipyrrolyl is more stable by 13.40 kJ/mol than the planar conformation ($\theta = 0°$).

Cyclic hetero-conjugated compounds are, as a rule, effective donors of electrons.

**Table 11.7.** Ionization potentials of amino- and alkoxypyridines

| Substituent | Substituent position | | | |
|---|---|---|---|---|
| | 2 | | 4 | |
| | $a_2$ | $b_1$ | $a_2$ | $b_1$ |
| H | 9.73 | 10.50 | 9.73 | 10.50 |
| $NH_2$ | 8.34 | 10.15 | 9.54 | 8.76 |
| $OC_2H_5$ | 8.78 | 10.20 | 9.40 | 9.25 |

(From Salakhutdinov and Koptyug [464])

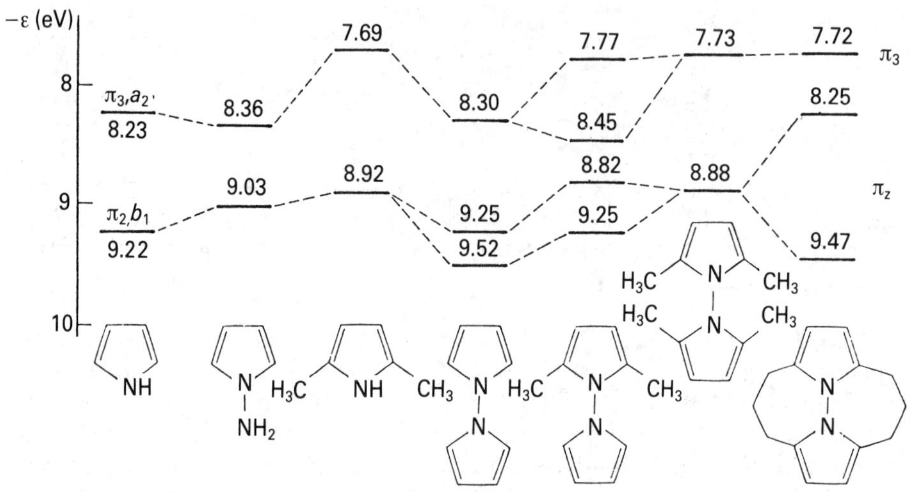


Fig. 11.11. Higher occupied electron levels in N,N′-dipyrrolyl and its analogs according to PES. (From Flitsch *et al.* [466])

They readily form CT complexes and are easily photo-ionized in solutions: e.g. dihydrophenazines form cation radicals by exposure of their alcohol solutions to radiation with the photon energy not exceeding 3.5 eV. It is believed that the ability to photo-ionize explains the high biological activity of many heteroaromatic molecules [33, 90]. As for other $\pi$ donors, the study of the interactions with $\pi$ acceptors gives data on the orbital structure of many heteroaromatic compounds [44]. The absorption spectra of TCNE complexes with pyrrole, furan, thiophene and their condensed analogues LXXXV have been studied in detail [467]:

S S  S Se  Se Se  S  Se

LXXXV: a b c d e

At least two CT bands can be seen in most spectra. The corresponding data are given in Table 11.8.

**Table 11.8.** Position of CT bands in absorption spectra ($\lambda_{CT}$) of TCNE complexes with compounds LXXXV, their formation constants ($K_{form}$), heats of formation ($\triangle H^0$) and ionization potentials of donors

| Donor | $\lambda_{CT}(CCl_4)$ (nm) | | $IP$ (eV) | | | | $K_{form}$, (l/mol) | $\triangle H^0$ (kJ/mol) |
|---|---|---|---|---|---|---|---|---|
| | | | $IP_1$ calcul. | $IP_2$ calcul. | $IP_1$ PES | $IP_2$ PES | | |
| | $\lambda_1$ | $\lambda_2$ | | | | | | |
| a | 500 | 333.3 | 8.51 | 10.16 | 8.61 | 10.04 | 3.90 | 24.58 |
| b | 527.4 | 363.6 | 8.34 | 9.75 | 8.39 | 9.86 | 5.37 | 28.09 |
| c | 548.3 | 357.1 | 8.22 | 9.83 | 8.20 | 9.63 | 6.31 | 32.99 |
| d | 555.6 | 471.7 | 8.18 | 8.71 | 8.20 | 8.76 | 3.23 | 16.87 |
| e | 576.0 | 475.3 | 8.07 | 8.68 | 8.03 | 8.64 | 3.57 | 18.63 |
| Naphthalene | 562.3 | 434.0 | 8.14 | 9.01 | 8.15 | 8.88 | 3.27 | 17.08 |
| Thiophene | 441.7 | 386.8 | 8.95 | 9.46 | 8.87 | 9.52 | 1.04 | 9.63 |
| Selenophene | 440.1 | — | 8.96 | | 8.80 | 8.95 | 1.42 | 11.64 |
| Benzene | 388.2 | — | 9.46 | | 9.31 | | 1.00 | 9.42 |

The given data are calculated from $h\nu_{CT} = aIP - b$ [467].
(From Cooper *et al.* [467])

Multiplicity of CT bands in the spectra of TCNE complexes of heterocycles is explained by possible electron transition to $\pi$ acceptor's frontier unoccupied MO not only from the HOMO but also from deeper occupied MOs of donors. It is noteworthy that $\Delta H^0$ and $K_{form}$ values seem to be the same for thiophene and benzene, and also for benzothiophene and naphthalene. This is of practical importance since the electron structure of the molecule remains the same with substitution of the —CH=CH— vinyl fragments by the sulphur atom.

As for polycyclic aromatic hydrocarbons, the objective character of PES and quantum chemical calculation data on the structure of occupied electron levels in polycyclic heteroaromatic molecules is confirmed by studies on their cation radicals [289]. There is a good correlation between electron transition energies measured from the absorption spectra of cation radicals of diphenylamine, acridine, and carbazole, and the energies of the same transitions estimated from the PE spectra. Figure 11.12 shows absorption spectra of cation radicals of acridine and carbazole. It should be noted that the difference between the first and subsequent ionization potentials for acridine is 1.50, 1.80, 3.53 eV and for carbazole 0.40, 1.41, 2.10 and 3.14 eV.

Electron absorption spectra of cation radicals of various biologically active phenazines and their PE spectra were studied in detail [468]. Thus, band energies (0.95 and 1.90 eV) in the electron absorption spectra of the anion radical of an effective neuroleptic 2-chloro-10($\gamma$-dimethylaminopropyl)phenothiazine were proved to be practically the same as the differences between the first and subsequent ionization potentials as determined from its PE spectrum. When comparing the PE spectrum of the compound with the electron absorption spectrum of its cation radical,

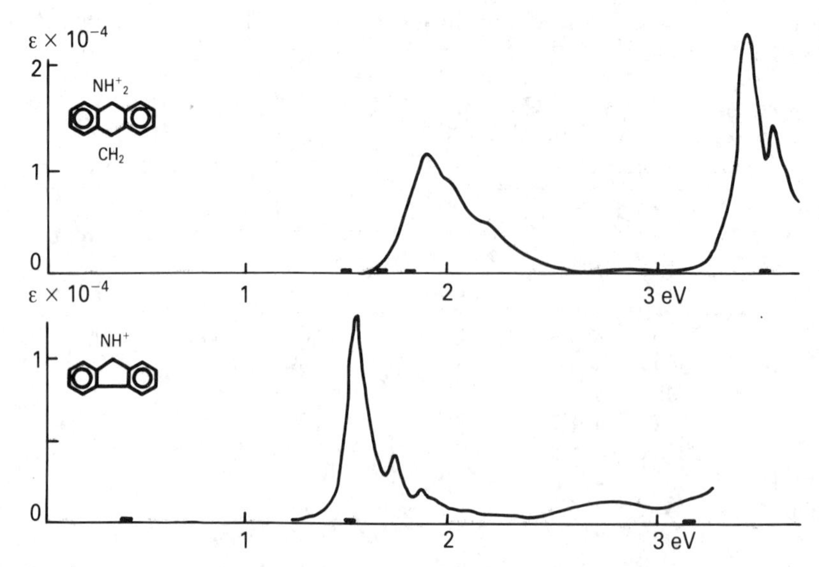


Fig. 11.12. Electron absorption spectra of cation radicals of acridane and carbazole (solid dots mark the position of the spectral bands as predicted by PES). (From Shida *et al.* [289])

it should be remembered that transitions between various electron states of the cation radical (recorded from the electron absorption spectrum of cation radicals in solution) obey the rule of selection by symmetry, as is the case with transitions between various electron states of neutral molecules.

The electron structure parameters determined by the study of the molecule in the gas phase were compared with similar parameters of the corresponding compounds in the solid state. PE spectra of solid films of phenothiazine, carbazole, acridine and anthracene were studied [469]. It was shown that low-polar heteroaromatic molecules (like the molecules of polycyclic aromatic hydrocarbons) largely preserve their electron structure in the solid phase in which van der Waals' forces are only effective.

Reactivity remains decisive for the estimation of the character of the frontier MOs in heteroaromatic compounds. It should, however, be noted that the application of the frontier orbitals concept to explain the reactivity of heterocycles was argued more strongly than that of aromatic hydrocarbons [68]. Thus, estimation of reactivity of heteroaromatic molecules in electrophilic substitution reactions is complicated by protonation of heteroatoms, while their molecules, on the whole, show a tendency to complexation. Thus, imidazole is protonated in water at the pH of about 7. At the same time, the basicity of pyrrole is much lower (pKa of the conjugated acid of pyrrole is about $-4$). In the interaction with weak electrophiles, the pyrrole molecule acts, most probably, as a neutral entity. This probably holds also for furan, thiophene, and related compounds.

If heteroaromatic compounds react with electrophiles as neutral molecules, the scheme of electrophilic aromatic substitution can be applied [470]. The substrate and position selectivity in substitution reactions (as in the reactions of aromatic hydrocarbons described in Section 6.3.5 and in Chapter 10) are, most probably, determined at the step of electron transfer between the heterocycle HOMO and electrophile LUMO. The important factors are thus the symmetry of the frontier occupied MO and its energy availability, which is determined by the first ionization potential of the heterocycle molecule.

The character of the HOMO of some nitrogen-containing heterocycles and their behaviour in electrophilic reactions were compared by Fukui within the framework of the frontier orbitals concept [64]. Electron densities in the HOMO (as determined by HMO calculations) are given in parentheses at the corresponding positions of the molecule. Written beneath the formulae are the isomers that are formed on nitration and halogenation:

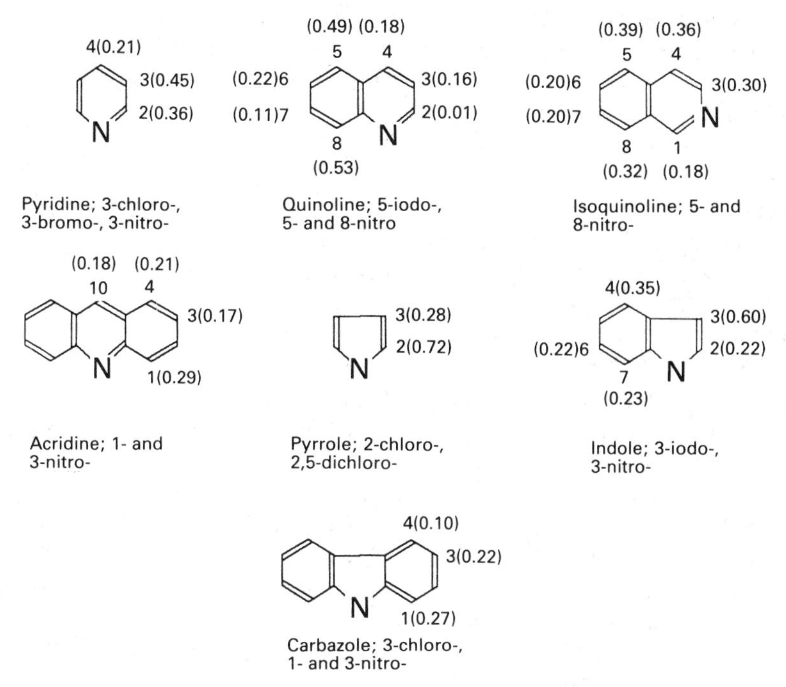


The frontier occupied orbitals of furan and thiophene have the $a_2$ symmetry and high electron density at position 2 (5). HOMO coefficients for thiophene, and the relative rate constants $k_{rel}$ of H/T exchange in its molecule (with trifluoroacetic acid as the solvent) are shown below:

0.37 $\qquad$ $1.9 \cdot 10^7$

0.60 $\qquad$ $3.2 \cdot 10^4$

S $\qquad$ S

It should be added that nitration of thiophene with acetyl nitrate gives mostly 2-nitrothiophene (70%). The 3-isomer is formed (less than 5%).

Since the $a_2$ orbital is the HOMO in five-membered heterocycles that are isoelectronic analogues of benzene (except for selenophene and tellurophene, in which $a_2$ and $b_1$ orbitals are either degenerate or inverted) their different relative reactivity can be explained by different energy availability of their frontier occupied MOs: the higher the HOMO level (the lower the $IP_1$ value), the higher is the rate. Below are $k_{rel}$ of bromination at position 2 for some heterocycles and their first ionization potentials [471]:

| | Benzene | Pyrrole | Furan | Thiophene | Selenophene |
|---|---|---|---|---|---|
| $k_{rel}$ | 1 | $3 \times 10^{18}$ | $6 \times 10^{11}$ | $5 \times 10^{9}$ | $2.4 \times 10^{11}$ |
| $IP_1$ (eV) | 9.24 | 8.23 | 8.83 | 8.87 | 8.92 |

The character of HOMO controls other properties of heteroaromatic molecules as well. Thus, the 'diene' character of the frontier occupied MOs, in accordance with their $a_2$ symmetry, explains why pyrrole, thiophene and furan act as dienes in the Diels–Alder reactions.

***Problems***

(1) Explain why the ability to release an electron is very similar in pyrrole, furan, and thiophene ($IP_1$ values are from 8.85 to 8.89 eV).

(2) Explain why the following reactions occur predominantly at position 2 of the substrate:

(a) thiophene + $CH_3COONO_2 \xrightarrow{BF_3}$ 2-nitrothiophene

(b) furan + $(CH_3CO)_2O \rightarrow$ 2-acetylfuran

(c) pyrrole + $\mathbf{C_6H_5N_2{}^+ Cl^-} \rightarrow$ 2-phenylazopyrrole

(3) Explain why cycloaddition of maleic anhydride to furan occurs at positions 2, 5 of the heterocycle.

(4) Explain why the ability to accept an electron is different in pyrrole and furan: $EA_1$ values are $-2.38$ and $-1.70$ eV respectively. (Note that the opposite is true for $IP_1$ values which are very similar!.)

(5) Explain why reactions of indole with electrophiles are regioselective. Suggest the most reactive position in its molecule.

(6) Explain why reactivity with electrophiles increases drastically in the series benzene < furan < pyrrole.

# Conclusion

Our knowledge of the electron structure of organic molecules becomes more definite and clear owing to its quantitative description within the framework of calculation and spectral methods of the MO theory. In accordance with the current interpretation of experimental findings in organic chemistry in terms of the orbital concepts, we have mostly discussed data concerning the structure of frontier orbitals of organic molecules. These orbitals often determine reactivity, biological activity, and the spectral characteristics of the molecule.

It can be expected that new experimental findings and further improvement of quantum chemical calculations will also give us an estimation of the role played by other electron levels of organic molecules. Their role seems to be quite important, mostly in charge-controlled reactions, and in understanding the properties of ion radicals and other intermediates.

# Answers to problems

## CHAPTER 4

(1) The substrate acts as a Lewis base in reactions with electrophiles. Alkanes are inert to electrophiles since they are very weak Lewis bases with low ability to release an electron (e.g. $IP_1$ of $CH_4$ is 13.0 eV).

(2) The substrate acts as a Lewis acid in reactions with nucleophiles. Alkanes are inert to nucleophiles since they have no properties of a Lewis acid, no ability to accept an electron to form a bond with a nucleophile.

(3) The octet rule gives no reasonable explanation. According to this rule, methane should have the only IP in the region of valence electron ionization. In terms of the MO theory, two of the methane valence electrons occupy the HOMO $\varphi_1$ and have lower energy than the other 6 electrons which occupy the degenerate HOMOs $\varphi_2$, $\varphi_3$ and $\varphi_4$.

(4) The C—C bonds of cyclopropane should be the first attacked by electrophiles, since its the frontier occupied MO has $\sigma$(C—C) character.

## CHAPTER 5

(1) Ethylene is more reactive to electrophiles than ethane since its frontier occupied MO is energetically and sterically more available for an attacking reagent ($IP_1$ values are 10.5 and 12 eV respectively).

(2) Ethylene is more reactive to nucleophiles than ethane since its frontier unoccupied MO is much lower in energy ($EA_1$ values are $-1.78$ and $< -6.0$ eV respectively).

(3) The direction of an alkene $A_E$ reaction (in the first step) is determined by the alkene HOMO coefficient; an electrophile attacks the carbon atom where the coefficient is the highest:

(a) $CH_3—CH(OSO_3H)—CH_3$

(b) $Cl_2CH—CH_3$

(c) $CH_3$—CH(OR)$OCH_3$
(d) HO—$CH_2$—$CH_2$—$NO_2$.

(4) The band at 185 nm in the UV spectrum of ethylene is due to the $\pi \rightarrow \pi^*$ electron transition and the band at 121 nm in the UV spectrum of ethane is due to the $\sigma \rightarrow \sigma^*$ electron transition. The $\pi$ and $\pi^*$ FMO of ethylene are much closer to each other in energy ($IP_1 = 10.51$ eV; $EA_1 = -1.78$ eV) than $\sigma$- and $\sigma^*$FMO of ethane ($IP_1 = 12.0$ eV; $EA_1 < -6.0$ eV).

(5) Ethylene is more reactive to electrophiles than acetylene, since it releases an electron much easier ($IP_1$ are 10.51 and 11.4 eV respectively).

(6) 'Hardness' values $[\chi = (IP_1 - EA_1)/2]$ of ethane, ethylene and acetylene are $>9.0$, 6.1, and 7.0 eV respectively.

(7) Buta-1,3-diene is more easily ionized and is more reactive to electrophiles than ethylene, since its frontier occupied $\pi$MO is higher in energy: $IP_1$ values are 9.09 and 10.51 eV respectively.

(8) Buta-1,3-diene attaches an electron much easier and is more reactive to nucleophiles than ethylene since its frontier $\pi$MO is lower in energy: $EA_1$ values are $-0.60$ and $-1.78$ eV respectively.

(9) In terms of MO theory, two buta-1,3-diene occupied $\pi$MOs have different energies and for this reason are ionized at different potentials.

(10) Buta-1,3-diene attaches an electron on the LUMO. Spin densities on the C(1) and C(2) atoms of butadiene anion radical (determined as $C^2_{LUMO,\mu}$) are 0.362 and 0.138 respectively.

(11) In terms of the frontier orbitals concept, the diene HOMO interacts with the dienophile LUMO during the Diels–Alder cycloaddition. Electron withdrawing substituents in alkene and electron donating substituents in diene make reacting orbitals closer in energy and increase the rate.

(12) (See Answer to Problem 11). Electron donating substituents in alkene and electron withdrawing substituents in diene make reacting orbitals greatly different in energy and decrease the rate.

(13) (a) The HOMO (2E, 4E)-hexa-2,4-diene is responsible for the thermal cyclization. It is similar to the butadiene HOMO in its coefficients. A new C(2)—C(5) bond can be formed only in a corotatory process, giving trans-3,4-dimethylcyclobutene as a product (see page 137).
(b) The LUMO (2E,4E)-hexa-2,4-diene is responsible for the photocyclization. It is similar to the butadiene LUMO in its coefficients. A new C(2)—C(5) bond can be formed only in a contrarotatory process giving cis-3,4-dimethylcyclobutene as a product (see page 137).

(14) (a) The (2E,4Z,6E)-octa-2,4,6-triene HOMO is responsible for the thermal cyclization. It is similar to the hexa-2,4,6-triene HOMO in its coefficients. A new C(2)—C(7) bond can be formed only in a contrarotatory process, giving cis-5,6-dimethylcyclohexa-1,3-diene as a product (see page 140).
(b) The (2E,4Z,6E)-octa-2,4,6-triene LUMO is responsible for the photocyclization. It is similar to the hexa-2,4,6-triene LUMO in its coefficients. A new C(2)—C(7) bond can be formed only in a conrotatory process giving trans-5,6-dimethylcyclohexa-1,3-diene as a product (see page 140).

(15) cis-1,2-Dimethylcyclobutane is the product, since this cycloaddition is a suprafacial process. On exposure to UV light, LUMOs (HOMOs*) of reactants are involved, providing bonding interaction at both sides of the reacting system:

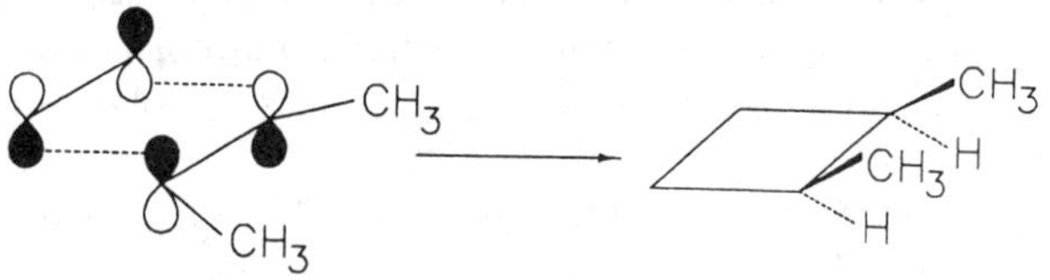

(16) The longest wavelength bands in UV spectra of ethylene and butadiene are due to electron transition between their $\pi$ FMOs. Frontier $\pi$MOs of butadiene are much closer in energy than those of ethylene (see Fig. 5.4).

(17) The Diels–Alder reaction is stereospecific since it proceeds as a cis-cycloaddition by a suprafacial mechanism:

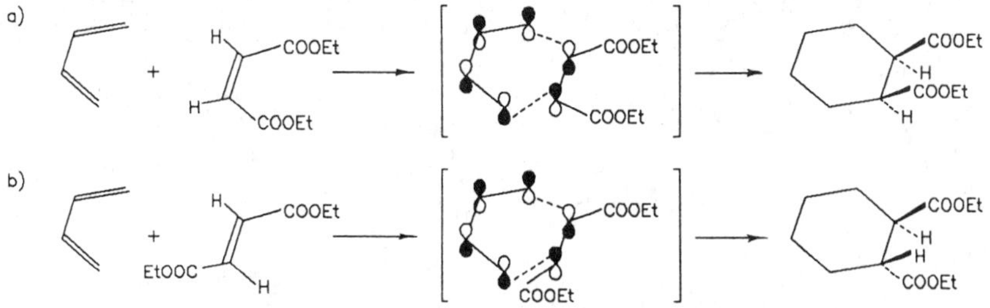

## CHAPTER 6

(1) All densities of unpaired electrons on carbon atoms of the benzene cation radical are equal to 0.167 (summate $C_\mu^2$ values of the degenerate benzene frontier occupied orbitals).

(2) Benzene should have the only $\pi$ ionization in terms of the octet rule and the model of hybridized orbitals. Benzene has two IPs of $\pi$ electrons since these electrons occupy the $\pi$MO $\varphi_1$ of lower energy and two degenerate $\pi$MOs $\varphi_2$ and $\varphi_3$ which have the highest energy.

(3) In terms of the frontier orbitals concept, the rate of PAH electrophilic nitration increases with diminishing PAH $IP_1$ values (this value is given in parentheses): naphthalene (8.15), chrysene (7.59), pyrene (7.41) and perylene (7.00).

(4) Pyrene, perylene, anthracene, phenanthrene and tetracene undergo electrophilic attack at positions 1, 3, 9, 9, 5 respectively since these positions have the highest values of HOMO coefficients (see Table 6.16).

(5) Highest values of unpaired electron density in anion radials of perylene, anthracene, pyrene, phenanthrene and tetracene are at the following positions: 3, 9, 1, 9, 5 (see Table 6.16) according to the LUMO coefficients.

(7) The unpaired electron occupied the PAH HOMO corresponding to the $C^2_{(\mathrm{HOMO})\mu}$ values when the PAH cation radical is formed. Highest values of unpaired electron density in cation radicals of naphthalene, anthracene, pyrene,

phenanthrene and tetracene are at the following positions: 1, 9, 1, 9, 5 (see Table 6.16).

(8) The longest wavelength band in UV spectra of PAH is due to the $\pi$(HOMO) → $\pi^*$(LUMO) electron transition. Frontier $\pi$MOs of PAH become closer with an increasing number of annelated benzene rings in its molecules (see Tables 6.13 and 6.16).

## CHAPTER 7

(1) The chlorine atom decreases the LUMO energy when it is present in a molecule. The $EA_1$ values (eV) illustrate this conclusion: $CH_4$, $<-6.0$; $CH_3Cl$, $-3.5$; $CH_2{=}CH_2$, $-1.78$; $CH_2{=}CHCl$, $-1.28$; $C_6H_6$, $-1.15$; $C_6H_5Cl$, $-0.75$.

(2) 'Hardness' values $\chi$(eV) are given below: $CH_4$, $>9.5$; $CH_3Cl$, 7.4; $CH_2{=}CH_2$, 6.1; $CH_2{=}CH{-}Cl$, 5.7; $C_6H_6$, 5.2; $C_6H_5Cl$, 4.9.

(3) In terms of the frontier orbitals concept, an alkyl halide acts as an acceptor in the $S_N^2$ reactions. $CH_3Br$ is more reactive than $CH_3Cl$, since it has higher electron affinity (compare $E_{1/2}^{red}$ values).

(4) 'Hardness' values $\chi$(eV) of $C_6H_5X$ are as follows: H, 5.2, F, 5.0; Cl, 4.93; Br, 4.87; I, $<4.7$.

## CHAPTER 8

(1) Hydroxy and alkoxy groups increase the substrate HOMO energy and therefore its availability for an electrophile attack. The $IP_1$ values (eV) illustrate this conclusion: $CH_4$, 13.0; $(CH_3)_2O$, 10.04; $CH_2{=}CH_2$, 10.51; $CH_2{=}CH{-}O{-}C_2H_5$, 9.15; $C_6H_6$, 9.24; $C_6H_5OH$, 8.70.

(2) Methanol is more easily oxidized than methane, since it has higher HOMO energy: $IP_1$ values (eV) are 10.94 ($n_O$) and 13.0 (C—H) respectively.

(3) Hydroxy- and methoxy groups as electron donating substituents increase ener→ gies of benzene frontier MOs $\varphi_2$ and $\varphi_5$, but do not touch FMOs $\varphi_3$ and $\varphi_4$ (see Figures 2.5 and 2.6).

(4) Elecrophiles attack phenol at ortho and para positions, which have the highest coefficients in its HOMO.

(5) An unpaired electron occupies the phenol LUMO in the phenol anion radical according to the $c^2_{(LUMO)\mu}$ values: ortho, 0.25; meta, 0.25; para, 0.

(6) An unpaired electron occupies the anisole HOMO in the anisole cation radical according to the $c^2_{(LUMO)\mu}$ values: ortho, 0.12; meta, 0.05; para, 0.28.

(7) Acrolein has higher electron affinity than formaldehyde and ethylene since its LUMO is formed by the interaction of two vacant $\pi^*$(C=O) and $\pi^*$(C=C) fragment orbitals (see Fig. 2.10).

(8) The benzaldehyde LUMO has lower energy compared with that of benzene. $EA_1$ values (eV) are $-0.76$ and $-1.15$ respectively.

(9) An unpaired electron occupies the benzaldehyde LUMO in the benzaldehyde anion radical according to the $c^2_{(LUMO)}$ values: ortho, 0.12; meta, 0.02; para, 0.19.

(10) The unpaired electron occupies the substrate LUMO when the substrate anion radical is formed. The energy of the LUMO decreases as follows: biphenyl, $-0.705$; benzaldehyde, $-0.617$; benzophenone, $-0.491$.

(11) 'Soft' donor–acceptor reactions occur according to the frontier orbitals concept. Ethylamine (donor) attacks acrolein (acceptor) at the terminal C atom which has the highest coefficient in the acrolein LUMO: 3-ethylaminopropanal is the product of this reaction.

(12) 'Hard' donor–acceptor reactions occur according to the charge-control concept. Ethylmagnesium bromide (donor) attacks acrolein (acceptor) at the C (carbonyl) which has the highest positive charge: 3-hydroxypentene is the product of the reaction.

(13) A possible mechanism is dimerization of anion radicals, which are formed at the first step. 2,2-di(1,4-naphthoquinonyl) is the product of dimerization since the 1,4-naphthoquinone anion radical has maximum spin density at position 2 (see Section 8.5.10).

(14) The 1,4-benzoquinone LUMO is mostly available for nucleophilic attack since this orbital has the lowest energy: $E^{red}_{1/2}$(V) values of 9,10-anthraquinone, 1,4-naphthoquinone and 1,4-benzoquinone are 0.155, 0.493 and 0.711 respectively.

## CHAPTER 9

(1) In terms of the frontier orbitals concept, $CH_3NH_2$ (donor) reacts with ethyl bromide (acceptor) at a high rate, since it has higher HOMO energy: $IP_1$ (eV) of $CH_3NH_2$ and $CH_3OH$ are 8.95 and 10.94 respectively.

(2) Amino group as an electron donating substituent substantially increases the energy of the benzene frontier MOs $\pi_2$ and $\pi_5$ but does not affect FMOs $\varphi_3$ and $\varphi_4$ (see Figs. 2.5 and 2.6).

(3) An unpaired electron occupies the aniline HOMO in the aniline cation radical according to the $c^2_{(\text{HOMO})\mu}$ values: ortho, 0.13; meta, 0.03; para, 0.25.

(4) An unpaired electron of the nitrobenzene anion radical occupies the nitrobenzene LUMO according to the $c^2_{(\text{LUMO})\mu}$ values: ortho, 0.11; meta, 0.005; para, 0.12 (see Figs. 2.5 and 2.6).

(5) An unpaired electron of the benzonitrile cation radical occupies the benzonitrile HOMO which is a slightly perturbed benzene orbital $\pi_3$ (see Figs. 2.5 and 2.6).

## CHAPTER 10

(1) An unpaired electron occupies the toluene HOMO in the toluene cation radical according to the $c^2_{(\text{HOMO})\mu}$ values: ortho, 0.12; meta, 0.06; para, 0.31.

(2) An unpaired electron of the benzoic acid cation radical occupies the benzoic acid HOMO which is a slightly perturbed benzene orbital $\pi_3$ (see Figs. 2.5 and 2.6).

(3) An unpaired electron of the toluene anion radical occupies the toluene LUMO which is a slightly perturbed benzene orbital $\pi_4$ according to $c^2_{(\text{LUMO})\mu}$ values: ortho, 0.25; meta, 0.25; para, 0.

(4) An anion radical of the substrate is formed as an intermediate of the Birch reaction:
   (a) hydrogenation of anisole takes place at position 2(3) and 5(6), which have the highest coefficients in the anisole LUMO (see Figs. 2.5 and 2.6);
   (b) see (a);
   (c) hydrogenation of benzoic acid takes place at positions 1 and 4, which have the highest coefficients in the benzoic acid LUMO (see Figs. 2.5 and 2.6).
   (d) naphthalene, anthracene, and phenanthrene have the highest LUMO coefficients at positions 1 and 4, 9 and 10, 9 and 10 respectively.

(5) An electrophilic addition to styrene starts with the attack on the terminal atom of the vinyl group. This position has the highest coefficient in the styrene HOMO.

(6) HOMO of the allyl $\beta$-naphthyl ether is responsible for the orientation in the Claisen rearrangement. The terminal atom of the allyl group attacks position 1 of the naphthalene ring, which has the highest coefficient in the ether HOMO (the HOMO of the ether is similar to the HOMO of 2-naphthol).

## CHAPTER 11

(1) The $IP_1$ of pyrrole, furan and thiophene are very close since the heteroatom is not involved in their HOMOs.

(2) In terms of the frontier orbitals concept, electrophilic substitution reactions with thiophene, furan and pyrrole are determined by their HOMOs which have the highest coefficients at positions 2 and 5.

(3) The Diels–Alder cycloaddition reactions of five-membered heterocycles are determined by their HOMOs which have the highest coefficients at positions 2(5).

(4) The furan LUMO is lower in energy than those of pyrrole: $-0.948$ and $-1.008$ respectively (in $\beta$ units).

(5) The position 3 of indole is the most reactive to electrophiles since it has the highest coefficient in the HOMO.

(6) Furan has the highest coefficients at position 2, both in HOMO and LUMO.

(7) Pyrrole is the most reactive since its HOMO is mostly available for an electrophilic attack: $IP_1$ values are 9.24, 8.83, and 8.23 eV respectively.

# Appendix

The HMO eigencoefficients and eigenvalues of frontier orbitals, the values of $EA_1$ and $IP_1$ of some key organic molecules are given as illustrations to the problems.

**LUMO** **HOMO**

Ethylene

| LUMO | |
|---|---|
| 1 | −0.707 |
| 2 | +0.707 |
| $\varepsilon =$ | $-1.000\beta$ |
| $EA_1 =$ | −1.78 ev |

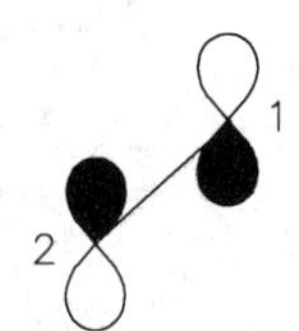


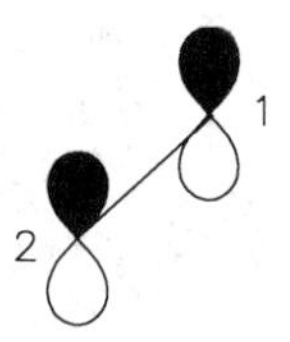


| HOMO | |
|---|---|
| 1 | +0.707 |
| 2 | +0.707 |
| $\varepsilon =$ | $+1.000\beta$ |
| $IP_1 =$ | 10.51 eV |

Propene (hyperconjugation model)

| LUMO | |
|---|---|
| $1(H_3)$ | +0.227 |
| 2 | +0.057 |
| 3 | −0.728 |
| 4 | +0.644 |
| $\varepsilon =$ | $-1.130\beta$ |
| $EA_1 =$ | −1.99 ev |

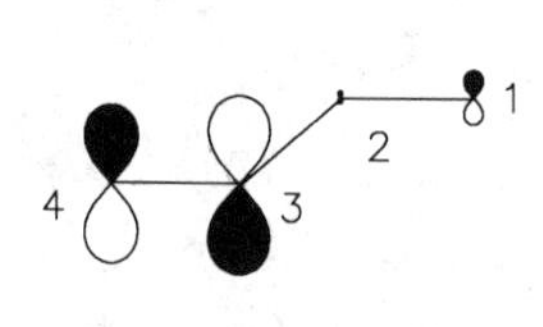


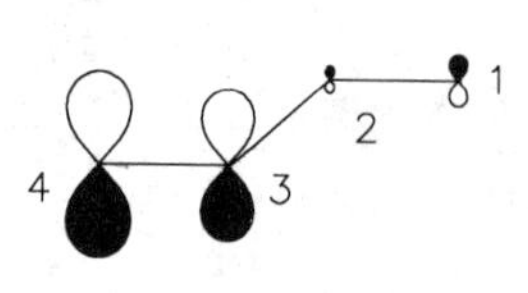


| HOMO | |
|---|---|
| $1(H_3)$ | +0.211 |
| 2 | +0.111 |
| 3 | −0.611 |
| 4 | −0.755 |
| $\varepsilon =$ | $+0.809\beta$ |
| $IP_1 =$ | 10.03 eV |

Buta-1,3-diene

| LUMO | |
|---|---|
| 1 | +0.602 |
| 2 | −0.372 |
| 3 | −0.372 |
| 4 | +0.602 |
| $\varepsilon =$ | $-0.618\beta$ |
| $EA_1 =$ | −0.62 ev |

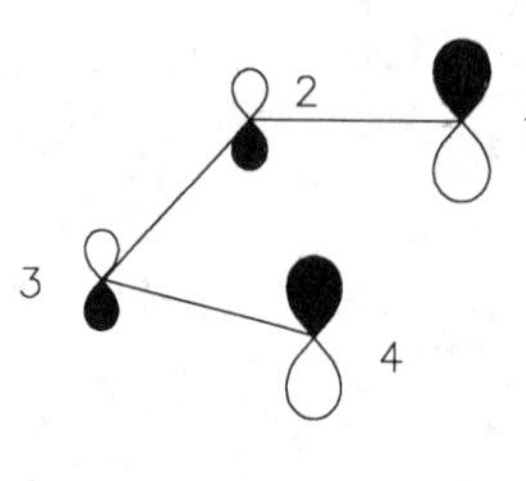


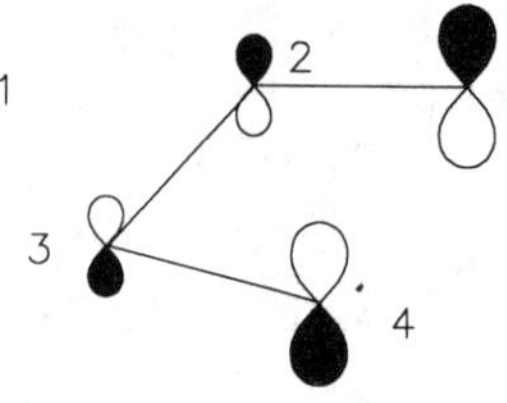


| HOMO | |
|---|---|
| 1 | +0.602 |
| 2 | +0.372 |
| 3 | −0.372 |
| 4 | −0.602 |
| $\varepsilon =$ | $+0.618\beta$ |
| $IP_1 =$ | 9.09 eV |

Hexa-1,3,5-triene

| LUMO | |
|---|---|
| 1 | +0.521 |
| 2 | −0.232 |
| 3 | −0.418 |
| 4 | +0.418 |
| 5 | +0.232 |
| 6 | −0.521 |
| $\varepsilon =$ | $-0.445\beta$ |
| $EA_1 =$ | — |

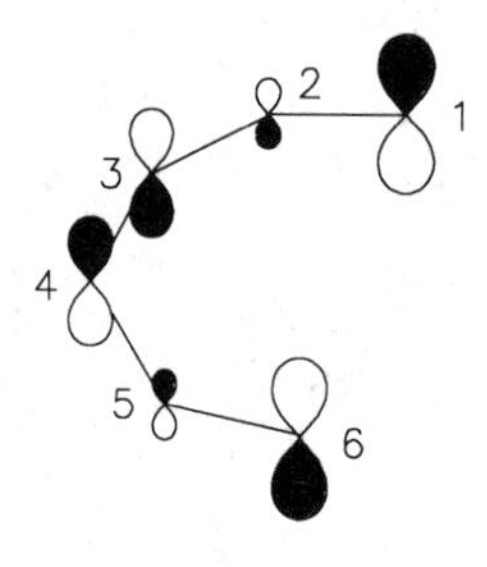


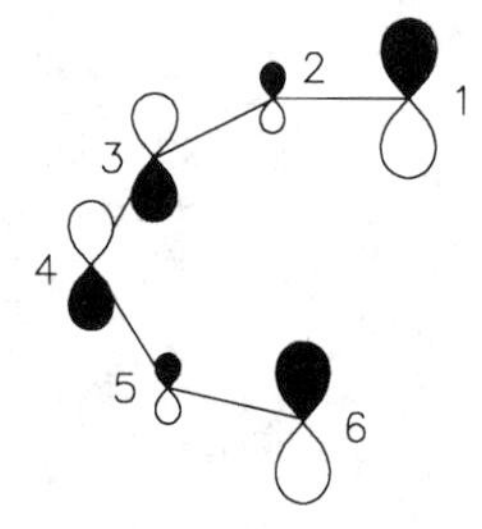


| HOMO | |
|---|---|
| 1 | +0.521 |
| 2 | +0.232 |
| 3 | −0.418 |
| 4 | −0.418 |
| 5 | +0.232 |
| 6 | +0.521 |
| $\varepsilon =$ | $+0.445\beta$ |
| $IP_1 =$ | 8.32 eV |

**LUMO** **HOMO**

Chloroethylene

| LUMO | |
|---|---|
| 1 (Cl) | +0.094 |
| 2 | −0.713 |
| 3 | +0.694 |
| $\varepsilon =$ | $-1.027\beta$ |
| $EA_1 =$ | −1.28 ev |

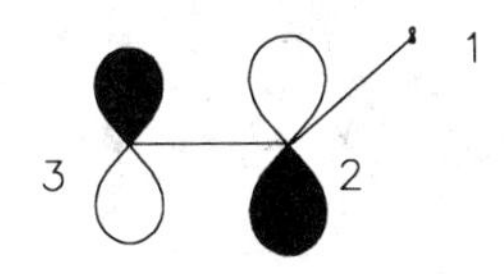


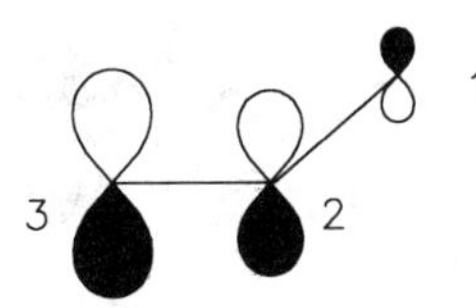


| HOMO | |
|---|---|
| 1 (Cl) | +0.246 |
| 2 | −0.659 |
| 3 | −0.710 |
| $\varepsilon =$ | $+0.928\beta$ |
| $IP_1 =$ | 10.20 eV |

Formaldehyde

| LUMO | |
|---|---|
| 1 (O) | +0.566 |
| 2 | −0.824 |
| $\varepsilon =$ | $-1.071\beta$ |
| $EA_1 =$ | −0.86 ev |

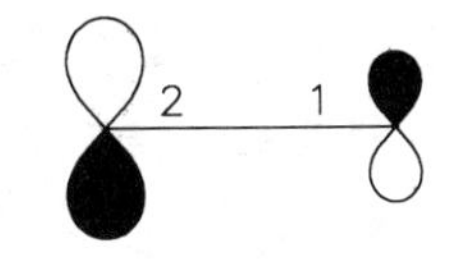


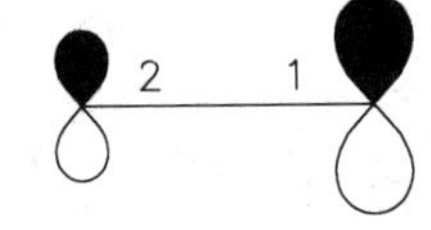


| HOMO | |
|---|---|
| 1 (O) | +0.824 |
| 2 | +0.566 |
| $\varepsilon =$ | $+2.271\beta$ |
| $IP_1 =$ | 10.88 eV |

Methoxyethylene

| LUMO | |
|---|---|
| 1 (OMe) | +0.160 |
| 2 | −0.731 |
| 3 | +0.663 |
| $\varepsilon =$ | $-1.104\beta$ |
| $EA_1 =$ | — |

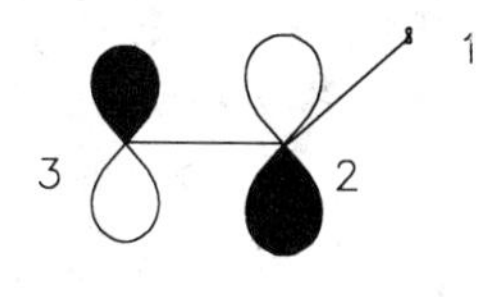


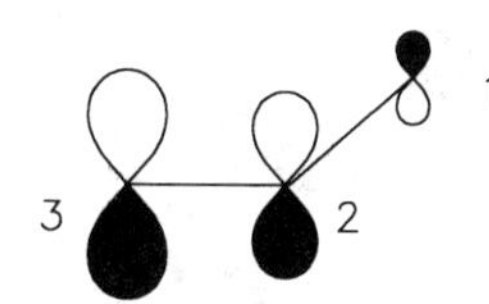


| HOMO | |
|---|---|
| 1 (OMe) | +0.256 |
| 2 | −0.618 |
| 3 | −0.744 |
| $\varepsilon =$ | $+0.831\beta$ |
| $IP_1 =$ | 9.14 eV |

Acrolein

| LUMO | |
|---|---|
| 1 (O) | −0.413 |
| 2 | +0.467 |
| 3 | +0.383 |
| 4 | −0.682 |
| $\varepsilon =$ | $-0.562\beta$ |
| $EA_1 =$ | — |

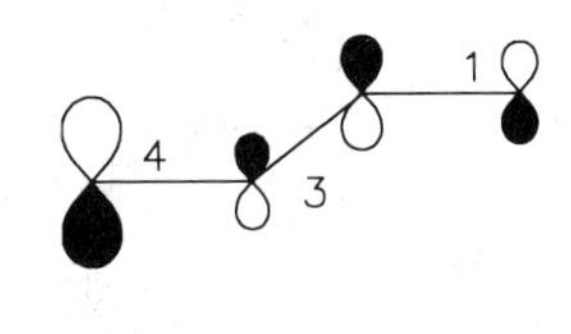


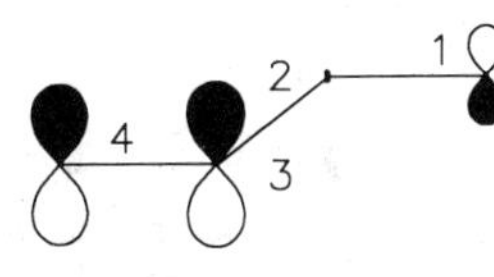


| HOMO | |
|---|---|
| 1 (O) | −0.395 |
| 2 | +0.042 |
| 3 | +0.659 |
| 4 | +0.639 |
| $\varepsilon =$ | $+1.033\beta$ |
| $IP_1 =$ | — |

**LUMO** **HOMO**

Nitroethylene

| LUMO | |
|---|---|
| 1(O) | −0.426 |
| 2(O) | −0.426 |
| 3(N) | +0.449 |
| 4 | +0.225 |
| 5 | −0.621 |
| $\varepsilon =$ | $-0.363\beta$ |
| $EA_1 =$ | — |

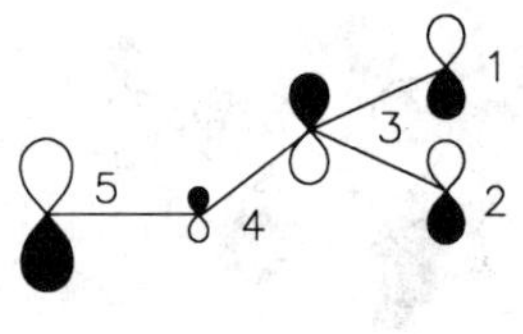


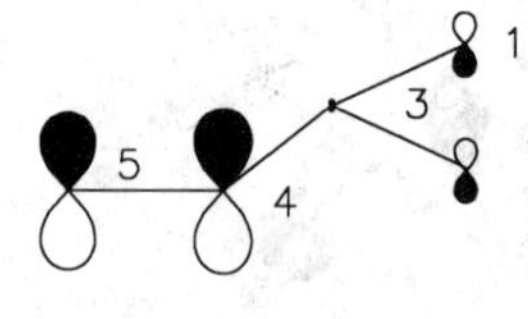


| HOMO | |
|---|---|
| 1(O) | −0.260 |
| 2(O) | −0.260 |
| 3(N) | +0.055 |
| 4 | +0.672 |
| 5 | +0.640 |
| $\varepsilon =$ | $+1.050\beta$ |
| $IP_1 =$ | — |

Toluene

| LUMO | |
|---|---|
| $1(H_3)$ | 0.000 |
| 2 | 0.000 |
| 3 | 0.000 |
| 4 | +0.500 |
| 5 | −0.500 |
| 6 | 0.000 |
| $\varepsilon =$ | $-1.100\beta$ |
| $EA_1 =$ | −1.11 ev |

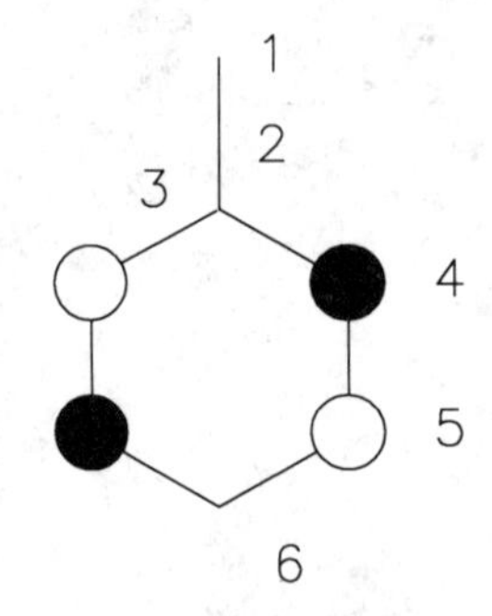


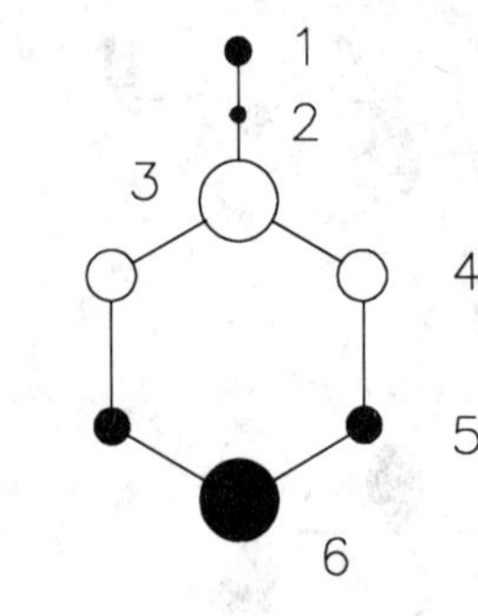


| HOMO | |
|---|---|
| $1(H_3)$ | +0.191 |
| 2 | +0.104 |
| 3 | −0.538 |
| 4 | −0.349 |
| 5 | +0.238 |
| 6 | +0.554 |
| $\varepsilon =$ | $+1.000\beta$ |
| $IP_1 =$ | 8.83 eV |

Styrene

| LUMO | |
|---|---|
| 1 | +0.586 |
| 2 | −0.364 |
| 3 | −0.367 |
| 4 | +0.312 |
| 5 | +0.147 |
| 6 | −0.416 |
| $\varepsilon =$ | $-0.706\beta$ |
| $EA_1 =$ | −0.25 ev |

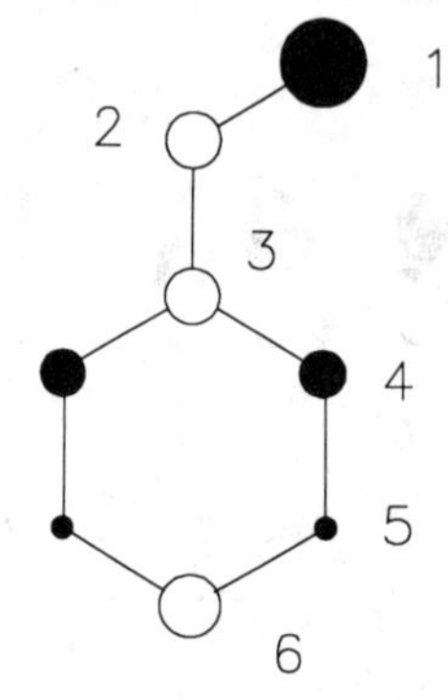


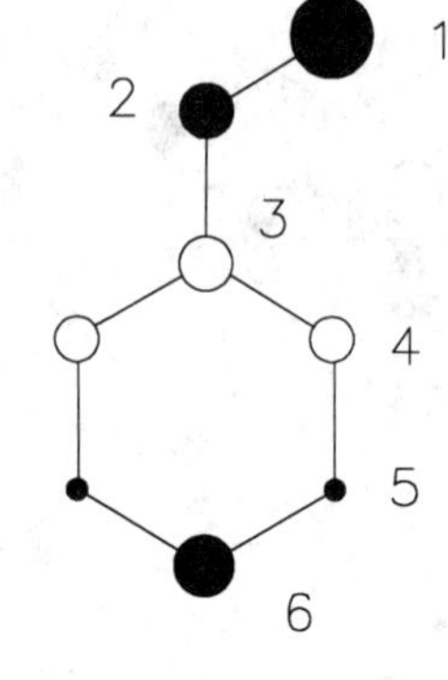


| HOMO | |
|---|---|
| 1 | +0.586 |
| 2 | +0.364 |
| 3 | −0.367 |
| 4 | −0.312 |
| 5 | +0.147 |
| 6 | +0.416 |
| $\varepsilon =$ | $+0.706\beta$ |
| $IP_1 =$ | 8.49 eV |

| | **LUMO** | | **HOMO** |
|---|---|---|---|

Naphthalene

| LUMO | | HOMO | |
|---|---|---|---|
| 1 | +0.425 | 1 | +0.425 |
| 2 | −0.263 | 2 | +0.263 |
| 3 | −0.263 | 3 | −0.263 |
| 4 | +0.425 | 4 | −0.425 |
| $\varepsilon =$ | $-0.618\beta$ | $\varepsilon =$ | $+0.618\beta$ |
| $EA_1 =$ | −0.19 ev | $IP_1 =$ | 8.15 eV |

2-Naphthol

| LUMO | | HOMO | |
|---|---|---|---|
| 1 | +0.391 | 1 | −0.475 |
| 2 | −0.236 | 2 | −0.305 |
| 3 (O) | +0.072 | 3 (O) | +0.172 |
| 4 | −0.305 | 4 | +0.160 |
| 5 | +0.033 | 5 | +0.398 |
| $\varepsilon =$ | $-0.633\beta$ | $\varepsilon =$ | $+0.581\beta$ |
| $EA_1 =$ | — | $IP_1 =$ | — |

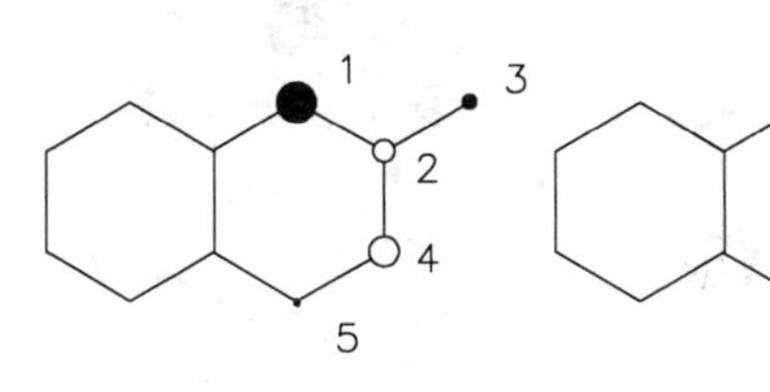


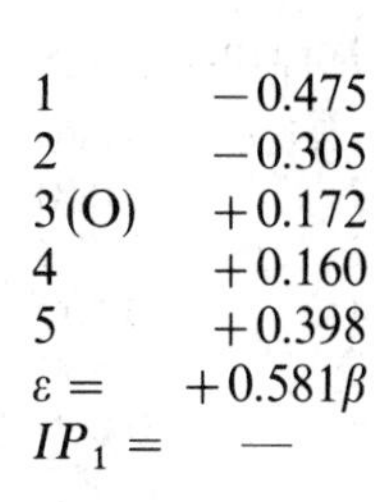


1,4-Naphthoquinone

| LUMO | | HOMO | |
|---|---|---|---|
| 1 (O) | +0.392 | 1 (O) | 0.000 |
| 2 | −0.331 | 2 | 0.000 |
| 3 | −0.375 | 3 | +0.408 |
| 4 | +0.157 | 4 | 0.000 |
| 5 | +0.178 | 5 | +0.408 |
| $\varepsilon =$ | $-1.118\beta$ | $\varepsilon =$ | $+1.000\beta$ |
| $EA_1 =$ | — | $IP_1 =$ | — |

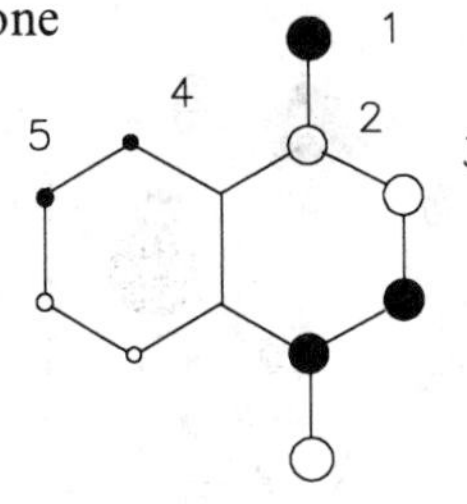


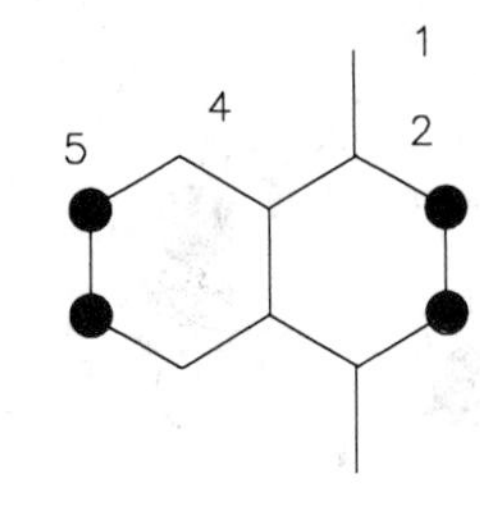

**LUMO** **HOMO**

Biphenyl

| LUMO | |
|---|---|
| 1 | +0.398 |
| 2 | −0.140 |
| 3 | −0.299 |
| 4 | +0.351 |
| 5 | +0.351 |
| 6 | −0.299 |
| 7 | −0.140 |
| 8 | +0.398 |
| $\varepsilon =$ | $-0.705\beta$ |
| $EA_1 =$ | −0.37 ev |

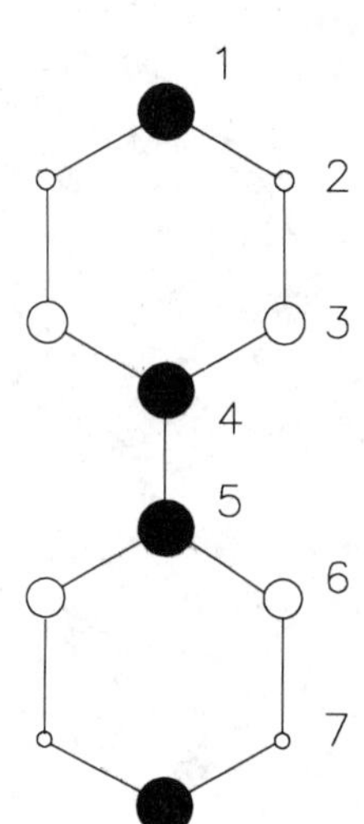


| HOMO | |
|---|---|
| 1 | +0.398 |
| 2 | +0.140 |
| 3 | −0.299 |
| 4 | −0.351 |
| 5 | +0.351 |
| 6 | +0.299 |
| 7 | −0.140 |
| 8 | −0.398 |
| $\varepsilon =$ | $+0.705\beta$ |
| $IP_1 =$ | 8.34 eV |

Chlorobenzene

| LUMO | |
|---|---|
| 1 (Cl) | 0.000 |
| 2 | 0.000 |
| 3 | +0.500 |
| 4 | −0.500 |
| 5 | 0.000 |
| $\varepsilon =$ | $-1.000\beta$ |
| $EA_1 =$ | −0.75 ev |

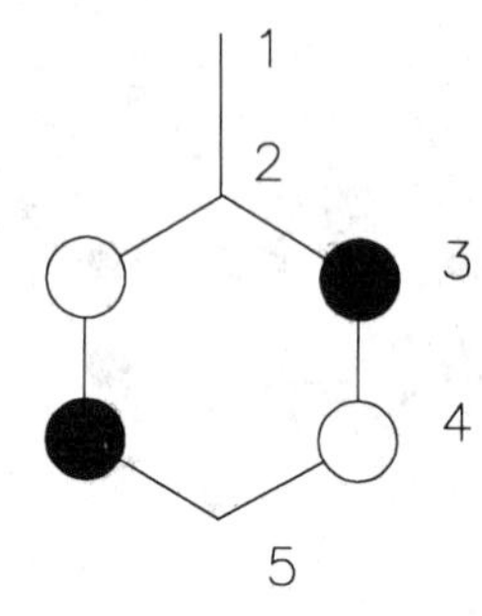


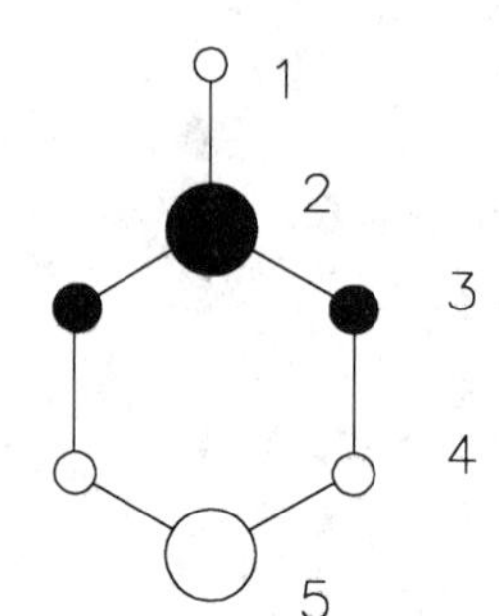


| HOMO | |
|---|---|
| 1 (Cl) | −0.212 |
| 2 | +0.558 |
| 3 | +0.307 |
| 4 | −0.266 |
| 5 | −0.560 |
| $\varepsilon =$ | $+0.950\beta$ |
| $IP_1 =$ | 9.10 eV |

Phenol

| LUMO | |
|---|---|
| 1 (O) | 0.000 |
| 2 | 0.000 |
| 3 | +0.500 |
| 4 | −0.500 |
| 5 | 0.000 |
| $\varepsilon =$ | $-1.000\beta$ |
| $EA_1 =$ | −1.11 ev |

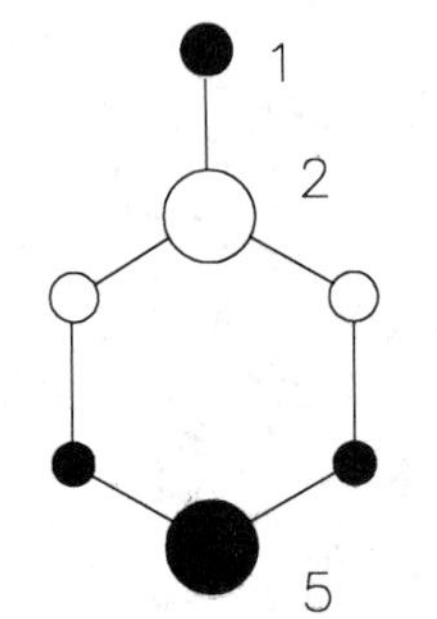


| HOMO | |
|---|---|
| 1 (O) | +0.347 |
| 2 | −0.508 |
| 3 | −0.349 |
| 4 | +0.219 |
| 5 | +0.531 |
| $\varepsilon =$ | $+0.827\beta$ |
| $IP_1 =$ | 8.67 eV |

**LUMO** **HOMO**

Benzaldehyde

| LUMO | | HOMO | |
|---|---|---|---|
| 1 (O) | +0.416 | 1 (O) | 0.000 |
| 2 | −0.485 | 2 | 0.000 |
| 3 | −0.350 | 3 | 0.000 |
| 4 | +0.351 | 4 | +0.500 |
| 5 | +0.134 | 5 | +0.500 |
| 6 | −0.433 | 6 | 0.000 |
| $\varepsilon =$ | $-0.617\beta$ | $\varepsilon =$ | $+1.000\beta$ |
| $EA_1 =$ | −0.76 ev | $IP_1 =$ | 9.59 eV |

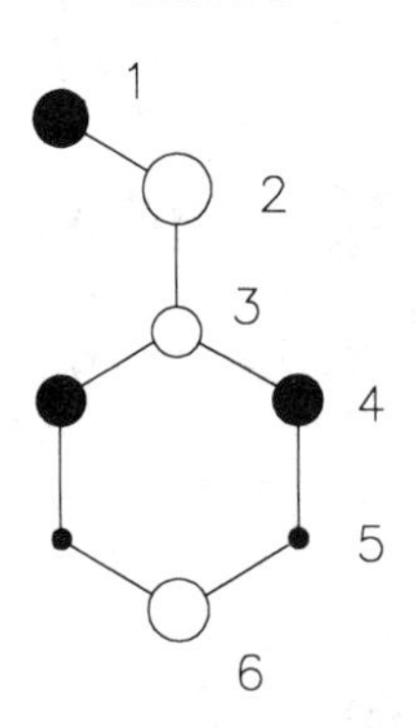


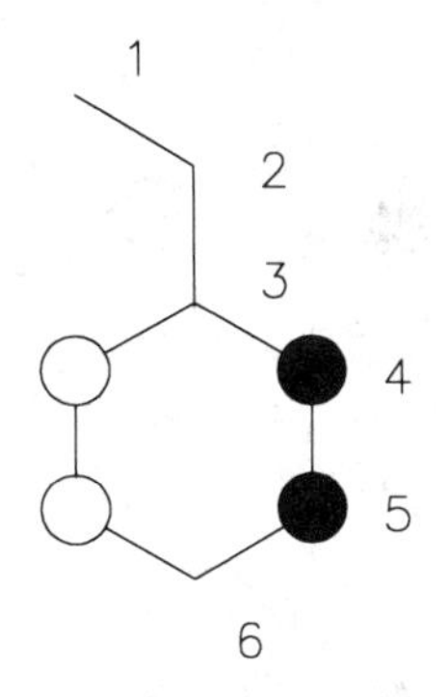


Nitrobenzene

| LUMO | | HOMO | |
|---|---|---|---|
| 1 (O) | +0.441 | 1 (O) | 0.000 |
| 2 (N) | −0.475 | 2 (N) | 0.000 |
| 3 | −0.200 | 3 | 0.000 |
| 4 | +0.325 | 4 | −0.500 |
| 5 | +0.070 | 5 | −0.500 |
| 6 | −0.353 | 6 | 0.000 |
| $\varepsilon =$ | $-0.398\beta$ | $\varepsilon =$ | $+1.000\beta$ |
| $EA_1 =$ | — | $IP_1 =$ | 9.96 eV |





Aniline

| LUMO | | HOMO | |
|---|---|---|---|
| 1 (N) | 0.000 | 1 (N) | +0.477 |
| 2 | 0.000 | 2 | −0.451 |
| 3 | +0.500 | 3 | −0.358 |
| 4 | −0.500 | 4 | +0.184 |
| 5 | 0.000 | 5 | +0.495 |
| $\varepsilon =$ | $-1.000\beta$ | $\varepsilon =$ | $+0.744\beta$ |
| $EA_1 =$ | −1.85 ev | $IP_1 =$ | 8.00 eV |

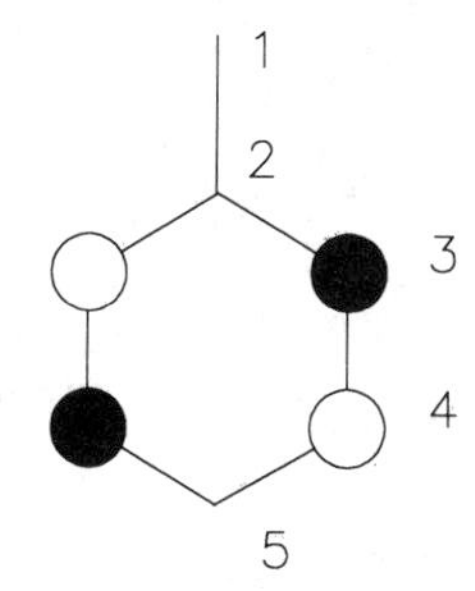


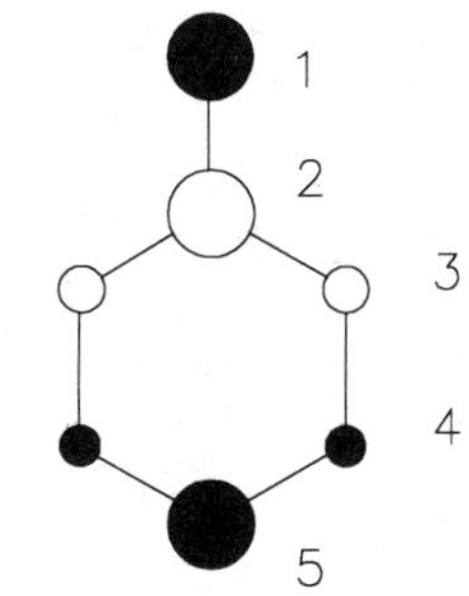

| | LUMO | HOMO | |
|---|---|---|---|

Furan

| | |
|---|---|
| 1 (O) | $-0.323$ |
| 2 | $+0.595$ |
| 3 | $-0.306$ |
| $\varepsilon =$ | $-0.948\beta$ |
| $EA_1 =$ | $-1.70$ ev |

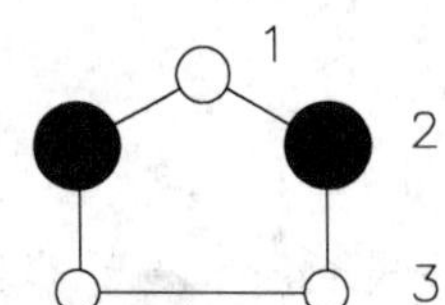


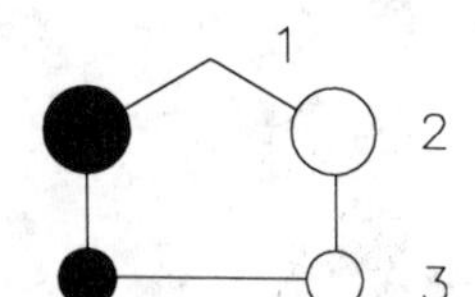


| | |
|---|---|
| 1 (O) | $0.000$ |
| 2 | $-0.602$ |
| 3 | $-0.372$ |
| $\varepsilon =$ | $+0.618\beta$ |
| $IP_1 =$ | 8.89 eV |

Pyrrole

| | |
|---|---|
| 1 (N) | $-0.347$ |
| 2 | $+0.587$ |
| 3 | $-0.292$ |
| $\varepsilon =$ | $-1.008\beta$ |
| $EA_1 =$ | $-2.38$ ev |

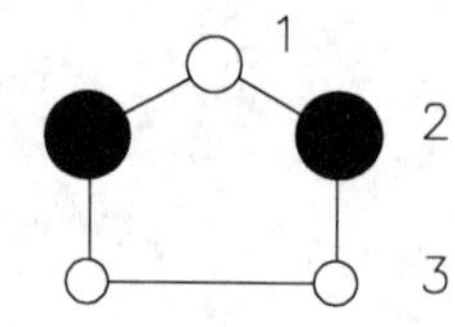


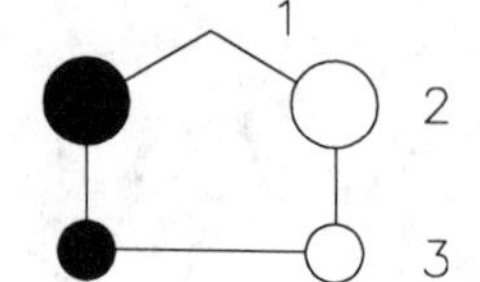


| | |
|---|---|
| 1 (N) | $0.000$ |
| 2 | $-0.602$ |
| 3 | $-0.372$ |
| $\varepsilon =$ | $+0.618\beta$ |
| $IP_1 =$ | 8.85 eV |

Indole

| | |
|---|---|
| 1 (N) | $-0.187$ |
| 2 | $+0.449$ |
| 3 | $-0.219$ |
| 4 | $+0.505$ |
| 5 | $-0.177$ |
| 6 | $-0.353$ |
| 7 | $+0.482$ |
| $\varepsilon =$ | $-0.863\beta$ |
| $EA_1 =$ | — |

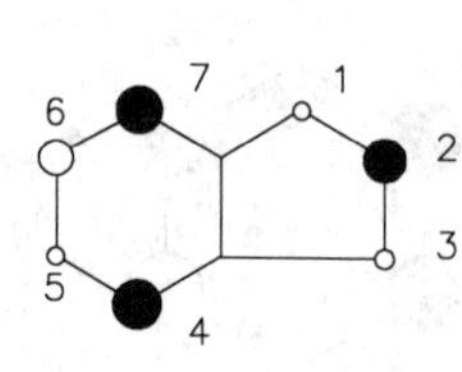


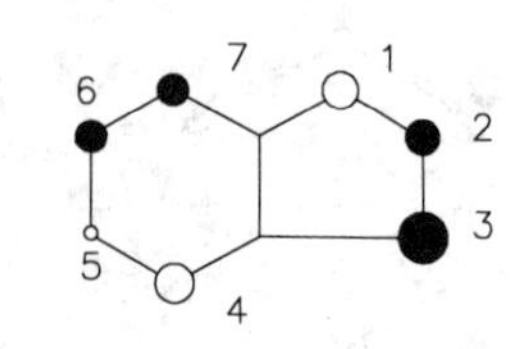


| | |
|---|---|
| 1 (N) | $-0.382$ |
| 2 | $+0.354$ |
| 3 | $+0.534$ |
| 4 | $-0.414$ |
| 5 | $-0.152$ |
| 6 | $+0.333$ |
| 7 | $+0.330$ |
| $\varepsilon =$ | $+0.534\beta$ |
| $IP_1 =$ | — |

# References


[1] Dunitz, J.D. (1979) *X-Ray Analysis and the Structure of Organic Molecules*, Cornell University Press, Ithaca, NY.

[2] Reutov, O.A. (1964) *Theoretical Principles of Organic Chemistry* (In Russian), Moscow University Press, Moscow.

[3] Ingold, C.K. (1969) *Structure and Mechanism in Organic Chemistry*, 2nd edn, Cornell University Press, Ithaca, NY.

[4] Streitwieser, A., Jr., Heathcock, C.H. (1985) *Introduction to Organic Chemistry*, 3rd edn, MacMillan, NY.

[5] March, J. (1977) *Advanced Organic Chemistry. Reactions, Mechanisms and Structure*, 2nd edn, McGraw-Hill, NY.

[6] Morrison, R.T., Boyd, R.N. (1983) *Organic Chemistry*, Allyn and Bacon, Boston.

[7] Saito, K. (ed.) (1979) *Chemistry and the Periodic Table*, Iwanami Shoten.

[8] Albert, A., Serjeant, E.P. (1973) *Determination of Ionization Constants*, 2nd edn, Chapman and Hall, London.

[9] Olah, G.A., Svoboda, J.J. (1973) *Synthesis*, 52–3.

[10] Hammett, L.P. (1970) *Physical Organic Chemistry: Reaction Rates, Equilibria and Mechanisms*, 2nd edn, McGraw-Hill, NY.

[11] Palm, V.A. (1967) *Fundamentals of Quantitative Theory of Organic Reactions*, Khimia, Leningrad.

[12] Chapman, N.B., Shorter, J. (1978) *Correlation Analysis in Chemistry: Recent Advances*, Plenum Press, NY.

[13] Shorter, J. (1982) *Correlation Analyses of Organic Reactivity with Particular Reference to Multiple Regression*. Research Studies Press, Chichester.

[14] Bryson, A. (1960) *J. Amer. Chem. Soc.*, **82**, 4862–71.

[15] Dewar, M.J.S., Grisdale, P.J. (1962) *J. Amer. Chem. Soc.*, **84**, 3546–53.

[16] Drago, R.S. (1970) *Physical Methods in Chemistry*, Vols 1, 2, W.B. Saunders Company, Philadelphia, PA.

[17] Minkin, V.I., Osipov, O.A., Zhdanov, Yu.A. (1968) *Dipole Moments in Organic Chemistry* (In Russian), Khimia, Moscow.

[18] Streitwieser, A.Jr. (1961) *Molecular Orbital Theory for Organic Chemists*, Wiley, NY and London.

[19] Lehmann, W.J. (1972) *Atomic and Molecular Structure: the Development of Our Concepts*, Wiley, NY.

[20] Tatevsky, V.M. (1973) *Classical Theory of Molecular Structure and Quantum Mechanics* (In Russian), Khimia, Moscow.

[21] Jolly, W.L. (1984) *Modern Inorganic Chemistry*, McGraw-Hill, NY.

[22] Ballard, R.E. (1978) *Photoelectron Spectroscopy and Molecular Orbital Theory*, Adam Hilger, Bristol.

[23] Purcell, W.P., Singer, J.A. (1967) *J. Chem. Eng. Data*, **12**, 235–46.

[24] Radom, L., Pople, J.A. (1972) Ab Initio Molecular Orbital Theory of Organic Molecules. In: Brown, W.B. (ed.) *Theoretical Chemistry*, Butterworth, London, pp. 71–112.

[25] Schaefer III, H.F. (1977) (ed.) *Methods of Electronic Structure Theory*, Plenum Press, NY.

[26] Schaefer III, H.F. (1977) (ed.) *Application of Electronic Structure Theory*, Plenum Press, NY, pp. 357–80.

[27] Zhidomirov, G.M., Chuvylkin, N.D. (1986) *Usp. Khimii* (In Russian), **55**, 353–70.

[28] Shchembelov, G.A., Ustynyuk, Yu.A., Mamaev, V.M. (1980) *Quantum Chemical Calculations of Molecules* (In Russian), Khimia, Moscow.

[29] Minkin, V.I., Simkin, B.Ya., Minyaev, R.M. (1979). *Theory of Molecular Structure* (In Russian), Vysshaya Shkola, Moscow.

[30] Peacock, T. (1965) *Electronic Properties of Aromatic and Heterocyclic Molecules*, Academic Press, London.

[31] Griffiths, J. (1982) *Dyes and Pigments*, **3**, 211–33.

[32] Kelemen, J., Moss, S., Glitsch, S. (1984) *Dyes and Pigments*, **5**, 83–108.

[33] Ghosh, P.K. (1983) *Introduction to Photoelectron Spectroscopy*, Wiley, New York.

[34] Shorygin, P.P., Burshtein, K.Ya. (1981) *Usp. khimii* (In Russian), **50**, 1345–75.

[35] Flurry, R.L. (1980) *Symmetry Groups. Theory and Chemical Applications*, Prentice Hall, Englewood Cliffs, NJ.

[36] Vovna, V.I., Vilesov, F.I. (1975) *Usp. fotoniki* (In Russian), Vol. 5, Leningrad University Press.

[37] Brundle, C.R., Baker, A.D. (eds.) (1977) *Electron Spectroscopy: Theory, Techniques and Applications*, Vol. 1, Academic Press, London.

[38] Kimura, K. (1981) *Handbook of He(I) Photoelectron Spectra of Fundamental Organic Molecules*, Japan Sci. Soc. Press, Tokyo.

[39] Nefedov, V.I., Vovna, V.I. (1987) *Electron Structure of Chemical Compounds* (In Russian), Nauka, Moscow.

[40] Koopmans, T. (1934) Phys., **1**, 104–13.

[41] Richards, W.G. (1969) *Int. Mass Spectr. Ion. Phys.*, **2**, 419–24.

[42] Wendin, G. (1981) *Breakdown of One-electron Pictures in Photoelectron Spectra*, Springer Verlag, Berlin.

[43] Hohlneicher, G. (1974) *Ber. Buns. Gesellsch.*, **78**, 1125–40.

[44] Foster, R. (ed.) (1979) *Molecular Association*, Academic Press, NY.
[45] Nenner, I., Schulz, G. (1975) *Chem. Phys.*, **62**, Re 5, 1747–58.
[46] Jordan, K.D., Burrow, P.D. (1978) *Acc. Chem. Res.*, **11**, 341–8.
[47] Mayranovsky, S.G., Stradyn, Ya.P., Bezugly, V.D. (1975) *Polarography in Organic Chemistry* (In Russian), Khimia, Leningrad.
[48] Miller, L.L., Nordblom, G.D., Mayeda, E.A. (1972) *J. Org. Chem.*, **37**, 916–18.
[49] Gassmann, P.G., Yamaguchi, R. (1979) *J. Amer. Chem. Soc.*, **101**, 1308–10.
[50] Gerson, F. (1970) *High Resolution ESR Spectroscopy*, Wiley, New York.
[51] Millefiori, S., Millefiori, A. (1981) *J. Chem. Soc., Far. Trans.* II, **77**, 245–58.
[52] Klessinger, M. (1982) *Electronenstructur Organischer Molekule*, Verlag Chemie, Weinheim.
[53] Bock, H., Ramsey, G. (1973) *Angew. Chem.*, **85**, 773–92.
[54] Heilbronner, E., Maier, J.P. (1977) Some aspects of organic photoelectron spectroscopy. In: Brundle, C.R. and Baker, A.D. (eds.) *Electron Spectroscopy: Theory, Techniques and Applications*, Vol. 1, Academic Press, London, pp. 205–92.
[55] Hoffmann, R. (1971) *Acc. Chem. Res.*, **4**, 1–9.
[56] Bock, H. *et al.* (1974) *Chem Ber.*, **107**, 1869–81.
[57] Modelli, A., Distefano, G. (1981) *Z. Naturforsch*, **36A**, 1344–51.
[58] Traven, V.F., Rodin, O.G., Shvets, A.F. (1987) *Zhur. Obsch. Khim.* (In Russian), **57**, 499–505.
[59] Mollere, P. *et al.* (1973) *J. Organomet. Chem.*, **61**, 127–32.
[60] Dewar, P.S. *et al.* (1974) *Tetrahedron*, **30**, 2455–9.
[61] Rodin, O.G., Traven, V.F., Redchenko, V.V. (1983) *Zhur. Obsch. Khim.* (In Russian), **53**, 2537–43.
[62] Traven, V.F., Stepanov, B.I. (1984) *Usp. Khimii* (In Russian), **53**, 897–924.
[63] Fukui, K., Yonezawa, T., Shingu, H. (1952) *J. Chem. Phys.*, **20**, 722–5.
[64] Fukui, K., Yonezawa, T., Nagata, C. *et al.* (1954) *J. Chem. Phys.*, **22**, 1433–42.
[65] Nicolas, D., Epiotis, N.D. (1978) *Theory of Organic Reactions*, Springer Verlag, Berlin–Heidelberg–New York.
[66] Fukui, K. (1982) *Angew. Chem.*, **94**, 852–61.
[67] Clar, E. (1964) *Polycyclic Hydrocarbons*, Vols 1 and 2, Academic Press, NY.
[68] Klopman, G. (1983) *J. Mol. Struct. Suppl. Teochem.*, **103**, 121–9.
[69] Ottolenghi, M. (1973) *Acc. Chem. Res.*, **6**, 153–60.
[70] Voight, E.M., Reid, C. (1964) *J. Amer. Chem. Soc.*, **86**, 3930–4.
[71] Farrell, P.G., Newton, J. (1965) *J. Phys. Chem.*, **69**, 3506–9.
[72] Traven, V.F., West, R., Donyagina, V.F., Stepanov, B.I. (1975) *Zhur. Obsch. Khim.* (In Russian), **45**, 824–30.
[73] Wagner, G., Bock, H. (1974). *Chem. Ber.*, **107**, 68–77.
[74] Traven, V.F., German, M.I., Stepanov, B.I. (1975) *Zhur. Obsch. Khim.* (In Russian), **45**, 707–8.
[75] Traven, V.F. *et al.* (1977) *Izv. AN SSSR*, series in chemistry (In Russian), 1042–7.
[76] Woodward, R.B., Hoffmann, R. (1965) *J. Amer. Chem. Soc.*, **87**, 395–7.
[77] Hoffmann, R., Woodward, R.B. (1965) *J. Amer. Chem. Soc.*, **87**, 2046–8.

[78] Woodward, R., Hoffmann, R. (1970) *The Conservation of Orbital Symmetry*, Academic Press, NY, Weinheim.

[79] Kochi, J.K. (1974) *Acc. Chem. Res.*, **7**, 351–60.

[80] Traven, V.F., Donyagina, V.F., Fedotov, N.S., Stepanov, B.I. (1976) *Zhur. Obsch. Khim.* (In Russian), **46**, 2761.

[81] Tabushi, J., Jamamura, K. (1975) *Tetrahedron*, **31**, 1827–31.

[82] Minkin, V.I., Simkin, B.Ya., Minyaev, R.M. (1986) *Quantum Chemistry of Organic Compounds: Reaction Mechanisms* (In Russian), Khimia, Moscow.

[83] Klopman, G. (1968) *J. Amer. Chem. Soc.*, **90** 223–34.

[84] Salem, L. (1968) *J. Amer. Chem. Soc.*, **90**, 543–66.

[85] Minkin, V.I. *et al.* (1985) *Zhur. Organ. Khim.* (In Russian), **21**, 926–39.

[86] Reents, W.D., Freiser, B.S. (1980) *J. Amer. Chem. Soc.*, **102**, 271–6.

[87] Fukuzumi, S., Kochi, J.K. (1981) *J. Amer. Chem. Soc.*, **103**, 7240–52.

[88] Fukuzumi, S., Kochi, J.K. (1981) *J. Org. Chem.*, **46**, 4116–26.

[89] Morkovnik, A.S. (1988) *Usp. Khimii* (In Russian), 254–80.

[90] Pearson, R.G. (1987) *J. Chem. Educ.*, **64**, 561–7.

[91] Pullman, B., Pullman, A. (1963) *Quantum Biochemistry*, Wiley–Interscience, New York.

[92] Richards, W.G. (1983) *Quantum Pharmacology* (2nd edn), Butterworths, London.

[93] Volkenshtein, M.V., Golovanov, I.B., Sobolev, V.N. (1982) *Molecular Orbitals in Enzymology* (In Russian), Nauka, Moscow.

[94] Ovchinnikov, A.A., Boldyrev, A.I. (1986) *Usp. Khimii* (In Russian), **55**, 539–54.

[95] Lebedeva, K.V., Pyatnova, Yu. B. (1984) *Zhur. Vses. Khim. Obsch. in D. I. Mend.* (In Russian), **29**, 54–63.

[96] Graedel, T.E. (1984) *J. Chem. Educ.* **61**, 681–6.

[97] Nizhniy, S.V., Epshtein, N.A. (1978) *Usp. khim.* (In Russian), **47**, 739–76.

[98] Luzhkov, V.B., Bogdanov, G.N. (1986) *Usp. khim.* (In Russian), **55**, 3–28.

[99] Klasinc, L., Ruščič, B., Sabljić, A. *et al.* (1979) *J. Amer. Chem. Soc.*, **101**, 7477–82.

[100] Domelsmith, L.N., Munchausen, L.L., Houk, K.N. (1977) *J. Amer. Chem. Soc.*, **99**, 4311–21.

[101] Domelsmith, L.N., Munchausen, L.L., Houk, K.N. (1977) *J. Amer. Chem. Soc.*, **99**, 6506–14.

[102] Deslongchamps, P. (1983) *Stereoelectronic Effects in Organic Chemistry*, Pergamon Press, Oxford.

[103] Boyd, R.J., Bünzli, J.-C.G., Snyder, J.P. (1976) *J. Amer. Chem. Soc.*, **98**, 2398–2406.

[104] Zefirov, N.S. (1965) *Usp. khim.* (In Russian), **34**, 1272–92.

[105] Wolfe, S. (1972) *Acc. Chem. Res.*, **5**, 102–11.

[106] Zefirov, N.S. (1977) *Tetrahedron*, **33**, 3193–3202.

[107] Kirby, A. (1982) *The Anomeric Effect and Related Stereoelectronic Effects of Oxygen*, Springer Verlag, NY.

[108] Friesen, D., Hedberg, K. (1980) *J. Amer. Chem. Soc.*, **102**, 3987–94.

[109] Zefirov, N.S., Samoshin, V.V., Subbotin, O.A., Sergeyev, N.M. (1981) *Zhur. Organ. Khim.* (In Russian), **17**, 1462–8.

[110] Zefirov, N.S., Samoshin, V.V., Palyulin, V.A. (1983) *Zhur. Organ. Khim.* (In Russian), **19**, 1888–92.
[111] Epiotis, N.D., Yates, R.L., Larson, J.R. *et al.* (1977) *J. Amer. Chem. Soc.*, **99**, 8379–88.
[112] Pierson, G.O., Runquist, O.A. (1968) *J. Org. Chem.*, **33**, 2572–4.
[113] Winter, C.E. (1987) *J. Chem. Educ.*, **64**, 587–90.
[114] Lehn, J.M., Wipff, G. (1978) *Helv. Chim. Acta.*, **61**, 1274–86.
[115] Stepanov, B.I. (1984) *Introduction to Chemistry and Manufacture of Organic Dyes* (3rd edn), Khimia, Moscow.
[116] Gordon, P., Gregory, P. (1983) *Chemistry in Colour*, Springer Verlag, Berlin–Heidelberg.
[117] Bock, H. (1977) *Angew. Chem.*, **89**, 631–55.
[118] Skala, V. *et al.* (1978) *Coll. Czech. Chem. Comm.*, **43**, 434–46.
[119] Klessinger, M. (1966) *Theor. Chim. Acta.*, **5**, 251–65.
[120] Kovshev, E.I., Blinov, L.M., Titov, V.V. (1977) *Usp. khimii* (In Russian), **46**, 753–98.
[121] Titov, V.V., Ivashchenko, A.V. (1983) *Zhur. Vses. Khim. Obsch. im D.I. Mend.* (In Russian), **28**, 176–87.
[122] Eltsov, A. (ed.) (1982) *Organic Photochromes*, Khimia, Leningrad.
[123] Wells, C.H.J. (1982) *Educ. in Chem.*, **19**, 38–9.
[124] Baum, S.J. (1978) *Introduction to Organic and Biological Chemistry* (2nd edn), MacMillan, NY.
[125] Samoshin, V.V., Zefirov, N.S. (1984) *Zhur. Vses. Khim. Obsch. im D.I. Mend.* (In Russian), **29**, 521–30.
[126] Tomoda, Sh., Kimura, K. (1983) *Chem. Phys.*, **82**, 215–27.
[127] Tomoda, Sh. (1986) *Chem. Phys.*, **110**, 431–45.
[128] Yeo, G.A., Ford, T.A. (1986) *South-Afr. J. Chem.*, **39**, 243–50.
[129] Taurian, O.E., Lunell, S. (1987) *J. Phys. Chem.*, **91**, 2249–53.
[130] Campbell, J. (1979) *J. Electr. Spectr. Relat. Phenom.*, **15**, 83.
[131] Carnovale, F., Gan, T.H., Peel, J.B. (1980) *J. Electr. Spectr. Relat. Phenom.*, **20**, 53–67.
[132] Aagren, H., Karlstroem, G. (1983) *J. Chem. Phys.*, **79**, 587–92.
[133] Shida, T., Haselbach, E., Bally, Th. (1984) *Acc. Chem. Res.*, **17**, 180–6.
[134] Harris, J.M., McManns, S.P. (eds.) (1987) *Nucleophilicity. Advances in Chemistry*, Ser. 215, Amer. Chem. Soc., Washington, pp. 181–94.
[135] Goodman, M., Morehouse, F. (1973) *Organic Molecules in Action*, Gordon and Breach, NY.
[136] Bartell, L.S., Kuchitsu, K., deNeni, R.J. (1961) *J. Chem. Phys.*, **35**, 1211–18.
[137] Dewar, M. (1962) *Hyperconjugation*, The Ronald Press Company, NY.
[138] Cambi, R. *et al.* (1981) *Chem. Phys. Lett.*, **80**, 295–300.
[139] Keim, W. (ed.) (1983) *Catalysis in $C_1$ Chemistry*, Reidel, Dordrecht.
[140] Chapman, N.B. (ed.) (1973) *Organic Chemistry*, Ser. One, Vol. 2, Butterworth, London–Boston.
[141] Goldshleger, N.F. *et al.* (1972) *Zhur. Fis. Kim.* (In Russian), **46**, 1353–4.
[142] Abramovich, R.A. (ed.) (1980) *Reactive Intermediates*, Vol. 1, Plenum, NY.

[143] Aleksandrov, Yu. A., Lebedev, S.A., Kuznetsova, N.V. (1986) *Zhur. Obsch. Khim.* (In Russian), **56**, 1667–8.
[144] Salahub, D.R., Sandorfy, C. (1971) *Theor. Chim. Acta.*, **20**, 227–42.
[145] Ikuta Shigeru (1987) *J. Mol. Struct. Theochem.*, **149**, 279–88.
[146] Lathan, W.A., Hehre, W.J., Pople, J.A. (1971) *J. Amer. Chem. Soc.*, **93**, 808–15.
[147] Iijima, T. (1973) *Bull. Chem. Soc. Jap.*, **46**, 2311–14.
[148] Topiol, S., Rather, M.A. (1977) *Chem. Phys.*, **20**, 1–7.
[149] Sovers, O.J. *et al.* (1968) *J. Chem Phys.*, **49**, 2592–9.
[150] Bartell, L.S., Kohl, D.A. (1963) *J. Chem. Phys.*, **39**, 3097–105.
[151] Shearer, H.M.M., Vand, V. (1956) *Acta Cryst.*, **9**, 379–84.
[152] Pitzer, K.S. (1940) *J. Chem Phys.*, **8**, 711–20.
[153] Ensslin, W., Bergmann, H., Elbel, S. (1975) *J. Chem. Soc. Far. Trans.* **II**, 913–20.
[154] Pople, J.A. (1970) *Acc. Chem. Res.*, **3**, 217–23.
[155] Hiraoka, K., Nara, M. (1981) *Bull. Chem. Soc. Jap.*, **54**, 1589–94.
[156] McQuilin, F.J., Baird, M.S. (1983) *Alicyclic Chemistry* (2nd edn), Camb. University Press.
[157] Plemenkov, V.V. *et al.* (1981) *Zhur. Obsch. Khim.* (In Russian), **51**, 2076–80.
[158] Plemenkov, V.V. *et al.* (1982) *Zhur. Organ. Khim.*, **18**, 1888–94.
[159] Sandorfy, C. (1955) *Can. J. Chem.*, **33**, 1337–51.
[160] Klopman, G. (1963) *Tetrahedron*, **19**, 111–22.
[161] Shorygin, P.P. *et al.* (1966) *Zh. teor. i eksp. khim.*, **11**, 190–5.
[162] Pitt, C.G., Jones, L.I., Ramsey, B.G. (1967) *J. Amer. Chem. Soc.*, **89**, 5471–2.
[163] Shorygin, P.P. *et al.* (1975) *Izv. AN SSSR*, series in chemistry (In Russian), 562–6.
[164] Fukui, K., Kato, H., Jonezawa, T. (1960) *Bull. Chem. Soc. Jap.*, **33**, 1197–1200.
[165] Bock, H., Ensslin, W. (1971) *Angew. Chem.*, **83**, 435–7.
[166] Traven, V.F., West, R. (1973) *J. Amer. Chem. Soc.*, **95**, 6824–6.
[167] Traven, V.F., Karelsky, V.N., Donyagina, V.F. *et al.* (1975) *DAN SSSR*, **224**, 837–40.
[168] Herman, A., Dreczewski, B., Wojnowski, W. (1983) *J. Organomet. Chem.*, **251**, 7–14.
[169] Sakurai, H., Kamiyama, Y. (1974) *J. Amer. Chem. Soc.*, **96**, 6192–4.
[170] Dubowchik, G.M., Gottschhall, D.W., Grossman, M.J. *et al.* (1982) *J. Amer. Chem. Soc.*, **104**, 4211–14.
[171] Reutov, O.A., Rozenberg, V.I. *et al.* (1977) *DAN SSSR* (In Russian), **237**, 608–11.
[172] Donyagina, V.F., Traven, V.F., Zaikin, V.G. *et al.* (1976) *Izv. AN SSSR*; series in chemistry (In Russian), 2117–19.
[173] Traven, V.F., Redchenko, V.V., Bartenev, V.Ya. *et al.* (1982) *Zhur. Obsch. Chim.* (In Russian), **52**, 358–62.
[174] Giordan, J.C., Moore, J.H. (1983) *J. Amer. Chem. Soc.*, **105**, 6541–4.
[175] Kuchitsu, K. (1968) *J. Chem. Phys.*, **49**, 4456–62.
[176] Cannington, P.H., Ham, N.S. (1983) *J. Electr. Spectr. Relat. Phenom.*, **31**, 175–9.

[177] Giordan, J.C. (1983) *J. Amer. Chem. Soc.*, **105**, 6544–6.
[178] Munakata, T., Kuchitsu, K., Harada, Y. (1979) *Chem. Phys. Lett.*, **64**, 409–12.
[179] Brundle, C.R. *et al.* (1972) *J. Amer. Chem. Soc.*, **94**, 1451–65.
[180] Sergeyev, G.B., Serguchev, Yu.A., Smirnov, V.V. (1973) *Usp. khimii* (In Russian), **42**, 1545–73.
[181] Bach, R.D., Henneike, H.F. (1970) *J. Amer. Chem. Soc.*, **92**, 5589–602.
[182] Volpin, M.E., Moiseyev, I.I., Shilov, A.E. (1980) *Zhur. Vses. Khim. Obsch. im D.I. Mend.* (In Russian), **25**, 515–24.
[183] Gritsenko, O.V. *et al.* (1985) *Usp. khim.* (In Russian), **54**, 1945–70.
[184] Basch, H. (1972) *J. Chem. Phys.*, **56**, 441–50.
[185] Baerends, E.J., Ellis, D.E., Ros, P. (1972) *Theor. Chim. Acta.*, **27**, 339–46.
[186] Vildavskaya, A.I., Rall, K.B., Petrov, A.A. (1971) *Zhur. Obsch. Chim.* (In Russian), **41**, 1279–85.
[187] Bieri, G. *et al.* (1977) *Helv. Chim. Acta.*, **60**, 2213–33.
[188] Masclet, P., Grosjean, D., Mouvier, G. (1973) *J. Electr. Spectr. Relat. Phenon.*, **2**, 225–37.
[189] Hanyu, Y., Boggs, J.E. (1973) *J. Chem. Phys.*, **43**, 3454–6.
[190] Zeelan, F.J. (1973) *Rec. Trav. Chim.*, **92**, 801–3.
[191] Jordan, K.D., Michejda, J.A., Burrow, P.D. (1976) *Chem. Phys. Lett.*, **42**, 227–31.
[192] Fock, J.H., Koch, E.E. (1984) *Chem. Phys. Lett.*, **105**, 38–43.
[193] Reutov, O.A., Beletskaya, I.P., Butin, K.P. (1980) CH. acids (In Russian), Nauka, Moscow.
[194] NgLily *et al.* (1982) *J. Amer. Chem. Soc.*, **104**, 7414–16.
[195] Desimoni, G., Tacconi, G., Barco, A. *et al.* (1983) *Natural Products Synthesis through Pericyclic Reactions*, Amer. Chem. Soc., Washington DC.
[196] Alston, P.V. *et al.* (1986) *Tetrahedron*, **42**, 4403–8.
[197] Dewar, M.J.S., Olivella, S., Stewart, J.J.P. (1986) *J. Amer. Chem. Soc.*, **108**, 5771–9.
[198] Kuchitsu, K. (1975) Gas electron diffraction. In: Allen, G. (ed.) *Molecular Structure and Properties*, Butterworths, London, 203–40.
[199] Masnovi, J.M., Seddon, E.A., Kochi, J.K. (1984) *Can. J. Chem.*, **62**, 2552–9.
[200] Hill, R.K., Bock, M.G. (1978) *J. Amer. Chem. Soc.*, **100**, 637–9.
[201] Fukui, K., Fujimoto, H. (1969) *Bull. Chem. Soc. Jap.*, **42**, 3399–409.
[202] Bock, H., Stafast, H. (1976) *Tetrahedron*, **32**, 855–63.
[203] Houk, K.N., Munchausen, L.L. (1976) *J. Amer. Chem. Soc.*, **98**, 937–46.
[204] Beez, M., Bieri, G., Bock, H., Heilbronner, E. (1973) *Helv. Chim. Acta.*, **56**, 1028–46.
[205] Heilbronner, E., Schmelzer, A. (1975) *Helv. Chim. Acta.*, **58**, 936–67.
[206] Bergmann, E.D. (1968) *Chem. Rev.*, **68**, 41–84.
[207] Clar, E., Schmidt, W. (1979) *Tetrahedron*, **35**, 2673–80.
[208] Mallion R.B. (1980) *Pure Appl. Chem.*, **52**, 1541–8.
[209] Lloyd, D. (1984) *Non-benzenoid Conjugated Carbocyclic Compounds*, Elsevier, Amsterdam.
[210] Glukhovtsev, M.N., Simkin, B. Ya., Minkin, V.I. (1985) *Usp. khimii* (In Russian),

**54**, 86–125.
[211] Garratt, P.J. (1986) *Aromaticity*, Wiley, NY.
[212] Masamune, S. *et al.* (1978) *J. Amer. Chem. Soc.*, **100**, 4889–91.
[213] Dunitz, J. *et al.* (1962) *Helv. Chim. Acta.*, **45**, 647–65.
[214] Pettit, R. (1975) *J. Organomet. Chem.*, **100**, 205–17.
[215] Bastiansen, O. *et al.* (1973) *J. Mol. Struct.*, **18**, 163–8.
[216] Cox, E.G. *et al.* (1958) *Proc. Roy. Soc. London*, Ser. A, **247**, 3–21.
[217] Cederbaum, L.S., Domcke, W., Schirmer, J. *et al.* (1978) *J. Chem. Phys.*, **69**, 1591–1603.
[218] Lindholm, E. (1972) *Trans. Far. Soc.*, **54**, 200–5.
[219] Case, D.A., Cook, M., Karplus, M. (1980) *J. Chem. Phys.*, **73**, 3294–3313.
[220] Price, W.C. (1977) Ultraviolet photoelectron spectroscopy: basic concepts and the spectra of small molecules. In: Brundle, C.R., Baker, A.D. (eds.) *Electr. Spectr. Theor. techn. Appl.*, Academic Press, London, 151–203.
[221] Brundle, C.R., Robin, M.B., Kuebler, N.A. (1972) *J. Amer. Chem. Soc.*, **94**, 1466–75.
[222] Duke, C.B., Yip, K.L., Ceasar, G.P. *et al.* (1977) *J. Chem. Phys.*, **66**, 256–68.
[223] Bloor, J.E., Sherrod, R.E. (1984) *Croat. Chem. Acta.*, **57**, 1011–30.
[224] Mines, G.W. *et al.* (1973) *Proc. Roy. Soc. London*, Ser. A., **333**, 171–81.
[225] Veszpremi, T. (1983) *Acta Chim. Hung.*, **112**, 307–18.
[226] Burrow, P.D., Michejda, J.A., Jordan, K.D. (1987) *J. Chem. Phys.*, **86**, 9–24.
[227] Evans, S., Green, J.C., Jackson, S.E. (1972) *J. Chem. Soc. Far. Trans.* II, **68**, 249–58.
[228] Vogel, E., Roth, H.D. (1964) *Angew. Chem. Int. Ed. Engl.*, **3**, 228–9.
[229] Effenberger, F., Klenk, H. (1976) *Chem. Ber.*, **109**, 769–76.
[230] Andrea, R.R., Cerfontain, H., Lambrechts, H.J. *et al.* (1984) *J. Amer. Chem. Soc.*, **106**, 2531–7.
[231] Zollinger, H. (ed.) (1973) *Aromatic Compounds*, Ser. 1, Vol. 3, Butterworths, London.
[232] Boschi, R., Schmidt, W., Gfeller, J.C. (1972) *Tetrahedron Lett.*, 4107–10.
[233] Dodler, M., Dunitz, J.D. (1965) *Helv. Chim. Acta.*, **48**, 1429–40.
[234] Farnell, L., Radom, L. (1981) *J. Amer. Chem. Soc.*, **103**, 7240–52.
[235] Gerson, F. *et al.* (1965) *Helv. Chim. Acta.*, **48**, 1494–1512.
[236] Kreiner, W.A., Rudolph, H.D., Tan, B.T. (1973) *J. Mol. Spectr.*, **48**, 86–99.
[237] Hehre, W.J., Radom, L., Pople, J.A. (1972) *J. Amer. Chem. Soc.*, **94**, 1496–1504.
[238] Klasinc, L. *et al.* (1978) *Croat. Chem. Acta.*, **51**, 43–53.
[239] Palmer, M.H. *et al.* (1978) *J. Mol. Struct.*, **49**, 105–23.
[240] Komatsu, T., Lund, A., Kinell, P.O. (1972) *J. Phys. Chem.*, **76**, 1721–6.
[241] Klopman, G. (1965) *J. Amer. Chem. Soc.*, **87**, 3300–3.
[242] Jordan, K.D., Michejda, J.A., Burrow, P.D. (1976) *J. Amer. Chem. Soc.*, **98**, 1295–6.
[243] Bolton, J.R. (1964) *J. Chem. Phys.*, **41**, 2455–7.
[244] Domenicano, A., Murray-Rust, P. (1979) *Tetrahedron Lett.*, 2283–6.
[245] Klessinger, M. (1972) *Angew. Chem.*, **84**, 544–5.
[246] Marschner, F., Goetz, H. (1974) *Tetrahedron*, **30**, 3451–4.

[247] Jonas, J.E., Schweitzer, G.K., Grimm, F.A. (1972/1973) *J. Electr. Spectr. Relat. Phenom.*, **1**, 29–66.
[248] Bischof, P.K. *et al.* (1974) *J. Organomet. Chem.*, **82**, 89–98.
[249] Ramsey, B.G. (1969) *Electronic Transitions in Organometalloids*, Academic Press, NY, London.
[250] Ramsey, B.G. (1977) *J. Organomet. Chem.*, **135**, 307–19.
[251] Pitt, C.G., Bock, H. (1972) *J. Chem. Soc., Chem. Commun.*, 28–9.
[252] Pitt, C.G. (1973) *J. Organomet. Chem.*, **61**, 49–70.
[253] Traven, V.F., West, R., Pyatkina, T.V., Stepanov, B.I. (1975) *Zhur. Obsch. Khim.* (In Russian), **45**, 831–7.
[254] Sakurai, H., Kira, M. (1975) *J. Amer. Chem. Soc.*, **97**, 4879–83.
[255] Traven, V.F., Eismont, M.Yu., Redchenko, V.V., Stepanov, B.I. (1980) *Zhur. Obsch. Khim.* (In Russian), **50**, 2007–16.
[256] Baidin, V.N., Kritskaya, I.I., Timoshenko, M.M. *et al.* (1983) *Usp. Fotoniki* (In Russian), Leningrad University Press, pp. 8–51.
[257] Baidin, V.N., Timoshenko, M.M. *et al.* (1981) *Izv. AN SSSR*, series in chemistry (In Russian), 2834–38.
[258] Lawler, R.G., Tabit, C.T. (1969) *J. Amer. Chem. Soc.*, **91**, 5671–2.
[259] Modelli, A., Jones, D., Distefano, G. (1982) *Chem. Phys. Lett.*, **86**, 434–7.
[260] Sipe, H.J., West, R. (1974) *J. Organomet. Chem.*, **70**, 353–66.
[261] Prins, I. *et al.* (1977) *Tetrahedron*, **33**, 127–31.
[262] Carreira, L.A., Towns, T.G. (1975) *J. Chem. Phys.*, **63**, 5283–6.
[263] Rabalais, J.W., Colton, R.J. (1972) *J. Electr. Relat. Phenom.*, **1**, 83–99.
[264] Kobayashi, T. *et al.* (1981) *Bull. Chem. Soc. Jap.*, **54**, 1658–61.
[265] Millefiori, S. *et al.* (1981) *Z. Phys. Chem.*, **128**, 63–72.
[266] Goldshtein, I.P., Fedotov, A.N. *et al.* (1984) *Zhur. Obsch. Khim.* (In Russian), **54**, 1363–73.
[267] Güsten, H., Klasinc, L., Ruščić, B. (1976) *Z. Naturforsch.*, **31A**, 1051–6.
[268] Burrow, P.D., Michejda, J.A., Jordan, K.D. (1978) *J. Amer. Chem. Soc.*, **96**, 6392–3.
[269] Kobayashi, T., Suzuki, H., Ogawa, K. (1982) *Bull. Chem. Soc. Jap.*, **55**, 1734–8.
[270] Haselbach, E., Klemm, U., Gschwind, R. (1982) *Helv. Chim. Acta.*, **65**, 2464–71.
[271] Munakata, T. *et al.* (1981) *Chem. Phys. Lett.*, **83**, 243–5.
[272] Elbel, S. *et al.* (1981) *Lieb. Ann. Chem.*, 1785–97.
[273] Palmer, M.H., Moyes, W., Spiers, M. (1980) *J. Mol. Struct.*, **62**, 165–87.
[274] Almenningen, A., Hartmann, A.O., Seip, H.M. (1968) *Acta Chem. Scand.*, **22**, 1013–24.
[275] Baca, A., Rossetti, R., Brus, L.E. (1979) *J. Chem. Phys.*, **70**, 5575–81.
[276] Gleiter, R., Schäfer, W., Eckert-Maskíc, M. (1981) *Chem. Ber.*, **114**, 2309–21.
[277] Klessinger, M., Rademacher, P., (1979) *Angew. Chem.* **91**, 885–96.
[278] Ruščić. B. *et al.* (1978) *Z. Naturforsch*, **33A**, 1006–12.
[279] Andrews, L., Arlinghaus, R.T., Payne, C.K. (1983) *J. Chem. Soc., Far. Trans. II*, **79**, 885–95.

[280] Carrington, A., dos Santos-Veiga, J. (1962) *Mol. Phys.*, **5**, 21–9.
[281] Hino, Sh., Seki, K., Inokuchi, H. (1975) *Chem. Phys. Lett.*, **36**, 335–9.
[282] Zagrubsky, A.A., Vilesov, F.I. (1974) *Usp. fotoniki* (In Russian), Leningrad University Press, 109–21.
[283] Zhong-Zhi, Y., Kovač, B. *et al.* (1981) *Helv. Chim. Acta.*, **64**, 1991–2001.
[284] Kovač, B., Mohraz, M., Heilbronner, E. (1980) *J. Amer. Chem. Soc.*, **102**, 4314–24.
[285] Lipari, N.O., Duke, C.B. (1975) *J. Chem. Phys.*, **63**, 1768–74.
[286] Dewar, M.J.S., Worley, S.D. (1969) *J. Chem. Phys.*, **50**, 654–67.
[287] Boschi, R., Clar, E., Schmidt, W. (1974) *J. Chem. Phys.*, **60**, 4406–18.
[288] Klasinc, L. (1980) *Pure Appl. Chem.*, **52**, 1509–24.
[289] Shida, T., Nosaka, Y., Kato, T. (1978) *J. Chem. Phys.*, **82**, 695–8.
[290] Andrews, L., Kelsall, B.J., Blankenship, T.A. (1982) *J. Chem. Phys.*, **86**, 2916–26.
[291] Notoya, R., Matsuda, A. (1981) *J. Res. Inst. Catal. Hokkaido Univ.*, **20**, 67–87.
[292] Kay, M.I., Okaya, Y., Cox, D.E. (1971) *Acta Cryst.*, Sec. B, **27**, 26–33.
[293] Schmidt, W. (1977) *J. Chem. Phys.*, **66**, 828–45.
[294] Bolton, J.R., Fraenkel, G.K. (1964) *J. Chem. Phys.*, **40**, 3307–20.
[295] Hilinski, E.F. *et al.* (1983) *J. Amer. Chem. Soc.*, **105**, 6167–8.
[296] Vartanyan, A.T. (1950) *DAN SSSR* (In Russian), **71**, 641–2.
[297] Chandrasekaran, K., Thomas, J. (1983) *J. Amer. Chem. Soc.*, **105**, 6383–9.
[298] Cameron, D.W. *et al.* (1964) *J. Chem. Soc.*, Sec. A, 62–79.
[299] Thompson, R.H. (1971) *Naturally Occurring Quinones* (2nd edn), Academic Press, New York.
[300] Laflamme, R.E., Hites, R.A. (1978) *Geochim. Acta.*, **42**, 289–303.
[301] Lee, M.L., Novotny, M.V., Bartle, K.D. (1981) *Analytical Chemistry of Polycyclic Aromatic Compounds*, Academic Press, NY–London.
[302] Nunome, K., Toriyama, K., Iwasaki, M. (1983) *J. Chem. Phys.*, **79**, 2499–503.
[303] Dewar, M.J.S. (1971) *Angew. Chem. Int. Ed. Engl.*, **10**, 761.
[304] Rabinovitz, M., Willner, I., Minsky, A. (1983) Acc. Chem. Res., **16**, 298–304.
[305] Blinov, L.M. (1983) *Usp. khimii* (In Russian), **52**, 1263–1300.
[306] Khidekel, M.L., Zhilyaeva, E.I. (1978) *Zhur. Vses. Khim. Obsch. im D.I. Mend.* (In Russian), **23**, 506–24.
[307] Nakamura, K. (1983) *J. Soc. Fiber Sci. Tech. Jap.*, **39**, 28–35.
[308] Misurkion, I.A., Ovchinnikov, A.A. (1977) *Usp. khimii* (In Russian), **46**, 1835–70.
[309] Moore, J.M., Ramamoorthy, S. (1984) *Organic Chemicals in Natural Waters: Applied Monitoring and Impact Assessment*, Springer, NY–Berlin.
[310] Müller, E. (1957) *Neuere Anschaungen der organischen Chemie*, Springer, Berlin.
[311] Clar, E., Schmidt, W. (1977) *Tetrahedron*, **33**, 2093–7.
[312] Clar, E., Schmidt, W. (1979) *Tetrahedron*, **35**, 1027–32.
[313] Sato, N., Inokuchi, H., Seki, K. *et al.* (1982) *J. Chem. Soc. Far. Trans.* II, **78**, 1929–36.
[314] Akiyama, I., Harvey, R.G., LeBreton, P.R. (1981) *J. Amer. Chem. Soc.*, **103**, 6330–2.

[315] Boschi, R., Murrell, J.N., Schmidt, W. (1972) *Farad. Discuss. Chem. Soc.*, **54**, 116–26.
[316] Clark, P.A., Brogli, F., Heilbronner, E. (1972) *Helv. Chim. Acta.*, **55**, 1415–28.
[317] Seki, K. *et al.* (1976) *Bull. Chem. Soc. Jap.*, **49**, 904–15.
[318] Holy, N.L. (1974) *Chem. Rev.*, **74**, 243–77.
[319] Tsubomura, H., Sunakawa, S. (1967) *Bull. Chem. Soc. Jap.*, **40**, 2468–74.
[320] Orgel, L.E. (1955) *J. Chem. Phys.*, **23**, 1352–3.
[321] Briegleb, G., Czekalla, J., Reuss, G. (1961) *Z. Phys. Chem.*, **30**, 316–32.
[322] Dewar, M.J.S., Rogers, H. (1962) *J. Amer. Chem. Soc.*, **84**, 395–8.
[323] Konovalov, A.I., Kiselev, V.D. (1966) *Zhur. Organ. Khim.* (In Russian), **2**, 142–4.
[324] Kiselev, V.D., Miller, J.G. (1975) *J. Amer. Chem. Soc.*, **97**, 4036–9.
[325] Thompson, C.C., Holder, D.D. (1972) *J. Chem. Soc. Perkin. Trans.*, **II**, 257–62.
[326] Fukuzumi, S., Kochi, J.K. (1982) *Tetrahedron*, **38**, 1035–49.
[327] Tuttle, T.R., Weissmann, S. (1958) *J. Amer. Chem. Soc.*, **80**, 5342–4.
[328] Paalme, L. *et al.* (1983) *Oxid. Commun.*, **4**, 27–34.
[329] Lowe, J.P., Silverman, B.D. (1984) *Acc. Chem. Res.*, **17**, 332–8.
[330] v. Szentpaly, L. (1984) *J. Amer. Chem. Soc.*, **106**, 6021–8.
[331] Buenker, R.J., Peyerimhoff, S.D. (1969) *Chem. Phys. Lett.*, **3**, 37–42.
[332] Cradock, S., Whiteford, R.A. (1971) *Trans. Farad. Soc.*, **67**, Pt. 12, 3425–434.
[333] Bernardi, F., Bottoni, A., Venturini, A. (1986) *J. Amer. Chem. Soc.*, **108**, 5395–400.
[334] Hehre, W.J., Pople, J.A. (1970) *J. Amer. Chem. Soc.*, **92**, 2191–7.
[335] Bazilevsky, M.V., Koldobsky, S.G., Tikhomirov, V.A. (1986) *Usp. khimii* (In Russian), **55**, 1667–98.
[336] Stepanov, B.I., Traven, V.F. (1968) *Zhur. Organ. Khim.* (In Russian), **4**, 1067–72.
[337] Burrow, P.D. *et al.* (1982) *J. Chem. Phys.*, **77**, 2699–2701.
[338] Kivelson, D., Wilson, E.B., Lide, D.R. (1960) *J. Chem. Phys.*, **32**, 205–9.
[339] von Niessen, W., Åsbrink, L., Bieri, G. (1982) *J. Electr. Relat. Phenom.*, **26**, 173–201.
[340] Wittel, K., Bock, H. (1974) *Chem. Ber.*, **107**, 317–38.
[341] Burrow, P.D. *et al.* (1981) *Chem. Phys. Lett.*, **82**, 270–6.
[342] Cvitaš, T., Klasine, L. (1977) *Croat. Chem. Acta.*, **50**, 291–7.
[343] Bethell, D., Gold, V. (1967) *Carbonium Ions. An Introduction.* Academic Press, London–New York.
[344] Maj, P., Hasegawa, A., Symons, M.C.R. (1983) *J. Chem. Soc. Far. Trans.* I, **79**, 1931–8.
[345] Fujisawa, S. *et al.* (1986) *J. Amer. Chem. Soc.*, **108**, 6505–11.
[346] Jones, D. (ed.) (1979) *Comprehensive Organic Chemistry. The Synthesis and Reaction of Organic Compounds*, Vol. 3, *Organometallic Compounds*, Pergamon Press, Oxford.
[347] Wilkinson, G., Stone, G.A., Abel, E.W. (eds.) (1982) *Comprehesive Organometallic Chemistry*, Vol. 8, Oxford University Press, Oxford.
[348] Jackson, W.R., Jennings, W.B. (1969) *J. Chem. Soc.*, Sec. B, 1221–8.

[349] Chowdhury, S. *et al.* (1986) *J. Amer. Chem. Soc.*, **108**, 3630–65.
[350] Stoddart, J.F. (ed.) (1979) Oxygen compounds. In: *Comprehensive Organic Chemistry. The Synthesis and Reactions of Organic Compounds*, Vol. 1, Pergamon Press, Oxford.
[351] Patai, S. (ed.) (1980) *The Chemistry of Ethers, Crown Ethers, Hydroxyl Groups and Their Sulphur Analogues*, Wiley, NY.
[352] Bock, H. *et al.* (1973) *J. Organomet. Chem.*, **61**, 113–25.
[353] Elbel, S., Bergmann, H., Ensslin, W. (1974) *J. Chem. Soc. Far. Trans.* II, **70**, 555–9.
[354] Cocksey, B.J., Eland, J.H.D., Danby, C.J. (1971) *J. Chem. Soc.*, Sec. B., 790–2.
[355] Knyazhevskaya, V.B., Traven, V.F., Stepanov, B.I. (1980) *Zhur. Obsch. Khim.* (In Russian), **50**, 606–13.
[356] Traven, V.F., German, M.I., Eismont, M. Yu. *et al.* (1978) *Zhur. Obsch. Khim.* (In Russian), **48**, 2232–8.
[357] Kira, M., Nakazawa, H., Sakurai, H. (1986) *Chemistry Letters*, 497–500.
[358] Sweigart, D.A., Turner, D.W. (1972) *J. Amer. Chem. Soc.*, **94**, 5599–603.
[359] Chan, S.I. *et al.* (1960) *J. Chem. Phys.*, **33**, 1643–55.
[360] Tschmutowa, G., Bock, H. (1976) *Z. Naturforsch.*, **31B**, 1616–20.
[361] Guryanova, E.N., Romm, I.P. *et al.* (1979) *DAN SSSR* (In Russian), **248**, 108–112.
[362] Gleiter, R., Larsen, J.-S. (1979) *Top. Curr. Chem.*, **86**, 139–95.
[363] Bernardi, F., Epiotis, N.D. *et al.* (1976) *J. Amer. Chem. Soc.*, **98**, 2385–90.
[364] Friege, H., Klessinger, M. (1977) *J. Chem. Res.*, 208–9.
[365] Hagen, K., Hedberg, K. (1973) *J. Chem. Phys.*, **59**, 158–62.
[366] Baker, A.D., May, D.P., Turner, D.W. (1968) *J. Chem. Soc.*, Sec. B., 22–34.
[367] Kobayashi, T., Nagakura, S. (1974) *Bull. Chem. Soc. Jap.*, **47**, 2563–72.
[368] Modelli, A. *et al.* (1983) *Chem. Phys.*, **77**, 153–8.
[369] Khelmer, B.Yu., Mazalov, L.N. *et al.* (1981) *Zh. struk. khim.* (In Russian), **22**, 135–63.
[370] Chmutova, G.A., Vtyurina, N.N., Bock, G. (1979) *DAN SSSR* (In Russian), **244**, 1138–41.
[371] Friege, H., Klessinger, M. (1979) *Chem. Ber.*, **112**, 1614–25.
[372] Schweig, A., Thon, N. (1976) *Chem. Phys. Lett.*, **38**, 482–5.
[373] Baker, A.D., Armen, G.H., Guang-di Yang (1981) *J. Org. Chem.*, **46**, 4127–30.
[374] Tschmutowa, G., Bock, H. (1976) *Naturforsch.* **31B**, 1611–15.
[375] Voronkov, M.G., Dolenko, G.N. *et al.* (1983) *DAN SSSR* (In Russian), **273**, 1406–10.
[376] Jordan, K.D., Michejda, J.A., Burrow, P.D. (1976) *J. Amer. Chem. Soc.*, **98**, 7189–91.
[377] Modelli, A. *et al.* (1984) *Wavefunct. Mech. Electr. Scattering Process*, Berlin, 19–23.
[378] Zabicky, J. (ed.) (1970) *Chemistry of the Carbonyl Group*, Vol. 2, Pt. 1, 5, Interscience, London.
[379] Nelson, R., Pierce, L. (1965) *J. Mol. Spectr.*, **18**, 344.
[380] Modelli, A. *et al.* (1984) *Tetrahedron*, **40**, 3257–62.

[381] Ashby, E.C., Laemmle, J.T., Neumann, H.M. (1974) *Acc. Chem. Res.*, **7**, 272–80.
[382] Beak, P., Worley, J.W. (1972) *J. Amer. Chem. Soc.*, **94**, 597–604.
[383] DeKock, R.L. (1977) Ultraviolet photoelectron spectroscopy of inorganic molecules. In: Brundle, C.R., Baker, A.D. (eds.) *Electron Spectroscopy: Theory, Techniques and Applications*, Vol. 1, Academic Press, London, 293–353.
[384] Ramsey, B.G. *et al.* (1974) *J. Organomet. Chem.*, **74**, 41–5.
[385] Wing-Cheung Tam, D. Yer, C. Brion (1974) *J. Electr. Spectr. Rel. Phen.*, **4**, 77–80.
[386] Von Niessen, W., Bieri, G., Schirmer, J., Cederbaum, L. (1982) *Chem. Phys.*, **65**, 157–76.
[387] Tandura, S.N., Voronkov, M.G., Alekseev, N.V. (1985) *Top. Cur. Chem.*, **131**, 99–189.
[388] Kurkovskaya, L.N., Negrebetsky, V.V., Traven, V.F. (1991) *Zhur. Obsch. Khim.* (In Russian), **61**, 2097–104.
[389] Cornovale, F., Gan, T., Peel, G., Franz, K. (1979) *Tetrahedron*, **35**, 129–33.
[390] Solodar, S.A. (1976) *Zhur. Vses. Khim. Obsch. im D.I. Mend.* (In Russian), **21**, 306–14.
[391] Traven, V.F., Safronov, A.I., Chibisova, T.A. (1991) *Zhur. Obsch. Khim.* (In Russian), **61**, 697–705.
[392] Traven, V.F., Safronov, A.I., Chibisova, T.A. (1991) *Zhur. Obsch. Khim.* (In Russian), **61**, 1448–53.
[393] Martin, H.D., Mayer, B. (1983) *Angew. Chem.*, **95**, 281–313.
[394] Gorelik, M.V. (1983) *Chemistry of Anthraquinones and Their Derivatives* (In Russian), Khimia, Moscow.
[395] Trotter, J. (1960) *Acta Cryst.*, **13**, 86–95.
[396] Lauer, G., Schäfer, W., Schweig, A. (1975) *Chem. Phys. Lett.*, **33**, 312–15.
[397] Von Niessen, W. *et al.* (1986) *Austral. J. Phys.*, **39**, 687–710.
[398] Dougherty, D., McGlynn, S.P. (1977) *J. Amer. Chem. Soc.*, **99**, 3234–9.
[399] Ady, E., Brickmann, J. (1971) *Chem. Phys. Lett.*, **11**, 302–6.
[400] Derissen, J.L. (1971) *J. Mol. Struct.*, **7**, 67–80.
[401] Baker, A.D., Betteridge, D. (1972) *Photoelectron Spectroscopy*, Pergamon, Elmsford, NY.
[402] Rudenko, A.P., Zarubin, M.Ya., Averyanov, S.F., Barsheva, N.S. (1979) *DAN SSSR* (In Russian), **249**, 117–20.
[403] Aarons, L.J. *et al.* (1973) *J. Chem. Soc. Far. Trans.* II, **69**, 643–7.
[404] Alder, R.W., Arrowsmith, R.J., Casson, A. *et al.* (1981) *J. Amer. Chem. Soc.*, **103**, 6137–42.
[405] Yoshikawa, K., Hashimoto, M., Morishima, I. (1974) *J. Amer. Chem. Soc.*, **96**, 288–9.
[406] Rozenboom, M.D., Houk, K.N. (1982) *J. Amer. Chem. Soc.*, **104**, 1189–91.
[407] Aue, D.H., Webb, H.M., Bowers, M.T. (1975) *J. Amer. Chem. Soc.*, **97**, 4136–7.
[408] Heilbronner, E., Muszkat, K.A. (1970) *J. Amer. Chem. Soc.*, **92**, 3818–21.
[409] Alder, R.W. *et al.* (1982) *J. Chem. Soc. Chem. Commun.*, 940.

[410] Spanka, G., Rademacher, P. (1986) *J. Org. Chem.*, **51**, 592–6.
[411] Sidorkin, V.F., Pestunovich, V.A., Voronkov, M.G. (1980) *Usp. khimii* (In Russian), **49**, 789–813.
[412] Voronkov, M.G., Dyakov, V.M., Kirpichenko, S.V. (1982) *J. Organomet. Chem.*, **233**, 1–147.
[413] Müller, K., Previdoli, F., Desilvestro, H. (1981) *Helv. Chim. Acta.*, **64**, 2497–2507.
[414] Lafon, C. *et al.* (1986) *Nouv. J. Chem.* (1986), **10**, 69–72.
[415] Müller, K., Previdoli, F. (1981) *Helv. Chim. Acta.*, **64**, 2508–14.
[416] Lister, D.G. *et al.* (1974) *J. Mol. Struct.*, **23**, 253–64.
[417] Nordgren, J. *et al.* (1984) *Chem. Phys.*, **84**, 333–6.
[418] Furin, G.G., Sultanov, A.Sh., Furley, I.I. (1986) *Izv. AN SSSR*, series in chemistry (In Russian), 474–8.
[419] Palmer, M.H. *et al.* (1979) *J. Mol. Struct.*, **53**, 235–49.
[420] Raevsky, O.A., Vereshchagin, A.N. *et al.* (1974) *Izv. AN SSSR*, series in chemistry (In Russian), 453–5.
[421] Tsvetkov, E.N., Kabachnik, M.I. (1971) *Usp. khimii* (In Russian), **40**, 177–225.
[422] Bokanov, A.I. (1978) *Trudi Mosk. Khim. Techn. Inst. of D.I. Mendeleev* (In Russian), **103**, 111–20.
[423] Zverev, V.V., Kitaev, Yu.P. (1977) *Usp. khimii* (In Russian), **46**, 1515–43.
[424] Puddephatt, R.J., Baneroft, G.M., Chan, T. (1883) *Inorg. Chim. Acta.*, **73**, 83–9.
[425] Ratovsky, G.V., Chuvashev, D.D., Panov, A.M. (1985) *Zhur. Obsch. Khim.* (In Russian), **55**, 571–6.
[426] Schmidt, H., Schweig, A., Vermeer, H. (1977) *J. Mol. Struct.*, **37**, 93–104.
[427] Effenberger, F. *et al.* (1978) *Tetrahedron*, **34**, 2409–17.
[428] Hallas, G. *et al.* (1984) *J. Chem. Soc. Perk. Trans.* II, 149–53.
[429] Morino, Y., Jijima, T., Murata, Y. (1960) *Bull. Chem. Soc. Jap.*, **33**, 46–9.
[430] Nelsen, S.F., Buschek, J.M. (1974) *J. Amer. Chem. Soc.*, **96**, 2392–7.
[431] Nelsen, S.F., Buschek, J.M. (1974) *J. Amer. Chem. Soc.*, **96**, 6982–7.
[432] Rademacher, P. (1974) *Tetrahedron Lett.*, 83–6.
[433] Rademacher, P., Koopmann, H. (1975) *Chem. Ber.*, **108**, 1557–69.
[434] Rademacher, P. *et al.* (1979) *Chem. Ber.*, **110**, 1939–49.
[435] Chong, D.P., Herring, F.G., McWilliams, D. (1975) *J. Electr. Spectr. Relat. Phenom.*, **7**, 445–55.
[436] Haselbach, E., Heilbronner, E. (1970) *Helv. Chim. Acta.*, **53**, 684–95.
[437] Haselbach, E., Schmelzer, A. (1971) *Helv. Chim. Acta.*, **54**, 1575–80.
[438] Houk, K.N., Yau-Miu Chang, Engel, P.S. (1975) *J. Amer. Chem. Soc.*, **97**, 1824–32.
[439] Bock, H. *et al.* (1976) *J. Amer. Chem. Soc.*, **98**, 109–14.
[440] Robin, M.B. *et al.* (1972) *J. Chem. Phys.*, **57**, 1758–63.
[441] Kobayashi, T., Yokota, K., Nagakura, S. (1975) *J. Electr. Spectr. Relat. Phenom.*, **6**, 167–70.
[442] Engel, P.S., Steel, C. (1973) *Acc. Chem. Res.*, **6**, 275–81.
[443] Traven, V.F., Kostyuchenko, E.E., Mkhitarov, R.A. *et al.* (1980) *Zhur. Organ.*

*Khim.* (In Russian), **16**, 1047–56.
[444] Bergmann, H., Elbel, S., Demuth, R. (1977) *J. Chem. Soc. Dalton Trans.*, 401–6.
[445] Compton, R.N., Reinhardt, P.W., Cooper, C.D. (1978) *J. Chem. Phys.*, **68**, 4360–7.
[446] Palmer, M.H. *et al.* (1979) *J. Mol. Struct.*, **55**, 243–63.
[447] Koptyug, V.A., Salakhutdinov, N.F. (1984) *Zhur. Organ. Khim.* (In Russian), **20**, 246–54.
[448] Koptyug, V.A. (1976) *Zhur. Vses. Khim. Obsch. im D.I. Mend.* (In Russian), **21**, 247–55.
[449] Koptyug, V.A., Rogozhnikova, O.Yu, Detsyna, A.N. (1982) *Zhur. Organ. Khim.* (In Russian), **18**, 2017–20.
[450] Koptyug, V.A., Salakhutdinov, N.F., Vasilyeva, V.G. (1986) *Zhur. Organ. Khim.*, **22**, 1127–33.
[451] Detsyna, A.N., Sidorova, N.V. *et al.* (1983) *Zhur. Organ. Khim.*, **19**, 1924–30.
[452] Koptyug, V.A., Salakhutdinov, N.F., Vasilyeva, N.V. (1984) *Zhur. Organ. Khim.*, **20**, 254–8.
[453] Schäfer, W., Schweig, A., Mathey, F. (1976) *J. Amer. Chem. Soc.*, **98**, 407–14.
[454] Munakata, T., Kuchitsu, K., Harada, Y. (1980) *J. Electr. Spectr. Relat. Phenom.*, **20**, 235–44.
[455] Modelli, A., Guerra, M., Jones, D. *et al.* (1984) *Chem. Phys.*, **88**, 455–61.
[456] Bart, J.C.J., Dely, J.J. (1968) *Angew. Chem.*, **80**, 843–4.
[457] Batich, C., Heilbronner, E. *et al.* (1973) *J. Amer. Chem. Soc.*, **95**, 928–30.
[458] Hodges, R.V. *et al.* (1985) *Organometallics*, **4**, 457–61.
[459] Burrow, P.D. *et al.* (1982) *J. Amer. Chem. Soc.*, **104**, 425–9.
[460] Von Niessen, W. Diercksen, G.H.F., Cederbaum, L.S. (1975) *Chem. Phys.*, **10**, 345–60.
[461] Von Niessen, W., Kraemer, W.P., Diercksen, G.H.F. (1979) *Chem. Phys.*, **41**, 113–32.
[462] Rodin, O.G., Redchenko, V.P., Kostitsyn, A.B., Traven, V.F. (1988) *Zhur. Obsch. Khim.* (In Russian), **58**, 1409–15.
[463] Traven, V.F., Chibisova, T.A., Kramarenko, S.S., Rodin, O.G. (1990) *Zhur. Obsch. Khim.* (In Russian), **60**, 414–21.
[464] Salakhutdinov, N.F., Koptyug, V.A. (1985) *Zhur. Organ. Khim.* (In Russian), **21**, 948–51.
[465] Heilbronner, E. *et al.* (1969) *Angew. Chem.*, **81**, 537–8.
[466] Flitsch, W. *et al.* (1978) *Tetrahedron*, **34**, 2301–4.
[467] Cooper, A.R., Crowne, C.W.P., Farrell, P.G. (1966) *Trans. Far. Soc.*, **62**, Pt. 1, 18–28.
[468] Dwivedi, P.C. *et al.* (1975) *Spectrochim Acta.*, Pt. A, **31**, 129–35.
[469] Karl, N. *et al.* (1982) *J. Chem. Phys.*, **77**, 4870–8.
[470] Gorb, L.G., Abronin, I.A. *et al.* (1984) *Izv. AN SSSR* series in chemistry (In Russian), 1079–85.
[471] Marino, G. (1971) *Adv. Heterocycl. Chem.*, **13**, 235–314.

# Index


acetaldehyde, 244–245
  charges on atoms, 247
  dipole moment, experimental and calculational, 247
  IP-values (PES), 244, 245
  MOs, *ab initio* (6-31G), 245
    graphic shapes, 243
  nucleophilic addition, relative reactivity, 243, 247–248
acetic acid, 263–265
  dimerization, 263–264
  geometry, 263
  IP-values (PES), 265
MOs, *ab initio* (4-31G), 265
  PE-spectra of monomer and dimer forms, 264
  protonation, 265
  effect of self-association on IP-values (PES), 99
acetone, 245–250
  charges on atoms, 247
  dipole moment, experimental and calculational, 247
  geometry, 246
  IP-values (PES), 244–245
  MOs, *ab initio* (4-31G), 246
  nucleophilic addition, relative reactivity, 243, 247–248
  tautomerism, 246
acetonitrile, 316
  $EA_1$-value, 316
  geometry, 316
  IP-values (PES), 316, 317
  MOs, *ab initio* (6-31G), 317
  pKa of conjugated acid, 316
acetophenone, 254–255
  conformation and barrier of rotation, 254
  electron densities, *ab initio* (STO-3G), 267
  IP-values (PES), 255
  isomer distribution, 267
  MOs, CNDO/2, 255
  nitration, 267
acetylacetone, 261
  IP-values (PES), 261
  pKa-value, 261
  tautomerism, 261
acetylene, 130–131
  C-H-acidity, 131
  $EA_1$-value (ETS), 131
  geometry, 130
  IP-values (PES), 130
  MOs, *ab initio* (6-31G), 130
  PE-spectrum, 126
  reactions with electrophiles, 130
acidity of carboxylic acids as a measure of inductive effect, 4
acridine, 351, 353
  HOMO electron densities (HMO), 353
  orbital control in reactions with electrophiles, 353
  PES of solid film, 352
acrolein, 250–253
  1,2- and 1,4-addition reactions, orbital and charge control, 251, 253
  charges on atoms, HMO, 253
  geometry, 251
  IP-values (PES),
  MOs, graphic shapes, 252
acrylonitrile, 316, 317
  $EA_1$-value, 317
  geometry, 316, 317
  IP-values (PES), 317
  MOs, 317
adamantane, 120
  CT-complexing with TCNE, EA-spectrum, 120
alcoholate anion, 5
  localization of electron density, 5
alcohols, 5, 219–230
  acidity in gas and liquid phases, 219
  charges on atoms, 224
  conformation and geometry, 220, 222
  IP-values (PES), 223
  MOs, *ab initio* (6-31G), CNDO/2, 223
alkadienes, 5, 132
alkenes, 5, 124
alkyl ethers, 219–230

conformation and geometry, 220
IP-values (PES), 223
alternant hydrocarbons, definition, 27
amines, 6, 269
1-amino-8-carboxy-naphthalene
derivatives, mixing of orbitals in them, 288, 289
geometry and mixing of orbitals in them, 288, 289
2-amino-pyridine, 350
IP-values (PES), 350
MOs, 350
4-amino-pyridine, 350
IP-values (PES), 350
MOs, 350
ammonia, 269–271
$IP_1$-value (PES), 271
geometry, 269
MOs, *ab initio*, 270
self-association and medium effects on, photoionization, 98
Walsh diagram, 270
aniline, 283–284
barrier of rotation, 284
basicity, 283
cation radical cf, 284
geometry and conformation, *ab initio*, 283
EA-value (ETS), 284
electron densities on atoms, *ab initio*, 267
IP-values (PES), 283
MOs, MINDO/3, 283
reactivity, 284
aniline, dipole moment, 10, 11
anion radical, definition, 42
anisole, 238–239, 241, 242
anion radical, ESR-spectrum, 242
cation radical, ESR-spectrum, 242
conformation, 238
IP-values (PES), 239
isomer distribution in electrophilic nitration, 326
homologues, 239
IP-values (PES), 239
relative rates of oxidation, 239
reactions with electrophiles, 238
transition energies in eA-spectra, 239
oxidation, 238
anthracene, 10, 40, 184, 201
absorption on $TiO_2$, 187
anion-radical, ESR-spectrum, 183
cation radical, EA-spectrum, transition energies, 183, 186
CT-complexes with TCNE, 186, 201
Diels–Alder reaction, 202
EA-spectrum, 52
EA-values (ETS), 181
energies of electron transitions in EA-spectrum, 40
ESR, anion radical, effect solvation on spin densities, 100
geometry, 184
IP-values (PES) 184, 323
MOs, CNDO/S, MINDO/2, HMO, PPP, EHMO, 185
PE-spectrum, 182
reactions with electrophiles, orbital control, 68, 180, 199
9-substituted ($CH_3$, $OCH_3$, Cl, F, $COCH_3$, COOH, CHO, CN, $NO_2$), CT-complexes with TCNE, 324
9,10-anthraquinone, 261–263
anion radical
ESR-spectrum, 263
spin densities, HMO, 263
$E_{1/2}^{red}$-values (polarography), 263
MOs, CNDO/S, 262
PE-spectrum, 262
aromatic hydrocarbons, alkyl, methoxy-, and halogen-substituted frontier MOs, reactions with electrophiles, 324–237
arsabenzene, 338–342
basicity, 341
EA-values (ETS), 341
IP-values (PES), 341
MOs, 341
proton affinity, 340
atomic orbital, 15–20
basis atomic orbital, 21
basis set of AO, 30
Gaussian functions, 30
Hartree–Fock functions, 20, 30
ionization potential of, 55
occupancy of, 15
overlap integral $S\mu\nu$, 22, 55
permitted combination of quantum numbers, 15
Slater functions, 30, 31
solution of Schrödinger equation, 18
1-azabicyclo[4,4,4]tetradecane, 277–278
basicity, 277
geometry, 277
IP-values (PES), 277
mixing of orbitals in it, 277
1-azabicyclo[4,4,4]tetradec-5-ene, 277
basicity, 277
geometry, 277
IP-values (PES), 277
mixing of orbitals in it, 277
oxidation, 273
*trans*-azoalkanes, linear and cyclic, 297–301
IP-values (PES), 300, 301
MOs, 300, 301
thermolysis and photolysis, 300, 301
EA-spectra, transition energies, 305
*trans*-azobenzene, 302
geometry, 302
IP-values (PES), 302, 303
*trans-cis*-isomerization, 302
MOs, INDO/S, 303
polarographic reduction, 303, 304
*trans*-azoethane, 297

IP-values (PES), 297
EA-spectra, transition energies, 305
*trans*-azomethane, 295
geometry, 295
MOs, *ab initio*, 295

basicity of amines as a measure of inductive effect, 4
benz[α]anthracene, 188, 190, 192, 199, 201
benzaldehyde, 253–254
anion radical, ESR-spectrum and spin densities, 254
barrier of rotation, 253–254
conformation, 253–254
EA-values (ETS), 254
electron densities, *ab initio* (STO-3G), 267
IP-values (PES), 254
isomer distribution, 267
nitration, 267
benzanthrone, 257–260
anion radical
ESR-spectrum, 260
spin densities, experimental and calculational, 259
basicity, 257
cation radical,
spin densities, experimental and calculational, INDO, 259
condensation in the presence of strong bases, 259
CT-complex with TCNE,
band positions in EA-spectrum, 260
$E_{1/2}^{red}$-values (polarography), 259
IP-values (PES), 258
MOs, INDO/S, 258
frontier, graphic shapes, 259
PE-spectrum, 258
reactions with electrophiles and nucleophiles, orbital control in, 259
benz[α]phenanthrene, 188, 199
benzene, 25, 148–155, 201
anion-radical, ESR-spectrum, 153
aromaticity, 149
EA-spectrum, transition energies, 53
$EA_1$-value (ETS), INDO/S, 42
EA-values (ETS), 152
effect of self-association on IP-values (PES), 100
'hardness' and electronegativity, 74, 88
geometry, 150
IP-values, 151
IP-values (PES), INDO/S, 53
MOs, $X\alpha$-calculations, 151
MOs, HMO method, 25
analytical form, 26
graphic, 27
PE-spectrum, 151
reactions with electrophiles, 154
removal of degeneracy of frontier MOs by substituents, 28
benzoic acid, p$K_a$-value, 6
$E_{1/2}^{red}$-values, polarography, 45
benzoic acid, 263–265
cation radical,
ESR-spectrum, 267
spin densities, experimental and calculational, 267
electron densities on atoms, *ab initio* (STO-3G), 267
geometry, 265
MOs, barrier of rotation, *ab initio* (STO-3G), 266
nitration, isomer distribution, 267
benzonitrile, 317–319
anion radical, ESR-spectrum, 319
cation radical, ESR-spectrum, 319
dipole moment, experimental and *ab initio*, 318
$EA_1$-value, 317
electron densities on atoms, *ab initio*, 319
geometry, 317
benzoperylene, 345
MOs, 345
1,12-benzoperylene, 192
1,4-benzoquinone, 261–263
anion radical,
ESR-spectrum, 263
spin densities, HMO, 263
$E_{1/2}^{red}$-values (polarography), 263
geometry in solid and gas phases, 262
IP-values (PES), 262
Koopmans defect, 262
MOs, CNDO/S, calculations with Green functions, 262
PE-spectrum, 262
benzoyloxymethyl(trifluoro)silane, 255
benzyl(trimethyl)germane, 165
EA-spectrum, transition energies, 165
hyperconjugation, 165
IP-values (PES), 165
MOs, 165
benzyl(trimethyl)silane, 165
EA-spectrum, transition energies, 165
hyperconjugation, 165
IP-values (PES), 165
MOs, 165
benzyl(trimethyl)silane, 323
CT-complex with TCNE, EA-spectrum, 323, 324
benzyl(trimethyl)stannane, 165
EA-spectrum, transition energies, 165
hyperconjugation, 165
IP-values (PES), 165
MOs, 165
1,1-biadamantyl, 120
CT-complexing with TCNE, EA-spectrum, 120
bicyclic amines
basicity, 274
cation radical, stability, 277
IP-values (PES), 273
bicyclic diamines

basicity, 274
cation radical, stability, 276
IP-values (PES), 275
mixing of orbitals in them, 273–275
TS- and TB-effects, in them, 274
bicycloalkadienes, 143–145
IP-values (PES), 145
TB- and TS-effects of orbital mixing, 145
bicyclo[4,2,2]deca-7,9-diene, 144
IP-values (PES), 144, 145
TB- and TS-effects of orbital mixing, 143
bicyclo[3,2,2]nona-6,8-diene, 144
IP-values (PES), 144, 145
TB- and TS-effects of orbital mixing, 143
bicyclo[2,2,2]octadiene, 144
IP-values (PES), 144, 145
TB- and TS-effects of orbital mixing, 143
bicyclo[4,4,4]tetradec-1-ene, 277
basicity, 277
geometry, 277
IP-values (PES), 277
biological activity and frontier orbitals of organic molecules, 76
agonist, 76
antagonist, 76
electrostatic potential, role in biological activity, 76
molecular pharmacology, 76
receptor, 76
biphenyl, 174–176, 201
anion-radical, ESR-spectrum, 177
cation radical, 176
EA-spectrum, 176
conformation, 174
EA-values (ETS), 177
MOs, 174–176
Birch reduction, orbital control in it
of benzene derivatives, 328
of PAH, 203
bis(benzene)chromium, 154
IP-values (PES), 154
MOs, 154
1,5,-bis(dimethylamino)naphthalene, 288
basicity, 288
geometry, 288
mixing of orbitals in it, 288
PES, 288
1,8-bis(dimethylamino)naphthalene, 287–288
basicity, 288
geometry, 288
mixing of orbitals in it, 288
PE, 288
1,2-bis(methylthio)ethylene, 234
conformation, 234
IP-values (PES), 234
*p*-bis(tert-butyl)benzene, 163, 166
EA-spectrum, transition energies, 165
EA-values (ETS), 166
hyperconjugation, 166
IP-values (PES), 166
bis(tert-butyl)diimine, 297
EA-spectrum, transition energies, 298, 301, 305
electron densities on atoms, CNDO, 297
IP-values (PES), 298
1,4-bis(trimethylsilyl)benzene,
anion radical, ESR-spectrum, 328
Birch reduction, 328
*p*-bis(trimethylsilyl)benzene, 163, 166
EA-values (ETS), 166
bis(trimethylsilyl)diimine, 297
EA-spectrum, transition energies, 298, 301
electron densities on atoms, CNDO, 297
IP-values (PES), 298
2,6-bis(trimethylsilyl)pyridine, 349
IP-values (PES), 349
MOs, 349
bond order, definition, 28
boron trifluoride, 'hardness' and electronegativity, 74
de Broglie principle, 16
bromobenzene, 212–218
bromocyclobutane, 208
bromocyclopropane, 208
bromoethylene
IP-values (PES), 211
mixing of orbitals, 211
bromomethane, 205–211
2-bromopropane, 208
N-brosylmitomycin, 278–279
geometry, 278
mixing of orbitals in it, 277–278
TS-effect, 278
buta-1,3-diene, 5, 134
anion radical of,
ESR-spectrum, 47
symmetry of state, 36
cation radical of, symmetry of state, 36
conjugation of double bonds in, 5
Diels–Alder reactions with, 134, 138
orbital control, 134
EA-values (ETS), 136
calculated by HMO, 42
experimentally determined by ETS, 42, 134
electron transitions
energies, 51
symmetry, 49
ET-spectrum, 43
geometry, 134
IP-values, 136
IP-values (PES), 134
mixing of $\pi$-orbitals in it, 139
MOs, 136
MOs (HMO)
shapes, 24, 25
symmetry, 35
PE-spectrum, 135
n-butane, 112, 118
conformations, 110
EA-spectrum, transition energies, CNDO, 113, 118

IP-values (PES), 112
MOs, *ab initio* (4-31G), 112
*cis*-butene, 128
$EA_1$-value (ETS), 129
IP-values (PES), 128
MOs, 128
N-(iso-butenyl)azetidine, 282
conformation, 282
mixing of orbitals in it, 282
PES, 282
N-(iso-butenyl)piperidine, 281
conformation, 281
mixing of orbitals in it, 281
PES, 281
N-iso-butenylpyrrolidine, 280–281
conformation, 280–281
mixing of orbitals in it, 280–281
PES, 280–281
tert-butyl alcohol, 219–227
tert-butyl chloride, 96
n-butyl vinyl ether, 232–235
conformation, *ab initio*, 232
cycloaddition with TCNE, 233
IP-values (PES), 233
tert-butyl vinyl ether, 232–235
conformation, *ab initio*, 232
cycloaddition with TCNE, 233
IP-values (PES), 233
tert-butylbenzene, 162–165
CT-complexes with TCNE, 164
EA-values (ETS, polarography), 164
hyperconjugation, 162–163
IP-values (PES), 163–164
isomer distribution in electrophilic nitration, 326
n-butyl(pentamethyldisilyl)sulphide, 227
tert-butyl(phenyl)ether, 227–239
tert-butyl(phenyl)sulphide, 240
4-(tert-butyl)pyridine, 349
IP-values (PES), 349
MOs, 349
n-butyl(trimethylsilyl)sulphide, 227
n-butyl(trimethylsilylmethyl)sulphide, 227
but-2-yne, 131
$EA_1$-value (ETS), 131

carbazole, 343, 351, 353
cation radical, EA-spectrum, 351
HOMO electron densities (HMO), 353
IP-values (PES), 351
MOs, 343
orbital control in reactions with electrophiles, 353
PES of solid film, 352
carbon tetrachloride, 210
carboxylate anion, 5
delocalization of electron density, 5
cation radical, definition, 25, 38, 45
charge control of chemical reaction, definition, 70, 75
charge transfer complexes, 68
dialkyl sulphides as donors, 69
permethylpolysilanes as donors, 69
saturated hydrocarbons as donors, 69
substituted benzenes as donors, 69, 72, 73
charge $Z$ on the atom, definition, 29
chloroacetaldehyde, 248
chlorobenzene, 212–218
isomer distribution in electrophilic nitration, 326
2-chloro-10-($\gamma$-dimethylaminopropyl)phenothiazine, 351
cation radical, EA-spectrum, 351
IP-values (PES), 351
chloroethylene,
EA-values (ETS), 212
geometry, 211
IP-values (PES), 211
$\beta$-chloroethyl vinyl ether, 232–235
conformation, *ab initio*, 232
cycloaddition with TCNE, 233
IP-values (PES), 233
chloroform, 210
chloromethane, 205–)211
'hardness' and electronegativity, 74
chloropromazine, 77
biological activity, 77
IP-values (PES), 77
chrysene, 188, 190, 192
clivorine, 278–279
geometry, 278
mixing of orbitals in it, 278
TS-effect, 278
colour and frontier orbitals of organic molecules, 88–92
HMO-calculations, 88
PPP CI-calculations, 89
conformational effects and frontier orbitals, 80–93
anomeric effect, 80, 81
*cis* effect, 81, 82
gauche effect, 80, 81
orbital overlap, 83
stabilization energy, 83
stereoelectronic effect, 85
conjugation, definition, 4
coronene, 187–188, 190, 200, 201
correlation of electron motion, 20
Coulomb integral, 23
covalent bond, definition, 3
o-cresole, IP-values (PES), 239
cryptopine, 278
geometry, 278
mixing of orbitals in it, 278
TS-effect, 278
cycloalkanes, 114–117
PE-spectra, 119
reactions, 114
strain effects, electron structure, 114
cycloalkanones, 117
EAS, 117

$\nu C = O$ frequencies, 117
cyclobutadiene, 148
MOs, 148
overlap of $\pi$-orbitals, 148
cyclobutane, 116
IP-values (PES), 116
MOs, *ab initio*, 116
PE-spectrum, 119
cyclobutene, 128
$EA_1$-value (ETS), 129
IP-values (PES), 128
MOs, 128
cyclohexa-1,3-diene, 140, 141
EA-values (ETS), 141
IP-values (PES), 140
mixing of $\pi$-orbitals in it, 55
MOs, *ab initio* (4-31G), 140
TS-effect of orbital mixing, 141
cyclohexane, 114, 120
conformations, 114
CT-complexing with TCNE, 120
PE-spectrum, 119
strain energy, 114
cyclohexene, 129
$EA_1$-value (ETS), 129
IP-values (PES), 128
MOs, 128
cyclohexyl vinyl ether, 232–235
conformation, *ab initio*, 232
cycloaddition with TCNE, 233
IP-values (PES), 233
cyclooctyne, 131, 132
$EA_1$-value (ETS), 131
PE-spectrum, 132
cyclopentane, 114
geometry, 114
IP-values (PES), 115
MOs, *ab initio* (4-31G), 115
PE-spectrum, 111, 119
strain energy, 114
cyclopropane, 114–117
geometry, 116
IP-values (PES), 116
MOs, *ab initio*, 116
PE-spectrum, 119
reactions, 117
cyclopropene, 128, 298
IP-values (PES), 128, 129, 299
MOs, *ab initio* (4-31G), 129, 299

decalin, 120
CT-complexing with TCNE, EA-spectrum, 120
decamethyltetrasilane, 121
CT-complexing with TCNE, 121
EAS, 121
IP-values (PES), 121
mass-spectrum, 122
reactions with $WCl_6$, $MoCl_6$, 121
n-decane, 118
CT-complexing with TCNE, 120
EA-spectrum, transition energies, 118
experimental and calculated, 118
diacetyl, 261
diagonal matrix, 23
1,4-diazabicyclo[2,2,2]octane, 274
basicity, 274
geometry, 274
IP-values (PES), 275
mixing of orbitals in it, 274–275
1,5-diazabicyclo[3,3,2]decane, 276
basicity, 274
geometry, 274
IP-values (PES), 275
mixing of orbitals in it, 274–275
1,6-diazabicyclo[4,4,0]decane, 291
conformation, INDO/2, 291
IP-values (PES), 290
MOs, *ab initio* (6-31G), 291
*trans*-diazene, 295
EA-spectrum, transition energies, 305
geometry, 295
IP-values (PES), 296
MOs, 296
diazirine, 298
IP-values (PES), 299
MOs, 299
thermolysis, 298
dibenzofuran, 343
MOs, 343, 345
dibenzoselenophene, 345
IP-values (PES), 345
MOs (HMO), 345
dibenzotellurophene, 345
IP-values (PES), 345
MOs (HMO), 345
dibenzothiophene, 344
1,2-dibromoethane, 82
anticonformer, 82
dibutyl ether, 227
dibutyl sulphide, 227
*trans*-1,2-dichlorocyclohexane, 96
conformation, solvent effect, 97
dichloromethane, 210
diethyldisulphide, conformation, IP-value, 231
di(iso-propyl) ether, 227
*trans*-di(i-propyl)diazene, 305
EA-spectrum, transition energies, 305
IP-values (PES), 305
N,N'-di(i-propyl)hydrazobenzene, 294
conformation, 294
IP-values (PES), 294
difluorodiazirine, 299
IP-values (PES), 299
MOs, 299
1,2-difluoroethane, 82
gauche conformer, 82
$\alpha$-diketones, 260–261
colour, 261
EA-spectrum, band positions, 261
IP-values (PES), 261

1,2-dimethoxyethane, conformation, 229
2,5-dimethoxy-4-methylamphetamine, 77
  biological activity, 77
  IP-values (PES), 77
dimethyl(phenyl)arsine, PES, CNDO/2, 285, 286
dimethyl(vinyl)arsine, PES, CNDO/2, 285
dimethyl ether, 220–224, 227
  'hardness' and electronegativity, 74
dimethyl selenide, 225–226
dimethyl sulphide, 225–226
  'hardness' and electronegativity, 74
dimethyl telluride, 225–226
N,N-dimethylaniline, 284
  EA-spectrum, transition energies, 286
  EA-value (ETS), 285
  geometry, 284
  IP-values (PES), 285
N,N-dimethyl-N',N'-diethylhydrazine, 290
  conformation, INDO/2, 291
  IP-values (PES), 290
  MOs, *ab initio* (6-31G), 291
2,5-dimethyl-N,N'-dipyrrolyl, 350
  conformation
  IP-values (PES), 350
dimethyldisulphide, conformation, IP-value, 231
N,N'-dimethyl-N,N'-di(tert-butyl)hydrazine, 290
  conformation, INDO/2, 291
  IP-values (PES), 290
  MOs, *ab initio* (6-31G), 291
N,N'-dimethylhydrazine, 289
  IP-values (PES), 290
  conformation, INDO/2, 291
  MOs, *ab initio* (6-31G),
N,N'-dimethylhydrazobenzene, 294
  conformation, 294
  IP-values (PES), 294
2,6-dimethylphenol, IP-values (PES), 239
P,P-dimethyl(phenyl)phosphine, 285–287
  anion radical,
    spin densities, ESR-spectrum, 285–287
  EA-spectrum, transition energies, 286
  geometry, 284
  IP-values (PES), 285
  MOs, CNDO/2, 285
2,6-dimethylpyridine, 349
  IP-values (PES), 349
  MOs, 349
2,5-dimethylpyrrole, 350
  IP-values (PES), 350
  MOs, 350
N,N-dimethyltryptamine, 77
  biological activity, 77
  IP-values (PES), 77
1,3-dioxane, 229, 230
  IP-values (PES), 229
  TB- and TS-effects in, 230
1,4-dioxane, 229, 230
  IP-values (PES), 229
  TB- and TS-effects in, 230
diphenylamine, 351
2,3-diphenyl-*trans*-decahydrophthalazine, 294
  conformation, 294
  IP-values (PES), 294
1,2-diphenylpyrazolidine, 292
  conformation, 294
  IP-values, 294
  PE-spectrum, 293
N,N'-dipyrrolyl, 350
  conformation, MINDO/2, 350
  IP-values (PES), 350
  MOs, 350
  di(tert-butyl)acetylene, 131
  $EA_1$-value (ETS), 131
di(tert-butyl) ether, 227
di(tert-butyl)disulphide, conformation, IP-value, 231
di(tert-butyl)ketone, reactions with nucleophiles, 248
di(tert-butyl)thioketone, reactions with nucleophiles, 248
1,2-dithiane, conformation, IP-value, 231
1,3-dithiane, 229, 230
  IP-values (PES), 229
  TB- and TS-effects in, 230
1,4-dithiane, 229, 230
  IP-values (PES), 229
  TB- and TS-effects in, 230
dodecamethylcyclohexasilane, 121, 122
  CT-complexing with TCNE, 121
  EAS, 121
  IP-values (PES), 121
  mass-spectrum, 122
  reactions with $WCl_6$, $MoCl_5$, 121
dopamine, 77
  biological activity, 77
  IP-values (PES), 77
dyes, colour of, 89
  azo for liquid crystals, 91
  cyanine, 89
  hydrogen bonding in, 91

electrocyclic reactions, 131–133
  role of frontier orbitals, 131–133
  stereoselectivity, 131–133
  Woodward–Hoffmann rules, 131–133
electron absorption spectroscopy, 37, 48
  electron transitions,
    configuration, 49
    multi-configurational interaction, 51
  energy of, 51
  oscillator strength, 50
  permitted and forbidden, 49
  selection rules, 49
    multiplicity selection rule, 49
    symmetry selection rule, 49
  symmetry, 52
  intensity of absorption, 49, 92
  polarization of absorption, 49, 92
electron affinity, definition, 3, 35, 42, 44
electron chemical potential, definition, 74

electron density, definition, 5, 18, 19, 28, 47, 67, 68, 77, 78
  partial, definition, 28
  total, definition, 28
electron donating substituents, 6
electron orbit, definition by Bohr's postulates, 14
electron repulsion, 19
electron shell, 16, 27
electron spin resonance spectroscopy, 37, 46
  anion radical of toluene, 48
  anion radical of trimethyl(phenyl)silane, 48
  hyperfine interaction, 46
  number of line in ESR-spectrum, 47
  spliting constant, 46
electron transmission spectroscopy, 36, 41–44
  electron affinity by
    adiabatic energy, 43
    vertical energy, 44
  vibrational structure of signal, 43
electron-withdrawing substituents, 6
electronegativity of,
  elements, definition, 3, 4
  chemical reagents, definition, 74
electrostatic repulsion, 20
energy operator, 16, 17, 19, 22
  Hamiltonian, 17
  Hartree, 20, 21
  Hartree–Fock, 20
ethane, 106–109
  barrier of rotation around C-C bond, 108
  conformations and geometry, 107, 109
EA-spectrum, transition energies, 109, 118
  IP-values (PES), 107
  MOs, *ab initio* (6-31G), 107
    graphic shapes, 108
  reactions with superacids, 108
2-ethoxypyridine, 350
  IP-values (PES), 350
  MOs, 350
4-ethoxypyridine, 350
  IP-values (PES), 350
  MOs, 350
ethyl alcohol, 219, 227
ethyl vinyl ether, 232–235
  conformation, *ab initio*, 232
  cycloaddition with TCNE, 233
  IP-values (PES), 233
ethyl-cation, conformation, mixing of orbital in it, 57
ethylene, 23, 57, 124–127
  correlation diagram, 24, 58
  cyclodimerization, 69
$EA_1$-value (ETS), 42, 126
ET-spectrum, 43
EAS, 127
  geometry, 124
  'hardness' and electronegativity, 74
  IP-values, 125
MOs, *ab initio* (6-31G), EHMO, HMO, MO LCFO, 23, 29, 57, 58, 125
    analytical form, 24
    graphic representation, 24, 58, 125
  PE-spectrum, 126
  reactions with electrophiles, 126–127
ethyl(phenyl)sulphide, 240
ethynylbenzene, 172
  geometry, 173
  IP-values (PES), 173
  MOs, HAM/3, SPINDO, 172, 173
    graphic shapes, 173
exchange energy, 20

field effect, definition, 4, 7
fluorene, 10, 177
  cation radical, transition energies in EA-spectrum, 177
  conformation, 177
  IP-values (PES), 177
fluorobenzene, 212–218, 326
  isomer distribution in electrophilic nitration, 326
fluoromethane, 205–211
formaldehyde, 31, 39, 43, 44, 243–244
  charges on atoms, 247
  dipole moment, experimental and calculational, 247
  EA-spectrum, transition energies, 244
  EA-values (ETS), 244
  ET-spectrum, 43
  IP-values (PES), 244
  hydration, 243
  MOs, *ab initio* (6-31G), 243–244
    graphic shapes, 243
  nucleophilic addition,
    relative reactivity, 243, 247–248
formic acid, acidity, 5
frontier orbitals, definition, 25, 36, 44–46
  concept of Fukui, 67–75
  energy gap, 69, 71, 75
  'hardness' and 'softness' of molecule, 75
  Klopman and Salem's consideration of interaction of, 70
  role in acids and bases interactions, 73
  role in CT-complexing, 68
  role in cyclodimerization of ethylene, 69
  role in nitration of naphthalene, 68
  rules of mixing, 67–68
fumaronitrile, 317
  $EA_1$-value, 317
  IP-values (PES), 317
  MOs, 317
furan, 332–334
  Diels–Alder reaction with, 333
  EA-values (ETS), 334
  geometry, 354
  IP-values (PES), 333
  MOs, *ab initio*, 333

haloalkanes, 205–211
  carbocations from 209, 210

dipole moments in gas phase, 205
EA-values (ETS), 210
geometry, 205
IP-value (PES), 208
MOs, *ab initio* (STO-3G), 208
orbital control, in $S_N2$ reactions, 70, 208
PE-spectra, 207, 208
halobenzenes, 212–218
cation radical, ESR-spectrum, 215
charges on atoms, CNDO/2, 214
dipole moments, 212
EA-values (ETS, polarography), 216
electron densities on atoms, *ab initio* (4-31G), 214
IP-values (PES), 214
MOs, *ab initio* (4-31G), 214
reactions with electrophiles, orbital control, 215
reactions with nucleophiles, orbital control, 217
Hammett equation, 6–10
constant $\rho$ of reaction, 6
constant $\sigma$ of substituents, 6–11
$\sigma,\rho$ method, 6, 9
hard and soft acids and bases (HSAB) principle, definition, 73, 74
absolute electronegativity of chemical reagent, definition, 74
absolute 'hardness' of chemical reagent, definition, 74
'hardness of chemical reagent, 74
'hardness' of reacting system, 75
heptacene, 194
n-heptane, 118
EA-spectrum, transition energies, 118
1,3,4,6,8,9-hexaazabicyclo[4,4,0]decane, 291
conformation, INDO/2, 291
IP-values (PES), 290
MOs, *ab initio* (6-31G), 291
(10*E*,12*Z*)10,12-hexadecadiene-1,ol, biological activity, 76
*cis*-hexafluoroazomethane, 295
geometry, 295
hexafluorobenzene, 150
IP-values (PES), 150
MOs, 150
hexamethyldisilane, 121
CT-complexing with TCNE, 121
EAS, 121
IP-values (PES), 121
mass-spectrum, 122
reactions with $WCl_6$, $MoCl_5$, 121
n-hexane, 118
EA-spectrum, transition energies, 118
hexa-1,3,5-triene, 42
IP-values (PES), 139
*cis*- and *trans*-isomers, 139
MOs, SPINDO, 139
Hoffmann rules for concerted reactions, 69
hydrazine, 289–292
barrier of rotation around N-N bone, 289
geometry and conformation, 289
IP-values (PES), 290
MOs, *ab initio* (6-31G), 290
hydrazobenzene, 292
conformation, 292
IP-values, 293
MOs, *ab initio*, 293
PE-spectrum, 293
hydrogen chloride, 'hardness' and electronegativity, 74
5-hydroxytryptamine, 79
biological activity, 78
IP-values (PES), 78
6-hydroxytryptamine, 78
biological activity, 78
IP-values (PES), 78
hyperconjugation, 6, 158, 167

indole, 353
HOMO electron densities (HMO), 353
orbital control in reactions with electrophiles, 353
inductive effect, definition, 4
iodobenzene, 212–218
iodoethylene,
IP-values (PES), 211
mixing of orbitals, 211
iodomethane, 205–211
'hardness' and electronegativity, 74
ion-radical, definition, 44, 46
ionic bond, definition, 3
ionization potential, definition, 3, 36, 37, 38, 44
IR, spectroscopy, 10, 11
study of electron structure of organic molecules, 10, 11
isopropyl(phenyl)ether, 239
isopropyl(phenyl)sulphide, 240
isoqinoline, 353

Koopman's theorem, definition, 36, 40, 42

liquid crystals, 92
5-amino-1,4-naphthoquinone, in liquid crystals, 93
norbornadiene, photoisomerisation, 94
vitamin $D_2$, 93
lysergic acid diethylamide, 77, 78
biological activity, 78
electrostatic potential, 78

manxine
conformation, 273
IP-values (PES), 273
mixing of fragment orbitals in it, 273
manxyl chloride, structure, reactivity, 273
mescaline, 77
biological activity, 77
IP-values (PES), 77
methane, 31, 102–106
EA-spectrum, transition energies, 106
geometry, 102

IP-values, 104, 106
MOs, MO LCFO, *ab initio*, 103, 104
PE-spectrum, 103
proton affinity, 106
reactions, 105
halogenation, 105, 106
with superacids, 105
1,6-methano[10]annulene, 155–158
anion-radical, ESR-spectrum, 158
conformation and geometry, 155
IP-values (PES), 157
MOs, HMO, CNDO/S, MNDO, *ab initio*, 157, 158
*p*-methoxyazobenzene, 302–305
$E_{1/2}^{red}$-values (polarography), 303
IP-values (PES), 302, 303
MOs, INDO/S, 303
*p*-methoxyazobenzenes, *m*' and *p*'-X-substituted ($X = NH_2$, $OCH_3$, OH, $CH_3$, Cl, COOH, $NO_2$)
$E_{1/2}^{red}$-values (polarography), 303
EA-spectra, transition energies, 308
IP-values (PES), 303
methyl alcohol, 219–225, 227
methyl bromide, 96
methyl vinyl ether, 232–235
conformation, *ab initio*, 232
cycloaddition with TCNE, 233
IP-values (PES), 233
methyl vinyl sulphide, 234
conformation, 234
IP-values (PES), 234
N-methylaniline, dipole moment, 11
methylbenzenes, isomers, 162–164
EA-spectra, transition energies, 163
geometry, 162
IP-values (PES), 162
MOs, 162
2-methylheptadecane, biological activity, 109
2-methylpyridine, 349
IP-values (PES), 349
MOs, 349
4-methylpyridine, 349
IP-values (PES), 349
MOs, 349
N-methyltetrahydroberberine, 278–279
geometry, 278
IR-spectrum, 278
mixing of orbitals in it, 278
methyl(trimethylgermyl)ketone, PES, 250
EA-spectrum, transition energies, 250
IP-values (PES), 250
methyl(trimethylsilyl)ketone, 250
EA-spectrum, transition energies, 250
IP-values (PES), 250
mixing of orbitals, rules, 55
MO LCFO-calculations, 54
molecular orbital, 13
antibonding, 24
bonding, 24
correlation diagram, 24
degenerate, 26
delocalized, 20
eigencoefficient, 21, 23
eigenvalue (eigenenergy), 21, 23, 24
frontier MO, 25, 36 44–46
graphic representation, 24
highest occupied (HOMO), 25
lowest unoccupied (LUMO), 25, 42, 45
nonbonding, 27
occupied, 24
symmetry, 35
unoccupied, 24
molecular orbital theory, 20
main concepts, 20
MO LCAO method, 20
MO LCFO method, 54
variational method, 22
molecular orbitals perturbation theory of Dewar, 320
monochloroacetic acid, $pK_a$-value, 4
monodeuterobenzene,
anion radical, ESR spectrum, 321, 322

naphthacene, EA-spectrum, 52, 195
naphthalene, 179
anion radical, ESR-spectrum, effect solvation on spin densities, 100
aromaticity, 179–180
cation radical, transition energies in EA-spectrum, 183
EA-spectrum, 52, 68
EA-values (ETS), 181
ESR-spectrum of anion-radical, 183
geometry, 180
IP-values (PES), 181
MOs, CNDO/S2, MINDO/1, INDO/S, *ab initio*, 180, 181
PE-spectrum, 182
reactions with electrophiles, orbital control, 68, 180, 199
naphthalenes
1-substituted ($CH_3$, $OCH_3$, $NH_2$, F, CN, $NO_2$), IP-values (PES), 323
2-substituted ($CH_3$, $OCH_3$, $NH_2$, F, CN, $NO_2$), IP-values (PES), 323
1,4-naphthoquinone, 261–263
anion radical,
ESR-spectrum, 263
spin densities, HMO, 263
$E_{1/2}^{red}$-values (polarography), 263
MOs, CNDO/S, 262
PE-spectrum, 262
neopentane, 122
EA-values (ETS), 122
IP-values (PES), 122
neopentylbenzene, 162–165
EA-spectrum, transition energies, 165
hyperconjugation, 164
IP-values (PES), 165

nitrobenzene, 314
  conformation, geometry, 314, 315
  dipole moment, 10
  electron densities on atoms, *ab initio*, 315
  ESR spectrum of anion-radical, 315
  IP-values (PES), 314, 315
  MOs, EHMO, *ab initio*, 314, 315
  polarographic reduction, 315
  reactivity, 315
nitrobenzenes,
  $E_{1/2}^{red}$-values, polarography, 45
*m*-nitrobenzoic acid, acidity, 6
nitrobutane, 312
  IP-values (PES), 312
nitroethane, 312
  acidity, 312
  dipole moment, 10
  IP-values (PES), 312
nitromethane, 311
  acidity, 312
  $EA_1$-value, 313
  geometry, 312
  IP-values (PES), 312
  MOs, *ab initio* (6-31G), 312
*o*-nitrophenol, H-bonding in it, 315, 316
*p*-nitrophenol, H-bonding in it, 315, 316
1-nitropropane, 312
  IP-values (PES), 312
2-nitropropane, 312
  acidity, 312
  IP-values (PES), 312
nitrosobenzene, 313
  charges on atoms, *ab initio*, 313, 314
  conformation geometry, 314
  IP-values (PES), 313
  MOs, CNDO/S, 313
nitrosomethane, 310
  EA-spectrum, transition energies, 311
  geometry, 310
  IP-values, 310
  MOs, *ab initio*, 310
  PE-spectra at different temperaturs, 311
NMR, study of electron structure of organic molecules, 10, 11
norbornadiene, 141–145
  EA-values (ETS), 143, 144
  IP-values, 143
  PE-spectrum, 143
  TS-effect of orbital mixing, 144
norbornene, 143
  IP-values (PES), 143
  MOs, 143

octamethyltrisilane, 121
  CT-complexing with TCNE, 121
  EAS, 121
  IP-values (PES), 121
  mass-spectrum, 122
  reactions with $WCl_6$, $MoCl_5$, 121
n-octane, 118
  EA-spectrum, transition energies, 118
  CT-complexing with TCNE, 120
  experimental and calculated, 118
Octet rule, definition, 3, 102
orbital control of chemical reaction, definition, 70, 75
orbital population, definition, 29
orbitals
  atomic, 15–20
  fragment, 53, 54
  group, 53
  interaction (mixing) of, 55
  molecular, 13
  non-orthogonal, 56
  orthogonal, 56
  overlap of, 22, 54, 55
*ortho*-effect, 7
ovalene, 187–188
oxetane, $IP_1$-value (PES), 228
oxyrane, $IP_1$-value (PES), 228

Pairing theorem, 27, 42, 184, 192
paracyclophanes, 179
  geometry, 179
  IP-values (PES), 179
parametrization of orbitals interaction, definition, 57, 60
*p*-bis(methylthio)benzene, 59, 61
  dimethyl ether, 63
  dimethyl sulphide, 61, 63
  dimethyl telluride, 63
  1,2-di)methylthio)ethylene, 59
  hydrogen sulphide, symmetry, 61
  thioanisole, 61
pentafluorophenetol, 238
  conformation, 238
  IP-values (PES), 238
pentamethyldisilyl(phenyl)ether, 227
pentamethylene sulphide, IP-values (PES), 229
n-pentane, 110–112, 118
  EA-spectrum, transition energies, CNDO, 113, 118
  IP-values, 111
  MOs, *ab initio* (4-31G), 112
PE-spectrum, 111
perylene, 188, 190, 194, 199, 201
phase effect on electron structure, 44
phenalenone, 255–257
  anion radical, spin densities, experimental and calculational, 257
  aromaticity, 255
  basicity, 256
  condensation in the presence of strong bases, 257
  geometry, 255
  MOs, HMO, CNDO, 256
  reactions with electrophiles, orbital control, 256
phenanthrene, 10, 40, 184, 188, 192, 199, 201
  EA-spectrum, transition energies, 40
phenetol, 239

phenol, 235–238
  acidity, 237
  EA-values (ETS), 238
  electron densities on atoms, *ab initio*, 237
  geometry, 236
  IP-values, 236
  MOs, *ab initio* (4-31G), 236
*p*-X-substituted (X = $CH_3$, $NO_2$, OH, I, Br, Cl, F), acidity, 237
phenothiazine, 352
  PES of solid film, 352
phenylacetic acids, 7
  substituted, $\sigma,\rho$-analysis, 7
1-phenylazo-2-naphthol, 308–310
  configurations of electron transitions, 309
  effect of substituents on colour, 308
  hydrogen bonds, 308
phenylpentamethyldisilane, 323
  CT-complex with TCNE, EA-spectrum, 323, 324
phenyl(trimethyl)germane, 165
  EA-spectrum, transition energies, 165
  hyperconjugation, 165
  IP-values (PES), 165
  MOs, 165
phenyl(trimethyl)silane, 165
  EA-spectrum, transition energies, 165
  hyperconjugation, 165
  IP-values (PES), 165
  MOs, 165
phenyl(trimethyl)stannane, 165
  EA-spectrum, transition energies, 165
hyperconjugation, 165
  IP-values (PES), 165
  MOs, 165
phosphabenzene, 338–342
  basicity, 341
  EA-values (ETS), 341
  geometry, 339
  IP-values (PES), 340
  MOs, 340
  proton affinity, 340
phosphine, 270
  geometry, 270
  $IP_1$-value (PES), 271
  MOs, 270
photochemical reactions and orbital structure of organic molecules, 88, 93, 94
photoelectron spectroscopy, 37
  correlation energy, 41, 60
  Franck–Condon transition, 39
  ionization potential
    adiabatic, 39
    vertical, 39
  Koopmans defect, 41, 60
  relaxation energy, 41, 60
  reorganization energy, 41, 60
  symmetry designation of cation-radical states, 39
  technique, 37
  vibrational structure of bonds, 38
piperidines
  2-,3-,4- and 6-methyl substituted, 273
    IP-values (PES), 273
    mixing of orbitals (STO-3G), calculations, 273
polarization energy, 190
polarography, 37, 44
  oxidation, $E^{ox}_{1/2}$-values, 45
    aromatic hydrocarbons, 46
    strained saturated cyclic hydrocarbons, 46
  reduction, $E_{1/2}^{red}$-values, 44–45
    aromatic hydrocarbons, 45
    aromatic ketones, 45
    linear conjugated polyenes, 45
    nitrobenzenes, 45
polycyclic aromatic hydrocarbons (PAH), 187
  Birch reduction, 202, 203
  carcinogenicity, 202
  cation radicals, transition energies in EA-spectrum, 194
  CT-complexes with TCNE, 200, 201
  EA-spectra, transition energies, 194–196
  EA-values (polarography), 192
  geometry, 187, 188
  IP-values (PES) in gas and solid phases, 188–190
  MOs, MINDO/2, HMO, PPP, EHMO, 189, 190
  orbital control, reactions with electrophiles, 199
  oxidation, polarographic, 45
  reduction, polarographic, 44, 46
polyenes,
  $E_{1/2}^{red}$-values, polarography, 45
propane, 113, 118
  EA-spectrum, transition energies, CNDO, 113
propene, 128
  IP-values (PES), 128
  $EA_1$-values (ETS), 129
  MOs, 128
N-propenylaziridine, 281
  conformation, 281
  mixing of orbitals in it, 281
  PES, 281
N-propenylpyrrolidine, 280–281
  conformation, PRDDO, 280–281
  IP-values (PES), 280–281
  mixing of orbitals in it, 280–281
iso-propyl alcohol, 219, 227
iso-propylidene norbornadiene, 142
  IP-values (PES), 143
  MOs, 143
  TB- and TS-effects of orbital mixing, 143
isopropyl(phenyl)ether, 239
isopropyl(phenyl)sulphide, 240
propyne, 131
  $EA_1$-value (ETS), 131
  IP-values (PES), 131
  MOs, 131
protopine, 278–279

geometry, 278
mixing of orbitals in it, 278
TS-effect, 278
[yrazine, 342–344
$E_{1/2}^{red}$-values (polarography), 344
IP-values (PES), 343
MOs, calculations with Green functions, 343
pyrene, 188, 192, 199, 201, 203
pyridazine, 344
$E_{1/2}^{red}$-value (polarography), 344
$EA_1$-value, 344
MOs, calculations with Green functions, 343
pyridine, 336–342, 353
basicity, 341
geometry, 337
HOMO electron densities (HMO), 353
IP-values (PES), 337–338
MOs, *ab initio*, 337, 338
orbital control in reactions with electropiles, 353
proton affinity (gas phase), 340
pyrimidine, 344
$E_{1/2}^{red}$-value (polarography), 344
$EA_1$-value, 344
MOs, calculations with Green functions, 343
pyrrole, 331–332, 353
Diels–Alder reaction with, 354
EA-values (ETS), 332
geomery, 331
HOMO electron densities (HMO), 353
EA-spectra, transition energies, 332
IP-values (PES), 333
MOs, CNDO/S, 332
orbital control in reactions with electrophiles, 353
rate of bromination, 353
resonance structures and aromaticity, 332

quantum chemical calculational methods, 29–32
*ab initio*, 29, 30, 40, 41, 53
Roothaan equations, 30, 60
self-consistent field (SCF) method, 30, 60
semiempirical, 29, 31
CNDO, 31
CNDO/S3, 32
HAM/3, 32
INDO/S, 52
MNDO, 32
NDDO, 31
PPP, 31
$X\alpha$, 32
simple, 20, 29
extended Huckel method (EHMO method), 29
Huckel method of MO (HMO method),2 3, 42, 47
quantum mechanics postulates, 14, 16
quantum numbers, 14, 15, 18
azimuthal, 15
magnetic, 15
principal, 14, 15
spin, 15
quinoline, 344, 353
$E_{1/2}^{red}$-value (polarography), 344
$EA_1$-value, 344
HOMO electron densities (HMO), 353
orbital control in reactions with electrophiles, 353
isoquinoline, 353
HOMO electron densities (HMO), 353
orbital control in reactions with electrophiles, 353
quinuclidine
basicity, 273
IP-values (PES), 273
mixing of orbitals in it, 273

resonance integral, 23
retazamine, 278–279
geometry, 278
mixing of orbitals in it, 278
TS-effect, 278
11-*cis*-retinal, photoisomerization of, 95

Schrödinger wave equation, 16–19, 22
secular determinant, 22
secular equations, 22
selenanthrene, 347
conformation, 347
IP-values (PES), 347
selenoanisole, 240
selenophene, 335
CT-complex with TCNE, EA-spectrum, 351
EA-values (ETS), $X\alpha$-method of calculation, 336
IP-values (PES), 336
MOs, 336
rate of bromination, 354
self-association of organic molecules, 95–101
effect on IP-values (PES) of benzene, 100
effect of MOs of
acetic acid, 99
ammonia, *ab initio*, 98
benzene, 100
serotonin, 77–79
biological activity, 77
IP-values (PES), 77
silatrane, 279
geometry, 279
TS-effect in it, 279
solvation, effect on electron structure, 96
acetaldehyde, dependence of $\nu C = O$ in IR-spectrum on, 96
acetone, $\mu$, 96
acetone, dependence of $\nu C = O$ in IR-spectrum on, 96
acetyl chloride, dependence of $\nu C = O$ in IR-spectrum on, 96
effect on spin densities in anion radicals, 100
energy of, 97

nitrobenzene, $\mu$, 96
nitromethane, $\mu$, 96
phosgene, dependence of $\nu C = O$ in IR-spectrum on, 96
spin density, definition, 28, 46
stibabenzene, 338–342
EA-values (ETS), 341–342
IP-values (PES), 341
MOs, 341
*trans*-stilbene, 170
comparison of MOs with azobenzene, 303
conformation and geometry, 170
EA-values (ETS), 171
ESR-spectrum of anion-radical, 172
IP-values, 172
MOs, 171
PE-spectrum, 171
styrene, 167–170
conformation, barrier of rotation, *ab initio* (STO-3GF), 167
EA-values (ETS), 169, 170
electron densities on atoms, CNDO/2, 168
IP-values (PES), 168
MOs, 168
symmetry of
molecules, 32, 36
operation of, 32
inversion, 32
mirror rotation, 32
reflection in plane sigma, 32
rotation around the axis, 32
orbitals, 35
point group of symmetry, 34
states, 35
symbols of symmetry, 33
table of characters, 33

telluranthrene, 347
conformation, 347
IP-values (PES), 347
telluroanisole, 240
tellurophene, 335
EA values (ETS), $X\alpha$-method of calculation, 336
IP-values (PES), 336
MOs, 336
*p*-terphenyl, 178–179
IP-values (PES), 178
PE-spectra in gas and solid phases, 178
tetrabenzoanthracene, 194
tetrabenzopentacene, 190
tetrabenzoperylene, 190
tetracene, 40, 188, 190, 199
EA-spectrum, transitions energies, 40
tetracyanoethylene, 317
CT-complexes of, 69
alkanes, 120
PAH, 200, 324
permethylpolysilanes, 69, 120, 121
substituted arenes, 69, 323
sulphides, 69
$EA_1$-value, 317
IP-values (PES), 317
tetradecamethylhexasilane, 121
CT-complexing with TCNE, 121
EAS, 121
IP-values (PES), 121
mass-spectrum, 122
reactions with $WCl_6$, $MoCl_5$, 121
tetrahydrofuran, $IP_1$-value (PES), 228
tetrahydropyran, IP-values (PES), 228, 229
2,2',5,5'-tetramethyl-N,N'-dipyrrolyl, 350
IP-values (PES), 350
MOs, 350
3,3,7,7-tetramethylcycloheptyne, PE-spectrum, 132
tetramethylgermane, 122
EA-values (ETS), 122
IP-values (PES), 122
tetramethylhydrazine, 290
conformation, INDO/2, 291
IP-values (PES), 290
MOs, *ab initio* (6-31G), 291
tetramethylsilane, 122
EA-values (ETS), 122
IP-values (PES), 122
tetramethylstannane, 122
EA-values (ETS), 122
IP-values (PES), 122
1,2,4,5-tetramethyltetrahydro-1,2,4,5-tetrazine, 291
conformation, 292
IP-values (PES), 292
tetra(methylthio)ethylene, 324
conformation, 234
IP-values (PES), 234
6,7-thiaperylene, 345
MOs, 345
thianthrene, 347
conformation, 347
IP-values (PES), 347
thioanisole, 240–242
conformation, 240
EA-values (ETS), 240
IP-values (PES), 240
EA-spectrum, transition energies, 241
thiobenzophenone, reactions with nucleophiles, 249
thiophene, 334–336
CT-complex with TCNE, EA-spectrum, 335, 351
Diels–Alder reaction with, 354
EA-values (ETS), 336
geometry, 334
IP-values (PES), 335
MOs, 336
PE-spectrum, 335
rate of bromination, 354
thiophene, 353
HOMO eigencoefficients (HMO), 353
orbital control in reactions with electrophiles, 353, 354

toluene, 158–162
anion radical, ESR-spectrum, 160
cation radical, ESR-spectrum, 160
conformation and geometry, 159
dipole moment, 158
hyperconjugation, 158
EA-spectrum, transition energies, 163
EA-values (ETS), 161
IP-values (PES), 160
MOs, *ab initio* (4-31G), 160
orbital and charge control of reactions with electrophiles, 160
*sym*-triazine,
$E_{1/2}^{red}$-value (polarography), 344
$EA_1$-value, 344
MOs, calculations with Green functions, 343
tri(n-butyl)amine, $IP_1$-value (PES), 272
trichloroacetic acid, p$K_a$-value, 4
triethylamine, $IP_1$-value (PES), 272
trifluoroanisole, 238
conformation, 238
IP-values (PES), 238
*trans*-1,1,1-trifluoroazomethane, 295
geometry, 295
trimethylamine, 271–272
barrier of inversion, 271
basicity in the gas phase, 340
geometry, 271, 340
'hardness' and electronegativity, 74
IP-values (PES), 271, 272
MOs, INDO, CNDO, EHMO, 271
proton affinity, 340
trimethylarsine, $IP_1$-value (PES), geometry, 271–272
trimethylpospine, 271–272
basicity and proton affinity, 340
geometry, 271, 272
'hardness' and electronegativity, 74
$IP_1$-value (PES), 271
trimethylsilylmethyl(phenyl) ether, 227
trimethylsilyl(phenyl) ether, 227
2-(trimethylsilyl)pyridine, 349
IP-values (PES), 349
MOs, 349
4-(trimethylsilyl)pyridine, 349
IP-values (PES), 349
MOs, 349
trimethylsilyl(tert-butyl)diimine, 297
EA-spectrum, transition energies, 298, 301
electron densities on atoms, CNDO, 297
IP-values (PES), 298
trimethylstibine, $IP_1$-value (PES), geometry, 271–272
tri(n-propyl)amine, $IP_1$ value (PES), 272
α,α,α-tris(trimethylsilyl)toluene, 323
CT-complex with TCNE, EA-spectrum, 323, 324
triphenylene, 188, 190, 194, 201

uncertainty principle, 16

valence basis approximation, 34
valence orbital, 16
valence shell, 16
vinylamine, PES, 280
N-vinylaziridine, 281
conformation, 281
mixing of orbitals in it, 281
PES, 281

water, 41, 222, 223
IP-values (PES), 41
geometry, 220
MOs, *ab initio* (6-31G), 223
PE-spectrum, 222
self-association and medium effects on, photoionization, 98

*p*-xylene, anion radical, ESR-spectrum, 328